CAMBRIDGE LIBRARY COLLECTION

Books of enduring scholarly value

Earth Sciences

In the nineteenth century, geology emerged as a distinct academic discipline. It pointed the way towards the theory of evolution, as scientists including Gideon Mantell, Adam Sedgwick, Charles Lyell and Roderick Murchison began to use the evidence of minerals, rock formations and fossils to demonstrate that the earth was older by millions of years than the conventional, Bible-based wisdom had supposed. They argued convincingly that the climate, flora and fauna of the distant past could be deduced from geological evidence. Volcanic activity, the formation of mountains, and the action of glaciers and rivers, tides and ocean currents also became better understood. This series includes landmark publications by pioneers of the modern earth sciences, who advanced the scientific understanding of our planet and the processes by which it is constantly re-shaped.

A Geological Manual

Sir Henry Thomas De la Beche (1796–1855) served as president of the Geological Society from 1847 to 1849, having contributed greatly to the development of geological science and surveying in the first half of the nineteenth century. He was also instrumental in the establishment of the Museum of Practical Geology in London. Reissued here in its 1831 first edition (which Darwin had with him aboard the *Beagle*), this work sought to help students to grasp the fundamentals of a rapidly advancing science. The first section considers the Earth's shape, density, temperature and other characteristics. The next part includes discussion of beaches, volcanos, and coastal processes. De la Beche then presents descriptions of various rock types, reflecting the state of contemporary geological knowledge. Highly successful, the book went through two further English editions; the expanded third edition is also reissued in this series.

Cambridge University Press has long been a pioneer in the reissuing of out-of-print titles from its own backlist, producing digital reprints of books that are still sought after by scholars and students but could not be reprinted economically using traditional technology. The Cambridge Library Collection extends this activity to a wider range of books which are still of importance to researchers and professionals, either for the source material they contain, or as landmarks in the history of their academic discipline.

Drawing from the world-renowned collections in the Cambridge University Library and other partner libraries, and guided by the advice of experts in each subject area, Cambridge University Press is using state-of-the-art scanning machines in its own Printing House to capture the content of each book selected for inclusion. The files are processed to give a consistently clear, crisp image, and the books finished to the high quality standard for which the Press is recognised around the world. The latest print-on-demand technology ensures that the books will remain available indefinitely, and that orders for single or multiple copies can quickly be supplied.

The Cambridge Library Collection brings back to life books of enduring scholarly value (including out-of-copyright works originally issued by other publishers) across a wide range of disciplines in the humanities and social sciences and in science and technology.

A Geological Manual

1831 edition

HENRY T. DE LA BECHE

CAMBRIDGE
UNIVERSITY PRESS

University Printing House, Cambridge, CB2 8BS, United Kingdom

Cambridge University Press is part of the University of Cambridge.
It furthers the University's mission by disseminating knowledge in the pursuit of
education, learning and research at the highest international levels of excellence.

www.cambridge.org
Information on this title: www.cambridge.org/9781108075121

© in this compilation Cambridge University Press 2014

This edition first published 1831
This digitally printed version 2014

ISBN 978-1-108-07512-1 Paperback

A

GEOLOGICAL MANUAL.

A

GEOLOGICAL MANUAL.

BY

HENRY T. DE LA BECHE, F.R.S., F.G.S.

MEM. GEOL. SOC. OF FRANCE, &c.

LONDON:

TREUTTEL AND WÜRTZ, TREUTTEL JUN. AND RICHTER.

PARIS AND STRASBURG:

TREUTTEL AND WÜRTZ.

1831.

PREFACE.

WHEN it is attempted, as in works of the following description, to sketch the actual state of a particular science, and at the same time to point out a few of the conclusions that may be hazarded from known facts, an author has always great difficulty in avoiding unnecessary and tedious detail on the one hand; while, on the other, he must notice such a number of facts as may convince a student, that he is not wandering in a wilderness of crude hypotheses or unsupported assumptions.

By some it will be considered that too much space has been allotted to lists of organic remains in the following pages. Considerable attention has certainly been paid to such catalogues, as the zoological character of certain rocks is now the subject of much research, and as the result of such investigations may be the knowledge of some of the principal conditions under which the fossiliferous rocks were produced: moreover, the author considered that, for practical purposes, there was no alternative between rendering them as perfect as his means of information would permit, or of omitting them altogether. It must however be confessed, that, though constructed from apparently the best authorities, these lists require severe examination; for, unfor-

b

tunately, the study of organic remains is beset with two
evils, which, though of an opposite character, do not neu-
tralize each other so much as at first sight might be anticipated : the one consisting of a strong desire to find similar
organic remains in supposed equivalent deposits, even at
great distances ; the other being an equally strong inclination to discover new species, often as it would seem for
the sole purpose of appending the apparently magical word
nobis.

There can be little doubt that from these and other
sources of error, the same organic remains, particularly
shells, often figure in our catalogues under two names ;
and that the exuviæ of certain animals are marked as discovered in situations where they have never been found.
Notwithstanding these difficulties, it will however be evident, from a glance at the catalogues of organic remains,
that a great mass of information has been gradually collected on this subject alone, from which the most important results must follow, even though the various lists may
require considerable correction.

As the author has endeavoured to address himself less
to the accomplished geologist than to the student, though
it is hoped that the former may also find matter interesting
to him, he has been particularly anxious to point out his
various sources of information, even when he has himself
visited the same countries ; that, independently of the fundamental principle *suum cuique*, the student should be
enabled more fully to avail himself of the labours of the
various authors cited, by referring to their published works
for greater detail than could be admitted into a volume of
this description.

In a rapidly advancing science like Geology, to which new facts are constantly added, and in which the chances of new views by their combination are consequently multiplied, it is almost impossible to avoid hazarding certain general conclusions, when the various known facts pass in review before us. In those which the author has ventured to bring forward, he has endeavoured always to follow that system of induction which can alone lead to exact knowledge ; but as truth, and truth alone, is the object of all science, he can sincerely declare,—that if from the discovery of new facts, or from more sound views respecting those already known, his conclusions should not appear tenable, he would not only be most ready to abandon them, but to rejoice that an untenable hypothesis may have been the means of leading to more exact knowledge, if it should have fortunately so happened that it promoted the requisite inquiry. Essentially it is of little importance, whose or what theory may in the end be found most accurate ; so long as we approximate towards the truth, we accomplish all that can be expected ; and it is clear, that the greater the amount of known facts, the greater the chance of accuracy, not only from the larger mass of information presented to the mind, but from the frequent checks offered to hasty conclusions.

Happily facts have become so multiplied that Geology is daily emerging from that state when an hypothesis, provided it were brilliant or ingenious, was sure of advocates and temporary success, even when it sinned against the laws of physics and facts themselves. It is not difficult to foresee, that this science, essentially one of observation, instead of being, as formerly, loaded with ingenious specu-

lations, will be divided into different branches, each investigated by those whose particular acquirements may render
them most competent to do so; the various combinations
of inorganic matter being examined by the Natural Philo-
sopher, while the Natural Historian will find ample occu-
pation in the remains of the various animals and vegetables,
which have lived at different periods on the surface of the
earth.

Excepting the lists of organic remains, general sketches
have been alone attempted in the following pages, even
though the temptations further to develope a given subject
were often sufficiently great, and the necessity of restraint
abundantly mortifying. It is however hoped that enough
has been done to assist the progress of those who may be
desirous of entering upon the important science of Geology;
and if fortunately this little work should fall into the hands
of any who may in consequence be induced to become fel-
low-labourers in that great work, the advancement of knowledge, the objects of the author will be most fully accomplished.

ABBREVIATIONS

OF

AUTHORS' NAMES

IN

THE LISTS OF ORGANIC REMAINS.

Bast.	Basterot.	Goldf.	Goldfuss.
Beaum.	Elie de Beaumont.	Jäg.	Jäger.
Blain.	Blainville.	Lam.	Lamarck.
Blum.	Blumenbach.	Linn.	Linnæus.
Bobl.	Boblaye.	Lons.	Lonsdale.
Broc.	Brocchi.	Mant.	Mantell.
Al. Brong.	Alex. Brongniart.	Munst.	Munster.
Ad. Brong.	Adolphe Brongniart.	Murch.	Murchison.
Brug.	Bruguière.	M. de S.	Marcel de Serres.
Buckl.	Buckland.	Nils.	Nilsson.
Conyb.	Conybeare.	Park.	Parkinson.
Cuv.	Cuvier.	Phil.	Phillips.
DeC.,or DeCau.	DeCaumont.	Raf.	Rafinesque.
Defr.	Defrance.	Rein.	Reinecke.
De la B.	De la Beche.	Schlot.	Schlotheim.
Desh.	Deshayes.	Sedg.	Sedgwick.
Des M.	Des Moulins.	Sow.	Sowerby.
Desm.	Desmarest.	Sternb.	Sternberg.
Desn.	Desnoyers.	Thir.	Thirria.
Dufr.	Dufrénoy.	Y. & B.	Young and Bird.
Fauj. de St. F.	Faujas de St. Fond.	Wahl.	Wahlenberg.
Flem.	Fleming.	Weav.	Weaver.

The localities marked A. in the lists of the carboniferous and grauwacke groups, are taken from a compilation on Swedish organic remains,
entitled: Esquisse d'un Tableau des Petrifications de la Svède; Stock
holm, 1829.

ERRATA.

Page 47, last line, *for* on *read* or.
—— 95, line 37, *for* Beyring's *read* Behring's.
—— 118, —— 28, *for* rent *read* vent.
—— 124, —— 15, *for* and rends off *read* and that the fiery mass rends off.
—— 129, —— 33, *for* were *read* was.
—— 131, last line, *for* is *read* afford.
—— 133, line 25, *for* Carne *read* Cane.
—— ib. —— 32, *for* rain water, resting on it *read* rain water resting on it.
—— 136, —— 2 from bottom, *for* though it is *read* though the latter is.
—— 184, —— 34, *for* by enchantment; after two or three days of rain, *read* by en-
      chantment, after two or three days of rain;
—— 209, —— 1, *for* Oysters Polypifers, *read* Oysters, Polypifers.
—— 222, —— 18, *for* Appendix (B.) *read* Appendix (C.)
—— 227, —— 3, *for* This is *read* These are.
—— 229, —— 3 from bottom, *dele* The probability.
—— 235, —— 9, *for* species *read* pieces.
—— 245, last line, *for* contains *read* contain.
—— 259, line 15, *for* the lower *read* its lower.
—— 260, —— 14, *for* these *read* their.
—— 261, —— 46, *for* cretaceous *read* arenaceous.
—— 264, —— 2, *for* Cerithium, Cornucopiæ, *read* Cerithium Cornucopiæ.
—— 268, —— 36, *for* 1. —— Caput Medusæ *read* 1. Pentacrinites Caput Medusæ.
—— 291, last line, *for* Grypha *read* Gryphæa.
—— 311, line 40, *for* of series *read* of the series.
—— 352, —— 14, *for* Peleolus *read* Pileolus.
—— 457, —— 2, *for* Steaschist for it, as a provisional term: steatite, however, it can
      read Steaschist for them, as a provisional term; steaschist, how-
      ever, they can.

TABLE OF CONTENTS.

————◆————

A

GEOLOGICAL MANUAL

Section I.

Figure of the Earth.

It has been concluded, both from astronomical and geodesical observations, that the figure of the earth is a spheroid. This spheroid has been considered as one of rotation, or such a figure as a fluid body would assume if possessed of rotatory motion in space.

The amount of the flattening of the poles, or the difference of the diameter of the earth from pole to pole, and its diameter at the equator, have been variously estimated; but it is commonly received that the polar axis is to the equatorial diameter as 304 to 305, the compression of the earth, or flattening at the poles, being thus considered as $= \frac{1}{305}$.

The equatorial diameter about	= 7924 miles*.
The polar axis	= 7898
Difference	26

Density of the Earth.

Various opinions have been entertained on this subject; but it appears certain that the internal density is greater than the solid superficial density. Daubuisson infers from the observations of Maskelyne, Playfair, and Cavendish, that " the mean density

* Considering the flattening of the poles as $= \frac{1}{305}$, M. Daubuisson has made the following calculations:—

Radius at the equator	6376851 metres.
Semi-terrestrial axis	6355943
Diff. or flattening of poles	20908
Radius in lat. 45°	6366407
A degree at same lat.	111115
A degree of long. in same lat.	78828
Surface of our earth	5098857 square myriametres
The volume	1082634000 cubic myriametres.

Traité de Géognosie, ed. 2me, tom. i.

B

of the earth is about five times greater than that of water, and, consequently, about double that of the mineral crust of our globe*." Laplace considered the mean density of our spheriod as = 1·55, the solid surface being 1. According to Baily, the density of the earth is 3·9326 times greater than that of the sun, and is to that of water as 11 to 2†.

Superficial Distribution of Land and Water.

The relative proportion of dry land to the ocean, as it at present exists, is such, that nearly three-fourths of the whole surface of the globe may be assigned to the latter. Of the former, the configuration is very various, presenting the greatest surface in the Northern hemisphere. Although the land sometimes rises high above the level of the sea, according to our general ideas on such subjects, it is, in reality, but slightly removed above that level, when considered, as it should be, with reference to the radius of the earth ‡. The superficies of the Pacific Ocean alone is estimated as somewhat greater that that of the whole dry land with which we are acquainted. Dry land can only be considered as so much of the rough surface of our globe as may happen, for the time, to be above the level of the waters, beneath which it may again disappear, as it has done at different previous periods. Laplace calculated that the mean depth of the ocean was a small fraction of twenty-five miles, the difference produced in the diameters of the earth by the flattening of the poles. It has been variously estimated at between two and three miles. The mean height of the dry land above the ocean-level does not exceed two miles, but probably falls far short of it; therefore, assuming two miles for the mean depth of the ocean, the waters occupying three-fourths of the earth's surface, the present dry land might be distributed over the bottom of the ocean, in such a manner that the surface of the globe would present a mass of waters;—an important possibility, for, with it at command, every variety of the superficial distribution of land and water may be imagined, and consequently every variety of organic life, each suited to the various situations and climates under which it would be placed.

The surface of the globe's solid crust is so uneven, that the ocean, preserving a general level, enters among the dry land in various directions, forming what are commonly termed inland seas; such as the Baltic, Red, and Mediterranean Seas, in which geological changes may be effected different from those in the open ocean.

* Traité de Géognosie, ed. 2me, tom. i. p. 18.
† Baily, Astronomical Tables.
‡ See the diagram in my Sections and Views illustrative of Geological Phænomena, pl. 40.

Masses of salt water are sometimes included in the dry land, which have been termed Caspians, from the Caspian Sea, the largest of them. These have no communication with the main ocean; indeed the level of the Caspian is represented as much lower than that of the Black or Mediterranean Seas. They have been variously accounted for, some supposing that they have been left isolated by a change in the relative level of land and water, while others imagine their saltness to arise from their occurrence in countries impregnated with saline matter. It is stated, in support of the latter opinion, that the Caspian, and the lakes Aral, Baikal, &c. are situated where salt springs abound. Whatever may be their origin, it will be obvious, that if the fresh water they receive be not equal to their evaporation, they will become gradually more saline, until, the water being saturated, the surplus salt will be deposited at the bottom, and strata of it will be formed of a size and depth proportioned to those of the lake or sea.

It would be out of place to attempt a general description of all the various combinations of land and water, with which all must be more or less familiar; but it may be useful to notice that fresh-water lakes cover very considerable spaces, and that thus, even at present, very extensive deposits may take place, which can only envelope the remains of terrestrial or fresh-water animals and vegetables.

Saltness and Specific Gravity of the Sea.

The whole body of the ocean is composed of salt water, which does not vary very materially in composition, as far as we can judge from the experiments made on it.

From evaporation and the fall of rain, the sea will be less salt at the surface than at some little depth beneath it.

According to Dr. Murray, sea-water collected from the Firth of Forth contained, in 10,000 parts,

Common salt	220·01
Sulphate of soda	33·16
Muriate of magnesia	42·08
Muriate of lime	7·84
	303·09

According to Dr. Marcet, 500 grains of sea-water, taken from the middle of the North Atlantic, contained,

Muriate of soda	13·3
Sulphate of soda	2·33
Muriate of lime	0·995
Muriate of magnesia	4·955
	21·580

According to the experiments of Dr. Fyfe (Edin. Phil. Journal, vol. i.), the waters of the ocean between 61° 52′ N. and 78° 35′ N. do not differ much in their saline contents, these being between 3·27 and 3·91 per cent. The waters were obtained by Scoresby.

Dr. Marcet instituted a series of experiments on the specific gravity of water, of which the following are the results:

	Sp. Gr.		Sp. Gr.
Arctic Ocean	1·02664	Sea of Marmora	1·01915
Northern Hemisphere	1·02829	Black Sea	1·01418
Equator	1·02777	White Sea	1·01901
Southern Hemisphere	1·02882	Baltic	1·01523
Yellow Sea	1·02291	Ice-Sea Water	1·00057
Mediterranean	1·0293	Lake Ourmia	1·16507

The same author concluded from his observations,

"1. That the Southern Ocean contains more salt than the Northern Ocean in the ratio of 1·02919 to 1·02757.

"2. That the mean specific gravity of sea-water near the equator is 1·02777, intermediate between that of the Northern and Southern hemispheres.

"3. That there is no notable difference in sea-water under different meridians.

"4. That there is no satisfactory evidence that the sea at great depths is more salt than at the surface*.

"5. That the sea, in general, contains more salt where it is deepest and most remote from land; and that its saltness is always diminished in the vicinity of large masses of ice.

"6. That small inland seas, though communicating with the ocean, are much less salt than the ocean.

"7. The Mediterranean contains rather larger proportions of salt than the ocean†."

The saltness of the sea, particularly that of its surface, would seem greatly to depend on the proximity of nearly permanent ice, and of large or numerous rivers. Thus, as is seen above, the Baltic, White, Black, and Yellow Seas are less salt than the main ocean, because they are supplied with comparatively large quan-

---

· * The author of the abstract of Dr. Marcet's observations in the Edin. Phil. Journal, cites the following observations of Mr. Scoresby in support of this conclusion.

		Sp. Gr.			Sp. Gr.
Lat. 76° 16′ N.	Surface	1·0261	Lat. 76° 34′ N.	Surface	1·0265
	At 738 feet	1·0270		At 120 feet	1·0264
	At 1380 feet	1·0269		At 240 feet	1·0266
				At 360 feet	1·0268
				At 600 feet	1·0267

† Phil. Trans. 1819; and Edin. Phil. Journal, vol. ii.

tities of fresh water. From the small proportion of salt contained in the Black Sea and Sea of Azof, the bays of the former frequently contain ice, and the latter is stated to be frozen over during four months in the year.

The superior saltness of the Mediterranean, though an inland sea, is attributed to the evaporation of its surface, which is supposed greater than the quantity of fresh water with which it is supplied. In consequence, two great currents, one from the Black Sea and the other from the Atlantic, flow into it to supply the waste caused by evaporation.

The saline contents of the sea are important, as all chemical changes or deposits, taking place in it, will be more or less affected by them. The gravity and pressure of the sea are of still greater consequence; for, as the pressure increases with the depth, effects, which would be possible at one depth, would be impossible at another. Thus, it is obvious from the ingenious experiments of Sir James Hall, that carbonate of lime may be fused by heat without the loss of its carbonic acid, if subjected to great pressure, such as exists at the bottom of the deep sea.

The compressibility of water, which was for a long time doubted, has been proved by experiment, and has been calculated at 51·3 millionths of its volume for a pressure equal to each atmosphere*. It follows, that at great depths, and beneath a great pressure of the ocean, a given quantity of water will occupy a less space than on the surface, and will, consequently, by this circumstance alone, have its specific gravity greatly increased.

Temperature of the Earth.

The superficial temperature of our planet is certainly very materially influenced by, if it may not be entirely due to, solar light and heat. That the difference of seasons, and of the climates of various latitudes, originates in the greater or less exposure to the sun, is obvious. That local circumstances cause great variations of superficial temperature, is also well known; yet the principle seems to prevail, that, under equal circumstances, the temperature decreases from the tropics to the poles.

It would be useless to increase the size of this little volume with a detail of the various temperatures that have been observed in different situations, or of the modifications arising from local causes; this will be found in various works devoted to the subject, more particularly in Humboldt's Treatise on Isothermal Lines.

Respecting the temperature of our globe, M. Arago has made the following remarks:—" 1st, In no part of the earth on land, and in no season, will a thermometer raised from two to three

* Turner's Elements of Chemistry; and Annales de Chim. et de Phys. tom. xxxvi.

metres above the ground, and protected from all reverberation, attain the 46th centigrade degree: 2ndly, In the open air, the temperature of the air, whatever be the place and season, never attains the 31st centigrade degree: 3dly, The greatest degree of
cold which has ever been observed upon our globe, with the thermometer suspended in the air, is 50 centigrade degrees below zero:
4thly, The temperature of the water of the sea, in no latitude, and
in no season, rises above + 30 centigrade degrees *."

Geologists have discovered that the superficial temperature of
the earth has not always remained the same, and that there is
evidence of a very considerable decrease. This evidence will be
found scattered over such parts of the following pages as treat of
organic remains, and therefore need not be adduced here. It may,
however, be right to remark, that it rests on the discovery of
vegetable and animal remains entombed in situations, where, from
the want of a congenial temperature, such animals or vegetables
would now be unable to exist. Undoubtedly this inference rests
on the supposed analogy between animals and vegetables now
existing, and those of a similar general structure found in various
rocks, and at various depths beneath the earth's surface: but as
we now find every animal and vegetable suited to the situations
proper for them, we have a right to infer design at all periods, and
under every possible state of our earth's surface; and therefore
to consider, that similarly constituted animals and vegetables have,
in general, had similar habitats.

This decrease in surface-temperature may arise either from external, superficial, or internal causes.

External Influence.—Heat, derived from the sun, producing
such great effects at present, it has been supposed that a difference in the relative position of our planet and our great lumi-
nary would cause a corresponding change in the surface-temperature of the globe. Theories have been invented which suppose
such a change in the earth's axis as would render the present poles
parts of the equator, and thus capable of having once supported a
tropical vegetation, which has gradually disappeared, and been
replaced by such plants as can exist amid masses of ice and snow.
Mr. Herschel, viewing this subject with the eye of an astronomer,
considers that a diminution of the surface-temperature might arise
from a change in the ellipticity of the earth's orbit, which, though
slowly, gradually becomes more circular. No calculations having
yet been made as to the probable amount of decreased temperature from this cause, it can at present be only considered as a
possible explanation of those geological phænomena which point
to considerable alterations in climates.

Superficial Influence.—A decrease of temperature may arise

* Ann. de Phys. et de Chim. tom. xxvii.; and Edin. Phil. Journ. 1825.

from such a variation in the relative position of land and water, and in the elevation and form of land, as may cause the climate, in any given position on the earth's surface, so to change, that a greater heat may precede a less heat and the land be capable of supporting the vegetables and animals of hot climates at one time, and be incapable of doing so at another. For this ingenious theory we are indebted to Mr. Lyell*. It supposes a combination of external and internal causes; the latter raising or depressing the land in the proper situations, the former supplying the necessary heat. It also supposes the possible recurrence of a cold climate, so that the same situations might alternately be placed under the influence of a raised and a depressed temperature. We have so few data for estimating the value of this theory, that it can only be considered as a possible explanation of a diminished temperature. It must, however, be admitted, that, in every state of the earth's surface, the relative disposition of land and water, and the form or elevation of the land, would always have had, as it now has, very considerable influence on climate.

Internal Influence.—From the earliest times an opinion has existed among philosophers that a central heat exists;—an opinion naturally arising from the phænomena of volcanos and hot springs. But, notwithstanding this opinion, it was not until a comparatively late period that direct experiments were instituted, for the purpose of determining whether the temperature does, or does not, increase with the depth, or from the surface downwards.

Various observations have been made on the temperature of mines in Great Britain, France, Saxony, Switzerland, and even Mexico. All those made previous to 1827 were collected, arranged, and commented on by M. Cordier†. Experiments on the temperature of mines have been made in various ways; sometimes by ascertaining the heat of air in the galleries, sometimes that of the stagnant water at various levels; at others, by observing the temperature of springs at different depths, or that of the waters pumped up from below; and sometimes, though rarely, by obtaining the temperature of the rock itself at various levels.

It soon suggested itself that, though these experiments pointed to an increase of temperature as we descended, the presence of the miners with their lamps or candles, and the explosions of gunpowder in some mines, would cause an increased heat of the air in galleries sufficient to produce exceedingly grave errors. M. Cordier endeavours to assign these and other objections their full value. It is calculated that a miner disengages, in an hour, a

* Principles of Geology.
† Essai sur la Température de l'Intérieur de la Terre: Mém. de l'Acad. tom. vii.

quantity of heat sufficient to raise the temperature of 542 cubic metres of air, one degree above a previous heat of 12° centigrade. It is also inferred that four miners' lamps will produce as much heat as three miners. It is further calculated, that the presence of two hundred miners and two hundred lamps, properly separated from each other, would elevate the temperature of a gallery whose dimensions are one metre by two, and 93,000 metres long, about one degree (centigrade) in one hour. M. Cordier also mentions, that in the coal-mine of Carmeaux "nineteen lamps and twenty-four miners, scattered through two levels, and continually employed during six days in the week, produced, by the hour, a heat sufficient to raise the temperature of the air in the galleries by 1°·66 cent." The air in these galleries was estimated at 12,560 cubic metres.

Another source of error arises from the circulation of air in mines, and its introduction from the surface. This will vary according to the local distribution of the galleries in a mine; but there will always be a tendency to replace expanded and heated air by that which is more dense and cold; consequently, from whatever cause the heat of a mine may be derived, if the air in it be, as usually happens, warmer than that at the surface, the cold air will always strive to get into the mine, and the heated air to escape from it. It follows, that the entrance of air from the exterior surface tends to lower the temperature of the mine, and in some measure to check the heat caused by the workings. M. Cordier observes, on this subject, that the mean temperature of *the mass of air*, introduced into a mine during a year, is lower than the mean temperature of the country for the same year, and estimates the difference between them at between 2° and 3° cent. for the greater part of the mines in our climate*.

The waters in mines may either give too high or too low a temperature, as they may be either derived from beneath or above. If waters descend from the surface into a mine, they will carry with them their original temperature, modified by the heat of the

* Essai sur la Température de l'Interieur de la Terre.

It has been supposed, the air in mines being under a greater pressure than that at the surface, and undergoing this change in a short time, that heat would be evolved sufficient to cause the appearance of an increase of temperature corresponding with an increased depth. But as the cold air will become expanded by the heated air of the workings, and as the change of pressure cannot be very sudden, this does not appear sufficient to account for the phænomena observed. According to Mr. Ivory (Phil. Mag. and Annals of Phil. vol. i. p. 94), one degree of heat, of Fahrenheit's scale, will be extricated from air when it undergoes condensation $= \frac{1}{180}$; and if a mass of air were suddenly reduced to half its bulk, the heat evolved would be $= 90°$.

substances through which they pass; so that their difference of temperature in the mine and on the surface will depend on their abundance or scarcity, and on their slowness or rapidity of motion. Moreover they will constantly tend to reduce the surfaces of rock through which they percolate to their own temperature. The same remarks apply to water derived from a lower level.

The temperature observed in the rock itself will be more or less affected, according to circumstances, by that of the water or air near it. So that the sides of a mine, to certain distances, might possess a heat not common to the mass of rock at the same level.

From these various sources of error, to which others might be added, the observations made under circumstances that might be influenced by them, can only be considered as approximations towards an estimate of the value of this mode of inquiry. To render each set of observations available for what they may be worth, M. Cordier has classed those made under different circumstances under different heads. His tables, thus formed, have also the great advantage of being reduced to common measures of heat and depth. From these the following have been selected as, perhaps, least liable to error.

Table of Observations made on the Springs in Mines.

Names, Authors, and Dates.	Mines.	Depth.	Temperature of the Springs.	mean of the Country.
		Metres.	Deg.	Deg.
Saxony. Daubu-isson. End of winter, 1802.	Lead and Silver of Junghohe-Birke . .	78	9·4	8·
	————	217	12·5	8·
	Beschert Glück	256	13·8	8·
	Himmelfahrt	224	14·4	8·
Britanny. Dau-buisson. 5th Sept., 1805.	Poullaouen	39	11·9	11·5
	. . . . .	75	11·9	11·5
	————	140	14·6	11·5
	Huelgoët	60	12·2	11·
	————	80	15·	11·
	————	120	15·	11·
	. . . . . .	230	19·7	11·
Cornwall. Fox. Pub. 1821.	Dolcoath—Copper . .	439	27·8	10·
Mexico. Humboldt	Guanaxuato—Silver .	522	35·8	16·

Tables of the Temperature of the Rock in Mines*.

I. Thermometer placed in a niche cut in the rock, distant from the principal workings :—the bulb in the rock ; the rest in a glass tube ;—the whole covered by a glass door, closing the niche, and only opened for observation.

			Depth. Metres.	Temperature of Rock.	of Country.
Saxony. De Trébra.	⎱ ⎰	⎰ Mine of Beschert ⎱	180	11·25	8
1805, 1806, 1807 .		⎱ Glück, lead & sil. ⎰	260	15	8
			71·9	8·75	8
Saxony. De Trébra.	⎱ ⎰	⎰ Mine of Alte Hoff-	168·2	12·81	8
1815.		⎱ nung Gotes . . .	268·2	15	8
			379·54	18·75	8

II. Thermometer plunged in the earthy matters at the bottom of galleries, which had been inundated two days†.

Cornwall. Fox.	⎱	United Mines	⎰ 348	30·8	10
Published 1821 . .	⎰		⎱ 366	311	10

III. Thermometer fixed in the rock of a gallery, for eighteen months, at a yard deep.

Cornwall. Fox. Pub- lished 1822 . . .	⎱ ⎰	Dolcoath	421	24·2	10

* The degrees of this, the preceding and following tables, are those of the Centigrade thermometer. When we consider the simplicity of this scale, and the facilities with which calculations can be made with it, it seems strange that its use should not be generally adopted in this country, where we continue to employ, from habit, the least philosophical of the three scales. The centigrade scale can easily be reduced to that of Fahrenheit, by considering that the latter is to the former, between the freezing and boiling points of water, as 180 to 100, or as 9 to 5. The degrees of Reaumur's scale are to those of Fahrenheit's as 4 to 9. As the zero of Fahrenheit's scale is 32° of that scale below the zero in the others, it is always necessary to make a proper allowance for it.

† M. Cordier remarks on the error that may, in this case, arise from the mixed temperature of the galleries, before inundation, produced by the usual causes in mines at work, and of the waters during inundation. On this subject he cites some observations of his own at Ravin, near Carmeaux, which show that the difference of temperature between the rubbish on the floor of the galleries, and that proper to the level, amounted to 2°·6, 20°·8, and even 3°·1 centigrade.

Table of the Temperatures of the Rock observed in the Coal-mines at Carmeaux, Littry, and Decise.

Carmeaux.

	Depth. Metres.	Temperature. Deg.
Water of the well Vériac	6·2	12·9
Water of the well Bigorre	11·5	13·15
Rock at the bottom of Ravin Mine	181·9	17·1
Rock at the bottom of Castellan Mine	192·	19·5

Littry.

Surface	0·	11·
Rock at the bottom of St. Charles Mine—mean of 2 obs. . . . . }	99·	16·135

Decise.

Water of the well Pélisson	8·8	11·4
Water of the Puits des Pavillons	16·9	11·67
Rock in the Jacobé Mine	107·	17·78
——————————————	171·	22·1

These observations were made with great care; "the thermometer was loosely rolled in seven turns of silk paper, closed at bottom, and tied by a string a little beneath the other extremity of the instrument, so that so much of the tube might be withdrawn as might be necessary for an observation of the scale, without fearing the contact of the air : the whole contained in a tin case." This was introduced into a hole from 60 to 65 centimetres in depth and 4 in diameter, inclined at an angle of 10° or 15°; so that the air once entered into the holes could not be renewed, because it became cooler, and consequently heavier, than that of the galleries. The thermometer was kept as nearly as possible at the temperature of the rock, by plunging it among pieces of rock or coal freshly broken off, and by holding it a few instants at the mouth of the hole, into which it was afterwards shut, a strong stopper of paper closing the aperture. The thermometer generally remained in this hole about an hour*.

Temperature of Water in Artesian Wells, and in neglected Mines.

Artesian wells are well known as borings, by which water, at different distances from the surface, rises to, and even above, that

* Where the investigation of the increase or decrease of temperature, beneath such a depth as may be out of atmospheric influences, is so easy, with a few necessary precautions, it is surprising, that n th e British collieries, which are so numerous, and many of which are very deep, so few direct experiments should have been made on the temperature of the rock itself.

surface, from its endeavour to escape. According to the observations of M. Arago, the greater the depth of these wells, the higher is the temperature of the waters that flow from them.

From experiments made by M. Fleuriau de Bellevue, in an Artesian well on the sea-side near Rochelle, the temperature increases with the depth.. The well, at the time of the first experiment, was 3½ inches in diameter, and 316 feet deep, and in it a column of brackish and stagnant water rose to the height of 294 feet. On February 14, 1830, he found the temperature at the bottom, after the thermometer had remained there 24 hours, to be $= 16°·25$ centigrade; the external air being $= 10°·6$. At 11 feet beneath the surface of the water, the temperature was found $= 13°·12$ cent. after the instrument had remained 17 hours. Common wells, varying in depth from 22 to 28 feet, afforded at the same time a mean temperature of $8°·75$. On March 22, MM. Emy and Gon made further experiments on the same well, which was then sunk to the depth of 125·16 metres, or 369½ metrical feet. They found the temperature at the bottom, after the thermometer had remained there 25 hours, $= 18°·12$ cent. Fearful of some inaccuracy in this experiment, they repeated it the next day, when, after the instrument had remained at the bottom for 15 hours, they obtained exactly the same result. M. Fleuriau de Bellevue estimates the mean temperature of the country at $11°·87$ cent.[*]

These experiments were conducted with great care, and seem highly illustrative of an increase of heat from the surface to the interior; for the column of water being subject to the usual laws, it would equalise its temperature by the descent of the cooler and the ascent of the warmer water, if a constant source of comparatively considerable heat did not exist at the bottom.

In the waters of neglected mines also there are numerous observations tending to show that the waters do not follow the laws of their greatest specific gravity in such situations, but that the temperatures greatly increase with their depth. Certainly, in many situations, such as in recently flooded mines, the water would be heated by the galleries in which work had been carried on; but such influence could not continue for a long period, and there are numerous observations which show an increase of temperature in neglected mines. On a subject of this kind, however, great caution is necessary in obtaining the true temperature, and it is very desirable that many of the experiments should be repeated[†].

[*] Fleuriau de Bellevue, Journal de Géologie, tom. i.
[†] A cold spring percolating rapidly from the surface to the deep waters of a neglected mine would tend to cool the waters at such depths.

Temperature of Springs.

The temperature of surface-springs has been supposed to give nearly, if not altogether, the mean temperature of the countries in which they appear. Their value in this respect would depend on whether the waters which supply them be derived from above or beneath, that is, whether they percolate from the surface through porous strata until thrown out by impervious beds, or are forced by some means from comparatively greater depths upwards. Many springs, we well know, come within the first class; but many, we are also certain, come within the second, for their temperatures are greatly above what they could have acquired by mere percolation downwards.

At Paris, the oscillations of the temperature of the earth do not quite cease at 28 metres. Professor Kupffer considers that 25 metres from the surface will afford a depth beneath which springs rise with a uniform temperature throughout the year, being sufficiently removed from atmospheric influences. Admitting this, it is clear that if surface-springs be small, and rise slowly, they may have their temperature somewhat changed during their passage through the 25 metres, while if they rise quickly, and their waters be copious, they will suffer little change in their traverse through that thickness. The question, however, of whence the waters may have been derived, remains the same.

Professor Kupffer has constructed the following table, principally from Von Buch's Treatise on the Temperature of Springs, and from Humboldt's Treatise on Isothermal Lines, with the view of corroborating the observations of Wahlenberg, that the temperature of springs in high latitudes is greater than that of the air, and of those of Von Humboldt and Von Buch, who found that in low latitudes the temperature of springs was lower than that of the air;—showing "that the temperature of the earth is sometimes very different from the mean temperature of the air, and that its distribution follows different laws*."

---

* Kupffer on the Mean Temperature of the Atmosphere and of the Earth in some Parts of Russia: Edin. New Phil. Journ. vol. viii.: and Poggendorf's Annalen, 1829.

Places.	Latitude.	Height above sea. metres.	Temp. of Earth. Fahr.	Temp. of Air. Fahr.	Observers.
Congo	9° S.	45	72·95	78·12	Smith.
Cumana	10¼N.	0	78·12	82·40	Humboldt
St. Jago (CapeVerde Isles)	15 —	0	76·10	77·00	Hamilton.
Rock Fort (Jamaica) . .	18 —	0	79·02	80·60	Hunter.
Havannah	23 —	0	74·30	78·12	Ferrier.
Nepaul	28 —	0?	73·85	77·00	Hamilton.
Teneriffe	28½—	0	64·40	70·92	Von Buch.
Cairo	30 —	0	72·5	72·5	Nouet.
Cincinnati	39 —	160	54·27	53·82	Mansfield.
Philadelphia	40 —	0	54·95	54·27	Warden.
Carmeaux	43 —	300?	55·40	57·87	Cordier.
Geneva	46 —	350	52·02	49·32	Saussure.
Paris	49 —	75	57·70	51·57	Bouvard.
Berlin	52½—	40	50·22	46·40	
Dublin	53 —	0	49·32	40·10	Kirwan.
Kendal	54 —	0	47·75	46·16	Dalton.
Keswick	54½—	0	48·65	47·97	
Konigsberg.	54¼—	0	46·62	43·25	Erman.
Edinburgh	56 —	0	47·75	47·75	Playfair.
Carlscrona	56¼—	0	47·30	47·30	Wahlenberg.
Upsal	60 —	0	43·70	42·12	———
Umeo	64 —	0	37·17	33·35	———
Giwartenfiäll	66 —	500	34·25	25·25	———

To this should be added Professor Kupffer's own observations in Russia.

Places.	Latitude.	Height. Metres.	Temp. of Earth.	Temp. of Air.
Kinekejewa	54½	300	39·87	34·7
Kasan	56	30	43·25	37·4
Nishney-tagilsk	58	200	37·17	31·55
Werchoturie	59	200	36·27	30·42
Bogoslowsk	60	200	35·37	29·30

The above tables, if correct, are sufficient to show that, though the terrestrial temperature, as deduced from springs, decreases from the equator to the poles, it does not decrease according to the mean temperature of the air above it. This seems to point out that there is some modifying cause in action independent of solar influence. Wahlenberg has noticed that many deep-rooted plants and trees only flourish because the temperature of the earth exceeds the mean temperature of the air; and Professor Kupffer remarks that he has often had occasion to confirm this observation in the northern Urals.

At the contact of the atmosphere and earth, we should expect, if they possessed different sources of temperature, that they would mutually act on each other, and that therefore the equal mean temperature of different parts of the earth's surface would, to a certain extent, correspond with equal terrestrial temperatures, as deduced from moderate depths. This may perhaps account for Professor Kupffer's conclusion, that " if we draw lines through all the points which have the same terrestrial temperature, these *isogeothermal lines* resemble the isothermal, as they are parallel to the equator, but diverge from it in several points*."

The temperature of the surface, as deduced from springs, is undoubtedly liable to many errors, as it rests on the assumption that they take the temperature of the earth at moderate depths. Those springs which percolate through porous strata, until thrown out, may take this temperature; but those which seem to come from beneath cannot be supposed, though cooled in their passage upwards, to do so.

The evidence that many springs rise from considerable depths, and possess a temperature independent of solar influence, rests on their great heat, which varies from the boiling point of water downwards to ordinary temperatures. It is impossible to account for this, otherwise than by supposing such heat communicated to the water in parts of the earth far beneath the surface, and removed from atmospheric influence.

The source of the heat in thermal waters has occupied the attention of Berzelius, Von Hoff, Keferstein, Bischoff, and others. The former remarks on those thermal springs which are charged with various salts of soda and carbonic acid, and attributes their origin to the percolation of atmospheric waters to volcanic regions, after which they are forced up to the surface, charged with the substances with which they have become combined in those situations. Von Hoff opposes the theory of a mere volcanic point supplying the necessary heat, and considers it much more probable that this is due to those processes in the interior of our globe which produce volcanos and earthquakes. Keferstein considers that hot vapours and springs are due to volcanic agency, which may be very deeply seated, even below the oldest formations. Bischoff, who details these various opinions†, does not appear to have adopted any decided one of his own on the subject, but directs attention to the possible increase of temperature in the

* Kupffer (memoir cited above).
† Uber die Vulchanischen Mineral quellen Deutschland und Frankreichs: and Edin. New Phil. Journal, 1830.

waters by the internal heat of the earth at great depths, independent of volcanic fires, and observes that if the channels through which the waters flow upwards become once heated, their walls would conduct little heat outwards, for rocks are bad conductors of heat, as is well shown in the case of lava streams, on the outside of which the hand may sometimes be placed, while the melted rock is still flowing inside*.

In support of the opinion that thermal waters may have their high temperature caused by a general internal heat, and not by mere volcanic points on the earth's surface, it may be remarked that thermal springs occur in almost all situations, some of which are far removed from any volcanic points on the surface.

The immediate connection of the Geysers and the volcanos of Iceland is so obvious that few will be found to doubt it; yet when hot springs have been found traversing cracks in strata not volcanic, theories have been invented to explain their origin by chemical combinations at small depths. The salts, however, usually held in solution in these waters do not afford support to this view, and Berzelius has shown it to be untenable with respect to the Carlsbad waters.

To show the various rocks among which thermal springs occur, we will select a few examples. In ranges of mountains they would appear to be far from uncommon, a circumstance which, supposing the ranges to have been elevated by a force acting from beneath, lends additional probability to a general heat beneath the surface. They have been observed in various places in the range of the Himalaya. Captain Hodgson notices them in the course of the Jumna river, so hot that the hand could not be kept in it many moments, and the temperature was too great to be measured by the short scaled thermometer usually employed to ascertain atmospheric heat. Again, at Jumnotri, very copious thermal springs rise through crevices in the granite. The heat was estimated at nearly the boiling point; the finger could not be kept in it two seconds. As the height of Jumnotri is estimated at 10,483 feet above the sea, the water would have the appearance of boiling at a lower temperature than in the plains below: moroever, the springs seem to evolve gas, for they rise with great ebullition; still, however, the temperature of the waters would appear to be very considerable†.

In the range of the Alps, there are also many thermal springs, as has been already remarked by Bakewell. The thermal waters of Bad-Gastein in the Salzburg country are well known.

---

* Monticelli and Covelli.
† Hodgson, Asiatic Researches, vol. xiv.: and Edin. Phil. Journ. vol. viii.

The following are Alpine warm springs noticed by Bakewell*:
Naters, Haut Valais;—temperature = 86° Fahr. Leuk, Haut
Valais,—twelve springs;—temperature varying from 117° to 126°.
Bagnes, in the valley of the same name;—the baths, village, and
one hundred and twenty inhabitants destroyed by the fall of part
of a mountain in the year 1545;—temperature unknown. Ther-
mal springs in the valley of Chamonix;—temperature unknown.
St. Gervais, near the Mont Blanc;—temperature from 94° to 98°.
Aix les Baines, Savoy;—two springs;—temperature from 112°
to 117°. Montiers, Savoy;—temperature not noticed. Brida,
Savoy;—temperature 93° to 97°. Sante de Pucelle, Savoy;—
temperature not noticed. Thermal springs at Cormayeur and St.
Didier, on the Italian side of the Pennine Alp;—temperature 94°.
Warm springs in the Alps near Grenoble.

Many of these thermal waters are of recent discovery, although
those of Aix were known to the Romans; therefore there may
be many in other parts of the Alps which remain unnoticed.

There are also warm springs in the Caucasus, to the N.W. of
the fortress of Constantinohor, with a temperature of from 110° to
114° F.; and there are, no doubt, numerous other thermal waters
in great mountain ranges, with which we are as yet unacquainted.

In the Pyrenees, we have the two celebrated thermal waters
of Barège and Bagnères; the former having a temperature of
120°, at the hottest spring, and the latter of 138° also at the
hottest spring.

The thermal springs at both these places are numerous. At the
latter place there are no less than thirty of them, the temperature
of the least hot of which is = 83¾ F.

There are also thermal waters in the valley of Barège, at St.
Sauveur, = 98°½; as also several springs at Cautieres not far
from the latter place, of which the temperatures vary from 98° to
131°. At Caberu, three leagues from Bagneres, there is a spring
= 80°.

It would be tedious to give a long list of thermal springs; they
occur in all parts of the world, as well remote from, as in the
vicinity of, active volcanos. A great burst of hot springs takes
place near the base of the south-eastern slope of the Ozark mountains, North America, and about six miles north from the Washita,
from which they take their name. They are about seventy in
number, and occur in a ravine between two slate hills. James
states the temperature of these waters at 160° Fahr. Major
Long gives that of several of them, as respectively, 122°, 104°,
106°, 126°, 94°, 92°, 128°, 132°, 151°, 148°, 132°, 124°, 119°,
108°, 122°, 126°, 128°, 130°, 136°, 140°. He also states that,

* On the Thermal Waters of the Alps, Phil. Mag. and Annals, 1828.

" not only confervas and other vegetables grow in and about the hottest springs, but great numbers of little insects are seen constantly sporting about the bottom and sides*.

Another example of the existence of animals and vegetables in thermal springs is to be found at Gastein, where the *Ulva thermalis*, and a fresh-water shell, the *Limneus pereger*, Drap., are found in waters at a temperature of 117° Fahr.

A very copious discharge of hot water takes place in an alluvial plain in a granitic district at Yom-Mack, about twenty miles from Macao, China. Three large springs have respectively the temperatures of 132°, 150°, and 186° Fahr. That with the temperature of 150°, is described as in a state of active ebullition, about thirty feet in diameter, and discharging at least fifteen gallons in a minute†.

The temperature of the waters of Carlsbad is also considerable, being, according to Berzelius, 165° Fahr. Those of Aix-la-Chapelle are = 143°; and at Borset, near Aix-la-Chapelle, there are two springs of which the temperatures are respectively 158° and 127° Fahr. At Balarue, department of Herault, there is one = 128° Fahr. The thermal springs of our own country are not very remarkable for their elevated temperature; for with the exception of those of Bath‡, which are at 116° Fahr., the others can only be considered as tepid, the waters at Buxton being at 82°, those of the Hotwells, Bristol, at 74°, and those of Matlock 68°§.

In the volcanic districts of Italy, thermal springs, as might be expected, are numerous. The waters of the Bagni di Lucca, are however sufficiently removed from a volcano to be here noticed: they rise on the sides of a hill, composed of a sandstone, the *macigno* of the Italians. The district is one of sandstone and limestone, and the hottest spring has a temperature of 131° F.

It may not be altogether out of place to notice the thermal waters of Bath, St. Thomas in the East, Jamaica, to show how widely distributed these heated springs are. They rise at the base of the Blue Mountains, in a valley composed of trap, limestone, and slate. I observed their temperature to be = 127° F.

The hot and cold springs of La Trinchera, three leagues from

* James, Expedition to the Rocky Mountains.
† Livingstone, Edin. Phil. Journal, vol. vi.
‡ These rise through lias, traversing probably red sandstone, carboniferous limestone, &c.
§ The thermal springs of the Hotwells, Matlock, and Buxton, appear among carboniferous limestone.
‖ Although no active volcanos exist in Jamaica, there are the remains of an extinct one on the north side of the island; and earthquakes are, as is well known, sufficiently common.

Valencia (America), may be cited to show, how differently derived waters may be, which make their appearance close to each other. According to Humboldt, there are two springs, only 40 feet asunder, the one cold, the other hot, the thermal waters having the great temperature of 90·°3 centigrade (194°·5 Fahr.). At Cannea, in Ceylon, a thermal spring is stated to exist which does not preserve a constant temperature, but varies from 38° to 41° cent. (100°·4 to 105°·8 F.).

Hot springs are common to the volcanic districts of different parts of the world, as also amid extinct volcanos, such as those of Central France ; to enumerate them would be useless ; but those of Iceland are so remarkable, that a short notice may not be unacceptable to the reader, particularly as they are the most extraordinary thermal springs with which we are acquainted.

Hot springs are numerous in Iceland, but those named the Geysers are the most singular. They are alternately in a state of rest and of violent activity, discharging, at intervals, immense quantities of hot water and steam.

Sir G. Mackenzie states that an irruption of the Great Geyser, which he witnessed, commenced with a sound resembling the distant discharge of a piece of ordnance. " The sound was repeated irregularly and rapidly ; and I had just," observes this author, " given the alarm to my companions*, who were at a little distance, when the water, after heaving several times, suddenly rose in a large column, accompanied by clouds of steam, from the middle of the basin, to the height of ten or twelve feet. The column seemed as if it burst, and sinking down it produced a wave, which caused the water to overflow the basin in considerable quantity. After the first propulsion, the water was thrown up again to the height of about fifteen feet. There was now a succession of jets to the number of eighteen, none of which appeared to me to exceed fifty feet in height ; they lasted about five minutes. Though the wind blew strongly, yet the clouds of vapour were so dense, that after the first two jets I could only see the highest part of the spray, and some of it that was occasionally thrown out sideways. After the last jet, which was the most furious, the water suddenly left the basin, and sunk into the pipe in the centre†." The water sunk in the pipe to the depth of ten feet, but afterwards rose gradually ; when sufficiently high, its temperature was observed, and found = 209° F.

A subsequent irruption of the same Geyser, is thus described by the same author. After an alarm given of its approaching activity, " in an instant," he says, " we were within sight of the

* Dr. Bright and Dr. Holland.
† Mackenzie's Travels in Iceland.

Geyser; the discharges continuing, being more frequent and louder than before, and resembling the distant firing of artillery from a ship at sea . . . . . . It raged furiously and threw up a succession of magnificent jets, the highest of which was at least ninety feet*."

One of the other fountains, which was formerly an insignificant spring, and now known as the New Geyser, alternates in like manner. The irruption commences, as at the Great Geyser, by short jets, which increase in size. When a considerable mass of water is thrown out, the steam rushes forth furiously, accompanied by a loud thundering noise, carrying the water, when Sir G. Mackenzie observed it, to at least seventy feet. He describes it as continuing in this magnificent play for more than half an hour. " When stones are dropped into this pipe, while the steam is rushing out, they are immediately thrown up, and are commonly broken into fragments, some of which are projected to an astonishing height†."

There are other alternating hot springs in Iceland, which are, however, of greatly inferior magnitude to the Geysers. The springs of Reikum, with a temperature of 212° Fahr., rise and fall, and dash up spray to the height of twenty or thirty feet. In the valley of Reikholt, there is a singular alternation of two boiling jets, one throwing the water up twelve feet, the other five‡.

Temperature of the Sea and of Lakes.

This temperature will probably be in part derived from that of the atmosphere, and partly from the earth; but water being, under certain circumstances, able to communicate heat with great rapidity, the temperature will be more speedily equalized in it, than in the solid earth beneath. Water, moreover, at a given temperature possesses a greater specific gravity than when that temperature is either increased or diminished, and will consequently, at that given temperature, sink to the lowest depths. Even if it should be heated there, on the presumption of an internal heat in the earth, the water will still obey the same laws, the newly heated water will ascend, and be replaced by that which is cooler and of greater specific gravity. For, in order that the water should sink to these depths in the first instance, it must be of such a temperature, or specific gravity, as shall enable it to do so, and any change in that temperature will cause it to rise.

* Mackenzie's Travels in Iceland.
† Travels in Iceland, where views of these fountains in full operation will be found.
‡ Waters actually at the boiling point seem exceedingly rare. The thermal waters of Urijino, in Japan, are stated to have a temperature of 212° Fahr., but it does not appear among what rocks they occur.

According to Dr. Hope, the maximum density of fresh water is at a temperature between 39½° and 40° Fahr.*, and this determination has been confirmed by Professor Moll. According to the experiments of Professor Hälloström, the maximum density of water occurs at the temperature of 4°·108 centigrade (39°·394 Fahr.).

It has been considered that the maximum density of sea-water approaches that of fresh water. On this head we have not any good experiments, but it may be supposed that the saline contents of sea-water would have considerable influence on its relative gravity at different temperatures.

In the years 1819 and 1820 I made numerous experiments, with great care, on the temperature of the Swiss Lakes at various depths, which are often considerable. The results of more than one hundred observations on the Lake of Geneva, in September and October 1819, were, that between the surface and a depth of 40 fathoms the temperature varied considerably. From 67° to 64° Fahr. was a common heat from one to five fathoms, and there was a general diminution of temperature downwards to the depth of 40 fathoms, whatever the surface-heat might be; in other words, there was a general increase of specific gravity downwards. From 40 fathoms to 90 fathoms the temperature was always 44°, with one exception near Ouchy, where 45° were observed at a depth of 40 fathoms. From 90 fathoms to the greatest depths, which amounted to 164 fathoms, between Evian and Ouchy, the temperature was invariably = 43°·5 Fahr. It will be observed, that in these experiments, made with a register thermometer constructed for the purpose, the water arranged itself according to the temperatures that would be expected, on the supposition of the maximum density of water being between 39° and 40°†.

After the severe winter of 1819, I made some further experiments, and found that the temperature of the lake still followed the same law.

In May, 1820, I tried the temperature of the lakes of Thun and Zug, and obtained the following results‡:

Lake of Thun.		Lake of Zug.	
Surface	60°	Surface	58°
At 15 fathoms	42	At 15 fathoms	42
At 50 fathoms	41·5	At 25 fathoms	41
At 105 fathoms . . .	41·5	At 38 fathoms	41

In these experiments also, the results are in accordance with the maximum density of water being between 39° and 40°, as

* Trans. Royal Soc. Edinburgh.

† A detailed account of these experiments, with a chart of soundings in the Lake, were inserted in the Bibliothèque Universelle for 1819, from whence they were copied, in part, into the Edin. Phil. Journal, vol. ii.

‡ See also Bibliothèque Universelle for 1820.

was also the case in some which I made in the Lake of Neufchatel, during very cold weather, so cold indeed, that the water froze on the oars of the boat, when the temperature increased towards the supposed maximum density of water.

If we now turn to the experiments that have been made by different navigators on the temperature of the sea at various depths, we shall observe that many point to a somewhat similar heat for the maximum density of sea-water. The following observations by Scoresby show an increase of temperature from the surface downwards, quite in accordance with this supposition.

Situation.	Depth.	Temp.	Situation.	Depth.	Temp.
	Surface ...	29°·0		Surface ...	28°·8
	13 fathoms	31·0		50 fathoms	31·8
Lat. 79° 4′ N.	37 fathoms	33·8	Lat. 76°16′ N.	123 fathoms	33·8
Long. 5° 4′ E.	57 fathoms	34·5		230 fathoms	33·3
	100 fathoms	36·0	Lat. 79° 4′ N.	Surface ...	29
	400 fathoms	36·0		730 fathoms	37

Again, in lat. 78° 2′ N. and long. 0° 10′ W., the same scientific navigator obtained 38° at 761 fathoms, the surface-water being 32°. In one situation, indeed, in lat. 76° 34′ N. the same observer obtained a temperature of 34° at 60 fathoms, and 34°·7 at 100 fathoms, after having had 35° at 40 fathoms; but when we reflect on the errors that may arise in experiments of this nature, even with the greatest care, this result can scarcely invalidate the general evidence, which, if we neglect the immediate surface-water, always liable to be acted on by the temperature of the air in contact with it, seems to point one way, whether observed by Scoresby, Parry, or Franklin*.

Kotzebue, in lat. 36° 9′ N. and long. 148° 9′ W. found the surface-water = 71°·9, the air being 73°; at 25 fathoms the water

---

* The experiments of Capt. Ross are, indeed, opposed to this view, for they give a decrease down to 25° at 660 fathoms, from 30° at 100 fathoms, 29° at 200 fathoms, and 28° at 400 fathoms; in lat. 60° 44′ N. and long. 59° 20′ W. According also to Dr. Marcet, the maximum density of sea-water is not at 40° Fahr. He states that this water decreases in weight to the freezing point, until actually congealed. In four experiments Dr. Marcet cooled sea-water down to between 18° and 19° Fahr., and found that it decreased in bulk till it reached 22°, after which it expanded a little, and continued to do so till the fluid was reduced to between 19° and 18°; when it suddenly expanded, and became ice with a temperature of 28°. It should always be recollected that a saturated solution of common salt does not become solid, or converted into ice at a less temperature than 4° Fahr.; and therefore if the sea should be, as is sometimes supposed, more saline at great depths, and as it appears to be in the Mediterranean from the experiments of Dr. Wollaston, ice could not be formed there at the same temperature as it could nearer the surface.

was at 57°·1 ; at 100 fathoms, 52°·8; and at 30 fathoms, 44° : showing a decrease of temperature towards 39° or 40°. In lat. 23° 3′ N. and 181° 56′ W. Krusenstern obtained, at the surface, 78°; at 25 fathoms, 75°; at 50 fathoms, 70°·5; and at 125 fathoms, 61°·5.

In latitudes south of the tropics, Kotzebue observed a temperature of 49°·5 at 35 fathoms, the surface being at 67°, the air at 68, in lat. 30° 39′ S. The same navigator found the temperature at 196 fathoms to be = 38°·8, in lat. 44° 17′ S. and long. 57° 31′ W.; the surface-water being 54°·9, and the air at 57°·6.

In the tropics, we have two observations at the great depths of 1000 fathoms each. One by Capt. Sabine, in lat. 20° 30′ N. and long. 83° 30′ W.; and the other by Capt. Wauchope, in lat. 3° 26′ S. and long. 7° 59′ E. Capt. Sabine found a temperature of 45°·5 at 1000 fathoms, the surface-water being at 83° ; and Capt. Wauchope obtained 42° at the same depth, the surface-water being at 73°. Other observations within the tropics, at inferior depths, show the same decrease of temperature downwards. Thus Kotzebue in lat. 9° 21′ N. obtained 77° at 250 fathoms, the surface-water being at 83°, and the air at 84°; and under the equator, in long. 177°·5 W., 55° at a depth of 300 fathoms, the surface-water being at 82°·5 and the air at 83°.

It will be observed, from what has been stated above, that the waters of lakes and the ocean (generally) arrange themselves according to certain temperatures, which seem to show that experiments made in the cabinet, and which fix the maximum density of fresh water at a temperature of between 30° and 40° Fahr., are correct, and that the greatest specific gravity of sea-water does very materially differ from that.

The probability of a central heat would appear to rest, first, on the experiments made in mines, which, notwithstanding their liability to error from various sources, still seem to show, particularly those made in the rock itself, an increase of temperature from the surface downwards; secondly, on thermal springs, which are not only abundant among active and extinct volcanos, but also among all varieties of rocks, in various parts of the world ; thirdly, on the presence of volcanos themselves, which are distributed over the globe, and present such a general resemblance to each other, that they may be considered as produced by a common cause, and that cause probably deep-seated; and fourthly, on the terrestrial temperature at comparatively small depths, which does not coincide with the mean temperature of the air above it.

The temperature at the bottom of seas and lakes is not at variance with this probability, as the waters merely arrange themselves according to their greatest specific gravity; and this would take place whether the earth was, or was not, heated towards the

centre. The temperature of the earth, to a small depth immediately beneath a mass of sea, is also likely to be the same as that
of the maximum density of the water, so constantly present to it.

Neither is the probability of internal heat at variance with the
figure of the earth or observed geological phænomena. The figure
of our planet being that which a fluid body would assume if re-
volving in space, it is as probable that this fluidity should be
igneous as aqueous. Geological phænomena attest the irruptions
of igneous matter from the interior at all periods ; as also elevations of mountains and great dislocations of the earth's surface,
caused by forces acting from beneath ; and, finally, a great decrease
of surface temperature. Should we be inclined to build a theory
on the probability of a central heat, we may suppose, as has often
been done, that our world is a mass of igneous matter in the act
of cooling.

Baron Fourier considered, from the form of our spheroid, the
disposition of the internal strata, shown by experiments with the
pendulum, to increase in density with their depth, and from other
considerations, that it seemed proved that a very intense heat formerly penetrated all parts of our globe. He concluded that this
temperature was dissipated into the surrounding planetary spaces,
the temperature of which he concluded from the laws of radiant
heat to be $= -50°$ cent. ($-58°$ Fahr.). He moreover inferred that the earth had nearly reached its limit of cooling. The
original heat contained in a spheroidal mass equal in magnitude to our globe, would diminish more rapidly at the surface than
at great depths, where the elevated temperature would remain for
a great length of time. He further inferred from these circumstances, and from the temperature of mines and springs, that there
is an internal source of heat, raising the temperature of the surface above that which the action of the sun could alone give it*.

Temperature of the Atmosphere.

The gaseous compound termed the Atmosphere, which surrounds the earth, has been calculated, from its powers of refrac-

---

* M. Svanberg, calculating what might possibly be the temperature of
the planetary spaces, proceeds upon another principle than that of the
radiation of heat. He supposes that the planetary spaces never undergo
any change of temperature, but that the capacity for elevation of temperature, above that which constantly reigns in the ethereal regions,
exists only within the limits of the planetary atmosphere. He obtains
for the result of his calculations a temperature $= -49°·85$ cent. Observing this near approach towards Baron Fourier's supposed temperature,
he had the curiosity to calculate the temperature according to Lambert's
statements, respecting the absorption which takes place in a ray of light
passing from the zenith through the whole atmosphere, and found that

tion, to extend upwards about forty-five miles. Dr. Wollaston considered, from the laws of the expansions of gases, that it might reach to at least forty miles, with its properties uninjured by rarefaction. On this head Dr. Turner observes, " that the tension or elasticity of gaseous matter is lessened by two causes, diminution of pressure, and reduction of temperature." And he further remarks, that the former alone has been taken into account by Dr. Wollaston, while it appears to him that the extreme cold at great heights would also be sufficient to limit the extent of the atmosphere*.

Though no part of the solid earth is so elevated above the general surface as to be exposed to a very considerable depression of temperature, yet numerous mountains are sufficiently high for being covered, at their summits, with what has been termed perpetual snow, the prolific parent of innumerable rivers, without which many regions would be uninhabitable. The line of perpetual snow differs generally according to the latitude, and is liable to very great variations from local causes. The following Table may serve to give a general idea of this line † :

Lat.	Height in English feet.	Lat.	Height in English feet.
0°	15,207	45°	7,671
5	15,095	50	6,334
10	14,761	55	5,034
15	14,220	60	3,818
20	13,478	65	2,722
25	12,557	70	1,778
30	11,484	75	1,016
35	10,287	80	457
40	9,001	85	117

It has been supposed that the temperature of the atmosphere diminished equally upwards in different latitudes; but the following Table, by Humboldt, will show that this is not the case, and that, on the contrary, the decrease is much more rapid in the temperate than in the equatorial zone.

he obtained — 50°·35 for the result. A curious coincidence between the results of the three modes of calculation.—Berzelius. Annual Progress of Chemical and Physical Science. Edin. Journ. of Science, vol. iii. New Series.

    * Turner, Elements of Chemistry, p. 221.
    † Supp. Encyc. Brit., art. *Climate.*

Height in English feet.	Equatorial zone; from 0° to 10°.		Temperate zone; from 45° to 47°.	
	Mean Temp.	Differ- ence.	Mean Temp.	Differ- ence.
0 . .	81·5	0	53·6	0
3,195 . .	71·2	10·3	41·0	12·6
6,392 . .	65·1	6·1	31·6	9·4
9,587 . .	57·7	7·4	23·4	8·2
12,792 . .	44·6	13·1		
15,965 . .	34·7	9·9		

The curve, representing the line of perpetual snow, will not be equal in the northern and southern hemispheres, as the latter is found to be colder than the former.

From the variable height at which perpetual snow commences, it follows, all other circumstances being the same, that the extent of dry land capable of sustaining animal and vegetable life, will decrease from the equator to the poles, and, consequently, that there is a greater probability of an abundance of terrestrial remains being entombed in any deposit now taking place in the tropics, than in similar deposits in high latitudes.

Valleys.

A classification of valleys cannot well be accomplished without some violence, as the various depressions of land, to which the term valley has been much too generally applied, pass into each other in such a manner as to produce compounds of no easy arrangement. No great value is therefore attached to the following sketch.

Mountain Valleys.—These are both longitudinal and transverse; ranging either in the direction of the mountain chain, or across that direction. Their sides are generally rugged, crowned by lofty pinnacles and broken masses, and are, for the most part, steep. Atmospheric agents, far from producing a milder outline, generally add to their broken appearance. The melting of ice, and snow, and the drain of rain-waters furrow their sides, bringing down detritus to the rivers, which, when levels are favourable, deposit it in situations, well suited to vegetation; so that in mountain regions patches of verdure occur amid the wildest scenes, presenting a singular contrast to the broken forms of the surrounding mountains. When levels are unfavourable, or the fallen blocks large, the masses accumulate in the water-courses, and produce innumerable cascades, adding to the desolate character of such regions.

Lowland Valleys.—These differ from the preceding in their

rounded form, which would render a section of them an undulating line, the undulations varying in the proximity of the higher parts and in depth, so that the more elevated portions may even be many miles asunder, and the depth inconsiderable. From the comparatively gentle slopes of these valleys, atmospheric agents, though still able to decompose the rocks beneath, do not transport the detritus to any considerable distance, except in climates and situations where heavy torrents of rain descend on land unfavourable to vegetation; yet, even in this case, the general rounded outlines of the hills are not very considerably impaired, though deep furrows are made in their sides.

Ravines and Gorges.—These are bounded by more or less perpendicular walls of rock, and are common both among mountain and lowland valleys, but more particularly the former. They frequently communicate between more open spaces, and their edges may often be approached without any suspicion that they exist, the country appearing as one continuous slope or level.

Broad Flat-bottomed Valleys.—Level plains of greater or less extent, bounded by hills or mountains on either side; such as the great valley of the Rhine below Basle, bounded on one side by the Swartzwald, and on the other by the Vosges.

Such a diversity of form would seem to suggest a diversity of origin. The mountain valleys for the most part resemble large cracks, produced when the strata were suddenly elevated and contorted, while the lowland valleys appear as if a large body of water had passed over them, rounding the inequalities, and acting on masses of strata in proportion to their power of resistance. The gorges or ravines would seem due to the cutting power of running waters, or to rifts in the rocks produced by violent convulsions. The flat-bottomed valleys have the character of drained lakes, or situations where the rivers or floods, not having any great velocity, deposit considerable quantities of sediment over a flat surface.

As we may suppose hill and dale, mountain and valley to have existed from the earliest geological periods, and that strata were by no means deposited in one even plane surface, we have now a very complicated system of depressions; though as a general fact it may be stated, that the superior stratified rocks have filled up and covered over numerous inequalities of the inferior stratified rocks, as is the case in Normandy, where the oolite group covers over the uneven surface of slates, limestones and grauwacke, the latter rocks here and there protruding through the stratification of the former, and becoming visible where rivers cut the superincumbent beds.

If we can imagine a violent disruption of strata, contorting or throwing them on their edges, large rents and fractures would be the natural consequences, producing longitudinal and transverse

C 2

fissures; but these would merely gape, and their origin would appear clear, if not modified by some subsequent action. If we suppose, with the advocates for no greater effects than we daily witness, that mountains have been raised gradually by a multi-tude of earthquakes acting always in the same line, we shall have great difficulty in explaining the position of strata in high ranges, more particularly those, (such are by no means uncommon in the calcareous Alps,) where whole mountains are contorted, and even appear as if thrown over, as at the Righi. Whereas, if we suppose that the elevations have been more violent, these dif-ficulties would appear to vanish, and the upturned, overthrown, and contorted strata, the longitudinal and transverse cracks or valleys, would be more in harmony with each other.

If we should suppose a violent disruption of strata to take place beneath the waters of an ocean, these waters would be greatly agitated and react upon the land, rushing into the cracks; sweep-ing away pinnacles; driving blocks and loosely aggregated strata before them; rounding off angles; and accumulating detritus at the bottom of hollows. Should such a sudden elevation be effected, partly in the ocean, and partly out of it, the reaction of the sea would only reach the lower portion of the upraised strata, and these only would present rounded forms. Should the strata be elevated only in the atmosphere, the modification of the original cracks would be effected by atmospheric agency alone.

Although lowland valleys generally present rounded forms, the strata composing such districts are often far from undisturbed; on the contrary, they are often upturned, contorted, and fractured, the lines of valleys being frequently the same with those of the faults or fractures. Often, however, no appearances of fracture are visible in the hills, though these are traversed by them in various directions. Of this fact the neighbourhood of Weymouth, in our own country, may be cited as affording good examples.

Valleys of Elevation are those which seem to have originated in a fracture of the strata, and a movement of the fractured part upwards, so that the strata dip from the valley on either side. Probably a very large proportion of mountain valleys might be arranged under this head; but at present geologists seem to have confined the application of the term to those which are bounded by hills of moderate height.

Prof. Buckland (in 1825) noticed valleys of this kind at New Kingsclere, Bower Chalk near Shaftesbury, and Poxwell near Weymouth. The annexed diagram is a section of that of Kingsclere.

Fig. 1.

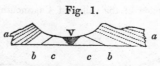

V, valley of Kingsclere : *a a*, chalk with flints : *b b*, chalk without flints : *c c*, green sand.

It will at once be observed, that the strata on either side were once continuous, and that they have been upheaved, producing a fracture, which, by subsequent denudation, has been formed into the valley we now see.

Subsequently to the observations of Prof. Buckland, similar valleys in Germany have occupied the attention of M. Hoffman, who endeavours also to show that they are connected with springs impregnated with carbonic acid gas. In support of this opinion he cites the valley of Pyrmont, of which he gives the following section, which will be seen closely to correspond, in its general characters, with that of Kingsclere.

Fig. 2.

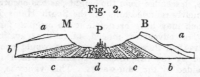

M, the Muhlberg, 1107 feet: B, the Bomberg, 1136 feet: P, Pyrmont, the bottom of the valley being 250 feet : *a a*, keuper (red or variegated marl) : *b b*, muschelkalk : *c c*, grès bigarré, broken into fragments at *d*, through which the acidulous waters are forced out.

As at Kingsclere, the strata have not been forced up to equal heights on either side. The grès bigarré rises to 850 feet on the Bomberg or north side ; while on the Muhlberg or south side it only reaches 540 feet, with an inferior dip. The theoretical opinions connected with these appearances will be noticed in the sequel; at present it is only necessary to point out the existence of such valleys.

M. Hoffman also notices similar appearances, with acidulous springs, in the valley of Dribourg, on the left bank of the Weser, and several other combinations of the like kind*.

Valleys of Denudation.—Although the valleys of elevation above noticed, may also be termed valleys of denudation, this name seems given, in preference, to those valleys where the strata are not far removed from an horizontal position on either side, and of which also the former continuity cannot be doubted. Of these,

---

* Journal de Géologie, tom. i.

the following section of the valley of Charmouth will afford an
example.

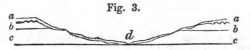

Fig. 3.

a a, summits of the hills composed of angular flint and chert
gravel, the remains of former superincumbent chalk and green
sand, which have been partially dissolved in place : *b b*, green
sand, with an uneven upper surface resulting from the causes that
have produced the gravel : *c c*, lias, in which the lower part of the
valley has been excavated : *d*, small river Char flowing at the bottom. If proportions had been strictly attended to, the stream
would have been invisible. On the sides of the hills, from *a* to *d*,
much chert and flint gravel is distributed over the rocks *b* and *c*,
and it may be questionable how much of it has, during a great
lapse of time, descended from the heights, as has occurred on the
slopes of similarly rounded hills, in the South Hams in Devon,
and how much may have been left at the original formation of
the valley. The advocates, indeed, of such excavations by no
greater powers than those we daily witness, would consider this
valley formed by the insignificant streamlet which now flows
through it, aided by the rain-waters. This valley is, however,
the sole channel of drainage for a district many miles in extent,
in which the actual river, with every assistance from floods, has
only effected a cut, varying from four to fifteen feet deep, bounded
by perpendicular walls; these walls not composed, for the most
part, of lias, but of gravel and drifted materials, such as are
strewed over the valley of all heights, from the bed of the river to
the tops of the hills. Such valleys are common in various parts
of the world, and not unfrequently are without running waters in
them, so that these could not have caused them. Even in Jamaica,
where heavy tropical rains are sufficiently common, there are
valleys, in which the waters are swallowed up by subterraneous
cavities, or sink holes, and no continuous streams are formed. In
England we have examples of dry valleys, in our chalk districts,
in the oolite of Yorkshire, and among the slates of the South
Hams, Devon * ; a covering of vegetation or turf most commonly
protecting the surface from removal, even during heavy rains.
On the west coast of Peru, where rain never falls, there are also
some remarkable examples of dry valleys, which, judging from

---

* These latter are due to the highly inclined position of the strata,
between the fissures of which the rain-water, after having been received
in a porous superficial gravel, percolates.

sketches, resemble many a lowland valley with rounded sides in Europe. The form of these valleys is also opposed to their production by running waters, for they are rounded and not bounded by perpendicular walls.

Sometimes the upper part of a hill being composed of harder materials than the lower portion, it advances with a somewhat bold escarpment.

The general form of these valleys would seem to suggest a mode of formation somewhat different from that of the mountain valleys; one which has permitted a very general removal of bold projecting points. There is scarcely a district of any considerable extent, composed of these valleys, which does not contain fissures, or faults, even when the strata are, as a mass, not far removed from an horizontal position. In other situations the strata are upheaved, contorted, and intruded by trap rocks, yet the general forms of the valleys are not considerably altered; the same rounded forms still prevail. From this general prevalence of the same character, it may not be unreasonable to conclude that some similar cause has produced them. They appear as if scooped out by large moving masses of waters; the least resisting parts first giving way. We might imagine this to have been effected by great disturbances beneath an ocean; such as would be caused by the elevation of long ranges of mountains near them, or a disruption of the strata of which they were actually composed;—in fact, by submarine earthquakes of much greater force than those which we now witness. Earthquakes of the present day frequently produce violent waves, which discharged on a coast ravage all within their reach. The sudden elevation of mountains to the height of several thousand feet would be accompanied by violent disturbance of the land, causing the waters of neighbouring seas to rise considerably, and overwhelm land within their reach; and these discharges of masses of water might have great scooping powers, more particularly if they acted on fractured strata, or small previously existing depressions. These valleys may also have been formed beneath agitated waters, in which currents moving with great velocity were produced; the land having been afterwards protruded above the level of the sea. These observations on the origin of lowland valleys should be considered as mere speculations, which future investigations may show to be either probable or improbable. One argument in their favour, in preference to the supposition that they have been scooped out by rivers, is, that in many instances the rivers quit the valleys, which would appear continuations of their natural channels, and pass through gorges or ravines cut in lands of considerable elevation on one side; the barrier to its natural passage onwards being merely a gentle rise of a few feet at the bottom of the valley, not easily observed.

Changes on the Surface of the Globe.

The present condition of our planet's surface is far from stable; on the contrary, if time enough could be allowed, a great change in the relations of land and water would be effected. This process is undoubtedly slow, but it is nevertheless certain, and so apparent, that many persons have been inclined to refer all geological phænomena to a continuance of those effects of existing causes which we daily witness. As far as we can judge from known facts, this opinion seems to have been somewhat hastily adopted, and not altogether in accordance with all those geological phænomena with which we are at present acquainted. As the student may, however, be supposed not to possess a knowledge of these phænomena, the consideration of their relative value must be waived until he becomes more familiar with the subject.

After geologists had ceased to amuse themselves by fabricating theories, without being at the trouble of examining the surface structure of that world which they made, modified, and broke to pieces at their own good will and pleasure, and when it was thought that a knowledge of facts was somewhat necessary to a knowledge of the subject, it was soon observed that changes had taken place on the world's surface. Facts being still few, hypotheses were easily formed, and were more or less plausible according to the knowledge of the day. These will be found in the various works which treat of the history of geology, and therefore need not be produced here; it will be sufficient to observe that the two prevailing theories of the present time are, 1st, That which attributes all geological phænomena to such effects of existing causes as we now witness; and, 2ndly, That which considers them referable to series of catastrophes or sudden revolutions. The difference in the two theories is in reality not very great; the question being merely one of intensity of forces, so that, probably, by uniting the two, we should approximate nearer to the truth.

Classification of Rocks.

The term Rock is applied by geologists, not only to the hard substances to which this name is commonly given, but also to those various sands, gravels, shales, marls, or clays which form beds, strata, or masses*.

Rocks were first divided into two classes, Primitive and Secondary, it being considered that they originated under different circumstances; the latter only containing organic remains. To this Werner added a third class, which he named Transition, considering that it exhibited a passage from the primary into the secondary. Subsequently, from observations made by MM. Cuvier

---

* For the terms used in geology, see Appendix A.

and Brongniart on the country round Paris, a fourth class was instituted, and called Tertiary, because the strata composing it occurred above the chalk, a rock considered as the highest of the secondary class. These divisions or classes are more or less in use at the present time, though it seems somewhat generally admitted that they are insufficient, and not in accordance with the present state of science. Numerous modifications and divisions have been proposed, which, though preferable to the preceding, have not been adopted, the force of habit, possibly, having prevailed.

To propose in the present state of geological science any classification of rocks which should pretend to more than temporary utility, would be to assume a more intimate acquaintance with the earth's crust than we possess. Our knowledge of this structure is far from extensive, and principally confined to certain portions of Europe. Still, however, a mass of information has gradually been collected, particularly as respects this quarter of the world, tending to certain general and important conclusions; among which the principal are,—that rocks may be divided into two great classes, the stratified and the unstratified;—that of the former some contain organic remains, and others do not;—and that the non-fossiliferous stratified rocks, as a mass, occupy an inferior place to the fossiliferous* strata, also taken as a mass. The next important conclusion is, that among the stratified fossiliferous rocks there is a certain order of superposition, apparently marked by peculiar general accumulations of organic remains, though the mineralogical character varies materially. It has even been supposed that in the divisions termed formations, there are found certain species of shells, &c. characteristic of each. Of this supposition extended observation can alone prove the truth; but it must not be supposed, as some now do, that in any accumulation of ten or twenty beds, characterised by the presence of distinct fossils in a given district, the organic remains will be found equally characteristic of the same part of the series at remote distances.

To suppose that all the formations, into which it has been thought advisable to divide European rocks, can be detected by the same organic remains in various distant points of the globe, is to assume that the vegetables and animals distributed over the surface of the world were always the same at the same time, and that they were all destroyed at the same moment, to be replaced by a new creation, differing specifically, if not generically, from that which immediately preceded it. From this theory it would also be inferred that the whole surface of the world possessed an uniform temperature at the same given epoch.

It has been considered, but has not yet been sufficiently proved, that the lowest rocks in which organic remains are found en-

* The term *fossiliferous* is here confined to organic remains.

tombed, show a general uniformity in their organic contents at points on the surface considerably distant from each other, and that this general uniformity gradually disappeared, until animal and vegetable life became as different in different latitudes, and even under various meridians, as it now is. How far this opinion may, or may not, be correct, can only be seen when geological facts shall have been sufficiently multiplied; but it is one which demands considerable attention, as the classification of fossiliferous rocks greatly depends upon it. Should it eventually be found to a certain degree correct, it would not be at variance with the theory of a central heat, which having diminished, permitted solar heat and light gradually to acquire an influence on the earth's surface.

Classifications of rocks should be convenient, suited to the state of science, and as free as possible from a leading theory. The usual divisions of Primitive, Transition, Secondary, and Tertiary, may perhaps be convenient, but they certainly cannot lay claim to either equality with the state of science, or freedom from theory.

In the accompanying Table, rocks are first divided into Stratified and Unstratified, a natural division, or at all events one convenient for practical purposes, independent of the theoretical opinions that may be connected with either of these two great classes of rocks. The same may, perhaps, also be said of the next great division; namely, that of the stratified rocks into Superior or Fossiliferous, and Inferior or Non-fossiliferous. The superior stratified or fossiliferous rocks are divided into groups. We are yet well acquainted with so small a portion of the real structure of the earth's exposed surface, that all general classifications seem premature; and it appears useless to attempt others than those which are calculated for temporary purposes, and of such a nature as not to impede, by an assumption of more knowledge than we possess, the general advancement of geology.

Stratified Rocks. Group 1. (*Modern*) seems at first sight natural and easily determined; but in practice it is often very difficult to say where it commences. When we take into consideration the great depth of many ravines and gorges, which appear to originate in the cutting power of existing rivers, the cliffs even of the hardest rocks which more or less bound any extent of coast, and the immense accumulations of comparatively modern land, such as those which constitute the deltas of great rivers, and the great flats, such as those on the western side of South America, there is a difficulty in referring these phænomena to the duration of a comparatively short period of time. Geologically speaking, the epoch is recent; but according to our ideas of time, it appears to reach back far beyond the dates commonly assigned to the present order of things.

Group 2. (*Erratic Block*) is exceedingly difficult to characterise. It may however be considered, merely for convenience, as comprising those superficial gravels, breccias, and transported materials which occur in situations where causes similar to those now in action could not have placed them. The most extraordinary feature of this group is the distribution of those enormous blocks or boulders found so singularly perched on mountains, or scattered over plains, far distant from the rocks from whence they appear to have been broken.

Group 3. (*Supercretaceous*) comprises the rocks usually termed tertiary: they are exceedingly various, and contain an immense accumulation of organic remains, terrestrial, fresh-water, and marine. This group has lately been shown to approach, more closely than was supposed, to the existing order of things on the one side, and to the following group on the other.

Group 4. (*Cretaceous*) contains the rocks which in England and the North of France are characterised by chalk in the upper part, and sands and sandstones in the lower. The term cretaceous' is perhaps an indifferent one; for probably, the mineralogical character of the upper portion, whence the name is derived, is local, that is, confined to particular portions of Europe, and may be represented elsewhere by dark compact sandstones and even sandstones. As, however, geologists are perfectly agreed as to what rock is meant when we speak of the ' chalk,' there seems no objection to retain it for the present. The Wealden rocks have been arranged, for the present, in this group, though their organic remains show a different origin, because they may be conveniently studied in connection with it.

Group 5. (*Oolitic*) comprises the various members of the oolite or Jura limestone formation, including lias. The term ' oolitic' has been retained upon the same principle as that of ' cretaceous.' In point of fact, this mineralogical character is found only in an insignificant part of the rocks known as the oolite formation in England and France; and moreover it is not confined to the rocks in question, but is common to many others. In the Alps and in Italy the oolite formation seems replaced by dark and compact marble limestones, so that its mineralogical structure is of little value.

Group 6. (*Red Sandstone*) contains the red or variegated marls (marnes irisées, keuper), the muschelkalk, the new red or variegated sandstone (grès bigarré, bunter sandstein), the zechstein or magnesian limestone, and the red conglomerate (rothe todte liegende, grès rouge). The whole is considered as a mass of conglomerates, sandstones, and marls, generally of a red colour, but most frequently variegated on the upper parts. The limestones may be considered subordinate : sometimes only one occurs, sometimes the other, and sometimes both are wanting.

There seems no good reason for supposing that other limestones may not be developed in this group in other parts of the world.

Group 7. (*Carboniferous.*) Coal-measures, carboniferous lime-stone, and old red sandstone of the English. The former would appear in the greater number of instances to be naturally divided from the group (6) above it; but the latter, though disconnected from the group (8) beneath in the North of England, is apparently so united with it in many other situations, that the old red sand-stone may be considered as little else than the upper part of the greywacke series in those places.

Group 8. (*Greywacke.*) This may be considered as a mass of sandstones, slates, and conglomerates, in which limestones are occasionally developed. Sandstones which mineralogically resem-ble the old red sandstone of the English, not only occupy the upper part, but frequently also other situations in the series.

Group 9. (*Lowest Fossiliferous.*) Slates of various kinds, among which stratified compounds, resembling some of the unstra-tified rocks, are by no means unfrequent. Organic remains very rare.

Inferior or Non-fossiliferous Stratified Rocks,—comprising slates of different kinds, and various crystalline compounds ar-ranged in strata, such as saccharine marble, in which other mine-rals may or may not be imbedded, gneiss, protogine, &c. From various circumstances, many rocks in the previous division so as-sume the mineralogical characters of those in this, as to be un-distinguishable from them, except by geological situation; but it may be assumed, that, as a mass, the strata in this division are greatly more crystalline than in those of the superior stratified rocks, whose origin seems chiefly mechanical.

Unstratified Rocks.—This great natural division is one of con-siderable importance in the history of our globe, as the rocks com-posing it seem to have caused, jointly with the forces that ejected them, very considerable changes on the earth's surface. They are very generally admitted to be of igneous origin, some of them in-deed, those produced by active volcanoes, never could have been doubted. Their great characteristic is a tendency to crystalline structure, though, in many, this cannot be traced. Every grada-tion from the crystalline to the non-crystalline structure can fre-quently be observed in the same mass. The minerals, felspar, quartz, hornblende, mica, diallage and serpentine enter largely into the composition of these rocks, more particularly the former.

In proposing this classification, I am fully aware that many just objections may be made to it, but it pretends to little beyond convenience: and if geologists could be induced to use something of this kind, or any other that would better answer the purpose of relieving us from the old theoretical terms, I cannot but imagine that the science would derive benefit from the change.

In the following part of this little volume, geological phænomena will be noticed in accordance with this classification. But to
enable those who prefer other arrangements to avail themselves
of any facts that may be brought forward, the equivalents of the
divisions or groups above noticed are given in the annexed Table,
where the classifications of Conybeare, Brongniart, and Omalius
D'Halloy, as also the improved Wernerian, are placed in parallel
columns with it.

Note to annexed Table.—It is by no means easy to show M.
Brongniart's classification in this Table, as the same things may
occupy two places in it. Thus the carboniferous or mountain limestone is found under both the heads of Izemian and Hemilysian
rocks, the old red sandstone being supposed interposed. Perhaps
M. Brongniart may have considered the limestones of the different localities which he cites in England, as different; they are,
however, the same.

STRATIFIED ROCKS.	SUPERIOR STRATIFIED, or FOSSILIFEROUS.	1. Modern Group.	{ Detritus of various kinds produced by causes now in action; Coral islands; Travertino, &c. } ...
		2. Erratic Block Group	{ Transported boulders and blocks; gravels on hills and plains, apparently produced by greater forces than those now in action. } ...
		3. Supercretaceous Group	{ Various deposits above the chalk, such as, in England, the Crag, Isle of Wight beds, London and Plastic clays. In France, the freshwater and marine rocks of Paris, &c. } ...
		4. Cretaceous Group.	{ 1. Chalk. 2. Upper greensand. 3. Gault. 4. Lower greensand. To which may be added, for convenience, 1. Weald clay. 2. Hastings sands. 3. Purbeck beds. }
		5. Oolitic Group......	{ The rocks usually known as the Oolite formation, including the Lias. } ...
		6. Red Sandstone Group............	{ 1. Variegated or Red marl. 2. Muschelkalk. 3. Red sandstone. 4. Zechstein; and 5. Red conglomerate. }
		7. Carboniferous Group.............	{ 1. Coal measures. 2. Carboniferous limestone. 3. Old red sandstone .... } ...
		8. Grauwacke Group.	{ Grauwacke, thick-bedded and schistose, sometimes red; Grauwacke limestones; Grauwacke clay slates, &c. } ...
		9. Lowest Fossiliferous Group......	{ Various slates, frequently mixed with stratified compounds resembling those of the unstratified rocks .... }
	INFERIOR STRATIFIED, or NON-FOSSILIFEROUS.	No determinate order of superposition....	{ Various schistose rocks, and many crystalline stratified compounds, such as Gneiss, Protogine, &c. .... } ...
UNSTRATIFIED ROCKS.		Volcanic, Trappean, Serpentinous, and Granitic rocks.....	{ Ancient and modern Lava, Trachyte, Basalt, Greenstone, Corneans, Augite and Hornblende Porphyries, Serpentine, Diallage rock, Sienite, Quartziferous Porphyry, Granite, &c. } ...

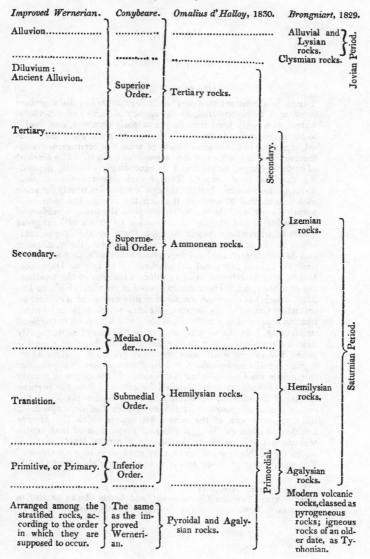

Improved Wernerian.	Conybeare.	Omalius d'Halloy, 1830.	Brongniart, 1829.	
Alluvion...............	....	....	Alluvial and Lysian rocks.	Jovian Period.
....	....	....	Clysmian rocks.	
Diluvium : Ancient Alluvion.	Superior Order.	Tertiary rocks.		
Tertiary...............	....	....		Secondary.
Secondary.	Supermedial Order.	Ammonean rocks.	Izemian rocks.	
....	Medial Order......	....		
Transition.	Submedial Order.	Hemilysian rocks.	Hemilysian rocks.	Saturnian Period.
....		....		
Primitive, or Primary.	Inferior Order.	....	Agalysian rocks.	Primordial.
....	....	....		
Arranged among the stratified rocks, according to the order in which they are supposed to occur.	The same as the improved Wernerian.	Pyroidal and Agalysian rocks.	Modern volcanic rocks, classed as pyrogeneous rocks; igneous rocks of an older date, as Typhonian.	

Section II.

MODERN GROUP.

Degradation of Land.

There is a constant tendency in all decomposed or disintegrated substances to be removed, by the agency of rains and superficial waters, to a lower level than they previously occupied, and finally to be transported into the sea.   There is no rock, even the hardest, that does not bear some marks of what has been termed weathering, or of the action of the atmosphere upon it.   The amount of surface-change, so produced, is exceedingly variable, depending much on local causes.   Thus, a rock may undergo complete disintegration in one situation, though composed of nearly the same materials as that in another, of which the change has been comparatively trifling.   When we contemplate the present surface of our continents and islands, we cannot but be struck with the great effects that have been produced upon them by the agents commonly known as existing causes; and among these, the weathering and degradation of land are very remarkable, attesting a lapse of time far beyond the usual calculations.   The tors of Dartmoor, Devon, may be referred to as excellent examples of the weathering of a hard rock.   These are composed of granite, which, as Dr. MacCulloch has observed, are divided into masses of a cubical or prismatic shape.   " By degrees, surfaces which were in contact become separated to a certain distance, which goes on to augment indefinitely.   As the wearing continues to proceed more rapidly near the parts which are most external, and therefore most exposed, the masses which were orginally prismatic acquire an irregular curvilinear boundary, and the stone assumes an appearance resembling the Cheese-wring (Cornwall). If the centre of gravity of the mass chances to be high and far removed from the perpendicular of its fulcrum, the stone falls from its elevation, and becomes constantly rounder by the continuance of decomposition, till it assumes one of the spheroidal figures which the granite boulders so often exhibit.   A different disposition of that centre will cause it to preserve its position for a greater length of time, or, in favourable circumstances, may produce a logging stone*.'' The weathering of these tors is so exceedingly slow that the life

---

* MacCulloch.   Geol. Trans. 1st series, vol. ii.; where there are views of Vixen Tor, the Cheese-wring, and Logan Rock; as also in Sections and Views illustrative of Geological Phænomena, pl. 20.

of man will scarcely permit him to observe a change; therefore
the period requisite to produce these present appearances must
have been very considerable. The surface of the whole country
round these districts attests the same great lapse of time. Whatever may be the nature of the rock, it is disintegrated to considerable depths; porphyries, slates, compact sandstones, trap rocks,
—all have suffered, but the valleys appear to have previously existed, and the general form of the land to have been much the
same as it now is. The following section will explain this decomposition of surface.

Fig. 4.

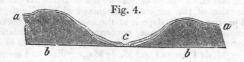

a a, decomposition of the rock *b b* following a line of previous
elevation and depression, the accumulation being greatest at the
bottom of the valley *c,* frequently cut through by a river or rivulet,
and sometimes exposing a stratified appearance, as if the disintegrated substances of the hill-sides had slipped over each other to
the bottom of the valley. The maximum quantity of detritus so
brought down to the bottom of a valley, sometimes amounts to
25 or 30 feet. This detritus, which is often very loosely aggregated, is now indeed protected from removal, at least to a great
extent, by grass and general cultivation. The various appearances
of this detritus are singular; for often larger pieces, perhaps of
twenty or thirty pounds weight, are included among small fragments and even sand. Of this the following section, exhibited on
the sea-shore at Black Pool, Dartmouth, affords an example.

Fig. 5.

a a, detritus from the grauwacke slates *b b,* more thickly accumulated at *e f. c c,* a high beach of small quartz shingles, defending
the bottom of the valley *d* (which is much lower than the crown of
the beach) and the cliffs on either side. The drainage of the valley
escapes in a serpentine manner by a rivulet at *e.* At *e* and *f,*
many large fragments are mixed with the smaller.

The slates in the South Hams, Devon, are frequently surmounted by a superficial covering of fragments, which, at their
union with the undecomposed rock, appear as if some force had
been exercised at the commencement; the slates being broken

and turned back in the manner
represented beneath.

Fig. 6.

a, vegetable soil : b, small fragments of slate resting in various
directions: c, portions of laminæ,
turned backwards, sometimes without fracture.

If we proceed to the eastward from the South Hams, the same
appearances present themselves, whatever may be the nature of the
rock, though they become somewhat more complicated upon Haldon Hill, and on the coast of Sidmouth and Lyme Regis, as this
decomposition of the surface seems mixed with a disintegration
effected previous to the deposit of the supercretaceous rocks. A
deep disintegration of surface, conforming to the undulations of
the country, is well observed in Normandy, where it has been described by M. de Caumont and M. de Magneville; and seems due
to the action of the same causes which have produced the decomposition of surface in the South of England.

This destruction of the surface is common to most countries;
and if the rock so weathered be limestone, there is, not unfrequently, a reconsolidation of the parts by means of calcareous
matter deposited by the water that percolates through the fragments, and which dissolves a portion of them. At Nice, the fractured surface thus reunited is so hard, that, if it occur on a line of
road, it must be blasted by gunpowder for removal. There are
some fine examples of this reconsolidation upon the limestone hills
of Jamaica, as for example near Rock Fort, and at the cliffs to
the eastward of the Milk River's mouth.

The felspar contained in granite is often easily decomposed,
and when this is effected the surface frequently presents a
quartzose gravel. D'Aubuisson mentions that in a hollow way,
which had been only six years blasted through granite, the rock
was entirely decomposed to the depth of three inches. He also
states that the granite country of Auvergne, the Vivarrais and the
eastern Pyrenees, is frequently so much decomposed, that the traveller may imagine himself on large tracts of gravel*.

Some trap-rocks, from the presence of the same mineral, are so
liable to decomposition that there is frequently much difficulty in
obtaining a specimen. The depth to which some rocks of this nature are disintegrated in Jamaica is often very considerable.

This decomposition is attributed to the chemical as well as mechanical action of the atmosphere. With the slow and quiet
changes effected by electricity on the surface we are very imperfectly acquainted; but all are familiar with the effects of a discharge
from a thunder-storm, shivering rocks, and hurling fragments from

* Traité de Géognosie.

the heights into the valleys beneath. In these electrical discharges the lightning often fuses the surface of rocks. Thus, De Saussure found a compound rock, on Mont Blanc, fused on the surface, white bubbles being on the felspar, and black bubbles on the hornblende. Similar observations have been made by other geologists in other parts of the world. The oxygen of the atmosphere produces considerable alteration in rocks, more particularly observed in those containing iron, which are thus often reduced from a hard to a soft substance.

At Peninis Point, St. Mary's, Scilly Islands, there is a curious example of that decomposition of granite, which antiquaries have termed *rock-basins*, and considered the work of the Druids. The Kettle and Pans, as these depressions are there named, occur in the large blocks of granite on the top of this promontory; they are generally three feet in diameter and about two feet deep; they are mostly circular and concave, but there are others much indented at the sides. "Some have perpendicular sides and flat bottoms, some are of an oval form, and others of no regular figure. Many of the blocks are six or seven yards high, eight or nine yards square, and several of them have four, five, six or more of these cavities in them. A large rock near the extremity of this group has two basins, of an immense size, besides several smaller ones. The upper and larger one appears to have been formed by the junction of three or more large basins. It is irregularly shaped and about eighteen feet in circumference, and six feet deep. When the water in this basin has attained the height of three feet, it discharges itself by a lip into a lower basin, more regularly formed, the back of which is about five feet high, but which is incapable of containing more than a depth of two feet of water, owing to the declivity of the surface of the rock*." As a proof that similar decomposition sometimes takes place on the sides of a block, the author above cited mentions an oval cavity, six feet long, five wide, and nearly four feet deep, thus situated. The following wood-cut will afford an idea of Kettle and Pans †.

There is scarcely a substance, which, having been exposed to

Fig. 7.

* Rev. G. Woodley; View of the present State of the Scilly Islands, 1822.
† From a sketch by Mr. Holland.

the action of the atmosphere for a considerable time, does not exhibit marks of weathering. It will even be observed on the hardest siliceous rocks. The action of the atmosphere on cliffs of sandstone, in which the cement varies in induration or otherwise, produces the most grotesque forms, which must be more or less familiar to the least observing. Variations in temperature much assist the chemical decomposing power of the air.

Water may be considered as the principal mechanical agent in the great work of atmospheric destruction, uniting at the same time the character of a chemical agent. By infiltration it tends to separate the particles of which the rocks are composed, uniting chemically with the cementing matter in some cases, and in others forcing it away mechanically; in both instances leaving the particles not previously acted upon, more easily disturbed by a continuation of infiltration. In those situations where the temperature descends sufficiently low to produce frost, the mechanical action of the atmospheric water becomes much more considerable. Having entered into the interstices of rocks when liquid, it assumes a greater volume when it becomes solid from a sufficiently diminished temperature, felt at greater or less depths in proportion to the amount of decreased heat of the climates where the rocks may be situate. Portions of rock are thus forced asunder, and fine particles so separated, that the mere return of the water to a liquid state, assisted by gravity, is sufficient to remove them. The large masses have their centres of gravity often so altered relatively to rocks on which they rest, that when no longer cemented by the ice, they fall from their situations to a lower level. The fall of rocks occasioned by this means is common in lofty mountains, where considerable heights are exposed to the alternations of frost and thaw.

By percolation through porous rocks the water attains others which are not so, such as clays. The water thus stopped in its course downwards, escapes as it best can to the sides of hills and other situations, producing springs. At the places where this discharge of water takes place, there is also a mechanical destruction of the parts through which the water delivers itself. Rocks are affected by this action of the water in proportion to their composition; which, though not porous, may still be acted on by the water. An argillaceous substratum will get gradually moist at the surface, and in favourable situations may become a wet clay. The stability of the mass above will depend upon the relative position of the strata. Thus in the wood-cut annexed, if on the mountain *a*, water percolate through the porous strata *b* to the impervious clay bed *c c*, the surface of the latter would become slippery, and the mass

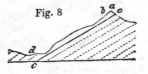

Fig. 8

above be launched into the valley *d.* Now, this is precisely what happened in the case of the Ruffiberg in Switzerland. This mountain, also known as the Rossberg, is 5196 feet above the level of the sea, and rises opposite the well-known Righi. Its upper part is composed of beds of a compound rock formed from the debris of the Alps at a previous geological epoch. These are to a certain extent porous, and the water percolates through them to a clay stratum on which they rest; the whole dipping at a considerable angle (about 45°). The clay becoming soft by the action of the water, and the thick superincumbent beds losing their support, the latter were launched over the slippery and inclined surface beneath, and the valley below was covered with their ruin.

This slide took place on the 2nd of September 1806, and covered a beautiful valley with rocks and mud. The villages of Goldau and Busingen, the hamlet of Huelloch, a large part of the village of Lowertz, the farms of Unter- and Ober-Rothen, and many scattered houses in the valley, were overwhelmed by the ruin. Goldau was crushed by masses of rocks, and Lowertz invaded by a stream of mud.

The torrent of rubbish and mud which rushed into the lake of Lowertz produced such a motion of the waters, that the village of Seven, situated at the other extremity, was inundated, and in great danger of being destroyed, two houses having been washed away. Live fish were found in the village of Steinen, thrown there by the flood. The lives lost were calculated from 800 to 900. Several travellers perished. It appears that there are traditionary accounts of former, though smaller, slides from the Rouffi or Rossberg*.

Large falls from mountains take place from the percolation of water to certain portions, which they mechanically loosen or chemically destroy without sliding over an inclined plane, as in the case of Rouffi, though the force of gravity still causes the fall. The Alps have afforded many examples of this fact, among others that of the great fall from the Diablerets in 1749.

Nothing is so common in mountainous regions as a talus of detritus brought to the foot of a cliff; this detritus composed of fragments of the rocks above, detached by decomposition from their surface, and brought down directly by their own gravity, or by the union of their own gravity and the force of surface-water, the latter derived from rains and the melting of snows. Avalanches of snow are great transporters of such fragments; and in the places where they fall there are always great accumulations of them, often borne from the greatest heights by the irresistible fury of the descending snow.

* For a view of this fall, taken four days after the catastrophe,—see Sections and Views illustrative of Geological Phænomena, pl. 33.

The under cliffs at Pinhay, near Lyme Regis, may be taken as an example of the destruction of a cliff by means of land springs, greater than that which is produced by the action of the sea in the same place.

Fig. 9.

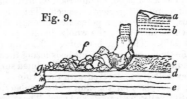

a, Gravel. *b*, Chalk. *c*, Green sand, both porous rocks through which the water percolates to the clay bed *d*, composed of the lower part of the green sand beds *c* and the upper part of the lias beds *e ;* being here arrested in its progress downwards, the water escapes by the easiest road, which is that presented by the cliff originally formed by the sea. It here gradually carries away the clay, first rendering it moist. The chalk and green sand lose their support, give way, and fall over into the sea. The lias *e* does not give way so fast before the sea at the cliff *g*, as the superincumbent mass, affected by the land springs; therefore the latter retreats until it has formed a great talus at *f*; but this talus tends constantly to move forward both by the destruction of the lias cliff at *g*, and by the tendency of the land springs to loosen its base, and to propel it into the sea. The chalk and green sand containing hard substances, often of considerable size, great protection is afforded to the cliff *g*, by their fall over its top, the fury of the breakers being greatly spent upon these masses.

Rivers.—These most frequently, though not always, take their rise among hills and mountains, and are supplied either by the melting of snows or glaciers, by the draining of rain-waters, or by springs. They transport the detritus formed either by the atmospheric agents, previously noted, or by themselves. The power of this transport depends upon their velocities. Now, the velocity of a river current is greatest in the centre, and least on the sides and bottom, being retarded by friction, water having a certain viscosity; consequently, the transporting power of a river is least where it comes in contact with the substances to be transported. These substances are generally angular if detached from simple rocks for the first time, such as pieces of limestone, granite, &c., and at the commencement present great obstacles to transportation; for the velocity of a current must be sufficient to move these angular fragments before they can suffer attrition. Rocks composed of fragments which have been previously rounded, such as conglomerates, will, if they decompose easily, contribute ready-formed gravel to the river, which might thus be able to carry them forward, while its velocity was insufficient to transport angular fragments of equal weight. The transport of sandstones will depend on their state of induration, and be easy where the particles are slightly aggregated; difficult, when so compact as to form angular fragments.

When the velocity of a river is sufficient to produce attrition of the substances which it has either torn up, collected by undermining its banks, or which have fallen into it, they gradually become more easy of transport, and would, if the force of the current continued always the same, be forced forward until the river delivered itself into the sea; but as the velocity of a current greatly depends on the fall of the river from one level to another, the transport is regulated by the inclination of the river's bed. Now, it is well known that this inclination varies materially, even in the same river; so that it may be able to carry detritus to one situation, but may be unable to transport it further, under ordinary circumstances, in consequence of diminished velocity. But this may be, and often is, so much increased further down, that its original transporting power may be, in a great measure, restored. It can now, however, only carry forward such detritus as it can receive or tear up in its course, and the pebbles which were left behind at the place of its first diminished velocity can only be brought within its power by floods, or, in other words, by extraordinary circumstances. As a general fact, it may be fairly stated that rivers, where their courses are short and rapid, bear down pebbles into the seas near them, as is the case in the Maritime Alps, &c.; but that when their courses are long, and changed from rapid to slow, they deposit the pebbles where the force of the stream diminishes, and finally transport more sand or mud to their mouths, as is the case with the Rhine, Rhone, Po, Danube, Ganges, &c.

It will follow that the form of the detritus will depend upon the length and velocities of rivers, all other circumstances being the same.

If in its course the form of the land be such that lakes are produced, the detritus borne down by a river will be deposited in their beds, which have thus a tendency to be gradually filled up, the quality of the detritus depending on the velocity of the river. Such inequalities, producing small lakes, are common in mountain valleys, and have evidently been once much more so. The velocity of the stream issuing from the lake will greatly depend upon the fall of land over which it flows. The stream will endeavour to cut down the barrier which produced the lake, but if it be slow or the rocks hard, it will effect little; while if it be rapid on the rocks easily cut, it will traverse the natural bar, drain

the lake, and permit the river to flow in an uninterrupted course.
Should the lake, while it existed, have been partially filled up by
the detritus from above, the river will cut through this also, and
the part thus cut away will be transported to a lower level. The
following diagram may assist the reader.

Fig. 10.

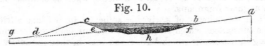

a b, course of a river flowing into the lake *b h c*, which is filled
with water to the level *b c*, the surplus falling over the slope *c d*,
and continuing its course in the direction *d g*. *e f*, deposit of detritus derived from the river *a b*, at the bottom of the lake
c h b. *b d*, bed of the river formed by cutting through the barrier
e c d, and part of the detritus *e h f*, so as to form a continuous
course with *a b* on the one side and *d g* on the other.

When lakes are large, such for instance as those of Geneva and
Constance, an immense lapse of time will be required to fill them
with detritus, so that, eventually, a continuous river may traverse
land occupying a space once filled by the water. Lakes of this
magnitude oppose great obstacles to the transport of pebbles. The
progress of a large proportion of detritus from the Alps is arrested
by lakes on their north and south sides. Thus, on the north, the
Rhine deposits its mountain detritus in the lake of Constance, and
the Rhone its transported pebbles and sands in the lake of Geneva.
Between these two great lakes, those of Zurich, Lucerne, &c., receive the gravels of other Alpine rivers. On the south the Lago
Maggiore receives the Alpine detritus of the Ticino, the lake of
Como, that of the Adda; and the lakes of Garda, &c. perform the
same office to other rivers. From these circumstances it will be
evident, that the detritus of a large portion of the Alps cannot travel, by the rivers, either into the ocean or the Mediterranean. The Po receives the waters of a large portion of the
Alps, and carries sand and silt into the sea; but the pebbles
are arrested before it receives the Ticino, which, though it transports rounded stones, does not bring them directly from the
Alps, but from its banks, after quitting the Lago Maggiore,
which banks contain the rounded Alpine fragments of a previous epoch. The same with the Rhone near Geneva, in which
Alpine pebbles occur, and which could not, in the actual state of
things, be derived from the Alps, because they would have been
stopped in the lake of Geneva. They are derived from its banks
and bed immediately on quitting the lake. Geological students,
in examining river-courses, should be very careful in distinguishing between pebbles from the immediate banks of rivers, and those
which might be derived from a distance, but to the transport of

which physical obstacles oppose themselves. From a want of attention to this circumstance many errors have arisen. It has been considered that the mode in which a river discharges itself into a lake, and pushes forward its detritus, would be such that the deposit would assume a nearly horizontal stratification. The angle of deposit must, however, depend upon the depth and the quality of the detritus discharged into it. Thus, if it be composed of sand and mud, it will be propelled further into the body of the lake than if it consist of pebbles. Examples of both cases will be found in the lake of Geneva. The ordinary deposit from the Rhone is sandy and muddy, which sinks in clouds, from its greater specific gravity, beneath the clear waters of the lake; yet the initial velocity is sufficient to transport a part of it about a league and a quarter, for I found a portion of it at the depth of 90 fathoms, raising the bottom of the lake between St. Gingolph and Vevey*. This would give a very slight dip from the embouchure of the Rhone. Off the mouth of the Drance, a torrent rushing into the lake near Ripaille, the pebbles, forced down, must arrange themselves at a much more considerable angle; for 80 fathoms are obtained at a short distance from the shore. The same variations in dip will also be observed in the lake of Como, where the turbid waters of the Adda have deposited a considerable quantity of sand and mud, which slopes gradually at a gentle angle; while the torrent-borne detritus at Bellano, Mandello, Abbadia, and other places, arranges itself with a much more considerable inclination. It would seem to follow that the stratification of lake deposits derived from the land around them, would not be uniform, but would depend on local circumstances, rivers or torrents propelling detritus before them, which would be as various as the rocks they respectively traversed; each collection would have a mode of deposit of its own, independent of the others, and they would tend to approach, and finally to unite with each other.

The higher part of the lake of Como is nearly filled up by the detritus transported by the Adda and Mera†. The former has divided the lake into two; the smaller portion (known by the name of the Lago di Mesola) being so shallow from the united deposit of the two rivers and some torrents, that aquatic plants grow through the water on the eastern part; while on the western, in which there is a greater depth, the process of filling up is hastened by means of stones, detached in such numbers, in particular seasons of the year, from the heights on that side, that a passage in a boat beneath the cliffs becomes exceedingly hazardous. Considering the many thousand revolutions of our planet

* For a map and sections of this lake, see Bibliothèque Universelle for 1819.

† See Sections and Views illustrative of Geological Phænomena, pl. 31.

D

that must have taken place since the land assumed its present general form, we should expect to find the barriers even of considerable lakes cut through under favourable circumstances, and accordingly we do discover appearances which would seem to warrant this conclusion.

It is by no means uncommon to find plains of greater or less extent, bounded on all sides by high land, and through which a principal river meanders, entering at one end by a valley, and passing out through a gorge at the other, augmented by tributary streams from the surrounding hills : sometimes these plains have no principal river passing through them ; but many small streams, descending from the mountains, unite in the plain and pass out also through a gorge. In such cases the plain presents the appearance of a drained lake, such as we may suppose would be exhibited in many now existing, if passages for the waters were cut through any part of the basins holding them. The gorge at Narni seems to have let out the waters of a lake supplied by the Nera, which now flows through the plain of Terni, the former bottom of the lake. The great fertile plain of Florence seems once to have been the bed of a lake, the drainage of which was effected by a cut through the high land that bounds it on the west. If this outlet were again closed, the waters of the Arno would again cover the plain and convert it into the bed of a lake. The period at which the break in the Jura was formed at the Fort de l'Ecluse, may perhaps be questionable ; but if closed, it would stop the course of the Rhone, and convert the lake of Geneva into a much larger body of water.

These appearances are not confined to one part of the world; they would appear, from the descriptions of intelligent travellers, to exist very commonly. I have myself observed examples in Jamaica. The district named St. Thomas in the Vale is a marked one. Here we have low land bounded on all sides by hills, which would form the banks of a lake, were not the waters let out by the gorge through which the Rio Cobre flows.

It would therefore appear, though large lakes collect mountain detritus, which is distributed over a large surface, enveloping, probably, animal and vegetable remains, that the barriers of the lakes may be cut through, and the rivers again act on a portion of the previous deposit.

The probability that many gorges originate from the cutting power of rivers discharged from lakes, is rendered stronger by examining those natural basins which are drained by subterraneous channels, and where gorges are not found. Thus Luidas Vale, in the island of Jamaica, is a district surrounded on all sides by high land, and would form a lake, were not the waters, derived from heavy tropical rains, carried off by sink-holes in the low grounds. A body of water, brought to turn the water-wheel of an estate's works, is swallowed up close to these works. A cavern, out of which water sometimes issues, near another estate, is speedily engulfed in a cave not far distant. In consequence of this escape of the waters, a gorge is not formed by means of a discharging river flowing over the lowest lip of the high land, as appears to have happened in the case of St. Thomas in the Vale, which adjoins Luidas Vale.

It is stated, " that a velocity of three inches per second at the bottom will just begin to work upon fine clay fit for pottery, and however firm and compact it may be, it will tear it up; yet no beds are more stable than clay when the velocities do not exceed this; for the water soon takes away the impalpable particles of the superficial clay, leaving the particles of sand sticking by their lower half in the rest of the clay, which they now protect, making a very permanent bottom, if the stream does not bring down gravel or coarse sand, which will rub off this very thin crust, and allow another layer to be worn off. A velocity of six inches will lift fine sand; eight inches will lift sand as coarse as linseed; twelve inches will sweep away fine gravel; twenty-four inches will roll along rounded pebbles an inch in diameter; and it requires three feet per second at the bottom to sweep along shivery angular stones of the size of an egg *."

The destructive power of rivers on solid rocks appears to act both chemically and mechanically. Chemically, by the affinity of water and of the air which it holds in solution for the various substances it encounters; and mechanically, by the friction of the detritus, independent of that of the water, upon the bottom and sides, but principally on the former. They may have thus effected a passage through the lake barriers previously noticed, and by these means they destroy the obstacles opposed to their courses. When a bank, a small hill, or the foot of a mountain, opposes their progress, they assail it, and form cliffs, the materials of which, if soft, fall into the stream, or make under cliffs, which are removed, and the work of destruction is slowly continued (Fig. 11. *a.*) ; or when the cliff,

Fig. 11.

thus formed, is of harder materials, blocks are accummulated in a talus at its base, and the cliff is secured, in a great measure, from

* Encyclopædia Britannica, art. *River.*

attack, until this protecting mass is removed (Fig. 11. *b*). There
is scarcely a river of any considerable length which does not afford
examples of cliffs thus produced; very frequently they overhang
flat or gently sloping land, on which the river has flowed while
employed in cutting the cliff. It is not a little curious to trace, in
countries where rivers wind considerably, the various obstacles
which have determined the course of the stream, causing it to attack the original more or less rounded forms of the bases of moderately elevated hills.

Rivers appear to be constantly striving to arrange their beds
in such a manner that they should suffer the least resistance in
their courses, cutting down obstacles and filling up depressions
which checked them. But the constant addition of new detritus
from the neighbouring highlands embarrasses this operation, causing accumulations in one situation which direct the waters in another. Thus the fall of a considerable quantity of rocks on one
side will throw the stream upon the opposite bank, which might
previously have been little attacked. This again forces the current in a direction that it did not previously follow; the bottom
becomes torn up by the new line of the principal stream, and the
effect of such a fall is felt far down the course of the river. In
consequence of this endeavour to avoid a new obstacle, continual
changes in a river's bed take place, as also from the destruction of
an old obstacle, which permits a new course in a direction that the
river has been striving to follow.

D'Aubuisson observed two rocks at the Falls of the Rhine,
near Schafhausen, isolated at the head of the precipice over which
the waters leap; these were observed corroded at their bases by
the action of the pent-up current between them. By gradually
diminishing their support, the rocks would finally be forced over
the cataract, and the waters having overcome this obstacle, would
fall in a different manner on the bottom beneath, producing a different effect from that which they had previously caused.

As all rivers must vary in their cutting power according to velocity, volume of water, and amount and quality of detritus in the
act of transport, it becomes exceedingly difficult to generalize on
the subject; but as barriers of even the hardest rocks have suffered, and as the destructive power of the same rivers on the same
obstacles is so exceedingly small as to be scarcely perceptible during the life of man, it seems fair to infer that this also tends to
confirm the opinion of the great age of the present general state
of the world.

Mr. Lyell indeed produces, as an example of the comparatively
quick cutting power of a river, a gorge in a lava-current at the
foot of Etna, formed by the erosion of the Simeto. The lava is
considered modern, and Gamellaro is cited as supposing it thrown
out in 1603. The lava is described as not porous or scoriaceous,
but as a compact homogeneous rock, lighter than common basalt,

and containing crystals of olivine and glassy felspar. Though there are two waterfalls, each about six feet, the general fall of the river's bed is stated as not considerable. The gorge is cut in some places to the depth of forty or fifty feet, and its breadth varies from fifty to several hundred feet *. It is therefore inferred that this is a good example of the speedy formation of gorges by running water; and this inference cannot be denied, if the date of the lava-current be correctly ascertained. It may be remarked that the present fall in the bed of the Simeto does not give that of the river during the great cutting operation. It must once have occupied a different level, or else the gorge could not have been commenced; and there must always have been a rapid fall, or, in other words, a cascade into the low land off the lava, equal to the height of the lava-current; the waters being raised to the top of the lava, at this place, by the formation of a lake behind, produced by the bar of lava. It would therefore follow, that the gorge in the lava-current has been principally formed by the cutting back of rapids or a cataract. Though this circumstance would facilitate the progress of destruction, and render it less remarkable than if the Simeto, with its present fall, had cut the gorge, it yet leaves this a good example of a ravine formed in hard rock during the course of two centuries, it being always understood that no doubt exists of the period when the lava-current was ejected, and crossed the previously existing valley.

The dates obtained by the well-known examples of the Auvergne rivers are only relative; but they are sufficient to show that a valley existed, through which a river kept its course, conveying detritus in the usual way, and that the progress of the river was barred by a lava-current (as in the instance just cited), which descending from a neighbouring volcano traversed the valley, and formed a lake. This lake, when full, discharged itself over the lower lip of its basin, which happened to be in the direction of the valley, and over the lava-current. This, by erosion, is cut down, not only to its original bed, but through it, into the rock which constituted the bottom of the original valley.

Notwithstanding appearances, there are numerous gorges or ravines through which rivers flow, which could not have been cut out by them, at least during the existence of the present general disposition of land; for the relative levels are such, that the rivers must be supposed to have run over land of much greater elevation towards their embouchures than they flowed over from their sources; in other words, such rivers must be supposed to have run up hill, if they be considered the agents which have formed these gorges. As a striking example of this fact, we may cite the course of the Meuse previous to, and during its traverse through the

* Principles of Geology.

Ardennes. M. Boblaye informs us, that previous to its passage
through these mountains, the Meuse is only separated from the
great basin of the Seine by hills or low *cols*, not more than thirty
or forty yards above the present bed of the river; while the Ardennes, through which it actually passes, rise to the height of several hundred feet above the same level. Now if all rivers had
really cut the beds or valleys through which they now flow, the
Meuse must have run up hill, and have cut a narrow channel
about three hundred yards deep; while nôthing prevented its flowing in the opposite direction into the Paris basin, when it had effected a rise of not much more than a tenth part of that height *.

At Clifton, near Bristol, we have also a striking example of the
same fact. The Avon here runs through a gorge or ravine, which
if closed would form a lake behind it; but this lake would exert no
action on the range of hill through which the present channel
passes; on the contrary, the lowest lip of the basin, and consequently the drainage, would be found in the direction of Nailsea,
to the sea beyond which the Avon would continue its course
from Bristol. The real rise of land between high water at Bristol
and the sea beyond Nailsea is trifling, and is bounded on the north
by the high ridge through which the Avon now finds its passage
to the Severn.

Other examples might easily be cited, but these are sufficient
to point out the fact. There are many gorges through which
rivers pass, the formation of which remains questionable from our
ignorance of the relative levels in their vicinity, and thus it becomes difficult to assign them any particular origin. They may
be either due to the same causes which have produced the ravines
of the Meuse in the Ardennes, and of the Avon near Bristol, or
to the cutting power of rivers discharging the surplus waters of
lakes. Under this head may be enumerated the celebrated Vale
of Tempe, in Thessaly; the tortuous course of the Wye, between
Monmouth and Chepstow; the famous Rheingau; the ravine by
which the Potomack traverses the Blue Mountains in the United
States; the Gates of Iron, through which the Danube escapes into
Wallachia, &c.

The Falls of Niagara may be adduced as an example of a river
discharging the surplus waters of a lake, and now cutting back a
gorge to that lake, which may eventually be drained by it. This
celebrated cascade is situated between the lakes Ontario and Erie.
For some distance above the embouchure of the river into the
former, the country is flat, and apparently alluvial, when suddenly
a plateau rises above it and continues to lake Erie. Over this
plateau the surplus waters of the latter lake have taken their course,
and appear to have originally fallen over the face of the plateau

* Boblaye, Ann. des Sci. Nat. t. 17, p. 37.

fronting lake Ontario. By degrees they have cut back their passage about seven miles, leaving about eighteen more to be worn away by future ages. When this shall have been accomplished, the gorge or ravine will be similar to those previously noticed. The manner in which the river cuts its passage is singular, and perhaps somewhat different from what, at first sight, might have been expected. It will be best explained by the following diagram.

a b, original level of the plateau. *a h*, river flowing over the plateau, and falling over to the abyss *c*, forming the cascade *h c*, after which the waters take their course in the direction *c g*. *d*, beds of limestone resting on beds of shale *e*, both being surmounted, in

Fig. 12.

the neighbouring flat country, by a mass of transported substances, varying from ten to one hundred and forty feet in depth, and containing large blocks. The rush of waters from *h* to *c* occasions violent gusts of wind, charged with water, to be driven against the shale *e* at *f*. The continued action of these water-charged whirlwinds displaces the shale, and throws it down in a talus at *k*. From the removal of this shale, the superincumbent limestone loses its support, falls from the combined gravity of itself and the water above, is dashed into the abyss beneath, and thus the falls are cut back so rapidly that they have considerably receded within the memory of man. The same operations are again renewed, and again the same results follow. So that unless some extraordinary circumstance should arrest their retreat, these falls will discharge the waters of lake Erie ; but not suddenly, as is sometimes supposed, so as to produce a violent deluge over the lower country further down the river, but much more gradually ; for the lake waters will only be lowered in proportion to the depth of the draining channel, as may be illustrated by the annexed wood-cut, in which

Fig. 13.

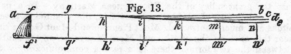

a b represents the level of the lake and of the plateau, rising but little above it. *h e*, the slope (exaggerated) of the lake bed from *h*, the spot where the surplus waters are delivered over the plateau. *f' n'* the level of the river below the falls. Supposing *g g'* to represent the falls which have approached the lake, by gradually cutting back the channel from *f f'* to *g g'*, it will appear that the

same kind of retreat may be effected to $h\,h'$ without discharging
more water than now passes down the river.   But the falls being
once at $h\,h'$, the retreat of every succeeding yard will occasion
more water to pass over them, by draining the waters of the lake
down to the point which now becomes its lowest lip ; so that when
the falls have cut their way back to $i\,i'$, the surface of the lake will
sink to the horizontal line $i\,c$, and the mass of water above the new
level will have passed over the falls in addition to the usual drainage.   Such an addition must add greatly to the velocity and cutting power of the falls, which will now retreat more rapidly and
effect their passage to $k\,k'$, reducing the level to $k\,d$ in less time
than it reduced it from $a\,b$ to $i\,c$.   After a certain time the water
forced over the falls would become less, because the superficies of
the lake would be diminished.   It would therefore appear that the
increased power of the falls, caused by the additional waters, would
decrease gradually, until, finally, there should not be more than
the waters of the river traversing the ancient bed of the lake.

The waters of a lake can only be suddenly let out, and produce
a debacle, when the hard barrier separating it from the land at a
lower level presents a perpendicular face to the whole depth of
the lake, which, even then, must be suddenly thrown down, in its
whole height, to produce the effect required.   Such rocky barriers
must be exceedingly rare ; and it must be still more rare, that
where they existed they were not cut down, to a certain extent,
by degrees.   The common character of lakes, as respects the inclination from their bottoms to the discharging outlet, varies materially, but in general the slope is very gradual, particularly in
lakes of considerable magnitude.

The often cited debacle caused by the bursting of a lake in the
Val de Bagnes was produced from a very different state of things
from that attending the drainage of a lake existing in a depression
of land, with a rocky barrier.

The Val de Bagnes, in the Vallais, is drained by the Dranse,
which, when unobstructed, is joined by the waters from the valley of Entremont, leading to the grand St. Bernard, and runs
into the great valley of the Rhone, near Martigny.   In a part
of the valley near the bridge of Mauvoisin, the channel is precipitous and much contracted.   Mont Pleureur and Mont Getroz rise
near this spot on the north, and Mont Mauvoisin on the south.
Between the two former there is a ravine communicating with
the Val de Bagnes, having a considerable glacier at its upper extremity.   Through this ravine blocks of ice and avalanches of snow
descend into the Val de Bagnes, and more or less obstruct the
channel of the Dranse, which is able, under ordinary circumstances, to remove the greater part, if not the whole, of such obstructions.   When however the blocks of ice are numerous, and the

avalanches are heavy, the force of the torrent is unable to contend
with them, and they accumulate. " For several years previous
to 1818," says M. Escher de la Linth, " the progress of the
Dranse had begun to be obstructed by the blocks of ice and avalanches of snow that descended from the glacier of Getroz; and as
soon as this accumulation was able to resist the heats of summer,
it acquired new magnitude during every succeeding winter, till it
became an homogeneous mass of ice of a conical form. The waters
of the Dranse, however, still found their way beneath the icy cone
till the month of April, when they were observed to have been
dammed up, and to have formed a lake about half a league in
length *.

The danger that threatened now became apparent, and accordingly the gradual drainage of the lake was attempted by means of
a gallery through the ice. This reduced its contents from about
800,000,000 cubic feet to 530,000,000 cubic feet. Finally, the
discharging waters attacked the debris at the foot of Mauvoisin,
and excavating a passage between the rocks and the ice, rushed
furiously out, carrying houses, trees, large blocks of rock, &c.,
before it. Escaping from the narrow valley, it desolated a large
portion of Martigny †, and passed with gradually diminished velocity down the Rhone into the lake of Geneva.

As might be expected, the velocity of the torrent varied materially in different parts of its course. M. Escher de la Linth calculates that from the glacier to Le Chable, a distance of 70,000
feet, the velocity was 33 feet per second; from Le Chable to Martigny, 60,000 feet, at the rate of 18 feet; from Martigny to St.
Maurice, 30,000 feet, at 11½; and from St. Maurice to the lake of
Geneva, 80,000 feet, with a diminished velocity of 6 feet per second ‡. The lake was drained in half an hour. Lakes may be
suddenly drained, if but a thin perpendicular partition divides
them from an inferior level; for this barrier may be rendered soft
by the percolation of water, and suddenly give way; but such cases
must be of very rare occurrence; and the lakes are not likely to
be of such magnitude as to cause appearances, by their sudden
discharge, that may be equal to those producible by the passage
of a more general mass of waters over land.

Mr. Strangways notices the bursting, or sudden considerable
drainage, of the lake Souvando, on the north of St. Petersburg.
Previous to 1818 this lake was separated from that of Ladoga
by the little isthmus of Taipala. The lake discharged its waters

* Edin. Phil. Journ. vol. i. p. 188.
† Among the debris transported to Martigny were many trees, resting
upright on their roots, the attached gravel and soil having kept them in
a position with the branches upwards.
‡ Edin. Phil. Journ. vol. i. p. 191.

into the Voxa at Keognemy, and so passed into the Ladoga at
Kexholm. In the spring of 1818, the water broke down the
isthmus and changed the direction of the discharging waters, by
presenting a lower lip in another direction. The water has been
lowered considerably, and continues to run through its new channel into the lake of Ladoga, having deserted the Voxa *.

The same author describes the falls or rapids of Imatra, about
six wersts below the point where the surplus waters of the lake
Saima first drain off by the Voxa. This river suddenly contracts
itself above the rapids, over which it runs, with great noise and
impetuosity, through a gorge that it has evidently cut for itself.
According to Mr. Strangways, we may consider the water to
have originally passed over a platform between two ranges of
hills, forming the bottom of a valley. The platform is composed
of gneiss, in very highly inclined strata; and into this the river
has cut a channel. "The surface of this platform is apparently
now about fifty feet above the level of the water, at the lower
extremity of the rapids. Its surface is in many parts quite bare
and deeply channelled in a direction parallel to the river. It is
covered with heaps of pebbles and boulders of great size, some
of which are hollowed and scooped into the most fanciful shapes.
One of the largest of the blocks now left dry, standing nearly in
the middle of the elevated platform, is worn through perpendicularly with a cylindrical hole †." It is stated that the level of the
lake Saima and its discharging river fall gradually.

Freshets.—These take place, more or less, in all rivers, greatly
augmenting their velocities and transporting power, carrying forward substances that could not have been moved under ordinary
circumstances. They are also geologically important, as they
surprise terrestrial animals in low situations, hurry them on with
trees and other matters into the sea, where they may be entombed
entire with estuary and marine animals in mud and silt.

It has been observed that, during freshets, a river tends
chiefly to widen its bed, "without greatly deepening it: for the
aquatic plants, which have been growing and thriving during
the peaceable state of the river, are now laid along, but not
swept away, by the freshes, and protect the bottom from their
attacks; and the stones and gravel, which must have been left
bare in a course of years, working on the soil, will also collect in
the bottom, and greatly augment its power of resistance ‡."
During these freshes, low lands on the sides of the river are frequently under water, and a deposit takes place; but notwithstanding all checks, a large quantity of detritus passes onwards to the
sea.

* Strangways, Geol. Trans. First Series, vol. v. p. 344.
† *Ibid.* p. 341.   ‡ Encyc. Brit. art., *River.*

We should be careful, in our estimates of the effects of a flood in a cultivated country, not only to separate the loss of lives and the destruction of property, which may affect the feelings, from the real physical change produced in the country; but also to remember, that the works of man greatly aid the destructive power of a flood. Instead of a body of water rushing into a plain, where from its diffusion over a more considerable space its velocity and transporting power are both diminished, all cross hedges and bridges, though they may check the waters for the moment, are the means of producing innumerable debacles, when they give way before the pressure exerted upon them. Suppose a bridge arrests the progress of the flood downwards, and, as very frequently happens on small plains, a causeway connects the bridge with the hills on either side, the waters will accumulate, and will finally burst through the least resisting part of the barrier, which will most probably be the bridge. Having once found a vent, the pent-up waters will issue forth with a velocity proportioned to the difference of the level and the mass of water, and a debacle will be produced, whose transporting power will be much greater than that of the general force of the flood if no such barrier had existed. It must also be recollected, that man, by his contrivances of ditches and drains, prevents the rain-water from remaining the time that it would otherwise do on the slopes of hills, conducting it as he does by numerous free channels into the valleys below; so that, in a given time, a much greater body of water is collected than could happen in an uncultivated country. He moreover, by dams and banks, often confines a body of river water within narrower channels than it would naturally take; and thus its dispersion over a larger surface being prevented during a freshet, its ordinary velocity is greatly increased, and with this its transporting power.

Glaciers.—These are large bodies of ice or indurated snow, formed upon land in the cold regions of the atmosphere, which descend into the valleys of mountainous countries; thus frequently presenting the singular appearance of desolation amid fertility, of ice amid vegetation. The levels to which glaciers descend depend greatly on the latitude of the place. Thus, in the arctic regions, where the line of perpetual snow approaches very nearly to the level of the sea, glaciers are produced in lower hills than could be the case in the Alps, where the line of perpetual congelation is much more elevated. So again in the Himalaya range the line of perpetual congelation being higher than in the Alps, the glaciers form at higher levels. Glaciers are instruments of the degradation of land, inasmuch as they drive before them and transport such substances as they may have the power to move. In front of glaciers there is usually a pile of rubbish composed of pieces of rock, earth, and trees, which they have forced forward, known in Switzerland by the name of *moraine*. If there be a line of moraine some distance from the front of the glacier, it is considered that the glacier has retreated to the amount of that distance; but if there be no other than that which the glacier immediately drives before it, it is considered to be on the increase. Glaciers assist the degradation of land by transporting blocks, often of very large dimensions, into lower regions than they could otherwise attain in so short a time. Many glaciers, particularly where they pass beneath precipices, are charged with fallen rubbish, which, as the ice constantly advances, are carried on with it; and should a precipice occur in the front of the moving mass, they are hurled over with it into the ravines beneath. Such falls are common in the high regions of the Alps, producing, with the rents suddenly formed in the glacier itself, the few interruptions to the dead silence which reigns in those lofty and wild regions. The velocity with which a glacier advances depends on the angle that it makes with the horizon, of course increasing with the steepness of the declivity.

A ladder, left by M. de Saussure at the upper end of a glacier, when he first visited the Col du Géant, has lately been discovered in the Mer de Glace, the continuation of the same glacier, and nearly opposite the aiguille named Le Moine. It must therefore have advanced about three leagues since the year 1787 *. From some experiments by Chamonix guides, mentioned by Capt. Sherwill, we learn that this rapid progress ceases, as might have been expected, where the declivity becomes less in the Mer de Glace itself; for it was there found that a block of rock advanced about two hundred yards in a twelvemonth †. No better proofs could be afforded of the advance of a glacier, the amount of which corresponds with the declivity. It hence appears to follow, that as the declivity remains nearly the same for a long period, the advance or retreat of the lower part of a glacier will correspond with the local variations in climate, which shall produce more or less ice in the higher, or destroy more or less of the glacier in the lower regions.

Almost all glacier waters are charged with detritus, the larger portions of which are deposited near the ice, but the lighter particles are transported to considerable distances; as is, for example, the case with the Arve, which having deposited its heavier burden in the valley of Chamonix, carries the lighter parts to its junction with the Rhone, near Geneva. Not unfrequently the turbid glacier waters are carried on, and deposit the detritus in some lake, as is the case with the Rhone, which transports silt,

* Phil. Mag. and Ann. of Philosophy, Jan. 1831.   † *Ibid.*

mud, and occasionally pebbles, into the lake of Geneva. The grinding of the glacier against the bottom over which it passes, may perhaps mechanically assist in the work of destruction.

In the northern regions glaciers have sometimes such a short distance to pass over before they reach the sea, that they project into it, as has been observed by northern navigators. The mass so forced into the sea will have a constant tendency to float, from its inferior specific gravity, and therefore when detached by any force from the glacier behind, it will be carried away ;—thus, forming those icebergs, so well known and so dangerous, in the Northern Atlantic Ocean.

Delivery of Detritus into the Sea.

We have seen above, that from the action of the atmosphere, the melting of snows and glaciers, landslips, and the cutting power of rivers, considerable destruction of dry land is effected. Local circumstances arrest a considerable portion of this detritus ; lakes are filled up, and again cut through ; low lands are occasionally flooded, and considerable deposits left upon them ; the velocity of the streams diminishes, and with it the power of transport ; so that, as previously observed, rivers when short and rapid may carry a large portion of their detritus forward, while, when long, they leave a considerable part of it in their courses. In favourable situations, such as in plains, they will raise their beds, if confined within bounds, that do not either permit a change of course, or a deposit in a new channel. This fact is well observed in Italy, where many plains have been under cultivation for a long period, during which it was always necessary to restrain the rivers within artificial banks, to prevent their range over the cultivated land, which would otherwise have been devastated by them ; so that, in travelling in that country, the road frequently passes up hill over high artificial ridges, upon which the rivers hold their course at a higher level than that of the surrounding country. These artificial ridges are particularly striking on the little plain of Nice, which has been under cultivation since the country was settled from the Phœnician colony of Marseilles. The height of the latter elevated river-courses is not only due to their antiquity, but to the loose nature of the conglomerate hills behind, which permits an easy transport of the pebbles.   Fig. 14.

The annexed diagram will illustrate this fact : *a b*, the level of the country, now cultivated, upon which the artificial banks have been gradually raised to *c d*, in order to protect the cultivated lands from being invaded by the detritus of the river or torrent *e*, which thus accumulated from *f* to *e*. There is a very general system of endeavouring to check this accumulation, and consequent rise of bed, by throwing, when the waters are low, the transported detritus out of the bed *e*, upon the protecting banks *c d*.

The Po affords a well-known example of this rise of bed, so that it becomes higher than the houses in the city of Ferrara. In Holland also the same phænomenon is observable, though not on so great a scale ; and may always be expected where artificial banks prevent detritus-bearing rivers from changing their beds on plains.

Although rivers, in certain situations, raise their beds, in others they deepen them.   This arises from two or more streams uniting into one river, when the water does not expose a surface equal to the two previous surfaces, but one very considerably less, the action of the united waters being to deepen their channel ; so that even with a diminished general inclination of the bed, the velocity continues the same, or is even increased.

This deepening of beds by the union of rivers is well exhibited by the following facts observed in the Po :—

" About the year 1600, the waters of the Panaro, a very considerable river, were added to the Po Grande; and although it brings along with it in its freshes a vast quantity of sand and mud, it has greatly deepened the whole Tronco di Venezia from the confluence to the sea. This point was clearly ascertained by Manfredi about the year 1720, when the inhabitants of the valleys adjacent were alarmed by the project of bringing in the waters of the Rheno, which then ran through the Ferrarese. Their fears were overcome, and the Po Grande continues to deepen its channel every day with a prodigious advantage to the navigations; and there are several extensive marshes which now drain off by it, after having been for ages under water : and it is to be particularly remarked, that the Rheno is the foulest river in its freshes of any river in that country *."

It might be supposed that all rivers would, by means of freshes, propel pebbles into the sea.   They certainly accomplish by these means a greater transport than could be effected in the same channels under ordinary circumstances ; but during freshes rivers can only be considered as of greater magnitude, and are therefore still subject to the general laws of rivers ; a greater body of water tending to deepen the channel ; the velocities, inclinations of beds, and the power of transport still being in proportion to each other.

In the beds of torrents, dry, or nearly dry, for the greater part of the year, we see examples of the deepening of river beds in proportion to the volume of water which passes through them, to the inclination of the beds, and to the resisting power of the bottoms and sides.   The transport of detritus will also be observed greater or less in proportion to these circumstances: the finer particles being more easy of transport, there are few rivers which, during freshes, do not convey a great quantity of such detritus into the sea: other kinds of detritus will be also transported, if levels permit; if not, they remain in the interior. Consequently, according to the circumstances already noticed will be the nature of the detritus conveyed to the mouths of rivers. But as circumstances vary in the same river, a deposit of such detritus in these situations also varies, and there may be alternations of clay or marl, and of sand or gravel.

If the mouths of rivers be tidal, the river detritus is committed to the charge of the estuary tides, and is dealt with according to the laws by which these are governed. If they be tideless, the whole mass of transported matter will be propelled without check into the seas at the embouchures. Between the extremes of great resistance and non-resistance the variations are so great, and depend so much on local circumstances, as to be of exceedingly difficult classification. The principal variations are produced by the difference in the volume of the discharging rivers, their velocities, and the quantity and quality of the substances they may transport. As a general fact, however, it may be stated that rivers tend to form deltas in tideless, or nearly tideless, seas, or where they can overcome the resistance of tides, currents, and the destructive action of the breakers; thus increasing the land by their deposit, and splitting into several channels; the superficial increase being in proportion to the depth of water into which the rivers discharge themselves.

In calculations of the advance of deltas, care has not always been taken to show the general depth of water into which they may have been protruded; so that a less quantity of transported detritus might expose a larger surface when thrown on a shallow bottom, than a larger quantity in deeper water.

The Nile, Danube, Volga, Rhone, and Po, afford us examples of deltas thrown forward into seas, which may, in common terms, be called tideless. As the Nile receives little atmospheric water from Egypt, on which rain seldom falls, the detritus which it brings down must be principally derived from above. This river begins to rise in June, attains its maximum of height—namely, twenty-four or twenty-eight feet—in August, and then falls till the next May. During a succession of ages, the Nile has transported a great mass of detritus into the Mediterranean, which has accumulated in a delta at the mouth, and is constantly on the increase. It has been calculated, that, as the sea deepens at the rate of a fathom in a mile, and supposing that the deposit is the same as in the Thebais, the addition would amount to a mile and a quarter since the time of Herodotus. According to Girard, the Nile has raised the surface of Upper Egypt about six feet four inches, since the commencement of the Christian æra. The quantity of water discharged per annum by this river is estimated at 250 times that of the Thames *. The delta is traversed by two main streams, which separate a few miles below Cairo; one descending to Rosetta, the other to Damietta. The present position of the latter city has led to very exaggerated ideas respecting the rapid increase of this delta. It was supposed that the present town was the same with that which during the first crusade of St. Louis was situated on the sea. Now, as Damietta is two leagues from the sea, it was calculated that this distance had been produced by deposits from the Nile within about 600 years. It now, however, appears, from the labours of M. Renaud, that after the departure of St. Louis, the Egyptian Emirs, wishing to prevent a new invasion on the same side, destroyed Damietta, and founded a new city in the interior, the present Damietta†. From the effect of the waves and currents, banks are thrown up on the outer edge of the delta, forming lakes, of which those of Menzalen, Bourlos, and that behind Alexandria, are the largest.

The delta of the Po advances at a rapid rate, in consequence of the shallow sea into which it is protruded. We are indebted to M. Prony for a very interesting collection of facts, which authorize him to conclude, " First, that at some ancient period, the precise date of which cannot now be ascertained, the waves of the Adriatic washed the walls of Adria. Secondly, that in the twelfth century, before a passage had been opened for the Po at Ficarrolo, on its left or northern bank, the shore had already been removed to the distance of nine or ten thousand metres from Adria. Thirdly, that the extremities of the promontories formed by the two principal branches of the Po, before the excavation of the Taglio di Porto Viro, had extended by the year 1600, or in four hundred years, to a medium distance of 18,500 metres beyond Adria; giving from the year 1200 an average yearly increase of the alluvial land of 25 metres. Fourthly, that the extreme point of the present single promontory, formed by the alluvions of the existing branches, is advanced to between thirty-two and thirty-three thousand metres beyond Adria; whence the average yearly progress is about seventy metres during the last two hundred years, being a greatly more rapid proportion than in former times‡."

The Mississippi, the great drain of so large a portion of North America, may be considered as delivering its waters into a nearly tideless sea. Its delta is very considerable, and little raised above the level of the ocean. During the greatest heights of flood, the fall of the river from New Orleans to the sea, a distance of about

---

* Supplement to Encyc. Brit., art. *Physical Geography.*
† Extraits des Historiens Arabes relatifs aux Guerres des Croisades.
‡ Prony, as quoted by Cuvier. Dis. sur les Rev. du Globe.

one hundred miles, has been calculated at only one inch and a half in a mile. When the waters are low, the fall is scarcely perceptible, the level of the sea being then nearly that of the river at New Orleans *.

This river affords a good example of a flood being higher at a distance from the embouchure of a river than at the mouth itself; for the rise of water, during the great freshets, is fifty feet at Natchez, three hundred and eighty miles inland, while at New Orleans it is only thirteen †.

Darby has furnished us with a mass of information respecting a large portion of the Mississippi's course, and of its delta, from whence very important geological information may be obtained ‡. It would appear that the Atchafalaya, which now, at a distance of about two hundred and fifty miles from the sea, conducts a large part of the Mississippi's waters into the Gulf of Mexico, did not always form a drain from that river, but that it once constituted a continuation of the Red River, which now flows into the Mississippi. During the autumns of 1807, 1808, 1809, Mr. Darby had frequent opportunities of examining the bed of the Atchafalaya, the waters in which were then at a low state. He found that " the upper stratum invariably consisted of a blueish clay, common to the banks of the Mississippi. This is usually followed by a stratum of red ochreous earth peculiar to the Red River, under which the blue clay of the Mississippi was again to be perceived §." From this we may infer, not only that the Red River flowed through the channel of the Atchafalaya, previous to the present course of the Mississippi, but that the latter river preceded the former, and that there have been alternations.

From the form of the Mississippi, where the Atchafalaya detaches itself, an immense quantity of trees brought down by the former are thrown into the latter. About fifty-two years since, these trees began to accumulate and form the " raft." " This mass of timber rises and falls with the water in the river, and at all seasons maintains an equal elevation above the surface. The tales that have been narrated respecting this phænomenon, its having timber of large size, and in many places being compact enough for horses to pass, are entirely void of truth. The raft is, in fact, subject to continual change of position, which, superadding its recent formation, renders either the solidity of its structure, or the growth of large timber, impossible. Some small willows and other aquatic bushes are frequently seen among the trees, but are too often destroyed by the shifting of the mass to acquire any considerable size. In the fall season, when the waters are low,

* Hall's Travels in North America.   † *Ibid.*
‡ Darby's Geographical Description of the State of Louisiana.
§ *Ibid.*

the surface of the raft is perfectly covered by the most beautiful
flora, whose varied dyes, and the hum of the honey-bee, seen in
thousands, compensate to the traveller for the deep silence and
lonely appearance of nature at this remote spot *.

Mr. Darby estimated the cubic contents of the raft, from observations made in 1808, at 286,784,000 cubic feet, considering the
breadth of the river = 220 yards, the length of the raft = 10 miles,
and the depth = 8 feet. The distance between the extremities of the
raft was actually more than twenty miles ; but, as the whole distance was not filled up by timber, he assumed ten miles as near
the truth.

Rafts of this description, but of less size, occur in other parts
of the Mississippi or its great tributaries. The banks are destroyed by the currents, and large collections of trees are suddenly hurled into the stream. Captain Hall was present when a
large mass of earth, loaded with trees, suddenly fell into the Missouri, and a larger mass had been detached a short time previous
to his arrival †.

There are few rivers whose course is more instructive than the
Mississippi, as man has not yet effected many changes on its banks ;
and we thus contemplate great natural operations, such as cannot
be so well observed in those which have been more or less under
his dominion for a series of ages. Its course is so long, and
through such various climates, that the freshets or floods produced
in one tributary are over before they commence in another :
hence arise those frequent deposits of detritus at the mouths of
the tributaries. These latter have their waters forced back, and
rendered, to a certain distance, stagnant by the rush of the
flood across their embouchures, and the consequence is a deposit,
which remains until the annual floods in the tributary remove
it ‡. When the Ohio is in flood, it stagnates the waters of the Mississippi for many leagues ; when the Mississippi is in flood, it
dams up the waters of the Ohio for seventy miles §.

Darby remarks that the Mississippi, in its long course from the
embouchure of the Ohio to Baton Rouge, washes the eastern
bluffs, which it tends to carry away and destroy, and that, even
to the sea, it does not come in contact with the western side of the
valley through which it flows. He attributes this, with great probability, to the deposits brought down by the great tributaries, which

* Darby's Louisiana, p. 65.
† Hall's Travels in North America.
‡ James, Exp. to Rocky Mountains.
§ Hall's Travels in North America, vol. iii. p. 370.   The same author
notices the curious mixture of the Missouri waters with those of the Mississippi, the former charged with detritus and wood, the latter beautifully
clear.

all enter the Mississippi from the west, and thus accumulate detritus on that side.

Notwithstanding the general tendency of the river to the eastward, innumerable smaller changes of channel take place. Thus winding courses shorten themselves, by cutting through isthmuses, the tendency of the winding currents being to destroy the barriers between them, as may be observed in numerous rivers flowing through plains. New obstacles present themselves; new sinuosities of channel are produced; trees growing upon old alluvial deposits of the river are carried away; and new vegetation springs up upon the recent alluvium, to be again removed by a new change of channel. During these various minor changes of bed, the degradation of the higher lands supplies a great abundance of detritus, which not only tends to raise the general level of the valley, by deposits over the low lands at floods, but is carried forward towards the sea, and forms an immense delta, composed of clay, mud, and silt, mixed with a large proportion of drifted trees and other vegetable substances.

The delta is divided into innumerable lakes, marshes, and streams, inhabited by a multitude of alligators. The main stream of the Mississippi will be observed to project forward, on all good maps, in a singular manner. The detritus brought down by it produces constant alterations, which require all the attention of the pilots. According to Captain Hall, millions of logs, or trunks of trees, are brought down during freshets, and carried several miles into the sea, so that it is difficult to navigate among them. When not carried to sea, these logs are bound together by a kind of cane, which retards the river and collects mud. The same author considers " that a belt of uninhabitable country, from fifty to one hundred miles in width, fringes the edge of the whole of that part of the coast *."

The mouth of the Ganges will afford us an example of the power of rivers to force forward deltas where no violent currents run across their embouchures, and where the body of water, particularly during freshets, is very considerable, even when such rivers are opposed to considerable tides. Major Rennell described this delta in 1781, so that probably, since his account was written, very material changes have been effected; yet as all these changes are likely to have been made in the same manner, Major Rennell's description will always be valuable, as showing the mode in which they have been carried on. The delta of the Ganges commences about two hundred and twenty miles from the sea in a direct line; or nearly three hundred, if the distance be reckoned along the windings of the river. The Ganges makes frequent windings, like many other rivers, and thus considerable changes of its

* Hall's Travels in North America, vol. iii. p. 340.

68   *Delivery of Detritus into the Sea.*

bed take place, the opposing bends cutting through the isthmus between them, as in the Mississippi. During the eleven years which
Major Rennell remained in India, the head of the Jellinghy river
was gradually removed three-quarters of a mile further down. He
also states, that "there are not wanting instances of a total change
of course in some of the Bengal rivers. The Cosa (equal to the
Rhine) once ran by Purneah, and joined the Ganges opposite
Rajenal. Its junction is now nearly forty-five miles higher up.
Gour, the ancient capital of Bengal, once stood on the Ganges."
It seems probable that the Ganges once ran in the line now occupied by the lakes and morasses between Nattore and Jaffier-
gunge *.

This delta is constantly on the increase. The quantity of detritus must be abundant, for the sea into which it is borne is
by no means shallow, the depths being considerable. The usual
checks are produced by the tide, but during the freshets the
ebb and flow are little felt, except near the sea. During these
times, therefore, the advance of the delta is most considerable,
the quantity of transported detritus being then greatest, and the
resistance of the sea at its minimum. The sea may ravage the
new lands, and apparently remove them for a time ; but eventually
they must gain, even from the accumulative power of the breakers themselves, which does but equalize the depths, by conveying
the detritus to a short distance : thus rendering the sea more
shallow, and consequently the easier filled up by river-borne detritus.

Coarse gravel transported by the Ganges does not approach
the sea within four hundred miles, and consequently does not
occur within one hundred and eighty miles of the commencement
of the delta ; therefore it would appear that during the present
order of things, the Ganges has not transported coarse gravel into
the sea at its present relative level. A great portion of the periodical inundations, represented as flowing on the level lands at
the rate of half a mile per hour, has been attributed to the rains
which fall on the low lands of India, as it has a blackish tint, from
being almost stagnant among vegetables of different kinds.
Small obstacles accumulate, as might be expected, very considerable banks and islands ; a large tree arrested in its progress downwards, or even a sunken boat, being sufficient for the purpose. As
these islands are quickly formed, so are they easily swept away by
any change in the mighty current, which is estimated to discharge
an annual quantity of water equal to 405,000 cubic feet per
second †.

At the junction of the Ganges and Burrampooter below Lucki-
poor, there is a large gulf in which the water is scarcely brackish,

* Rennell, Phil. Trans. 1781.   † *Ibid.*

even at the extremity of the islands, some of which are described by Major Rennell as equalling the Isle of Wight in size and fertility. The sea is represented as perfectly fresh to the distance of several leagues from this place during the rainy season.

It will be seen that deltas not only occur in situations where there is neither tide nor considerable current to prevent a great accumulation of new land, as at the embouchures of the Nile and Po, but also where the tides are small (Mississippi), and even where they are considerable (Ganges). The deltas thus produced are no doubt large, and the amount of animal and vegetable matter which they may entomb very considerable; but we must not be led away by measurements and comparisons with the length, breadth, or superficies of districts with which we may be familiar, and which we may, from habit, consider important. They should be regarded with reference to their relative importance as portions of dry land, when it will be seen that they do not expose so considerable a surface as might at first be supposed. The augmentation of deltas will correspond with the detritus carried forward to the embouchures of rivers, and it will be obvious that the facility of the transport will depend, all other circumstances being the same, on the length and fall of the channel. Now the course will be shortest and the declivity greatest at the commencement of the delta, and therefore it might be concluded that deltas would accumulate heavier materials, and increase most rapidly at the first periods of their formation, and that this increase would gradually diminish as the fall of the river channel became less, and its length increased; without reckoning on the innumerable checks given to the stream by the increasing divisions in the delta. It may also be supposed that the detritus from the high lands would become gradually less, from the equalization of levels, and the fewer asperities that meteoric agents have to act on. Should these remarks, made under the supposition of the non-interference of man, be correct, it will follow that the increase of deltas would gradually diminish if these were the only circumstances which regulated them. But it must be admitted that heavy rains, more particularly in tropical countries, would tend to cut up and destroy the delta itself, (still accumulating at its highest parts,) and force the detritus into the sea. The dense aquatic vegetation, common at the extremities of deltas, would render this transport difficult, yet still some detritus would escape. The amount of such additions to the outskirts of the new land would not, perhaps, be considerable, but it would correspond with the size of the delta, and consequently the larger this was, the greater would be the increase thus derived. Such an access of detritus would probably not compensate for the diminished quantity borne down by the river, and the increase of the delta would gradually diminish, if the operations of man, by dams and other works, did not cause a temporary increase.

Between those rivers, such as the Ganges, which obtrude deltas into tidal seas, and those which have large open embouchures, such as the Maranon, St. Lawrence, Tagus, and Thames, there are such variations, produced by local causes, that it would be exceedingly difficult, even if useful, to classify them. In the delivery of their detritus, therefore, such rivers will either produce deltas or estuaries at their embouchures, as they either partake of the characters of the Ganges or the St. Lawrence; if of the latter, the detritus will be dealt with according to the mode of deposit or transport in estuaries.

Action of the Sea on Coasts.

Breakers, or the waves falling on sea beaches or coasts, are continual and powerful agents of destruction in some situations; while in others they pile up barriers against themselves. Their destructive influence is principally felt when the rocks on which they are discharged are composed of soft materials, and rise somewhat abruptly above the level of the sea. Their protecting influence is most commonly experienced in front of low level lands, and across the mouths of valleys, on each side of which a hard rocky point supports the ends of a beach.

The destruction of coasts of equal hardness almost always bears a proportion to the extent of open sea to which such coasts are exposed, all other circumstances being the same. The configuration of most coasts will be seen to be determined by the hardness of the rocks composing them; the softer strata giving way before the battering power of the breakers, while the harder rocks preserve their places for a greater length of time. If the rocks forming a coast be stratified, much depends on the dip of the strata relatively to the breakers. Thus, in many situations on the southern coasts of Devon and Cornwall, the slaty rocks dip in such a manner towards the sea, that the waves have never effected more than the removal of some loose superficial matter, the same that covers all the hills in the vicinity. In fact, a skilful engineer could not have protected the coast better than has been accomplished by the dip of the strata. The destructive power in other situations is well known; and of this, the eastern coast of our island presents abundant proof, where very considerable encroachments of the sea have been recorded within the lapse of a few centuries. The substances so forced away by the action of the breakers will be acted on according to their weight, form, and solidity. The tides will remove so much of them as they are able to transport, and the rest will remain on the shore within the immediate influence of the breakers, which constantly tend to grind them down into smaller portions, and finally into sand.

In the destruction of a cliff of unequal hardness, it not unfrequently happens, that the harder portions, when large, such as many concretions in sandstones and marls, or blocks of indurated strata, remain at the base of the cliff, and in a great measure protect it from the more powerful effects of the breakers, as will be seen in the annexed figure.

Fig. 15.

a, a defence of blocks, derived from the hard strata *b*, and the concretions *c*.

Among the unstratified rocks, great variety of hardness prevails, so that they frequently present an uneven front to the sea, resulting from the quicker decomposition and destruction of some parts than of others. Veins of one substance, or rock, traversing another are generally of different textures and solidity from that which they cut, and consequently nothing is more frequent, on sea shores, than to observe them either standing out in relief or hollowed into coves.

When a shingle or sandy beach, but more particularly the former, is partly torn up and held in temporary mechanical suspension by the breakers during a heavy gale, the action of the waves is very considerable, even on the hardest rocks, so as to scoop them out near the ordinary level of the sea. In exposed situations, the hardest rocks are often drilled into holes or caverns, from the force of the broken wave being driven, by local circumstances, more in one direction than another, or from the inferior hardness of different portions of the rock. The most beautiful of ocean caverns, Fingal's Cave in Staffa, owes its existence to the circumstance of the basaltic columns being jointed in that place, while the general character is to be without divisions in the columns *.

After the sea has formed a cavern, the vault of which does not rise above high water, it sometimes works its way upwards at the inmost extremity, partly by means of the compressed air, held between each wave as it rolls into the cave. Of this kind of cavern Bosheston Mere in South Wales is an example on the large scale. It is formed through strata of carboniferous limestone, and the noise caused by the blast of compressed air and sea water upwards is heard at a considerable distance.

The protecting influence of breakers is shown in long lines of shingle and sandy beaches, which often protect low and marshy land, particularly at the mouths of valleys, from the destructive power of the sea.

* Macculloch, Western Islands of Scotland.

Shingle Beaches.

In the case of shingle beaches, it will be observed, that during a heavy gale every breaker is more or less charged with the materials composing the beach; the shingles are forced forward as far as the broken wave can reach, and in their shock against the beach drive others before them that were not held in momentary mechanical suspension by the breaker. By these means, and particularly at the greatest height of the tide, the shingles are projected on the land beyond the reach of retiring waves. Heavy gales and high tides combined seem to produce the highest beaches; they do indeed sometimes cause breaches in the rampart they have raised against themselves, but they quickly repair them. The great accumulation of beach upon the land being effected at high water, the ebb tide, it is clear, cannot deprive the land of what it has gained. In moderate weather, and during neap tides, various little lines of beach are formed, which are swept away by a heavy gale; and when these little beaches are so obliterated, it might be supposed, by a casual observer, that the sea was diminishing the beach; but attention will show that the shingles of the lines, so apparently swept away, are but accumulated elsewhere. These remarks do not apply to situations where the sea, during gales, has access to cliffs or piers, from whence there might be a retiring wave carrying all before it; but to such situations—and they are abundant—where the breakers meet with no resistance, and strike nothing but the more or less inclined plane of a shingle beach. Even in cases where the waves in heavy gales and high tides do reach cliffs, and for the time remove shingle beaches, it is curious to see how soon these latter are restored, when the weather moderates, and when the breakers, in consequence of a diminished projecting force, cease to recoil from the cliff behind.

Shingle beaches travel in the direction of the prevalent winds, or those which produce the greatest breakers: of this there are abundant examples on our own southern coast, where the prevalent winds being W. or S.W., the beaches travel eastward until arrested by some projecting land, when the sea forms a barrier against itself, and not unfrequently leaves a space between it and the cliff which it formerly cut: this space, under favourable circumstances, is covered by vegetation, suited to such a situation, even the cliff being sometimes studded with sea-side plants, when they can find root. Works are sometimes constructed to arrest beaches, either to protect land behind or to prevent their passage round pier heads into artificial harbours; and thus engineers are practically aware of their travelling power in the direction of certain winds. This progressive march of beaches is far from rapid, and can only be in proportion to the greater power or duration of one wind to another; moreover, the pebbles become comminuted in their passage, and thus the harder can only travel to considerable distances.

The Chesil Bank, connecting the Isle of Portland with the main land, is about sixteen miles long, and, as a general fact, it may be stated that the pebbles increase in size from west to east. It protects land which has, evidently, never been exposed to the destructive power of the Atlantic swell and seas, which break with great fury against the Bank; for the land behind is composed of soft and easily disintegrated strata, which would speedily give way before such a power. Perhaps a gradual sinking of the land might produce the present appearances; for though the sea would have attacked the land when the relative levels were different, the form of the bay, and the projection of the Isle of Portland, would soon cause a beach to be formed, which would rise as the land sunk, so that, finally, no traces of a back cliff could be observed. Under this hypothesis, Portland would not have formed an island, but merely the projecting point of a bay, which, with its exposure, would soon have accumulated the beach required. It may be remarked, that this supposed gradual sinking of the land is in accordance with appearances more westward on the same coast, where the facts presented seem to require this explanation. The sea separates the Chesil Bank from the land for about half its length, so that, for about eight miles, it forms a shingle ridge in the sea. The effects of the waves, however, on either side are very unequal; on the western side the propelling and piling influence is considerable, while on the eastern, or that part between the bank and the main land, it is of trifling importance. The following is a section:

Fig. 16.

a, the Chesil Bank: *b*, the water called the Fleet: *c*, small cliffs formed by the waves of the Fleet and land springs: *d*, various soft rocks of the oolite formation, protected from destruction by the Chesil Bank *a*.

Another curious example of land protected by a shingle bank occurs on the southern coast of Devon, and is remarkable, as it shows that the sea, at its present relative level with the land, has never reached the land behind the beach,—a fact that will admit of the same explanation as that previously given for the Chesil Bank. At the bottom of Start Bay, and for the distance of about five or six miles, a considerable bank, principally composed of small quartz pebbles, has been thrown up by the sea. The line of coast faces

E

the east. Between Tor Cross and Beeson Cellar, a point of land
comes within the reach of the breakers; but here, as well as elsewhere behind the bank, the land has evidently gained on the sea,
or, in other words, the latter has piled up a barrier which prevents
its reaching the cliff, as it once did, even during heavy gales. This
bank, generally known as Slapton Sands, though composed wholly
of small pebbles, protects and blocks up the mouths of five valleys.
Between Slapton Sands (properly so called) there is a fresh-water
lake, divided into two at Slapton Bridge, where the waters of the
northern lake drain into the southern. The northern portion is
nearly silted up by the detritus borne down by a river that drains
a few miles of country, and is nearly covered by bullrushes and
other aquatic plants. The southern and larger portion is open,
and of many acres in extent. The waters are supplied by the
rivers behind, and commonly percolate through the pebbles into
the sea. When, however, the tides are high, and the waters kept
up by heavy gales, it sometimes happens that, the relative levels
being altered, the sea-water passes through the shingles into the
lake, and renders it to a certain extent brackish. This usually
happens in winter; but, generally speaking, the relative levels are
such, that the lake drains into the sea and remains perfectly fresh.
It contains a great abundance of trout, perch, pike, roach, and
flounders. The presence of the latter, a marine or estuary fish,
shows that it can be gradually accustomed to fresh water. The
percolation of the sea through the pebbles, during heavy gales,
does not seem to injure the fresh-water fish; but when a breach
was made through this beach during the gale of November 1824,
they were nearly all killed by the sudden influx of the sea. Those
which escaped up the streams were sufficient, in five years, again
to stock the lake abundantly.

The breach made through Slapton Sands continued open for
nearly a year, becoming gradually smaller. The complete re-
storation of the sands was hastened by throwing a few bags, filled
with shingles, into the gap, upon which two or three gales soon
piled up a heavy beach.

The old bank must have remained undisturbed for a long period; for vegetation had become active upon it, as we see by those
portions which remain uninjured, where turf and even furze-bushes
have established themselves upon the shingles.

Fig. 17.

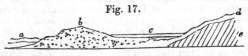

The above exhibits a section of the beach and lake.—*a*, the sea
which throws up the beach *b* : *c*, the fresh-water lake behind the

beach : *d,* several feet in depth of pieces of slate and sand derived from the slate-rocks *e.*

This diagram shows that the sea could not have acted upon the hill *d e* since the accumulation of the loose substances *d,* which it would have instantly removed.

The great size of rock fragments moved by the action of the breakers attests their power. During heavy gales, blocks of many tons in weight have been forced from their places; and others, even squared and bolted together in the form of piers and jetties, have been torn asunder by the battering power of the waves. During the gale of November 1824, which ravaged a considerable part of the southern coast of England, a squared block, from a ton and a half to two tons in weight, strongly trenailed down, was torn away from a jetty at Lyme Regis, and tossed upwards by the force of a breaker. The same gale tore away part of the Breakwater at Plymouth, and threw blocks of three tons or more in weight far over on the other side*.

At the Scilly Islands the blocks of granite that fall from the cliffs are ground by attrition into great boulders, which become the sport of the heavy Atlantic seas in tempestuous weather.

The effect produced by a heavy sea must depend considerably on the form of the block on which the sea acts. Thus, a flat front would present the greatest resistance to the shock, and the mass so struck would have a tendency to be more easily moved than a rounded mass, if it were not that the resistance to removal offered at its base, is very considerably greater than in a rounded mass.

The wedging power of the breakers is also very considerable where heavy blocks of difficult removal are mixed with smaller stones easily transported. A beach of this nature sometimes acquires much solidity, as the smaller pieces are often forced among the larger so tightly as to require very great force, and even fracture, before they can be taken out.

* Mr. Harris, of Plymouth, informs me, that, during the severe gales of November 1824, and at the commencement of 1829, blocks of limestone and granite, from 2 to 5 tons in weight, were washed about on the Breakwater like pebbles; about 300 tons, in blocks of these dimensions, being carried a distance of 200 feet, and up the inclined plane of the Breakwater. These blocks were thrown over on the other side, where they remained, after the gale, scattered in various directions. A block of limestone, weighing 7 tons, was washed round the western extremity of the Breakwater, and carried 150 feet. Two or three blocks of this size were washed about. At the Pier in Bovey Sand Bay, on the east side of Plymouth Sound, a piece of masonry may be now seen, which was washed back about 10 feet, being, at the time it was struck, 16 feet above the level of an 18-feet spring tide. This piece of masonry weighs about 7 tons, and consists of a few blocks of limestone cemented together and covered by a large block of granite. The mass was dovetailed into, and formed part of, a parapet facing the sea.

It would appear that, though shingle beaches, or those composed partly of pebbles, and partly of larger masses, may be moved in the direction of the predominating and heaviest breakers, we have no evidence of their being transported outwards, or into the depths of the ocean, but that, on the contrary, the waves of the sea strive to throw them upon the land; and this, not only in the case of substances derived from the land, but also in that of corals, shells, and marine plants which have been produced in the sea itself. In tropical countries it is found that many coral reefs and islands are defended on their windward sides by beaches of coral shingles, and even large fragments of coral. Lieut.-col. Hamilton Smith informs me that, during a hurricane which he witnessed at Curaçoa in September 1807, large pieces of coral were torn up from a depth of ten fathoms, and thrown on the bank uniting Punta Brava with the land. Beaches composed wholly or entirely of comminuted marine shells are not uncommon, and will be noticed in the sequel.

The seaward front of most shingle beaches, particularly when they defend tracts of flat country, is bounded by a line along the edge of the beach; above this line the beach generally makes a considerable angle with the sands, in cases of sandy flats. In cases where shingle beaches are not entirely quitted by the tide, sandy, shelly, or very fine gravel soundings are commonly obtained at a short distance from the shore, unless the bottom be rocky. It would appear that, if the present continents or islands were elevated above, or depressed beneath, the present ocean-level, shingle beaches would be found to fringe the land, but not to extend far seaward *.

Sandy Beaches.

The observations made respecting shingle beaches apply, in a great measure, to those composed of sand. The sand is derived either from the detritus borne down by rivers, from the attrition of sea-shore shingles against each other, or immediately from the

* We should be careful, when we obtain shingles in various soundings, to consider that the probability is as great of finding pebbles at the bottom of the sea as on the dry land; and that their presence there, is no proof that they have been transported by existing currents, unless it can be shown that the velocity of the existing current is sufficient to transport such detritus, and that the direction of the current is that which would carry the fragments from the known place of the parent rock. Without attention to this circumstance, it might be supposed that the small shingles, covering the bottom of the newly discovered bank off the north-west coast of Ireland, were carried there by the present currents, when they are quite as likely to have been otherwise produced. That they are not now rolled about to any extent, is evident from the serpulæ and other marine productions attached to some of them brought up by Captain Vidael, during his survey, by the arming of the sounding lead.

sand and sandstones of the land. The breakers have the same
tendency to force sand upon the land, as was observed in the case
of shingles; but, being so much lighter than the latter, sand can be
transported by coast tides or currents whose velocity would be
insufficient to move shingles. On the other hand, however, smaller
forces and bodies of water can throw sand on the shore. The
spray that could not transport a pebble can carry sand, and thus
this substance can be, and is, conveyed far beyond situations
where the reflux of a wave can be felt. When the tide is low, or
the sea less agitated, sand, dried by the sun or winds, is transported by the latter to great distances, so that whole districts of
once fertile land have been overwhelmed by it.

Such transported sand, when sufficient to form hills, is known
by the name of *dunes*, more or less common behind sandy shores
or beaches over the globe. A striking example of the progress of
such drifted sand inland, is to be found in the Bay of Biscay, on
the eastern shore of which the sands have overwhelmed and are
continuing to cover large tracts of country. Cuvier states the
advance of these dunes as perfectly irresistible, forcing lakes of
fresh water before them, derived from the rains which cannot find
a passage into the sea. Forests, cultivated lands, and houses disappear beneath them. Many villages noticed in the middle ages
have been covered, and, in the department of the Landes alone,
ten are now threatened with destruction. "One of these villages,
named Mimisan, has been striving for twenty years against them;
and one sand-hill, more than sixty feet high, may be said to be
seen advancing. In 1802, the lakes invaded five fine farms belonging to Saint Julien; they have long since covered a Roman
causeway which led from Bourdeaux to Bayonne, and which was
seen, about forty years since, when the waters were low. The
Adour, which was once known to flow by Vieux Boucaut, and fall
into the sea at Cap Breton, is now turned aside more than a
thousand toises *."

M. Bremontier calculated that these dunes advance at the rate
of sixty, and even seventy-two, feet per annum.

Under favourable circumstances, sands, transported from a
beach into the interior, become consolidated: of this a good example is found on the north coast of Cornwall, where the matter
thrown up is formed from comminuted sea-shells, and the consolidation is principally effected by means of oxide of iron. From
the drift having taken place at different times, this recent calcareous sandstone is stratified, with occasionally interposed vegetable
remains. Houses have been overwhelmed, and human remains
entombed where churchyards have existed. Mr. Carne describes
a pot of old coins dug out of it. The induration of this rock is so
considerable, that holes are drilled in it at New Kay, for the pur-

* Cuvier, Dis. sur les Rev. du Globe.

pose of securing vessels to the cliff. It is also used for architectural purposes, and according to Dr. Paris the church of Crantock is built with it. The same author states that the high cliffs of this recent rock, which extend several miles in Fistrel Bay, are occasionally intersected with veins of breccia. " In the cavities, calcareous stalactites of rude appearance, opaque, and of a gray colour, hang suspended." " The beach is covered with disjointed fragments, which have been detached from the cliff above, many of which weigh two or three tons *."

Indurated dunes occur in various parts of the world : they have been noticed by Peron in New Holland; and the rock in which the human remains of Guadaloupe have been found would appear to be similar. These latter are discovered at the Port du Moule, in an indurated beach composed of comminuted shells and corals. The specimen in the British Museum is formed of coral and small pieces of compact limestone, and in it Mr. König has observed *Millepora miniacea*, madrepores, and shells referred to *Helix acuta* and *Turbo Pica*. According to Cuvier, the specimen in the Jardin du Roi, at Paris, exhibits a gangue of travertine containing shells of the neighbouring sea, and terrestrial shells, especially the *Bulimus guadaloupensis* of Férussac. Near Messina, loose sand becomes consolidated on the beach, and is used for building. It is stated that the cavities thus made are again filled up by sand, which becomes consolidated and used in its turn.

Dr. Clarke Abel describes a large bank, rising from the sea to the height of about a hundred feet to the eastward of Simon's Town, Cape of Good Hope, formed of shell and sand, thrown up by the S.E. wind. In this he discovered singular cylindrical bodies, which resembled bones bleached by the air. " On a closer examination, many of them are found to be branched ; and others are discovered rising through the soil, and ramifying from a stem beneath, thicker than themselves. Their vegetable origin imme-

---

* Paris, Geol. Trans. of Cornwall. Not only sands but shingle beaches are sometimes indurated.—Captain Beaufort describes a plain, several miles in length, near Selinty, coast of Karamania, as bounded by a gravel beach, which has become consolidated from the top of the crest to some distance into the sea; the consolidation extending to the depth of from one to two feet, and being generally covered with loose sand and gravel, so that it is not easily observed. The pebbles are cemented by a calcareous paste, and the whole is so hard, that a blow "more frequently fractures even the quartz pebbles than dislodges them from their bed." Other beaches of the like kind, but on a smaller scale, were observed on other parts of the coasts of Asia Minor and of Greece. Rocky ledges of a similar nature occur to the westward of Sidé, partly above and partly under the water. They contain broken tiles, shells, bits of wood, and other rubbish. They are very hard, and are cemented by calcareous matter, probably derived from some calcareous slate in the vicinity.— Beaufort's Karamania, p. 182 and 185.

diately suggests itself, and is confirmed by a further inquiry. They are seldom solid, their centres being either hollow or filled with a blackish granular substance, which in many specimens, except in colour, resembles the substance called roestone by mineralogists. Their outer crust is chiefly composed of a large proportion of sand, and a small proportion of calcareous matter, and in many specimens contains fragments of ironstone and quartz an inch square. That they are really incrustations formed on vegetables which have afterwards decayed, is proved by the different degrees of change which the internal parts of different specimens have undergone. In some the organization of the plant sufficiently remains to leave its nature unequivocal ; and near the sea the very commencement of the process of incrustation may be witnessed on the large *Fuci* which strew the shore *.

Peron's previous description of the change undergone by vegetable substances in similar situations on the coasts of Australia, is nearly the same. He considers that the shells undergo decomposition, and form a cement with the sand ; and that the vegetables become altered and finally replaced by this sandstone, leaving nothing to show its origin but its general form. On our coasts the sands thrown on shore by the action of the sea, and afterwards drifted by the winds, are often comparatively considerable. Mr. Ritchie describes a district of ten square miles in Morayshire, once termed the Granary of Moray, as having been overwhelmed. "This barren waste may be considered as hilly ; the accumulation of sand composing these hills frequently varying in their height, and changing their situation †."

The following account by Mr. Macgillivray affords an additional example of the tendency of coast-seas to throw even the substances formed in them upon the land. "The bottom of the sea, along the whole west coast of the Outer Hebrides, from Barray Head to the Butt of the Lewis, appears to consist of sand. Along the shores of these islands this sand appears here and there in patches of several miles, separated by intervals of rock of equal or greater extent. In some places the sandy shores are flat, or very gently sloping, forming what are here called Fords ; in others, behind the beach, there is an accumulation of sand to the height of from twenty to sixty feet, formed into hillocks. This sand is constantly drifting ; and in some places islands have been formed by the removal of isthmi. The parts immediately behind the beach are also liable to be inundated by the sand ; and in this manner most of the islands have suffered very considerable damage . . . . . . The sand consists almost entirely of comminuted shells, apparently of the species which are found in the neighbouring seas. It is rather coarse in the grain ; but during high winds, by the rubbing of its

* Clarke Abel, Voyage to China, p. 308.
† Notes appended to Cuvier's Theory of the Earth, by Jameson.

particles on each other, a sort of dust is formed, which at a distance resembles smoke, and which, in the island of Berneray, I have seen driven into the sea to the distance of upwards of two miles, appearing like a thin white fog *."

It would be useless to accumulate notices of these various sand-drifts, which often contain seams of vegetable matter that have been successively covered up, and of which sections are afforded †. The action of the waves round coasts tends to disturb the bottom at certain depths, and to move the shells, sands, and other substances, of which this bottom is composed, towards the land. The exact depth to which the moving action of waves extends, seems never to have been very accurately estimated; indeed, when we consider that the power of the wave is continually varying, such an estimate becomes exceedingly difficult. Ninety feet, or fifteen fathoms, has been sometimes considered as the limit, in depth, to which this disturbing power extends; but this requires confirmation. Around coasts and on shores which do not much exceed ten or twelve fathoms, the action of the waves is very apparent in the discoloration of the water during heavy gales. This turbid character of the sea is due to the moving power of the waves on the bottom, and becomes more marked as the water becomes more shallow, either in approaching the land or over shoals. The transporting power of the waves will therefore be in proportion to the depth of water beneath them, the transport being greatest in the shallowest places. The waves will tend to throw substances on coasts, because the off-shore wind produces smaller waves than the wind blowing upon the land. On shoals distant from the land, the effect will be somewhat different, and the piling or propelling power will be greatest on the side of the prevalent or more violent winds. Shoals will be also liable to shift, as the turbid waters on the crown of a shoal will be forced over on the lee side. Accordingly, we do find, that shoals shift, more particularly when near the surface, unless there be an equal counteracting effect in a current or tide. We may, in some measure, learn the effects of waves at different depths, from the form of the outer

---

* Notes to Cuvier's Theory of the Earth, by Jameson.

† Not only are sand-hills thrown up by the sea, but also by the waves of extensive fresh-water lakes. Dr. Bigsby states (Journal of Science, vol. xviii.) that large quantities of sand are thrown up between the Crags and the Otter's Head, on the east side of Lake Superior; and that from seven to eleven miles eastward of the last-mentioned place, there are sand-hills 150 feet in height. In the same vicinity also, angular fragments, torn from the neighbouring rocks, are thrown up in vast heaps, and scattered among the trees. This operation must be greatly assisted by the rise of water consequent on a westerly gale; for Dr. Bigsby informs us, that if a gale from that quarter lasts more than a day, it raises the water on the eastern side of the lake to the amount of twenty or thirty feet.

talus of the *Digue,* or Breakwater, at Cherbourg, where they have, to a certain extent, arranged the stones, four-fifths of which are small, in the manner best fitted to resist themselves. According to M. Cachin, there are four kinds of taluses, arranged one beneath the other. The upper line of talus, being only touched by the higher break of the waves, presents a height proportioned to its base, as 100 is to 185. The second line, comprising the whole distance between the line of high and low water at the equinoxes, and thus exposed to the battering power of the breakers during the whole flood and ebb, is consequently the most inclined, and its height is to its base as 100 to 540. The third line, being below the lowest water at the equinoxes, is only acted upon during the first of the flood or the last of the ebb. Its height is to its base as 100 to 302. The fourth line, or the base of all, not being acted on by the waves, maintains a talus, of which the height is to the base as 100 to 125*.

The action of waves on coasts is not only exhibited by piling up detritus in the direction of their greatest force on the shore, by which embouchures of rivers are turned on one side, but also by heaping up bars, as they are termed, even at their mouths, rendering their navigation dangerous, and in many instances preventing it altogether; though, behind these barriers, the rivers may have considerable depth and breadth. In some situations these bars are partially dry at low water, at others they are never uncovered, though rendered visible by the breaking of a furious surf. To produce examples would be useless, as they are common in all parts of the world. In many cases, the bars are liable to shift, particularly after a gale of wind, so that vessels are frequently lost by keeping the direction of the old channels; and it requires the constant attention of pilots to be aware of the exact position of the new passages.

When the rivers are small, the force of the waves frequently blocks up their embouchures, and artificial means are necessary to permit the escape of the pent-up waters, that would otherwise form a lake in the low country behind. If the dam be a shingle beach, the water usually percolates through it; but if composed of sand, the water will accumulate until its level enables it to cut a passage through the barrier and escape. This done, the breach will be again repaired, and another accumulation of water take place behind, and so on. But, in the mean time, the level of the low land would rise, first, by deposition from the river waters; and, secondly, from the sand blown over the bank. In such an alluvial land there would probably be found remains of terrestrial, fresh-water, and even marine shells, the latter worn or broken.

Rivers are deflected from their courses into the sea by beaches

* Mém. de l'Académie, tom. vii. p. 413.

extending from one side, and produced by the winds and breakers;
both forcing detritus before them, if it be composed of sand or
comminuted shells, while the latter acts upon the shingles alone,
except when light pebbles are caught up in the heavier spray,
and are thus driven by the wind. Examples of this deflection
may be seen in many situations, and the harbour of Shoreham,
on our southern coast, is a marked one*.

Rivers, when thus deflected from their courses by beaches,
generally escape into the sea by the sides of cliffs, which seem to
give them such support that they can cut channels.

In tropical countries the breakers throw up barriers against the
advance of the mangrove trees, either from a deep bay or creek,
or at the mouths of rivers, if they come within their influence.
Capt. Tuckey remarks, that "the peninsula of Cape Padron and
Shark Point, which forms the south side of the estuary (of the
Zaire), has been evidently formed by the combined depositions of
the sea and river, the external or sea shore being formed of quartzy
sand constituting a steep beach; the internal or river side, by a
deposit of mud overgrown by mangroves; and both sides of the
river towards its mouth are of similar formation, intersected by
numerous creeks (apparently forming islands), in which the water
is perfectly torpid." This mangrove tract appears to extend inland,
on both banks about seven or eight miles, and is represented as impenetrable. Did not the sea pile up a barrier against it, and thus
afford it protection from its own attacks, it would be destroyed†.
Similar phænomena, though on a much smaller scale, are seen at
the mouths of the Rio Minho, and other rivers in Jamaica.
Beaches are accumulated in front of mangrove trees, under somewhat similar circumstances, in the same island, on the south side
of which, particularly near Albion estate, lakes are formed on the
inside of a shingle beach thrown up by the sea. The lake near
Albion has a small opening in the protecting bank, permitting
the surplus water to escape; this water being apparently derived
from the drain of the mountains behind, and the splash of the
sea during gales. The mountain drainage has carried much mud
into the lake, upon which mangrove trees have established themselves. These by their roots entangle various substances, and
form land, the accumulation being a compound of mineral, vege-

* See Geological Notes, pl. 1. fig. 2.; and Phil. Mag. and Annals of
Philosophy, N. S. vol. vii. pl. 11. fig. 2.

† Expedition to the Zaire or Congo, p. 85. This author further remarks, that "small islands have in many places been formed by the
current (of the river); and doubtless in the rainy season, when the
stream is at its maximum, these islands may be entirely separated from
its banks, and the entwined roots keeping the trees together, they will
float down the river, and merit the name of floating islands."

table and animal substances *. A much larger lake of the same
description is found under Yallah's Mountain, the most projecting
part of the beach forming Yallah's Point †.

The bank called the Palisades, at the end of which stands Port
Royal, Jamaica, seems thrown up by the action of the prevalent
breakers, caused by the sea breezes, or winds from the east and
south-east, which propel the materials of the beach from east to
west. This bank is between eight and nine miles long, of little
elevation above the sea, having a beach on the seaward front,
with mangrove trees on many parts of the inward side. If the
passage between the western end of this bank and the land opposite to it should be barred up by a continuation of the bank, a
large lake would be inclosed, into which the Rio Cobre would
discharge itself. The mangrove trees would assist in the formation of new land, in which a mixture of marine, fresh-water, and
terrestrial remains might be entombed.

Mangrove trees afford support to beaches thrown up by the sea;
and if such a beach originated from a shoal, there is always a
tendency to increase land to leeward by their agency. Protection
being once afforded, the mangrove trees establish themselves, and
accumulate silt, mud, and drift-rubbish about their roots. Thus,
support is afforded to the original bank, and new materials are
piled upon it to windward by the action of the breakers, additional
consolidation being produced by the tropical sea-side creepers.
Meanwhile the advance to leeward continues, until the land immediately against the beach becoming too dry for the support of
the mangrove trees, others, more suited to the new land, establish
themselves; and, finally, a grove of cocoa-nut trees may gradually
appear ‡.

Tides and Currents.

The principal motions in the waters of seas and oceans are produced by tides and currents; the former due to the action of the
sun and moon, the latter probably caused by the winds and the
motion of the earth.

The streams of water caused by tides are chiefly felt on coasts,
while the currents produced by winds are more or less experienced
over the whole surface of the ocean. It must frequently happen
that the direction of a tide and a current being the same, they add
mutually to the velocity of each other, while the contrary arises
with opposed courses.

The streams of water produced by tides and currents are geo-

* For a section of this lake, see Sections and Views illustrative of
Geological Phænomena, pl. 35. fig. 6.
  † These waters contain a multitude of alligators.
  ‡ For a section of such an island near Jamaica, see Sections and Views
illustrative of Geological Phænomena, pl. 36. fig. 2.

logically important, as they may be the means of distributing the
detritus derived from the land over spaces at a greater or less
distance from the shore; their power of effecting this being proportioned to their velocity and depth.

Tides.

The velocity of a stream of tide depends on the obstacles it
encounters. These obstacles generally present themselves in the
form of projecting headlands, a gradually diminishing channel, or
a group of islands and shoals. In the former case the velocity of
the tide is considerably increased round the opposing capes, gradually diminishing to its usual rate at a short distance on either
side, or in the offing. The English Channel will present us with
many examples, more or less striking, according to circumstances.
Round the Start and the Bill of Portland the tides run exceedingly
strong, causing dangerous races when opposed to the winds. But
these considerable streams of tide are merely local; for in the bays,
and at a short distance out at sea, the velocity of the tides does
not exceed a mile and a half or two miles; while at the headlands
above noticed, it frequently flows at the rate of four or five miles *.
Generally speaking, the increased velocity of the tidal stream
round capes is in proportion to the body of water forced into the
bays of which they form the extreme points.

The greatest obstacle opposed to the tidal wave flowing up the
English Channel, is the great bight on the west of Cap la Hague,
where we find innumerable islands and rocks, of which the principal are Guernsey, Jersey, and Alderney. The stream of flood
being completely opposed to the line of coast, and pent-up by the
islands and rocks, it rises to a very considerable height, and escapes
through the Race of Alderney, between the island of the same
name and the main land, with a velocity of seven miles an hour.
It continues to run with great rapidity round Cap Barfleur, gradually decreasing in strength until the general level is restored.
Some idea may be formed of the variation in the Channel level,
caused by this obstacle, by the differences in the rise of tide observed between the mouth of the Channel and the Straits of
Dover.

The perpendicular rise of tide on each side of the mouth of the
Channel is nearly the same, being twenty-one feet at Ushant, and
twenty feet at the Land's End. In the great bight or bay west
of Cap la Hague, the tide rises forty-five feet between Jersey and
St. Maloes, and thirty-five feet at Guernsey. At Cherbourg this
great elevation of the level is diminished; the tide there rising
about twenty-one feet. On the opposite side of the Channel, on

* All the miles mentioned in the following notice of tides and currents are nautical, sixty being equal to one degree.

the English coast, the perpendicular rise of the tidal wave is comparatively trifling, being thirteen feet at Lyme Regis, seven feet in Portland Road, fifteen feet at Cowes, and eighteen feet at Beachy Head. Therefore, the elevated level of the Guernsey and Jersey waters produces no perceptible effect on the English coast opposite. Between Beachy Head and Dover, there is a rise of twenty-four feet on the west of Dungeness, and twenty feet at Folkestone. On the opposite coast there is a rise of twenty feet at Havre, nineteen feet at Dieppe, and nineteen feet at Boulogne. The tides are twenty feet at Dover, and nineteen feet at Calais.

The Bristol Channel is a familiar example of a high rise of tide caused by a gradually contracted channel, at the end of which there is no outlet. At St. Ives, Cornwall, the perpendicular rise of the spring tides is eighteen feet, of the neap tides fourteen feet *. At Padstow the tide rises twenty-four feet; at Lundy Island, thirty; at Minehead, thirty-six; at King Road, near Bristol, from forty-six to fifty; and at Chepstow, about the same.

The difference of level, produced by obstacles to the tide, is remarkably exhibited on each side of the isthmus separating Nova Scotia from the main land of North America. In the Bay of Fundy, on the south side, the tides have a very considerable rise, amounting, according to Des Barres, to sixty and seventy feet at the equinoxes; while on the northern side, in Baie Verte, they only rise and fall eight feet. The tidal stream is, as might be expected, very rapid in these gradually diminished channels, particularly where the rise and fall is most considerable. This unusual rapidity ceases by degrees as we approach the mouths of such channels, and arrive at the more common levels.

From the great diversity in the line of coasts, innumerable modifications are effected in tidal streams, causing them to flow with augmented or diminished velocity. As such streams are only visible on coasts, it seems fair to infer that the effects produced by them do not extend to any considerable distance beyond the land.

The tide in the offing, and the tide along shore, do not exactly correspond, the flood tide continuing in the offing some time after the ebb has commenced on shore; the ebb tide the same. It has been stated that "the length of time between the changes of the tide on the shore and the stream in the offing, is in proportion to the strength of the current and the distance from the land; that is, the stronger the current, and the greater the distance that the current is from the land, the longer it will run after the change on the shore †."

* The rise of tide at St. Ives is sometimes stated at twenty-two feet.
† Purdy, Atlantic Memoir, 1829. In the same work it is stated that "the time which the flood-stream runs in the middle of the English

Among the small islands of the Pacific Ocean the tide rises about two feet, there being no great range of coast near them to produce a greater elevation. At the islands of the Atlantic Ocean the rise is greater, probably because the body of water is less and the coasts nearer. At the Azores the rise is from six to seven feet; at Madeira, eight or nine; among the Canaries, eight or ten; at the Cape Verde Islands, from four to six; and at the Bermudas, five or six.

The stream of tide along a coast is greatly increased at the time of full and new moon, so that at spring tides the current often runs at double the rate experienced at neap tides. The transporting power of tidal streams is therefore perpetually changing, independent of the variations produced by winds upon them.

From various circumstances the tides of flood and ebb are sometimes unequal. Thus, at the Land's End the flood runs nine hours to the north, and the ebb three to the south. In the expedition under Captains Parry and Lyon, it was found that in the higher part of Davis's Straits the flood tide set from the north at the rate of three miles an hour for nine hours, the tide of ebb making only three hours.

A current setting into the Straits of Malacca, during part of the year, causes the tide to run nine hours one way and three hours the other. The tides are irregular through the Straits of Banca, with an easterly wind. The ebb sets to the northward for sixteen hours, while the flood only lasts eight hours. In common tides there are two floods and two ebbs in twenty-eight hours in these straits, the duration of which is in some sort regulated by the winds: the flood lasts six hours, and the ebb eight hours; or there are five hours flood, and nine hours ebb.

The tides are very trifling and irregular in the West Indies, perhaps owing to the accumulation of water pent up by the equatorial current and trade winds. At Vera Cruz there is only one tide in twenty-four hours, and that irregular. Among these islands the tide varies in perpendicular rise from a few inches to two feet or two feet and a half. The stream or current produced by them must consequently be very trifling.

Theoretically, all bodies of water, even large fresh-water lakes, have tides; but they are so insignificant that inland seas, such as the Mediterranean and Black Seas, are generally termed tideless. The current setting into the Mediterranean from the Atlantic

Channel after the time of high water on shore, is, westward of the meridian of Portland, about three hours; but to the eastward, off Beachy Head, only one hour and three quarters. In the offing, between the meridians of Dungeness and Folkestone, the North Sea and Channel tides seem to meet; and the ebb of the one uniting with the flood of the other, set in an easterly direction off the French coast, more than four hours after high water on the western shore of Dungeness." p. 88.

is somewhat modified by the tides. In the middle of the Straits
of Gibraltar the current sets eastward; on each side, however,
the flood tide sets to the westward.

"On the European side, west of the island of Tarifa, it is high
water at 11^h, but the stream without continues to run until 2^h.
On the opposite shore of Africa, it is high water at 10^h, and the
stream without continues to run until one o'clock; after which
periods it changes on either side, and runs eastward with the general current. Near the shore are many changes, counter currents and whirlpools, caused by and varying with the winds. Near
Malaga the stream runs along shore about eight hours each way.
The flood sets to the westward*."

The strongest tides of which I can find mention, occur among
the Orkney and Shetland Isles, and through the Pentland Frith,
between the main land of Scotland and the former. The flood
comes from the north-west, and is not of unusual strength until it
encounters the obstacles of the islands and main land. The tides
change near the shores sooner than at a distance from them. The
difference of time varies according to situation, amounting in some
places to two or three hours. The velocity of the tide through
Stronsa Frith is about five miles an hour during spring tides, and
a mile or a mile and a half at neaps. In North Ronaldsha Frith,
the springs run at five miles an hour; the neap tides at one mile
and a half. The flood divides near the shore at Fair Isle, forming
a large eddy on the east side. The springs here run six miles an
hour, the neaps two. These tides increase in velocity when supported by the winds. The most rapid stream of tide occurs at the
Pentland Frith, its velocity being nine miles an hour during the
springs, though it runs only three miles an hour at neap tides.

Tides in Rivers and Estuaries.—These are necessarily much
modified by circumstances; but, generally speaking, the tide of
ebb is stronger than the flood, from the body of fresh water being
pent up by the flood, to which the rivers must always present a
certain resistance, proportioned to their velocity and abundance of
water;—the greatest resistance to the flood, and increased velocity
of the ebb, being during freshets, or when the rivers have a sur-
charge of water produced by rains in the interior.

When the flood tide takes place in rivers of sufficient depth, the
first operation of the tide appears to be that of a wedge, elevating
the fresh water from its inferior specific gravity to a higher level.
The flood gradually opposes greater resistance to the outflow of
the river, and in the end succeeds in damming it up. I have
found many fishermen aware of this "creeping," as they have
termed it, of the salt water beneath the fresh at the commencement of the flood, and have seen a rise of two or three feet caused

* Purdy, Atlantic Memoir, p. 90. The tide rises three feet at Malaga.

in water in the higher parts of tidal rivers, while the water so raised has continued perfectly fresh at the surface.

At the ebb, if the fresh or river waters be abundant, they will, after the salt water has been discharged, flow over the salt water to greater or less distances from the shore according to circumstances. After the rains, a strong freshet sets down the Senegal, and a powerful current of fresh water runs some distance out at sea. Masters of vessels crossing this stream have been surprised by the sudden increased draught of their ships, caused by their entrance into a fluid of inferior specific gravity.

Captain Sabine states, that while proceeding in his voyage from Maranham to Trinidad, on September 10, 1822, the general current running at the great rate of ninety-nine miles in twenty-four hours (more than four miles per hour), they crossed discoloured water in 5° 08′ N. lat, and 50° 28′ W. long. He considers this water as that of the river Amazons or Maranon, which had preserved its original impulse three hundred miles from its embouchure, having flowed over the waters of the ocean, from its less specific gravity. The line between the ocean water and discoloured water was very distinct, and great numbers of gelatinous marine animals were floating on the edge of the river water. The temperature of the ocean water is stated as $= 81°\cdot1$, and that of the supposed river water $= 81°\cdot8$, both near the division line; "the specific gravity of the former was $1\cdot0262$, and of the latter $1\cdot0204$." From experiments made, the depth of the discoloured water was superficial, and did not amount to 126 feet. There was no bottom at 105 fathoms. In this discoloured water the ship was set N. 38° W., sixty-eight miles in twenty-four hours, or rather less than three miles per hour. The western side of the fresh water was gradually lost in that of the sea. Captain Sabine attributes the unusual velocity of the ocean current, of ninety-nine miles per day, to the obstacle which this fresh-water current opposes to it*.

* Experiments to determine the Figure of the Earth.

We have other accounts of discoloured waters in the Atlantic, which would render it necessary that the specific gravity and relative freshness of simply discoloured waters should always be ascertained, as was done by Captain Sabine, before we can be certain that waters even flowing in the necessary direction were derived from rivers. Captain Cosmé de Churruca states, that 128 leagues to the eastward of St. Lucia, and 150 to the N.E. of the Orinoco, there is always discoloured water as if on soundings, but there is no bottom at 120 fathoms. The same appearances are observed about seventy or eighty leagues to the eastward of Barbadoes. Humboldt notices a place in the latitude of Dominica at about 55° W. longitude, where the sea is constantly milky, although it is very deep ; and seems to think that there may possibly be a volcano beneath it. Captain Tuckey observed the same kind of milkiness upon entering the

In the river St. Lawrence we have a striking example of the superior velocity of the ebb tide to the flood. "At the Isle of Coudre, in spring tides, the ebb runs at the rate of two knots. The next strongest tide is between Apple and Basque Isles; the ebb of the river Saguenay uniting here, it runs full seven knots in spring tides; yet, although the ebb is so strong, the flood is scarcely perceptible; and below the Isle of Bic there is no appearance of a flood tide *."

The great difference in the ebb and flood of river tides must depend on many local causes, but be principally in proportion to the perpendicular rise of tide on the one side, and the mass of fresh water on the other. The flood tide sets up many rivers so suddenly, as to cause a wave of greater or less magnitude, according to circumstances, called the *bore*, appearing as if the flood suddenly overcame the resistance of the ebb. The bore of the Ganges is very considerable. According to Major Rennell, it "commences at Hughly Point, below Fulta, the place where the river first contracts itself, and is perceptible above Hughly Town; and so quick is its motion, that it hardly employs four hours in travelling from one to the other, although the distance is near seventy miles. At Calcutta, it sometimes causes an instantaneous rise of five feet; and both here and in every other part of its track, the boats on its approach immediately quit the shore, and make for safety to the middle of the river †."

According to Romme, there is a considerable *bore* at the mouth of the Amazons or Maranon during three days at the equinoxes. It is observed between Maraca and the North Cape, and opposite the mouth of the Arouary. A wave of twelve or fifteen feet in height is suddenly formed, and is followed by three or four others. The advance of this *bore* is exceedingly rapid, and the noise caused by

---

Gulf of Guinea; but considered it due to multitudes of crustacea which were caught, and which produced great luminosity at night.

Sir Gore Ouseley mentions that on February 12, 1811, when off the Arabian shore, a partial line of *green water*, such as generally indicates shallows, and perfectly different from the *blue* of a deep sea, was perceived extending considerably. It appeared eight or nine miles from the land. The change from the blue to the green waters was sudden, so that the ship was in green and blue waters at the same time. Having entered the green water they sounded, and found bottom at seventy-nine fathoms; proving that the change of colour was not due to a shoal; for previous to entering this water they sounded in the blue water, and found sixty-three fathoms, so that the blue was more shallow than the green water. This was observed not far from the Persian Gulf.—Sir Gore Ouseley, Travels, vol. i.—In this case there was no great river near to produce the difference of colour. "Green Sea" is the name given to the Persian Gulf by eastern geographers.

* Purdy, Atlantic Memoir, p. 91.   † Phil. Trans.

it is stated to be heard at the distance of two leagues. It occupies the whole breadth of the river, and in its progress carries all
before it, until it has passed the banks into deeper and wider water,
where it ceases. M. De la Condamine has described this phænomenon, and has observed that there are two opposing currents
during the flood, one superficial, the other deep. There are also
two superficial currents, one setting by the shore on each side,
while a central but retarded current descends. Tides are stated
to be felt two hundred leagues up the Amazons, so that there are
several in the river at the same time, and the surface of the water
for that distance forms an undulating line.

The most curious *bore* which I find recorded, was observed by
Monach, Port Commandant at Cayenne: he states, that the sea
rises forty feet in less than five minutes in the Turury Channel,
river Arouary; that this suddenly elevated water constitutes the
whole rise of tide, the ebb immediately taking place, and running
with great velocity *."

In the Zaire or Congo we have an example of the comparatively
small effect of the tide upon a large body of fresh water discharged
with sufficient velocity. Notwithstanding the aid of Massey's
machine, bottom was not found in Tuckey's expedition at 113
fathoms in mid-channel and at the mouth, and the stream ran at the
rate of four and five miles an hour †. This stream became checked
but not overcome in mid-channel, and the tide only produced
counter currents near the shore. The rise of water is felt between
thirty and forty miles up the river. Alluvial land is continually
forming into flat islands, which are covered by mangrove trees
and papyrus, and are often partially or wholly carried by the river
into the ocean ‡. Professor Smith describes a floating isle of this
kind which he saw further north off the coast of Africa; it was
"about 120 feet long, and consisted of reeds, resembling the
Donax, and a species of *Agrostis*? among which were still growing
some branches of *Justicia* §."

Currents.

Currents are sometimes classed as constant, periodical, and
temporary.

The great current which flows from the Indian Ocean round
the Cape of Good Hope, up the coast of Africa to the equatorial
regions, whence it strikes across the Atlantic to the West Indies,
is considered a constant current, produced by the tropical or trade
winds, assisted by the motion of the earth. The current having
driven, by these means, a body of water to the continent of Ame-

* Romme: Vents, Marées et Courants du Globe, tom. ii. p. 302.
† It has been since supposed that this stream had greater velocity.
‡ Tuckey's Expedition to the Zaire or Congo.   § *Ibid.* p. 259.

rica, through which it cannot escape, passes up through the channel
offered it at the Straits of Florida, flows considerably to the northward, and then bends to the eastward and south-east, taking its
course to the west coast of Europe and the upper part of Africa.
It is considered that the latter division of the current again unites
with the northern portion of the equatorial current, and again
traverses the Atlantic.

Between Cape Bassas in Africa and the Laccadives or Lakdivas,
there is a constant current to the westward, mostly to the S.W.
or W.S.W. Its rate is supposed to be from eight to twelve miles
per day. The current south of the equator, in the Indian Sea,
runs to the west. During the N.E. monsoon the currents of the
Mosambique Channel run to the south along the African coast,
and even in the offing; their usual velocity being about seven or
eight leagues in twenty-four hours. On the coast of Madagascar
the currents take an opposite direction, and set towards the north.
At the southern extremity of Africa, the currents set round the
bank of Agulhas, or Lagullas as it is more commonly termed, a
bank of considerable extent, the soundings in which are described
as *mud* to the westward of Cape Lagullas, and *sand* to the eastward, the latter containing numerous small shells. Rennell informs us that this current is strongest during the winter, and that
the outer verge of the stream runs into 39° S. before it turns to
the northward, after which it proceeds slowly along the western
coast of Africa to, and even beyond, the equator*. The general
velocity of the current round the bank is not stated; but it appears
that one vessel was carried by it one hundred and sixty miles in
five days, or thirty-two miles per day †.

Beyond St. Helena, the current above noticed unites with
the equatorial current of the Atlantic, and sets across from the
Ethiopic Sea to the West Indies. The velocity of this current has
not been well ascertained, but is generally considered as about
one mile and a half per hour, increasing as it proceeds westward,
and setting off the coast of Guyana at the rate of two or three
miles per hour. Captain Sabine states that, sailing from Maranham in 1822, and entering the current, he estimated it as running at the rate of ninety-nine miles in twenty-four hours, or a
little more than four miles per hour. The central direction of
this current is W.N.W.

* Captain Tuckey in his expedition to the Zaire or Congo, found a
current setting to the N.N.W. after making St. Thomas off the African
coast. Its velocity was thirty-three miles in twenty-four hours.

† As the current round the Lagullas Bank evidently conforms to the
bank, we may, perhaps, consider that it there has considerable depth,
that is, a depth equal to about sixty or seventy fathoms. But of this
we cannot be quite certain, for we do not know to what distance water
thrown off by the bank at lesser depths may be carried round it.

"On the Colombian coast, from Trinidad to Cape la Vela, the currents sweep the frontier islands, inclining something to the south, according to the strait they come from, and running about a mile and a half an hour with little difference. Between the islands and the coast, and particularly in the proximity of the latter, it has been remarked, that the current at times runs to the west, and at others to the east. From Cape la Vela, the principal part of the current runs W.N.W.; and, as it spreads, its velocity diminishes: there is, however, a branch which runs with the velocity of a mile an hour, directing itself towards the coast about Cartagena. From this point, and in the space of sea comprehended between 14° of latitude and the coast, it has however been observed, that in the dry season, the current runs to the westward, and in the season of rains to the eastward *."

It is asserted, that there is a constant stream entering the Mexican Gulf by the western side of the channel of Yucatan; and that there is commonly a re-flow on the eastern side of the same channel around Cape Antonio †.

On the northern coasts of St. Domingo and Cuba, in the windward passages, at Jamaica, and in the Bahama passages, the currents appear variable, their greatest observed velocity being about two miles per hour.

The accumulation of water in the Caribbean and Mexican seas does not raise the level of those seas so much as was, perhaps, once supposed. The difference of level observed by Mr. Lloyd, in his researches on the Isthmus of Panama, between the Mexican Sea and Pacific Ocean, was in favour of the greater height of the Pacific Ocean by 3.52 feet,—an unexpected result; but the measurements were conducted with such care, that we can scarcely doubt it. The high-water mark at Panama is 13.55 feet above high-water mark of the Atlantic at Chagres; but from the difference in the tides on each side the isthmus, the Pacific is lower than the Atlantic at low water by 6·51 feet ‡. If we consider the body of water pent up by the effects of currents over so large a space as the Mexican Sea at five feet, or even less, above the Atlantic Ocean, we need not be surprised at the velocity of the current produced by its escape through the Straits of Florida.

If the temperature of the waters, heated in the Gulf of Mexico and Caribbean Sea, be greater, as we know it is, than the waters north of the tropics through which the Gulf stream flows, their specific gravity will be less, and consequently they will flow onwards over the colder waters or those of greater specific gravity, precisely as river-water flows out to sea over those of the ocean, and

---

* Purdy, Atlantic Memoir, translated from the " Derrotero de las Antillas."

† Purdy, Atlantic Memoir.   ‡ Phil. Trans. 1830.

will continue to do so until their progress be gradually checked and finally stopped.

From a mass of information that has been collected, it appears that the Gulf stream varies considerably in breadth, length, and velocity. It has been found that winds much affect the current, diminishing its breadth, and augmenting its velocity, or augmenting its breadth and diminishing its velocity.

In mid-channel, on the meridian of the Havanna, the direction is E.N.E., and the velocity about two miles and a half per hour. Off the most southern parts of Florida, and at about one third over from the Florida Reefs, it runs at the rate of about four miles per hour. Between Cape Florida and the Bemini Isles it runs to the N. by E., with a velocity of more than four miles an hour. The stream is weak on the Cuba side, and sets to the eastward.

A re-flow or counter current sets down by the Florida Reefs and Kays to the S.W. and W., and by its aid many small vessels have made their passages from the northward *. To the northward of Cape Canaveral there is no stream of tide, along the southern coast of the United States, further from the shore than in ten or twelve fathoms of water; from that depth to the edge of soundings, a current sets to the southward at the rate of a mile an hour; out off soundings, the Gulf stream is found setting to the northward †. It is also stated that there is a re-flow or counter current on the eastward of the stream.

Capt. Sabine remarks, that in the latter part of 1822 the velocity of the current after passing Cape Hatteras was seventy-seven miles per day ‡. Rennell, considering the force of the stream as determined at different points, calculates that the water requires about eleven weeks to run in the summer, when its rapidity is greatest, from the Gulf of Mexico to the Azores, a distance of about 3000 miles. Capt. Livingston, however, observes that the calculations of the velocity of the Gulf stream are not to be depended on. He found it setting at the rate of five knots and upwards on the 16th and 17th of August 1817. On the 19th and 20th of February 1819, it seemed to be almost imperceptible. In September 1819, it set at about the rate described in the charts §.

Lieut. Hare has found in the meridian of 57° W., that the stream ranges to 42¾° N. in the summer, and even to 42° N. in the winter.

It would appear, that the waters, after issuing through the

* Purdy, Atlantic Memoir.   † *Ibid.*
‡ Capt. Livingston observes that the current set him, off Cape Hatteras, 1° 8 to the northward of his dead reckoning; this he ascertained by stellar and solar observations.—Atlantic Memoir.
§ Purdy, Atlantic Memoir.—These observations appear to have reference to the stream between Cape Florida and the Bemini Isles.

Straits of Florida, run off from the eastern edge of the stream to
the eastward, as might be expected from their tendency to equalize their level, particularly in those parts not carried forward
with considerable velocity.

A strong current sets from the Polar Seas, and through Hudson's
Bay and Davis's Strait, commonly denominated the Polar or
Greenland current. It sets southerly down the coast of America
to Newfoundland, bringing down large icebergs beyond the Great
Bank. Captains Ross and Parry found the velocity of the current
from three to four miles per hour in Baffin's Bay and Davis's
Strait.

A current from the polar regions sets into the North Atlantic
between America and Europe : it produced such a drift of the ice
to the south in Capt. Parry's attempt to reach the North Pole
over the ice, that the expedition was finally abandoned in consequence of it.

The Polar current coming from Davis's Strait, may be said to
unite with the Gulf stream, and then to set eastward, directing
its course to the coasts of Europe and Africa. Off the coast of
Newfoundland, the current sometimes runs at the rate of two
miles an hour, but is much modified by winds. About five degrees
to the westward of Cape Finisterre the current has a velocity of
thirty miles in twenty-four hours.

Between Cape Finisterre and the Azores there is a tendency of
the surface waters to the S.E., being variable in winter. Lieute-
nant Hare, in September 1823, found a current setting E.S.E.
with a velocity of a mile and a half per hour between N. latitude
45° 20' and 43° 40', and W. longitude 22° 30' to 16°. Rennell
remarks, respecting the currents between Cape Finisterre and the
Canary Islands, that " it may be taken for granted, that the whole
surface of that part of the Atlantic from the parallel of 30° to 45°
at least, and to 100 or 130 leagues off shore, is in motion towards
the Straits of Gibraltar."

" Near the coasts of Spain and Portugal, commonly called The
Wall, the current is always very much southerly (as it is more
easterly towards Cape Finisterre), and continues as far as the
parallel of 25°, and is, moreover, felt beyond Madeira westward ;
that is, at least 130 leagues from the coast of Africa ; beyond
which a S.W. current takes place, owing, doubtless, to the operation of the N.E. trade wind." The same author observes, that the
velocity of the current varies considerably, being from twelve to
twenty, or more, miles in twenty-four hours. He considers sixteen as below the mean rate.

A current sets along the coast of Africa from the Canaries to
the Gulf of Guinea, running westerly out of the Bight of Biafra.
The rainy seasons, and Harmattan wind, interrupt this stream.
From Cape Bojador and the Isles de Los, the velocity of the cur-

rent has never been found to exceed a mile and a half per hour
on the coast and on the outer edge of the bank. Its more common rate is less than a mile. At the distance of four leagues from
the coast it becomes half a mile, and even less. In the meridian
of 11° W. the current runs twenty-five miles to the E.S.E. in
twenty-four hours. Off Cape Palmas it sets to the E. at forty
miles; off Cape Three Points, and thence to the Bight of Benin,
at from fifteen to thirty miles. It then decreases in strength, runs
to the southward, turns to the S.W between 6° and 8° S., and
thence flows N.W. to the Cape Verde Islands. It is considered
that the portion flowing eastward into the Gulf of Guinea, is not
altogether continuous with that which comes from Cape Bojador
to the south.

A current is described to pass round Cape Horn and Terra del
Fuego, from the Pacific into the Atlantic, during the greater part
of the year *. From the Straits of Magellan to the equator, a current sets northward along the western coast of South America.
At eighty leagues from the coast, between 15° S. latitude and the
equator, and even to 15° N. latitude, the currents generally run
westward. Captain Hall found a constant current setting off the
Galapagos, to the N.N.W. At Guayaquil a strong current sets
out of the gulf at the rate of forty miles in twenty-four hours.
Between Panama and Acapulco, and at about 180 miles from the
latter place, Captain Hall met with a steady current running E. by
S. at rates varying from seven to thirty-seven miles per day.
Great quantities of wood are drifted from the continent of America
to Easter Island by the force of a current setting in that direction.
Currents have been found at Juan Fernandez, and 300 leagues to
the westward of it, running W.S.W. at sixteen miles per day. At
the Marquesas they flow with a velocity of twenty-six miles in
twenty-four hours. Between the Marquesas and the Sandwich
Islands they have been found to run westward at the rate of
thirty miles a day, in April and May. A southerly current has
been observed at California; and a northerly current along the
N.W. coast of America, from Cape Orford, the latter having a
velocity of a mile and a half per hour.

A northerly current sets through Beyring's Straits †, and is sup-

* Captain Hall states, that he did not meet with any current round
Cape Horn. A naval officer, however, assures me that a current runs
out of the Pacific into the Atlantic during nine months; and this is rendered probable from the prevalence of strong westerly winds during the
greater part of the year, which would drive the waters before them.
Kotzebue found a current which turned rapidly to E.N.E. near Staaten
Land, having had another direction (S.W.) off Cape St. John.

† Kotzebue describes this current as setting through the straits with
a velocity of three miles per hour to the N.E. At Anchorage, near East

posed to run along the north coast of America, and deliver itself
through Baffin's Bay and Hudson's Straits, into the Atlantic.

King found a current setting N.E. near the Japanese Islands,
the velocity five miles per hour; but he also found it to vary considerably in direction and strength.

Among the Philippine Islands a current comes from the N.E.,
and runs with considerable force among the passages dividing the
islands; it has been found with a strength of twenty miles a day
near these isles. This current varies.

Cook found a southerly current, in August, flowing ten or fifteen
miles a day, between Botany Bay and 24° S. On the same side
of Australia a vessel was set forty miles to the southward in twenty-
four hours, in the month of March; and in July another vessel
was carried thirty miles in two days in the same direction.

A constant current sets eastward into the Mediterranean, with
a velocity of about eleven miles in twenty-four hours. It has been
considered that there is an under or counter current setting westward, and carrying out the dense water, rendered more than
usually saline from evaporation within the Straits of Gibraltar;
but this has lately been controverted. It was remarked by Dr.
Wollaston, that the salt carried into the Mediterranean by the
current from the Atlantic must remain there after the evaporation
of the water which held it in solution, unless it could escape by
some means. He inferred its escape to be by an under-current,
usually thought to exist, and this he considered proved by experiment; for water brought up from the depth of 670 fathoms
about fifty miles within the Straits, by Captain Smyth, was found
to contain about four times the usual quantity of saline matter.
Water taken from depths of 450 and 400 fathoms, at 680 and 450
miles within the Straits, did not exceed in its saline contents many
ordinary examples of sea-water. He further observed, that if the
under current moved only with one fourth the velocity of the
upper current, and was of the same depth and breadth as it, the
former would convey out as much salt as the latter brought in *
Mr. Lyell infers that this dense water cannot pass out because
the bottom of the sea rises between Capes Spartel and Trafalgar,
and has only 220 fathoms of water upon it; and therefore, if
the under and more saline water be as deep as is supposed, it
would be impossible for it to escape, and it would deposit great
quantities of salt in the bed of the Mediterranean†. It is much

---

Cape, the current was found to set at the rate of one mile per hour; but
shortly afterwards, notwithstanding a brisk wind, the expedition under
Kotzebue made but little way against it, though going, by the log, at the
rate of seven miles per hour.

* Wollaston, Phil. Trans. 1829.
† Lyell, Principles of Geology, vol. i.

to be regretted that we do not possess better information on this
subject, and that direct experiments have not been made on thsi
supposed under-current. That this has not been done is the more
remarkable, when we consider the numerous opportunities afforded
by the continual passage of ships, and the proximity of such esta-
blishments as those of Gibraltar. Mr. Lyell's theory of a great
deposit of salt at the bottom of the Mediterranean, though very
ingenious, can scarcely be true; for, supposing it to be so, the sea
would, as the depth increased, be more and more charged with
saline matter, until it finally became mere salt, the density increasing at the same time. This being the case, we should bring
up salt with the sounding-lead, and little else. But the fact is,
that the deep soundings, as shown by Captain Smyth, are mud,
sand, and shells. Sand and shells form the bottom, beneath 980
fathoms of water, a little east of the meridian of Gibraltar; and
the same bottom is found in the Straits beneath 700 fathoms of
water. Now these places are near where the sea-water, so highly
charged with saline matter, was brought up; and where, according
to the theory, there should be a bottom of salt. The same may
be said of other situations *.

The current entering the Mediterranean passes along the
southern shores of that sea, and is felt at Tripoli and the Island of
Galitta. At Alexandria there is a stream flowing east, as well as
between the coast of Egypt and Candia: arrived on the coast of
Syria, it runs north, and then advances between Cyprus and the
coast of Caramania. A strong current flows from the Black Sea
into the Mediterranean, through the Dardanelles.

A constant current flows out of the Baltic, through the Sound and
Cattegat, into the German Ocean. Its velocity in the narrowest
part of the Sound is about three miles per hour; but the ordinary
rate, in fine weather, is about one mile and a half or two miles.
The currents out of the Sound and two Belts are directed towards

---

* In all our remarks on the changes that may be supposed to occur
at the bottom of the Mediterranean, we should be careful to remember
that this bottom is divided into two great basins (See Smyth's Charts)
by a winding shoal, which connects Sicily with the coast of Africa.
This shoal, known as the Skerki, has the following line of soundings
upon it, proceeding from the African to the Sicilian coast; namely, 34,
48, 50, 33, 74, 20, 70, 52, 91, 16, 15, 32, 7, 32, 48, 34, 54, 70, 72, 38,
55, and 13 fathoms, from whence an idea of its inequalities may be
formed. There are soundings in 140, 157, and 260 fathoms, on either
side, as also places where 190 and 230 fathoms of line have been run
out, without finding bottom. It may be here remarked, that, at the entrance of the Dardanelles into the Mediterranean, there are only thirty-
seven fathoms of water; so that the quantity of matter requisite to bar
the communication between the Black and Mediterranean seas, would
not be very considerable.

F

the Scau or Skagen, and flowing thence, turn N.E. towards Marstrand, at the rate of about two miles per hour. It is not impossible that a counter- and under-current setting into the Baltic from the ocean may exist; for Captain Patton observed, when at anchor a few miles from Elsinore, in an upper-current setting at the rate of four miles an hour outwards, that in sounding in fourteen fathoms, he found the line continue perpendicular to his hand, when the lead was raised a little from the ground. Hence he concluded that there was an under-current that prevented the line from being carried away.

In the Indian and Chinese seas we have good examples of periodical currents, evidently referable to the periodical winds or monsoons.

From St. John's Point to Cape Cormorin there is a nearly constant current in the direction of the coast from N.N.W. to S.S.E. ; except that between Cape Cormorin and Cochin, it flows from S.E. to N.W. from October to the end of January.

The current sets from the ocean into the Red Sea from October to May, and runs out of that sea from May to October. A current commonly sets from the Gulf of Persia towards the ocean during the whole time that the current flows into the Red Sea, and runs into the Gulf from May to October.

In the Gulf of Manar, between Ceylon and Cape Cormorin, the current flows northward from May to October, setting the remaining six months to the S.W. and S.S.W. From Pedro Point on the north of Ceylon, to Pointe de Galle on the south, the current runs S.E., S.S.E., S., S.W., and W., according to the nature of the coast, uniting at the Pointe de Galle with the current that comes out of the Gulf of Manar. The ordinary velocity of the stream on the south coast of Ceylon is about a league an hour. The Ceylon currents are weak in June and November. In the Bay of Bengal the currents run with the wind towards the N.E. during the S.W. or W. monsoon, and slacken in September. On the coast of Orissa, about eight days before the equinox their direction is towards the south, and they become strong at the end of the month. During the N.E. and E. monsoon, the currents are, as before, with the wind, and strong in proportion to it.

In the S.W. monsoon the current between the coast of Malabar and the Lakdivas sets to the S.S.E. with a velocity of twenty, twenty-four, or twenty-six miles in twenty-four hours. Between the Lakdivas its direction is to the S.S.W. and S.W., its rate being from eighteen to twenty-two miles in twenty-four hours. The current, setting W. or W.S.W. to the westward of these islands, varies in velocity from eight to eleven miles per day. There is a strong current among the Maldives: among the southern isles the direction is generally to the E.N.E. in March and April. In May it sets to the eastward; in June and July the currents often run to the W.N.W., particularly to the south of the equator. Between these isles and Ceylon they frequently set strongly to the westward during the months of October, November, and December.

The currents in the China seas, at a distance from shore, generally flow, more or less, towards the N.E. from the middle of May to the middle of August, and have a contrary direction from the middle of October to March or April. The velocity of the currents from the N.E. is usually greater in October, November, and December, along the adjacent shores, than that of the opposite set in May, June, and July. Their strength is most felt among the islands and shoals near the coast.

The strongest currents of these seas are experienced along the coasts of Cambodia, during the end of November. They run with a velocity of from fifty to seventy miles to the southward, in twenty-four hours, between Avarella and Poolo Cecir da Terra. Some part of the stream setting into the Straits of Malacca, causes the tide to run nine hours one way and three hours the other. The currents to the northward commence running in April through the Straits of Banca, past the Straits of Malacca, and along the west coast of the Gulf of Siam, setting along the north-east side of the same Gulf to the E.S.E. until to the eastward of Point Ooby; they then bend to the N.E., running along the coasts of Cambodia, Cochin-China, and China, till September, when the opposite monsoon and currents prevail from the N.E., and continue to March or April.

Periodical currents occur, according to M. Lartigue, along the west coast of South America, from Cape Horn to latitude 19°. The S. and E.S.E. winds produce a current, setting to the N.W., off the coast of Peru, of which the maximum velocity is fifteen miles in twenty-four hours, and the mean velocity about nine or ten miles. Between this current and the coast, there is a counter-current flowing to the S.E. During the prevalence of the wind from N. to W. the current flows S.E., but is only sensible near the land *.

Temporary currents are innumerable, every severe gale of any duration producing one. Nothing is more common than these partial currents, which are more particularly felt along coasts and through channels.

The direction and velocity of the currents above enumerated may be considered rather as approximations to the truth, than as the truth itself, for the determination of currents is liable to many errors. The usual manner of ascertaining them is by comparing the true place of a ship, determined by means of chronometers and astronomical observations, with the position of the same ship

* Lartigue, Description de la Côte du Pérou.

F 2

as deduced by dead reckoning. The latter is a calculation of the vessel's way through the water in a given direction. The rate of the vessel's way is estimated by means of a contrivance called a log-line, or a line at the end of which there is a float. According to the quantity of line run out in a given time, with allowances for the agitation of the sea, &c., is the rate of the vessel's way calculated. This operation is liable to numerous errors; and even with the line and glasses in the highest order, requires a nicety of execution seldom practised. The direction of the vessel's course is estimated by the compass, with allowances for magnetic variation. Here we have a most fruitful source of error, for until lately no allowance whatever was attempted for the local attraction of the ship. It is now well known that the disposition of iron in a vessel is such, that no two ships will be found to have the same local attraction, consequently no rules can be adopted for correcting the error of aberration by means of placing the magnets in any particular situation, though situations have been found more favourable for true observations than others. It was not until Mr. Barlow invented his plate of iron for counteracting the effect of aberration, that the error arising from it could be fully known. Now nearly all the preceding observations, as to the direction and velocity of currents, were made before this great source of error was understood; consequently many of them are erroneous, and require that re-examination which the advance of science has rendered necessary. It is clear, that if a vessel is steering one course, and those on board consider they are taking another, the position deduced from dead reckoning must wander from the truth in proportion to the amount of aberration, even supposing the rate of way through the water and other necessary observations correct.

If, in the annexed diagram, a vessel, without any allowance for aberration, be supposed to hold her course from *a* to *b*, while in reality her course, with proper allowance for aberration, is from *a* to *c*, the distance from *b* to *c* will, according to the usual practice, be referred to current, after an observation shall have shown that her true place is at *c*. It will be clear that in this case no such current exists, and that the difference between the true and calculated situations of the ship arises solely from want of attention to local attraction.

Fig. 18.

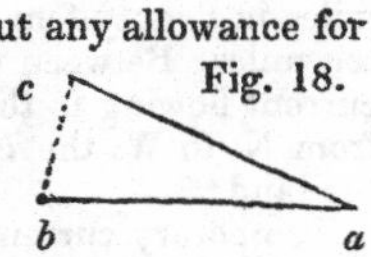

Another great source of error in estimating the value of currents has been noticed by Captain Basil Hall. This author observes, that the usual method of laying down ships' tracks by two lines, one representing the course as estimated from the dead reckoning, and the other as deduced from chronometers and lunar observations, leads to no information as "to where the current began, or where it ceased, or what was its set, or its velocity."

He proposes instead of this, that the position of the ship found at each good observation should form the point of departure, both for the line representing the distance and direction to the next observed true position, and for that representing the ship's course as estimated by dead reckoning. A very superior plan, and one that should supersede the old method *.

Although these causes of error render the exact velocity and course of currents heretofore observed, vague and uncertain, so that many minor streams may be found imaginary, and that the navigator may be exposed to great danger from implicitly depending upon them; yet to the geologist, perhaps, they may not be so formidable; as, probably, the general velocity of currents will not be found greatly increased; and as it is with their velocity and consequent transporting power that he is principally concerned.

Transporting Power of Tides.

The stream caused by tides varies much in strength, but a common velocity appears to be one mile and a half per hour, when head-lands, shallow banks, and other obstacles are not opposed to it; and therefore, even supposing the superficial velocity to extend to the bottom, which would not be the case except in comparatively shallow seas, the general transporting power of such tides would appear, judging from the effects we witness near shores, to be but small. This the unchanged character of soundings for a great length of time, though principally composed of mud and sand, seems to attest.

Where obstacles are opposed to the tides, the transporting power will be increased, and the changes produced more rapid. The tide through the Pentland Firth having a velocity of nine miles per hour, would scour out pebbles of considerable size from its channel; but its power to do this would cease at each extremity, where the tides flow at the rate of two or three miles per hour, and the local cause would merely produce a local effect. The same with the Race of Alderney, and other similar places.

Changes in the shape of sand-banks frequently take place when they approach the surface; but as they then come within the influence of another cause, the action of the waves, the transporting power of which is very considerable, too much must not be attributed to the mere force of a tidal stream.

The transporting power of tidal rivers outwards, or into the waters of the sea, is considerable, more particularly during the time of freshets or floods. As has been seen, the tide of ebb in rivers is always greater than the flood; therefore, although estuary waters are very turbid, and a great proportion of them merely carried backwards and forwards, detritus will escape into the open sea in proportion to the difference of velocity between the ebb and flood. It should be remarked, that all estuaries have a tendency to be filled up by deposition of the matters held mechanically in solution by their waters. The heads of estuaries are very frequently alluvial plains, formed of the same kind of mud and silt as are at present brought down by the rivers; and it often appears as if the tides had flowed up to much greater distances than they now do, the higher parts having been gradually silted up *. These appearances are so common, that it is useless to insist upon them; but the extent of flat lands, evidently accumulated in this way on the sides and heads of estuaries, is often very remarkable, and would seem to have required a long lapse of ages for their formation; more particularly when the present deposits of the same estuaries are considered.

Notwithstanding this deposit in the estuary itself, and the bars and banks accumulated at the mouths of so many tidal rivers, above noticed, mud and silt escape into the sea, and are transported by the tides to greater or less distances from the rivers; as may often be seen at low water, in coasts where tidal rivers discharge themselves.

The transporting power of tides and currents being proportioned to their velocity, and this being greatest when obstacles are opposed to either, it is in these situations where we should look for the greatest transporting power.

The difference between the velocity of tides on the surface and at moderate depths must be very considerable, otherwise the previously noticed power of water to tear up different kinds of substances at given velocities must be incorrect; for if the velocities were nearly as great at moderate depths as on the surface, tidal streams would be little else than a mass of turbid waters.

The discoloration of the sea to greater or less distances from the shore, according to depth, is well known to be effected during heavy gales, and is due to the action of the waves, and not to that of the tide merely passing over sand or mud with a certain strength, and therefore must not be confounded with it.

To take an example of tidal waters running over a certain bottom :—At the Shambles, a well-known bank near the island of Portland, the tides run at the rapid rate of from three to four nautical miles per hour, over soundings of gravel which do not alter. Now, if the calculations above noticed were correct, and the inferior velocity not very considerably different from that on the

---

* If we could always give implicit confidence to old maps and charts, great deposits of this nature would seem to have taken place within historical times.

surface, stones, the size of eggs, could be torn up by water with a velocity of three feet in a second, or 3600 yards in an hour; consequently the pebbles on the bank would be carried away, and nothing but bare rock or masses of stone would be left; but the soundings on the Shambles are the same at present as they are represented to have been, by the charts, many years ago.

The preservation of the same kind of bottoms or soundings, over which tides or currents pass with considerable velocity without their being altered, is familiar to most mariners; and it would seem that we are far from being acquainted with the respective velocities required to tear up mud, sand, and pebbles at various depths in the sea. Tidal streams flow over mud banks in some estuaries at the rate of two miles and even more per hour without removing them; though, if the above-noticed calculations were always applicable, the current would be sufficiently strong to remove pebbles of some size. The same remark applies to innumerable sand-banks *.

Transporting Power of Currents.

In estimating the transporting power of currents, we should consider the causes which produce them, and the nature of the fluid in which they are produced. The motion of the earth, although it would seem to give a certain general movement to the waters of our globe, would not appear capable, taken by itself, to produce currents of geological importance. The great cause of ocean currents would seem to be prevalent winds; and accordingly we find that in the equatorial regions of the world, over which the more or less easterly winds, commonly called the Trade Winds, prevail, there is a tendency of the waters to flow westward in the Pacific Ocean, in the Atlantic, and in those parts of the Indian seas free from the monsoons. That the winds are the great cause of ocean-currents, is a fact sufficiently proved by the velocity and direction of such currents in the Indian and Chinese seas, varying

---

* While on the subject of soundings, it may be noticed that the British Islands are in reality united to the continent beneath the sea by banks of various kinds, at greater or less depths; the principal soundings on which are mud or sand. The whole is more or less known by the name of soundings, because bottom can be easily obtained by a line of ninety or one hundred fathoms in length. The boundary of these soundings is traced on all good charts, and is seen to commence at the bottom of the Bay of Biscay, then to run round the British isles, and to communicate with the shallows of the German Ocean.

The bed of the sea in these soundings can only be considered as so much of the continent, which happens to be at no great depth beneath the ocean level.

The course of the tides round our islands is represented in Dr. Young's Natural Philosophy, vol. i. pl. 38. fig. 521.

with the force and direction of the monsoons. On this subject Major Rennell observes, " It is well known how easily a current may be induced by the action of the wind, and how a strong S.W., a N.W., or even a N.E., wind on our own coasts raises the tide to an extraordinary height in the English Channel, the river Thames, the East Coast of Britain, &c., as those winds respectively prevail. The late ingenious Mr. Smeaton ascertained, by experiment, that in a canal of four miles in length, the water was kept up four inches higher at one end than at the other, merely by the action of the wind along the canal. The Baltic is kept up two feet at least by a strong N.W. wind of any continuance; and the Caspian Sea is higher by several feet, at either end, as a strong northerly or southerly wind prevails. It is likewise known that a large piece of water, ten miles broad, and generally only three feet deep, has, by a strong wind, had its waters driven to one side, and sustained so as to become six feet deep, while the windward side was left dry. Therefore, as water pent up so that it cannot escape acquires a higher level, in a place where it can escape, the same operation produces a current, and this current will extend to a greater or less distance according to the force by which it is produced or kept up * "

It is also considered that the moon exercises an influence on the waters of the tropical regions, increasing their velocity by drawing them from E. to W. The current setting six hours one way and six hours the other through the Straits of Messina, though there is no rise or fall of water with it, is attributed to the influence of the moon, and may be considered as a tide. It has also been inferred that the sun, by its attraction, increases the velocity of the Gulf-stream. Capt. Livingston observes, that "when the sun's declination is N., the N.E. trade wind blows fresher, and extends further to the northward than when the sun's declination is S., thus forcing a greater body of water into the Caribbean Sea†."

The current setting into the Mediterranean through the Straits of Gibraltar is commonly attributed to the evaporation of that sea, which also receives a large supply of water from the Black Sea through the Dardanelles. The easterly indraught from the Atlantic is stated to commence nearly one hundred leagues to the westward of the Straits of Gibraltar. It has been supposed that an under- and counter-current sets outwards ; but this, as has been above noticed, has been lately controverted‡. That undercurrents do, however, occur in the Mediterranean, Capt. Beaufort affords us sufficient proof. After remarking that from Syria to the Archipelago there is a constant current to the westward,

* On the Channel Current.   † Purdy's Atlantic Memoir.
‡ Lyell's Principles of Geology.

slightly felt at sea, although very perceptible on shore, amounting
to three miles per hour, between Adratchan Cape and the opposite island, he observes, "the counter-currents, or those which
return beneath the surface of the water, are also very remarkable:
in some parts of the Archipelago they are sometimes so strong as
to prevent the steering of the ship; and in one instance, on
sinking the lead, when the sea was calm and clear, with shreds
of buntin, of various colours, attached at every yard of the line,
they pointed in different directions all round the compass *."

These observations of Capt. Beaufort are of the highest importance when we consider the transporting power of currents, because
they seem to show that we cannot judge of the force or direction
of under-currents from those known to flow on the surface.

The winds being, generally speaking, the cause of the great
ocean-currents, and effects being only in proportion to their causes,
the streams of water thus produced will not extend deeper than
the propelling power of the winds can be felt. Now, as the ocean
varies in density according to its depth, the cause sufficient to
move waters on the surface, and to certain depths beneath it, will
constantly meet with opposition, at an increasing ratio; until
finally, the moving power and the resistance being equal, no effect
whatever is produced; and all water beneath a certain depth would
be, as far as respects surface causes, immovable, and consequently
would have no transporting power.

Hence it would appear that the transporting power of currents
will depend on the depth of the sea, all other things being equal,
and that the smaller the depth the greater the transporting power.
Consequently, coasts are the situations where we may look for
this power.

If the current entering into the Mediterranean from the Atlantic be due to the evaporation of the former, this also is a
superficial cause, and its effects will gradually become less, until,
in deep water, it ceases altogether.

We have seen that tides as well as currents have their greatest
velocity in shallow water, across headlands, or in contracted
channels; consequently, their greatest transporting power exists
in the same situations, and will be local. Tides exert their transporting power in two directions, for the most part opposite to
each other, except in the case of rivers, where this power is greater
on the ebb than at the flood. Unless the rivers be very considerable, the detritus brought through their embouchures by the superior velocity of the ebb, enters into the power of the coast tides,
and is carried backwards and forwards by them until deposited.
But in the case of great rivers, such as the Maranon, St. Lawrence,
and Orinoco, the unchecked detritus is borne forward, until stop-

* Beaufort's Karamania.
F 5

ped and turned by the ocean-currents. Large additions are daily
made to the coast of South America by the deposit from the waters
of the Maranon, which are carried toward the shore by the prevailing current*.

Upon a review of what has been stated respecting the streams
of water caused by tides and currents, it would appear that
their geological importance will depend upon the relative depth
of water which they traverse, and their proximity to land, by
which their velocity is increased. Round coasts they have a
transporting power, which varies according to circumstances, being
greatest, all other things remaining the same, nearest the land.
In great depths we have no reason to suppose that this transporting power exists; or if it does, the causes must be different from
those which produce motion on the surface. It does not appear
that we are acquainted with the velocities which could tear up
mud, sand, or gravel; for currents pass over the bottom in shallow
water, composed of mud and sand, without mixing them, with a
considerable surface velocity. The changes produced on the bottom
are scarcely perceptible, within the periods we should consider
long, unless in shallow water, and near the mouths of great rivers,
the deposits from which must gradually accumulate, and diminish
the depth of the water. In the soundings round coasts, we do
not generally find any great inequalities, but in the ocean these
must exist to a very great extent, as is shown by the rocks, shoals,
and small islands scattered over it, the tops of mountains emerging from the water, which is generally of great depth close to
them.

Active Volcanos.

The surface of the earth is irregularly marked by orifices, through
which various gases, cinders, ashes, stones, and streams of red hot
melted rocks are projected. From this continued propulsion of
matter through a vent or vents, a conical mass is accumulated,
to which the name of *volcano* is applied. Volcanos differ materially
in the quantity of matter ejected, but agree in such a general
resemblance to each other, that they seem all referable to the
operations of the same causes.

Various theories have been formed for the explanation of volcanic phænomena; but it must be confessed, that they are all
more or less defective, and that the real causes of such phænomena are mere subjects of conjecture. With some of the effects
we are familiar; though with the districts most ravaged by erupted
matter, we are far from being well acquainted; our principal

* The water upon this coast is so shallow, that the land is dangerous
to approach without great care, the only harbours being the mouths of
rivers.

knowledge of volcanos being derived from the two largest active vents of Europe, Etna and Vesuvius, but principally from the latter. Etna certainly covers a considerable surface, but Vesuvius sinks into insignificance before some of the great volcanos of the world.

From their general proximity to, or occurrence in, the sea, it has been supposed that the active state of volcanos is produced by the percolation of sea-water to certain metallic bases of the earths or alkalies at various depths beneath the surface, which metallic bases being thus inflamed, cause the phænomena observed in volcanic eruptions. The volcanos in the interior of Mexico, as also the supposed volcanos of Tartary, have been accounted for by the advocates of this theory; the former, by supposing a connection between the vents of Colima, Jorullo, Pococatepetl, and Orizaba, all situated on the same line;—the latter, by considering that the waters of salt lakes may percolate to their foci. As the first chemical operation, if this theory were true, would be the union of the oxygen with the metallic base, and the escape of an immense quantity of hydrogen, M. Gay Lussac has objected to it, that pure hydrogen gas is not evolved from volcanos; and as a proof of it, observes, that if it were present, it would be inflamed by the red hot matter ejected from the craters. Dr. Daubeny endeavours to meet this objection, by supposing the hydrogen " to have combined in its nascent state with sulphur, and the two bodies to have been evolved in the form of sulphuretted hydrogen gas." He also considers that the presence of large quantities of muriatic acid would destroy the inflammability of the hydrogen *

According to the same author, the gases evolved from volcanos consist of muriatic acid gas, sulphur combined with oxygen or hydrogen, carbonic acid gas, and nitrogen; to which must be added a great quantity of aqueous vapour †.

Volcanic eruptions are usually preceded by detonations in the mountain, and agitations of the earth, or earthquakes in the vicinity, after which the mountain vomits forth an abundance of ashes, cinders, and stones; and streams of melted lava flow from apertures made in the side of the cone, the resistance of which becomes unequal to the pressure of the melted mass within. The lava very rarely seems to proceed from the lip of the crater.

The following is a summary, from various authorities, of the heat and appearances of a lava-current. " Lava, when observed as near as possible to the point from whence it issues, is for the most part a semifluid mass of the consistence of honey, but sometimes so liquid as to penetrate the fibre of wood. It soon cools externally, and therefore exhibits a rough unequal surface; but as it is a bad conductor of heat, the internal mass remains liquid

* Description of Volcanos, p. 377.   † *Ibid.* p. 376.

long after the portion exposed to the air has become solidified. The temperature at which it continues fluid is considerable enough to melt glass and silver, and has been found to render a mass of lead fluid in four minutes; when the same mass, placed on red hot iron, required double that time to enter into fusion." The heat does not, however, appear to be always equal; for it is stated, that when bell-metal was thrown into lava (of 1794), the zinc was melted and the copper remained unfused *.

The volcanic eruption which produced the greatest quantity of lava known to have been thrown out at one time, is that recorded as having proceeded in 1783, from the low country near Skaptar Jokul, in Iceland. The lava burst out, according to Sir G. Mackenzie, at three different points, about eight or nine miles from each other, and spread in some places to the breadth of several miles†.

The whole of Iceland may be considered little else than a volcanic mass, in which there are many apertures through which lava, ashes, and other products have been ejected. The igneous matter struggles to escape in various places, and, consequently, many single eruptions from different points have been recorded since historical times; nevertheless, volcanic discharges have taken place at various times through the same apertures. Thus, there have been twenty-two eruptions from Hecla since the year 1004 ; seven from Kattlagiau Jokul since 900 ; and four from Krabla since 1724.

As might be expected in such a region as that of Iceland, the eruptions are not confined to the immediate dry land, but have pierced through the sea in the vicinity. In January 1783, a volcanic eruption, described as flame, rose through the sea, about thirty miles from Cape Reikianes; several islands were observed, as if raised from beneath, and a reef of rocks exists where these appearances occurred. "The flames lasted several months, during which, vast quantities of pumice and light slags were washed on shore. In the beginning of June, earthquakes shook the whole of Iceland; the flames in the sea disappeared; and the dreadful eruption commenced from the Skaptar Jokul, which is nearly two hundred miles distant from the spot where the marine eruption took place‡."

Another submarine eruption occurred near the same island, in June 13, 1830. An island was produced, and consequent eruptions were feared in the interior, as in the case above cited§.

An example of a volcano forcing its way from beneath the sea

---

* Daubeny, Description of Volcanos, p. 381.
† Sir George Mackenzie. Travels in Iceland, 2nd edit.
‡ *Ibid.*
§ Journal de Géologie, tom. i.

into the atmosphere was observed off St. Michael's, Azores, in 1811. It was first seen above the sea on June 13th. On the 17th it was observed by Capt. Tillard and some other gentlemen from the nearest cliff of St. Michael's. The appearances were exceedingly beautiful, the volcano shooting up columns of the blackest cinders to the height of between 700 and 800 feet above the surface of the water. When not ejecting ashes, an immense body of vapour or smoke revolved almost horizontally on the sea. The bursts are described as accompanied by explosions resembling a mixed discharge of cannon and musketry, and by a great abundance of lightning*. By the 4th of July, a complete island was formed, described by Capt. Tillard (who landed upon it) as nearly a mile in circumference, almost circular, and about 300 feet in height. In the centre there was a crater, then full of hot water, which discharged itself through an opening facing St. Michael's. To this island, which afterwards disappeared, Capt. Tillard gave the name of Sabrina, from that of the frigate which he commanded.

It can only have been since historical times, and by mere accident, that instances of volcanos so forcing themselves from beneath the sea could have been recorded. Now, the power of man to do this is so recent, that we may conclude such occurrences to have been far from rare, and that, even in the present day, they may happen in remote regions, into which civilized man rarely, if ever, enters, and therefore they remain unknown.

There are numerous islands in the ocean, composed almost entirely of volcanic matter, and in which active volcanos still exist, that may have been thus formed, the dome or cone not giving way before the pressure of the water, but gradually accumulating a mass of lava, cinders and ashes, so that the islands have become firm, and even of considerable size. Owhyhee, or Hawaii, is perhaps a magnificent example of such an island. The whole mass, estimated as exposing a surface of 4000 square miles, is composed of lava, or other volcanic matter, which rises in the peaks of Mouna Roa and Mouna Kaah, to the height of between 15,000 and 16,000 feet above the level of the sea. Mr. Ellis describes the crater of Kirauea as situated in a lofty elevated plain, bounded by a precipice fifteen or sixteen miles in circumference, apparently sunk from two hundred to four hundred feet below its original level. "The surface of this plain was uneven, and strewed over with loose stones and volcanic rock; and in the centre of it was the great crater, at the distance of a mile and a half from the place where we were standing. We walked on to the north end of the ridge, where, the precipice being less steep, a descent to the plain below

---

* For a view of this scene, and a plan and elevation of the island, see Sections and Views illustrative of Geological Phænomena, pl. 34 & 35.

seemed practicable. After walking some distance over the sunken plain, which in several places sounded hollow under our feet, we at length came to the edge of the great crater, where a spectacle sublime, and even appalling, presented itself before us. Immediately before us yawned an immense gulf, in the form of a crescent, about two miles in length, from N.E. to S.W., nearly a mile in width, and apparently 800 feet deep. The bottom was covered with lava, and the S.W. and northern parts of it were one vast flood of burning matter, in a state of terrific ebullition, rolling to and fro its 'fiery surge' and flaming billows. Fifty-one conical islands of varied form and size, containing so many craters, rose either round the edge, or from the surface of the burning lake; twenty-two constantly emitted columns of gray smoke, or pyramids of brilliant flame; and several of these at the same time vomited from their ignited mouths streams of lava, which rolled in blazing torrents down their black indented sides into the boiling mass below." Mr. Ellis concluded, from the existence of these cones, that the mass of boiling lava resulted from the streams poured from the craters into this upper reservoir, which appeared to vary in its level; for there were marks on the rocks bounding it, which showed that the great crater had been recently filled up 300 or 400 feet higher to a black ledge, from whence there was a slope to the hot fluid mass*.

It will be obvious that this crater by no means resembles those with which we are more familiar. Instead of the more or less rounded orifice usually found, we have a semicircular crack in a level of considerable extent, and, by the description, this level does not appear to have been ravaged by lava streams flowing from the crater over it. The depth of water round Owhyhee, and indeed round the Sandwich Islands generally, is so great, that they are somewhat dangerous to approach in stormy weather, as anchorage cannot be obtained except close to the land; seeming to show that these volcanic masses rise from considerable depths, and are only partly out of the water.

The number of volcanos which fringe the Pacific Ocean, or occur in it, or in that part of the Indian Seas which contain Java and the neighbouring islands, far exceeds that of any other part of the world. From Terra del Fuego they occur northerly through the range of the Andes, often attaining very considerable elevations. In Mexico the northerly line is met by an east and west line, connecting it with the volcanos in the West Indian Islands. In California there are three volcanos, of which one, Mount St. Elia, is variously estimated from 13,000 to 17,000 feet in height. America is connected with Asia by means of the volcanic vents of the Aleutian Isles. From Kamtschatka southwards we observe volcanos

* Ellis, Tour through the Sandwich Islands.

in the Kurule Islands, Japan, the Loo Choo Isles, Formosa, and the Philippines. From the latter islands a range of volcanic vents proceeds to nearly lat. 10° S., ranges westward along this parallel for about twenty-five degrees of longitude, and then turns up N.W. diagonally through about twenty degrees of latitude. This line, which when represented in maps* resembles an enormous fish-hook, passes from the Philippines, by the N.E. point of Celebes, Gilolo, the volcanic isles between New Guinea and Timor, Floris, Sumbawa, Java, and Sumatra, to Barren Island.

Active volcanos are by no means relatively so abundant in, or on the shores of, the Atlantic. Indeed the shores of this ocean in Europe, Africa and America, appear free from them, if we except Mexico and the land connecting the main body of North America with the Southern continent, and which may be considered as common both to the Atlantic and Pacific Oceans†.

Teneriffe affords the greatest volcanic elevation in the Atlantic, the Peak rising 12,216 feet above its surface. Iceland, though its volcanos do not attain any considerable elevation, presents the largest accumulation of volcanic matter above the level of the same mass of waters.

We have seen that in Iceland high cones or elevations of land do not always accompany volcanic eruptions, for the lava of 1783 seems to have flowed from comparatively low apertures. Elevations seem more especially formed when the erupted matter consists of cinders, ashes, or stones, which being ejected, arrange themselves in a conical manner around the central aperture, where the amount of melted rock or lava may vary. The escape of this melted rock will, in a great measure, depend on its relative proportion to the cinders, ashes, or stones thrown out. If these be in comparatively small quantity, the lava will have the less difficulty to escape, and may easily break down its barrier and rush forth. But when the proportions are inverted, a large cone may be raised without the escape of any lava-current. Between the two extremes there will be every kind of variation, and lava-currents will flow from various apertures and at various heights. By repeated action a volcano acquires considerable solidity at its base, for the loose erupted matter is, independently of the consolidation produced by other causes, bound together by lava radii proceeding from the central aperture. Rents are often produced in the base, particularly when the great vent has accumulated matter to

* See Von Buch's Canary Islands, pl. 13 ; and a corrected reduction of this in Lyell's Principles of Geology, pl. 1.

† Mr. Scoresby notices a volcano off the main land of Greenland. This volcano is situated in the island of Jan Mayen, presented marks of recent eruption, and had a crater about 500 feet deep, and 2000 feet in diameter. Edin. Phil. Journal.

a considerable height, and through these, lava is protruded; the streams so thrown out serving to brace the lower parts of the mountain more firmly together. The occurrence of such apertures is precisely what we should expect in a volcano, which had accumulated materials upon it nearly equal to the average force of the elastic vapours propelling igneous matter upwards; for the pressure of the elevated column being very considerable, and in proportion to its height, it will always struggle to free itself in the direction of the least resistance. Now the sides of a volcanic mountain are not likely to be homogeneous, but to vary much in their resisting powers, being most solid where crossed by lava-currents, and weakest where merely formed of ashes or substances of the like nature. If to these causes of unequal resistance to pressure we add the fractures and rents produced by shocks in the mountain itself, we should always expect to find lateral discharges of lava common, while similar streams from the mouth would be rare.

M. von Buch is the author of a theory respecting the elevations of volcanos, which has been adopted by many geologists, while it has been combated by others. He observes, that the appearances of many craters are such, that we can scarcely consider them as erupted in the ordinary way; because they do not seem to present either lava-currents, or such an arrangement in the deposit of other volcanic substances as to justify such a conclusion. To these craters he has given the name of Craters of Elevation (*Erhebungs Cratere*). It has been opposed to this theory, that it presupposes an horizontal accumulation of lava or other volcanic matters, previously to the propulsion of elastic vapours through it, which should elevate the flat mass in a dome or cone, and burst through the highest part, presenting the appearance of a crater of eruption. How far this objection may be valid would seem to depend on the possibility of forming sheets of volcanic matter, which heat might soften and elastic vapours force up, so that the necessary forms should be produced. It may be questionable whether, under a great pressure of the sea, there is the same tendency to produce cinders and ashes as in the atmosphere; and if the superincumbent weight would not so act upon the solid matter ejected, that it would be forced into fusion, and sheets of melted matter be the result, if the elastic vapours beneath a column of melted rock were sufficiently powerful to overcome the resistance both of the column of lava and the superincumbent water. If such would be the state of things beneath a considerable depth of water, the tendency to produce ashes and cinders in a volcanic vent would increase with its approach to the surface of the water; and therefore all the phænomena of eruptions from beneath the surface of the sea would differ but little from those observed in the atmosphere. Another objection to the theory of craters of elevation

is, that the stratification of such supposed craters is precisely that
of craters of eruption ; and that therefore the inference from this
circumstance would be in favour of the latter, because we now have
daily examples of such modes of formation, while of the other we
have none. Data on this subject are so few, that it seems difficult to estimate the value of this objection. The fact, however,
that solid rocks can be raised by elastic vapours, is shown in the
case of the Little and New Kameni, (Island of Santorino,) where
brown trachyte, of a resinous lustre and full of crystals of glassy
felspar, was upraised ; the former in 1573, and the latter in 1707
and 1709. The elevation of the Little Kameni was " accompanied by the discharge of large quantities of pumice, and a great
disengagement of vapour*." By terming this rise an earthquake,
we merely seem to be using two names for the same thing. That
there were elastic vapours it is clear, and that these vapours were
the propelling power may fairly be inferred; therefore the fact is
the same, whether we call it an earthquake or a volcanic elevation, and it would be somewhat difficult to draw fine lines of
distinction between the two. The trachyte of New Kameni was
observed to have shells upon it when raised, and limestone and
marine shells are described as composing a part of these otherwise
igneous islands†. These occurrences at Santorino are quite sufficient to show, that volcanic rocks, with shells upon them, may be
raised bodily to the surface. Langsdorff notices a trachyte rock
3000 feet high, which appeared in 1795 near the Island of Una-
laschka, and which seemed to have been thrown up as a mass
from the bottom of the sea‡.

Ingenious explanations have been given to account for the large
orifices which have been termed craters of elevation. Mr. Lyell
considers that the crater resulting from the destruction of the
summit of Etna in 1444, was as large as those noticed in other
places and named craters of elevation; and supposes that a series
of great explosions might so reduce the cone, that finally there
would be a circular bay, forty or fifty miles round, in an island seventy or eighty miles in circumference, wholly composed of volcanic rocks which should dip outwards. But supposing such appearances to have been produced, the whole base of Etna, a kind
of circular island, would still show its lava-currents, sections of
which would be observed in the interior bay, or might be exposed
outside, and no doubt would remain that it was a crater of eruption. How far the so called " craters of elevation" may resemble
the supposed case of Etna remains to be seen ; yet if they should
not, as is considered they do not, present traces of lava-currents,
radiating from a centre or centres, but large envelopes of trachyte

* Lyell, Principles of Geology, vol. i. p. 386.   † *Ibid.*
‡ Daubeny, Description of Volcanos, p. 310.

or other fused volcanic rock, they can scarcely be referred to the
same origin. There does seem a possibility of producing craters
of elevation by the action of heat and elastic vapours on a sheet of
lava, therefore the subject should be fairly investigated, without
bias, with proper caution, and in the necessary detail.

It is supposed that after the craters of elevation were formed,
the eruptive action poured forth the usual volcanic substances,
which, when it was continued sufficiently long, produced a cone like
the Peak of Teneriffe ; but when such eruptive action was small,
or the crater comparatively recent, the appearances were such as
we now observe at Barren Island in the Bay of Bengal, where a
central cone, in activity, in the midst of a basin of water, is surrounded by a circular range of volcanic ground, which, according
to the figure given by Mr. Lyell*, rises at an angle of about 45°
from the sea. The height of the central cone is about 1800 feet
above the water, and the elevation of the surrounding volcanic
circle being nearly the same, the interior is only viewed through a
break in it. It would appear that the rocks of this island are extremely hot, for Capt. Webster landing upon it in March 1822 or
1823, found the water almost boiling at one hundred yards from
the shore ; the stones upon the beach, and the rocks exposed by
the ebb tide, hissing and steaming, and the water bubbling around
them†.

Von Buch adduces the Caldera in the Isle of Palma, Canaries,
as a good example of the craters of elevation. A large precipitous cavity or crater exists in a lofty range sloping outwards,
which incloses it on all sides but one, where a gorge forms the
only communication from the exterior to it. The sides of this
great cavity expose a section of beds of basalt, and conglomerates
composed of basaltic fragments, dipping regularly outwards. Now
if the beds be so regular, and not composed of scoriaceous matter
or ashes, their formation would seem not to have taken place in
the air or beneath a small pressure of water, but under different
circumstances, which would permit the basalt to be flattened into
tabular masses, not presenting the appearance of lava-currents
which have flowed in the atmosphere.

Jorullo affords a striking example of the outburst of volcanic action in the interior of dry land, where no active volcanos then existed, though the rocks in the vicinity would seem to indicate their
previous presence. Judging from the direction of the vents, a cleft
seems to extend east and west across Mexico to the Revillagi-
gedo Isles in the Pacific. Previous to June 1759, the space where
the volcano of Jorullo now stands was covered by indigo and sugar-
canes, bounded by two brooks, the Cuitimba and San Pedro. In

---

* Principles of Geology, vol. i. p. 390.
† Edin. Phil. Journal, vol. viii.

June, hollow subterranean noises were heard, accompanied by earthquakes, which lasted from fifty to sixty days. Tranquillity seemed re-established at the commencement of September, but on the 28th and 29th of this month, the subterraneous noises again commenced, and, according to Humboldt, the ground, with a superficies of three or four square miles, rose up like a bladder. The extent of this movement is considered to be now marked by an elevation round its edges of 39 feet, gradually acquiring a height of 524 feet towards the centre of the present volcanic district. The eruption appears to have been very violent, fragments of rock were hurled to great heights, ashes were thrown up in clouds, and the light emitted was seen at considerable distances. The Cuitimba and San Pedro are described as having precipitated themselves into the volcanic vent, and to have assisted, by the decomposition of their waters, the fury of the eruption. " Eruptions of mud, and especially of strata of clay, enveloping balls of decomposed basalt in concentric layers, appear to indicate that subterraneous water had no small share in producing this extraordinary revolution. Thousands of small cones, from 6 to 10 feet in height, called by the natives Hornitos (ovens), issued forth from the Malpays. Each small cone is a fumirole, from which a thick vapour ascends to the height of from 22 to 32 feet. In many of them a subterraneous noise is heard, which appears to announce the proximity of a fluid in ebullition." From amid these cones, six volcanic masses, varying from 300 to 1600 feet in height above the old plain, were ejected from a chasm having a direction N.N.E. and S.S.W. The most elevated mass is named Jorullo, and from its north side a considerable quantity of lava, containing fragments of other rocks, has been thrown out. The great eruptions ceased in February 1760, and afterwards became gradually less frequent.—The opponents to the theory of craters of elevation consider the raising of the ground in the form of a bladder as not altogether proved, resting on Indian accounts of appearances, which have been considered with reference to a particular theory.

The well-known Monte Nuovo near Naples was thrown up in a day and a night, in 1538. This is also described as ejected from a fissure. The present height of this volcanic elevation is 440 feet above the sea, and its circumference about a mile and a half.

Various descriptions of volcanic eruptions will be found in works dedicated to the subject, and could not be admitted within the necessary limits of this volume. The following account, however, obtained by the exertions of Sir Stamford Raffles, of a great eruption from Tomboro, in the island of Sumbawa, is too important to be omitted. The first explosions were heard at various distant places, where they were very generally mistaken for discharges of artillery. They commenced on the 5th of April 1815, and continued more or less until the 10th, when the eruptions became more violent; and such a great discharge of ashes took place that the sky was obscured, and darkness prevailed over considerable distances. It appears that a Malay prow, while at sea on the 11th, far from Sumbawa, was enveloped in utter darkness, and that, afterwards passing the Tomboro mountain at the distance of about five miles, the commander observed that the lower part appeared in flames, while the upper portion was concealed in clouds. Upon landing, for the purpose of procuring water, he found the ground covered to the depth of three feet by ashes, and " several large prows thrown on shore by the concussion of the sea." Quitting Sumbawa, he with difficulty sailed through a quantity of these ashes floating on the sea, which he described as two feet thick, and several miles in extent. This person also stated that the volcano of Carang Assam, in Bali, was convulsed at the same time. The most interesting account is that presented us by the commander of the East India Company's cruiser Benares, which is nearly as follows :—At the commencement of the explosions this vessel was at Macasar, and the reports so closely resembled those of cannon, that it was supposed there was an engagement of pirates somewhere in the neighbourhood. Troops were consequently embarked on board the Benares, and the vessel stood out to sea in search of the supposed pirates. On the 8th of April she returned, without having found any cause for alarm. On the 11th, the apparent discharges of cannon were again heard, sometimes shaking the ship and Fort Rotterdam. The vessel proceeded southward to ascertain the cause of these explosions. At eight o clock on the morning of the 12th, " the face of the heavens to the southward and westward had assumed a dark aspect, and it was much darker than when the sun rose; as it came nearer it assumed a dusky red appearance, and spread over every part of the heavens ; by ten it was so dark that a ship could hardly be seen a mile distant ; by eleven the whole of the heavens was obscured, except a small space towards the horizon to the eastward, the quarter from which the wind came. The ashes now began to fall in showers, and the appearance was altogether truly awful and alarming. By noon the light that remained in the eastern part of the horizon disappeared, and complete darkness covered the face of day. This continued so profound during the remainder of the day that I (the commander of the Benares) never saw any thing to equal it in the darkest night; it was impossible to see the hand when held close to the eyes. The ashes fell without intermission throughout the night, and were so light and subtile that, notwithstanding the precautions of spreading awnings fore and aft as much as possible, they pervaded every part of the ship."

" At six o'clock the next morning it continued as dark as ever, but began to clear about half-past seven, and about eight o'clock objects could be faintly observed on deck. From this time it began to clear very fast. . . . . The appearance of the ship when day-light returned was most singular; every part being covered with the falling matter. It had the appearance of calcined pumice-stone, nearly the colour of wood-ashes; it lay in heaps of a foot in depth in many parts of the deck, and several tons weight of it must have been thrown overboard; for though an impalpable powder or dust when it fell, it was, when compressed, of considerable weight. A pint measure of it weighed twelve ounces and three quarters; it was perfectly tasteless, and did not affect the eyes with a painful sensation; had a faint smell, but nothing like sulphur; when mixed with water it formed a tenacious mud difficult to be washed off."

The same vessel left Macasar on the 13th, and made Sumbawa on the 18th. Approaching the coast she encountered an immense quantity of pumice-stone, mixed with numerous trees and logs with a burnt and shivered appearance. When arrived at Bima Bay, the anchorage was found to be altered, as the vessel grounded on a bank where a few months previously there had been six fathoms of water. The shores of the bay were entirely covered with the ashes ejected from Tomboro, which is distant about forty miles. The explosions heard at Bima were described as terrific, and the fall of ashes so heavy as to break in the Resident's house in many places. There was no wind at Bima, but the sea was greatly agitated, the waves rolling on shore, and filling the lower parts of the houses a foot deep. When off the Tomboro mountain, about six miles distant on the 23rd, the commander of the Benares observed the summit to be enveloped in smoke and ashes, while the sides showed lava-currents, some of which had reached the sea.

The explosions were heard at very considerable distances. Not only were they noticed at Macasar, which is 217 nautical miles from Tomboro, but also throughout the Molucca islands; at a port in Sumatra, distant about 970 nautical miles from Sumbawa; and at Ternate, distant 720 miles.

Lieut. Phillips being dispatched to relieve the wants of the inhabitants, who were perishing from famine and disease, learned from the Rajah of Saugar, that about seven o'clock in the morning of the 10th of April, there was an appearance of three distinct columns of flame, all within the crater, which united at a great height upwards; and that, subsequently, the whole mountain appeared like a mass of liquid fire. How far the appearance of flame may be correct, it would be difficult to say, as nothing is so common as deceptive appearances of this kind; its character, however, would seem remarkable.

The Rajah's account proceeds :—" The fire and columns of

flame continued to rage with unabated fury, until the darkness caused by the quantity of falling matter about eight P.M. Stones at this time fell very thick at Saugar, some of them as large as two fists, but generally not larger than walnuts." Soon after ten P.M. a violent whirlwind arose, "which blew down nearly every house in the village of Saugar, carrying the tops and light parts along with it. In the part of Saugar adjoining Tomboro its effects were much more violent, tearing up by the roots the largest trees, and carrying them into the air, together with men, houses, cattle, and whatever else came within its influence." The sea was agitated, rising twelve feet higher than it was ever known to do before. The water rushed upon the land, sweeping away houses and all within its influence, and destroying the few rice-grounds which previously existed at Saugar. As might have been expected amid such a convulsion, a great destruction of life was effected, and many thousand inhabitants were killed. The vegetation on the north and west sides of the peninsula was completely destroyed, with the exception of a high point of land where the village of Tomboro previously stood, and where a few trees still remained *.

The changes produced by such eruptions as that here recorded, would, independently of the alteration in the shape of the volcano itself, and of the streams of lava which flowed from it, extend to very considerable distances. On the dry land, vegetables and animals would be entombed beneath stones and ashes, the quantity of the covering matter probably increasing with the proximity to the volcano. And if it should chance, as sometimes happens, that the aqueous vapour discharged from the volcanic rent were suddenly condensed, the torrents produced would sweep away not only the looser parts of the volcano, but also the plants and animals which they might encounter, embedding them in a thick mass of alluvial matter.

The vegetable and animal substances enveloped by the discharged ashes, cinders, and stones falling into the sea, would be both marine and terrestrial, and a very curious mixture, as far as regarded its organic contents, would be observed; trees, men, cattle, fish, corals, and a great variety of marine remains, would be encased, and it might so happen that both on the land and in the sea a bed of lava might cover such accumulations.

In the case of the great discharge of lava in Iceland, in the year 1783, many terrestrial remains might have been covered by the igneous matter, possibly some in such situations as to preserve their form. Should a similar eruption take place in the sea, where, as before observed, the conditions are more favourable for the production of a sheet of lava, sands and clays, perhaps full of

* Life of Sir Stamford Raffles.

marine remains, would be covered over, and very considerable changes might be produced by such a superincumbent mass of heated matter. Upon which, after a certain time, sands and clays, again charged with organic remains, might be accumulated, when a new eruption might again cover them. Thus producing an alternation of igneous and aqueous rocks.

Mr. Henderson notices an alternation of fossil wood, clay, and sandstone in Iceland, surmounted by basalt, tuff, and lava. When this accumulation of vegetable matter was so covered is not so clear; but if Mr. Henderson be right in considering many of the fossil leaves as those of the poplar, it is not, probably, very recent, for it supposes a change of climate, as poplars do not now grow in Iceland.

During great explosions, volcanos cannot be approached sufficiently near for the purposes of very minute observation; therefore we can only judge of some of the probable effects from appearances at their calmer periods, and consequently a minor state of activity is very favourable for such examinations. After ineffectual attempts to observe the workings of the fluid mass within the crater of Vesuvius at the commencement of 1829, when that mountain was somewhat active, I was fortunate enough on the 15th of February to have ascended on a calm day, when the vapours darted majestically upwards as they were propelled from the small cone in the middle of the grand crater *, and the incandescent matter in the vent was at times distinctly visible,—a rare circumstance, as when there is the slightest movement in the air, the vapours obscure every thing. After the more continued detonations there was a lull or calm, succeeded by a violent explosion, throwing up stones to a considerable height, mixed with pieces of red-hot lava, which latter fell like lumps of soft paste on the sides of the small cone. When the vapour cleared away, the red-hot mass appeared as if in ebullition from the passage of the gaseous matters through it. The light emitted varied exceedingly in intensity, being brightest at the moment of the great explosion, when a great volume of vapour suddenly forced its way through the fiery mass, darting up with great velocity, and carrying all before it. Wishing to profit by my good fortune, I continued many hours on the mountain, until night closed in, hoping that objects might be perceived within the crater not previously observed. In this I was disappointed, appearances being the same, though more distinctly visible. The picturesque effect, however, was greatly heightened; the solid ejected substances darted upwards like a grand discharge of red-hot balls, while the reflection of the incandescent matter within, on the vapour above, was at times exceed-

* For a sketch of the crater at this time, see Sections and Views illustrative of Geological Phænomena, pl. 22.

ingly brilliant, producing, at a distance, those false appearances of flames, which, there are very good reasons for supposing, do not issue from volcanos; the often recorded appearances of this nature being merely reflected light, varying in intensity according to the activity of the mountain.

The products of active volcanos, though man seems to exhaust his language in finding terms to express his horror and dismay at their mode of ejection, do not constitute such an addition to dry land as at first sight would appear probable,—for their mass must be regarded relatively to the mass of dry land generally, and not with reference to particular districts. Moreover, cavities corresponding with the quantity of matter thrown out will sometimes occur not far beneath the surface; and when the weight above shall overcome the resistance below, either suddenly from a violent convulsion, or slowly from gradual change, the mass above will fall into the abyss beneath, and matter be, in some measure, restored to its place. Among volcanic changes it is by no means uncommon to hear of hills disappearing, and being converted into lakes. The most memorable example, perhaps, of the disappearance of a volcano, is that which took place in Java in 1772. The Papandayang, on the south-western part of the island, reputed one of its largest volcanos, was observed at night, between the 11th and 12th of August, to be enveloped by a luminous cloud. The inhabitants being alarmed, betook themselves to flight, but before they could all escape, the mountain fell in, accompanied by a sound resembling the discharge of cannon. Great quantities of volcanic substances were thrown out, and carried over many miles. The extent of ground thus swallowed up was estimated at fifteen miles by six. Forty villages were engulphed or covered by the substances thrown out, and 2957 persons were reported to have been destroyed *.

Extinct Volcanos.

From a similarity of appearances, rocks existing under certain circumstances where there are at present no active vents, have been attributed to a volcanic origin. To draw fine lines of distinction between volcanos now in activity and those which appear extinct, would be almost impossible, for there is no certainty that the one may not soon be converted into the other. Of this we probably have a good example in Vesuvius, which after being, as far as we can judge from historical records, for a long period extinct, became convulsed in the year 79, destroyed the higher part of its old cone, part of which now remaining is named Monte Somma, and overwhelmed Herculaneum, Pompeii, and Stabiæ, entombing not only men, but theatres, temples, palaces, and innumerable works of art, which have afforded by their

* Horsfield, as quoted by Daubeny.

disinterment more real knowledge of the manners and customs of the ancient inhabitants of these beautiful regions of Italy than all the writings which have escaped destruction.

Solfataras, as they are termed, are usually considered as semi-extinct volcanos, emitting only gaseous exhalations and aqueous vapour; but there can be no certainty that they also may not again enter into activity. According to Dr. Daubeny, sulphuretted hydrogen and a small portion of muriatic acid are contained in the steam which rushes out of the fumaroles at the Solfatara near Naples. The rocks of the crater and vicinity are greatly decomposed by the action of these gaseous exhalations; and, among other salts thus formed, the muriate of ammonia is the most abundant. Solfataras, variously modified, are by no means rare in volcanic countries.

Not only do extinct volcanic vents occur in regions where active volcanos now exist, so that we may imagine a mere change of fiery orifice, but they are also found in districts where all trace of activity has been lost since the earliest historical times, if we except the presence of mineral and thermal springs. In central France and Germany such appearances are particularly remarkable, and it has been attempted to draw a line of distinction between those volcanos which have existed in a state of activity since the establishment of the present order of things, and those whose activity was previous to this state. The subject is full of difficulty, more especially as respects Central France, where volcanic ejections have taken place at different periods; so that there is no ready mode of making geological distinctions between the ejections, which would seem little else than productions from new orifices opened for the discharge of volcanic matter in the same region. We may be able to observe the extremes, but to mark striking and easily distinguishable points intermediate between them would be exceedingly difficult. Volcanic ejections were probably continued through nearly the same orifices for a long period of time, during which many and great geological changes were taking place around them, and on the surface of the earth generally. It has been attempted to determine the relative ages of volcanos by the absence or presence of craters; as also on the supposition that some have existed prior to the excavation of valleys, while others have been produced after their formation, their lava-currents having been discharged into them. Such distinctions can scarcely be made; for craters may be easily obliterated, and relative age, from the excavation of valleys, cannot be very satisfactorily established amid circumstances which could so easily produce changes in this respect. A more direct mode has been to try their relative antiquity by means of the mineral structure of their lavas; and if this should hold good, it would be the safest guide; but it may be doubted how far our knowledge of volcanic products authorises so general a conclusion.   That there is a great difference in the mineral character, generally, between the igneous rocks of the older periods of the world and those at present formed, few will doubt.   We know of no granite or serpentine streams thrown out from modern volcanos; but when igneous rocks so closely allied in geological dates as those produced by active and extinct volcanos are under consideration, such distinctions should not be too hastily adopted.

Dr. Daubeny considers that the more modern volcanic products of Auvergne are more cellular, have in general a harsher feel, and possess a more vitreous aspect than the more ancient *.

In Auvergne and the Vivarais there are numerous examples of the more modern extinct volcanos, the craters of which are frequently very perfect, or merely broken down by the discharge of large lava-currents from them.   Details respecting them will be found in works written expressly on the subject, and pictorial representations among the views contained in Mr. Scrope's work on Central France†.

In the district of the Eyfel, near the Rhine, there are also extinct volcanos which have been considered as comparatively recent, from the situations which they occupy; having been apparently produced after the formation of the valleys in the neighbouring country.   In the volcanic district of Central France the lava-currents have in some places traversed valleys, and dammed up the waters that passed through them.   The waters so dammed up accumulated into a lake, which was subsequently drained through a gorge cut in the rocky barrier by means of the surplus water; which not only effected this, but also cut, by continual erosion, into the rock beneath, forming a part of the original valley.

Many other examples of extinct volcanos have been noticed in districts where active volcanos do not now exist.   Their relative antiquity is however so little understood, that a general classification of them cannot be attempted.

Mineral Volcanic Products.

Various classifications of volcanic substances have been proposed, among which the division into Trachytic and Basaltic seems to be that most commonly adopted; trachyte being considered as essentially composed of felspar, and containing crystals of glassy felspar; while basalt is supposed to be essentially composed of felspar, augite, and titaniferous iron.   Lavas, however, present such various mixtures of different minerals, that exact classifications of them would appear exceedingly difficult; and, when we con-

---

* Description of Volcanos.

† Part of one of the most striking of these views is copied in " Sections and Views illustrative of Geological Phænomena," pl. 24.

sider that these different compounds may be infinitely modified by circumstances, such classifications cannot be of much value. These products are of such a compound nature, consisting of felspar, augite, leucite, hornblende, mica, olivine, and other minerals, that definite names can scarcely be attached to them. Mr. Poulett Scrope has distinguished the rocks termed trachyte, basalt, and graystone (the latter a name proposed by himself) under the following heads:—1. *Compound trachyte*, with mica, hornblende, or augite, sometimes both, and grains of titaniferous iron. 2. *Simple trachyte*, without any visible ingredient but felspar. 3. *Quartziferous trachyte*, when containing numerous crystals of quartz. 4. *Siliceous trachyte*, when apparently much silex has been introduced into its composition. 1. *Common graystone*, consisting of felspar, augite, or hornblende, and iron. 2. *Leucitic graystone*, when leucite supplants the felspar. 3. *Melilitic graystone*, when melilite supplants the felspar, &c. 1. *Common basalt*, composed of felspar, augite, and iron. 2. *Leucitic basalt*, when leucite replaces the felspar. 4. *Olivine basalt*, when olivine replaces the felspar. 5. *Hauyine basalt*, when hauyine replaces the felspar. 6. *Ferruginous basalt*, when iron is a predominant ingredient. 7. *Augite basalt*, when augite composes nearly the whole rock *.

As all fused substances will tend to crystallize, or arrange their component parts more compactly, where their liquidity continues the longest, and their loss of temperature is the slowest, we find that lava-currents are always more crystalline or compact in their interior parts, and that dykes cutting volcanic cones are generally more compact and crystalline than the lavas which flow from them; such dykes being also more crystalline towards their interior parts than towards their walls or sides. It has been inferred from the appearance and distribution of the ejected matters, that many volcanic rocks have not been formed in the atmosphere, but beneath seas, and that they have been subsequently elevated. The ashes and pumice ejected from volcanos seem merely, if I may so express myself, the frothy part of the great fused and incandescent matter within, produced by the action of elastic vapours, or by the intumescence of that matter under diminished pressure. The force required to eject such light substances is evidently far inferior to that necessary for the propulsion of the more solid lava, and consequently the one is in general more common than the other. As might be expected from the nature of such mineral productions, volcanic substances vary, from the lightest ash to a highly crystalline rock, the intermediate states being vitreous, and of the character of obsidian. The quantity of minerals detected in volcanic products is exceedingly great, a circumstance by no means surprising when we consider the various elementary substances acted

* Quarterly Journal of Science, vol. xxi. 1826.

on by heat in the bowels of a volcano, and striving to combine with each other in various ways *.

Not only are fused substances ejected, but also various portions of rocks traversed by the volcanic vent; and as this is very variously situated, so are the rocks various which are thrown out. Vesuvius having been under observation for so long a period, its products have received greater attention than falls to the lot of most volcanos; and it has been observed, though no doubt volcanos vary most materially in this respect, that such ejected substances are far from being either rare or of one kind. The Chevalier Monticelli's invaluable collection of Vesuvian products at Naples contains a great variety of these substances, among which may be seen fragments of the compact limestones of the district, with organic remains in them, seeming to show that the vent traverses the limestones, and rends off portions of them, as indeed might be expected from the nature of the country. The limestones so ejected are often impregnated with magnesia, supposed to have been acquired in this great natural crucible.

Volcanic Dykes, &c.

Dykes or fissures in the sides of volcanos, subsequently filled by melted lava, are sufficiently common. M. Necker de Saussure mentions numerous dykes which traverse the beds of Monte Somma. These veins are nearly all of the same composition, differing somewhat from the lava-beds they cut; augite being more abundant, while leucite, so common in the beds, occurs rarely in the dykes, with the exception of one vein of Monte Otajano, and another near the foot of the Punte del Nasone, which contain large crystals of leucite. The lava of the dykes also contains minute crystals of felspar (?), with a considerable abundance of a yellow substance, which may be olivine. The rock composing the veins is fine-grained on the sides, and more crystalline in the middle. These veins vary from one to twelve feet in width.

One remarkable dyke, different from the rest, occurs at Otajano. It is about ten and a half feet wide, and rises perpendicularly to the crest of the mountain, having apparently turned up the alternating beds of porous and compact lava which it traverses. Another singular dyke cuts the rocks of the Primo Monte. It rises perpendicularly, and is formed of a slightly greenish gray and homogeneous rock. At its base (that of the mountain) it is only eleven inches wide, and for twelve feet of its height is bordered by a line of vitreous lava, half an inch thick, separating it from the porous volcanic breccia which it cuts. Above the twelve feet,

---

* Sulphur is exceedingly common, and is often sublimed in such quantities as to be carried away for economical purposes.

the vitreous lava ceases entirely, the solid rock occupying the whole vein *.

Dr. Daubeny notices tuff traversed by dykes of a cellular trachytic lava at Stromboli, and at Vulcanello in the island of Lipari †. Dykes, described as resembling greenstone, were noticed by Sir George Mackenzie traversing alternate beds of tuff and scoriaceous lava in Iceland.

Dykes of porphyry traverse the older lavas of Etna. The formation is by no means difficult of explanation, by supposing fissures which sometimes have, and sometimes have not, penetrated to the surface, injected with incandescent lava. Of fissures extending to the surface, the cleft twelve miles long and six feet broad, which opened on the flank of Etna, between the plain of St. Lio and a mile from the summit, at the commencement of the great eruption of 1669, is an example‡. This fissure gave out a vivid light; from which Mr. Lyell with great probability concludes that it was filled to a certain height with incandescent lava. After the formation of this, five other fissures were produced, and emitted sounds heard at the distance of forty miles §.

While on this subject it may be as well to notice the probable effects of a column of lava passing through stratified rocks, insinuating melted matter among the strata, or through fissures formed in them.

Let *a b* in the annexed diagram represent a column of liquid lava, traversing horizontal strata. It is obvious that it will strive to overcome the resistance of the sides, and such resistance will always be less between the strata than elsewhere. If it obtain an aperture in that direction, it will endeavour to separate one stratum from another; and it will the more readily accomplish this, as to the pressure of the column of lava will be added the mechanical action of the wedge; and eventually an injection of liquid lava may be made, and carried laterally, so far as the pressure will permit. Thus, if a separation of the strata can be commenced at *d*, it will be carried on in the direction *d c* as far as the pressure of the column *a d* will permit. If, instead of this kind of injection, we consider the strata to have been fractured, as is very likely to be the case near volcanic action, the fissure

Fig. 19.

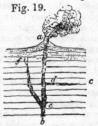

* Necker, Mémoire sur le Mont Somma, Mém. de la Soc. de Phys. et d'Hist. Nat. de Genève, 1828.

† Daubeny's Description of Volcanos, p. 185—187, where there are views of these appearances.

‡ Lyell, Principles of Geology.   § *Ibid.* vol. i. p. 364.

will be filled, and forced asunder as far as resistances will permit. Thus, if a fracture *e f* be made, it will be filled by liquid lava as far as can be effected by the pressure of the column *a e.* The strata have here been supposed horizontal, for the sake of illustration, but as they might occur in all modes, the effects would be varied accordingly, the principle remaining the same.

Earthquakes.

The connection between volcanos and earthquakes is now so generally admitted that it would be useless to enumerate the various circumstances that point to this conclusion. They both seem the effects of some cause as yet unknown to us. The motion of the ground produced by earthquakes is not always the same; sometimes resembling the undulatory movement of a heavy swell at sea, though much quicker, and being at others tremulous, as if some force shook the ground violently in one spot. The former of these is far the most dangerous, as it forces walls and buildings off their centres of gravity, crushing whatever may be beneath them.

It has been considered that earthquakes are presaged by certain atmospheric appearances, but it may be questionable to what extent this supposition is correct. Historians of earthquakes seem to have been generally desirous of producing effect in their descriptions, adding all that could tend to heighten the horror of the picture. They have not always, moreover, been anxious or able to separate accidental from essential circumstances. As far as my own experience goes, which is however merely limited to four earthquakes, the atmosphere seemed little affected by the movement of the earth; though I would be far from denying that it may be so; for we can scarcely imagine such movements to arise in the earth, without some modification or change of its usual state of electricity which would affect the atmosphere. If animals be generally sensible of an approaching shock, it might arise as well from electrical changes as from the sounds which they may be supposed capable of distinguishing.

Earthquakes very frequently precede violent volcanic explosions, even though they may be felt far from a fiery vent. Thus, the great earthquake which destroyed the Caraccas, March 26, 1812, was followed by the great eruption of the Souffrier in St. Vincents, on April 30th of the same year; when, according to Humboldt, subterranean noises were heard the same day at the Caraccas and on the banks of the Apure.

Earthquakes are felt over very considerable spaces, and of this no better example has yet been recorded than the celebrated earthquake of Lisbon in 1755, the shock of which was felt over nearly the whole of Europe, and even in the West Indies. The force capable of causing such extensive vibrations must have been very considerable; and, with every allowance for the easy transmission of motion and sound laterally through rocks, must have required considerable depth for its production. Motion seems always to be communicated to water during earthquakes, the vibratory movement being very frequently felt by vessels at sea, and waves of greater or less magnitude, according to the force of the shock, being commonly driven on shore. The wave produced during the great Lisbon earthquake rose sixty feet high at Cadiz, and eighteen feet at Madeira, causing various movements of the water on the coasts of Great Britain and Ireland. Similar waves, though of proportionally less size, are common during volcanic eruptions; motion being produced in the surrounding water, which being unable to rend and crack like the land, communicates the impulse it has received to the waters around, and thus a wave is propagated which will diminish in height in proportion as it recedes from the disturbing cause. In almost all ports irregularities in the motion of the sea are at times observable, which cannot be reconciled with the tides or motions communicated to water by temporary currents or winds in the offing. The movement is generally a quick flow or reflow of the water, and is often so trifling as to escape the attention of all but seamen or fishermen, who, constantly engaged with their vessels or boats in harbours, are surprised to find them suddenly floated or left dry, and this sometimes several times repeated. May not these movements be caused by earthquakes beneath the depths of the sea, or too trifling to escape observation on land? If, as it seems reasonable to conclude, earthquakes are propagated laterally through considerable distances, in the same manner as sound is conveyed through the air, the intensity of the shock will depend on the medium through which it is conveyed; and if this view should be correct, earthquakes will not be equally felt on every description of rock. I once observed a fact, which, though it struck me much at the time, cannot in itself form the basis of any reasonable hypothesis, but as it may be the means of exciting inquiry it may be as well to mention it. While sitting in a house in Jamaica, situated on a hill, near the verge of the white limestone of that island, where the large gravelly, sandy, and clay plain of Vere and Lower Clarendon meets it, I experienced a slight shock of an earthquake. Having occasion about half an hour afterwards to descend to some houses at the foot of the hill, and on the gravel plain, I inquired if the inhabitants had felt the earthquake, when they ridiculed the idea, stating that if any such had occurred they must have known it, as they also had been sitting quietly, and were too much accustomed to shocks not to have observed the earthquake if it had really occurred. I then considered that I had deceived myself, and thought no more of the subject until the evening; when some negroes, who had been employed on their own account, a few miles distant in some mountains composed of the white limestone, reported that they had felt the shock of an earthquake; and it subsequently appeared, that a much more considerable shock had been felt in the vicinity of Kingston, about forty miles distant. The importance of this fact certainly rests on the little apparent sensation produced at the lower house; and, therefore, as the shock may have escaped the attention of those then present, this circumstance is in itself of no great value, and is merely stated to promote inquiry. It may, however, be remarked, that gravel would transmit a vibration less readily than compact limestone, though it might more easily give way before a vertical movement. Humboldt has remarked that during the Caraccas earthquake of 1812, the Cordilleras were more shaken than the plains. This may have arisen from the more easy transmission of the vibration through the gneiss and mica slate, than through the rocks in the plains; or, as might be also the case in Jamaica, the inferior rocks might be more shaken through their continuity than the superior rocks, being nearer the disturbing cause.

It may also be remarked that rocks would transmit sounds unequally from variations in their texture and continuity, and that subterranean noises might be audible while the shock which produced them could not be distinctly felt. Various sounds are recorded as accompanying earthquakes, but the most general seems a low rumbling noise like that of a waggon passing rapidly along. The first shock I ever experienced was, during a beautiful night, on the north side of Jamaica, when it appeared as if a waggon, rolling rapidly to the house, gave it a smart rap and then passed on.

It has been considered, and with much probability, that the very great distances at which volcanic explosions from surface-vents have been heard, arises from the transmission of the sound through the rocks. The great explosion at Sumbawa above noticed is described as having been heard in Sumatra, a distance of 970 geographical miles, and at Ternate, 720 miles in another direction *. It is also stated that the eruption from the Aringuay, in the island of Lucon, Philippines, in 1641, was heard in Cochin-China†.

Earthquakes produce changes in the level of the land, raising and depressing ground, and causing clefts, slips or faults, and various other modifications of surface. The raising of the surface implies either an expansion of the solid matter beneath, or a separation of parts, which should form a cavity, filled either by gaseous or liquid substances. We are not aware of anything that could

* Life of Sir S. Raffles.
† Chamisso, Kotzebue's Voyage.

produce the expansion required but heat, so that if the temperature were again diminished, contraction would ensue. If a separation of parts were effected, and the upper portion raised, the gaseous or liquid support could scarcely be considered permanent, unless the injected matter became solid, as might happen with liquid lava, and the hollow produced by such injection be far removed from the surface.

The best example of the bodily elevation of land with considerable surface appears to be that recorded by Mrs. Maria Graham, as having taken place during the Chili earthquake of 1822. The shock extended along the coast for more than a thousand miles, and the land was raised for a length of one hundred miles, with an unknown breadth, but certainly extending to the mountains. The beach was raised about three or four feet, as was also the bottom near the shore; on the former shell-fish were still adhering to the rocks on which they grew. It was also observed that there were other lines of beach, with shells intermixed, above that newly elevated, attaining in parallel lines a height of about fifty feet above the sea; seeming to show that other elevations of the same land had been effected by previous earthquakes. During this earthquake the sea flowed and ebbed several times. No visible change in the atmosphere was produced previous to the shocks, but it is supposed that some effect, perhaps electrical, may have been caused by the earthquake, for the country was subsequently deluged by storms of rain *.

Mr. Lyell has accumulated a considerable mass of evidence to show that such elevations have been the consequence of earthquakes in other places, and that considerable depressions have also occurred †. Thus, during the Cutch earthquake of 1819, the eastern channel of the Indus was altered, the bed of which was in one place deepened about seventeen feet, so that a spot once fordable became impassable.

A variety of surface-changes were effected during the great earthquake in Calabria in 1783. Of these a summary has been given from various authorities by Mr. Lyell, whose account will be perused with interest, however little we may feel inclined to adopt the theoretical conclusions that have been deduced from it. The earth had a waving motion; numerous and deep rents were formed; faults were produced, even through buildings; large landslips took place; lakes were formed,—one about two miles long by one broad, from the obstruction of two streams; the usual agitation of the neighbouring sea was produced, and heavy waves broke upon the land, sweeping all before them.

The great earthquake in Jamaica of 1692, generally described as having swallowed up Port Royal, has been adduced as an ex-

---

* Journ. of Science; Geol. Trans. vol. i.   † Principles of Geology.

ample of great derangement. By a careful perusal of the statements made, and an examination of the places said to have been most affected, the accounts appear to have been much exaggerated; nor need this surprise us, when we reflect how difficult it is to elicit truth respecting natural phænomena, from those who have been dreadfully alarmed by them. The narrow spit of land, terminated by the present town of Port Royal, is a sand-bank some miles long, thrown up, apparently, by the sea. The great mischief done at Port Royal seems to have been occasioned by a heavy wave, such as we have seen common in great earthquakes, which rushed into the town, then, as at present, elevated but a few feet above the level of the water, sweeping away all within its influence *.   Rents also appear to have been formed, and there may have been subsidence, though it must have been somewhat difficult to find chimney-tops in Port Royal for the masts of wrecked vessels to appear among, such conveniences being confined to little low and detached kitchens where their presence may be required.   The accounts also of submerged standing houses must be received with some caution, particularly when it is recollected that a frigate was driven by the force of the water over their tops.

Funnel-shaped, or inverted conical cavities are by no means unfrequent on plains after earthquakes, and are so much alike wherever they occur, that they must have some common cause for their production. Circular apertures were produced in the plains of Calabria by the earthquake of 1783 : they are described as commonly of the size of carriage-wheels, but often larger and smaller; they were often filled by water, but more frequently by sand.   Water seems to have spouted through them †.   During the earthquake in Mercia in 1829, numerous small circular apertures were produced in a plain near the sea, which threw out black mud, salt water, and marine shells ‡.   After the earthquake at the Cape of Good Hope, in December 1809, the sandy surface of Blauweberg's Valley is described as studded with circular cavities, varying from six inches to three feet in diameter, and from four inches to a foot and a half in depth. Jets of coloured water are stated, by the inhabitants of the valley, to have been thrown out of these holes to the height of six feet during the earthquake §.   It seems somewhat difficult to account for these appearances, though the common aqueous discharges through rents or chasms can be more readily understood. During the Chili earth-

* If a great earthquake were to produce a wave from ten to twenty feet high, it would sweep away the greater part of the present Port Royal.

† Lyell, Principles of Geology; where a view and section of these curious cavities are given, pp. 428, 429.

‡ *Ibid;* and Ferussac's Bulletin, 1829.

§ Phil. Mag. and Annals, January 1830.

quake, previously noticed, sands were forced up in cones, many of which were truncated with hollows in their centres*

The courses of springs are, as would be anticipated, often deranged amid such motions of the ground; and flashes of light, or bright meteors, are so frequently mentioned that we can scarcely doubt their occurrence, and they may, perhaps, be considered as electrical.

If we now withdraw ourselves from the turmoil of volcanos and earthquakes, and cease to measure them by the effects which they have produced upon our imaginations, we shall find that the real changes they cause on the earth's surface are but small, and quite irreconcileable with those theories which propose to account for the elevations of vast mountain-ranges, and for enormous and sudden dislocations of strata, by repeated earthquakes acting invariably in the same line, thus raising the mountains by successive starts of five or ten feet at a time, or by catastrophes of no greater importance than a modern earthquake. It is useless to appeal to time: time can effect no more than its powers are capable of performing: if a mouse be harnessed to a large piece of ordnance, it will never move it, even if centuries on centuries could be allowed; but attach the necessary force, and the resistance is overcome in a minute.

Gaseous Exhalations.

In several situations removed from any volcanic action, so far as is visible on the surface, natural jets of inflammable gases are seen to issue, affording decisive evidence of chemical changes that are taking place at various depths beneath. Of these, some have served the purpose of the priest to delude mankind, while part of the others have been more usefully employed.

Carburetted hydrogen gas is well known to be the "fire-damp" of the coal districts, and to issue from the coal strata; collecting in the ill-ventilated galleries of collieries, and, when sufficiently mixed with atmospheric air, exploding with great violence if approached incautiously with an unprotected flame, spreading mourning and misery among the families of the miners. If the genius of Davy had only produced his safety-lamp, it would alone have entitled him to the applause and thanks of mankind.

As carburetted hydrogen is so freely liberated in coal-mines, it would be expected that it should occasionally be detected on the surface, and accordingly it has been so discovered. Inflammable gas also occurs in other situations, where there is no reason to suspect the presence of coal strata. Of this, the well-known jets of gas in the limestone and serpentine district of the Pietra Mala, between Bologna and Florence, is an example.

* Journal of Science.

Captain Beaufort describes an ignited jet of inflammable gas, named the Yanar, near Deliktash, on the coast of Karamania, which perhaps once figured in some religious rites. He states that, "in the inner corner of a ruined building, the wall is undermined, so as to leave an aperture of about three feet in diameter, and shaped like the mouth of an oven: from thence the flame issues, giving out an intense heat, yet producing no smoke on the wall." Though the wall was scarcely discoloured, small lumps of caked soot were found in the neck of the opening. The hill is composed of crumbly serpentine and loose blocks of limestone. A short distance down the hill there is another aperture, which from its appearance seems once to have given out a similar discharge of gas. The Yanar is supposed to be very ancient, and is possibly the jet described by Pliny*.

Colonel Rooke informed Captain Beaufort that high up on the western mountain at Samos, there was an intermittent flame of the same kind; and Major Rennell stated that a natural jet of inflammable gas, inclosed in a temple at Chittagong, in Bengal, is made use of by the priests, who also cooked with it.

The village of Fredonia, in the State of New York, is lighted by a natural discharge of gas, which is collected by means of a pipe into a gasometer. The quantity obtained is about eighty cubic feet in twelve hours. It is carburetted hydrogen, and is supposed to be derived from beds of bituminous coal. The same gas is discharged in much larger quantities in the bed of a stream about a mile from the village.

According to M..Imbert, gaseous exhalations are employed at Thsee-Lieou-Tsing, in China, to distil saline water obtained from wells in the neighbourhood. "Bamboo pipes carry the gas from the spring to the place where it is to be consumed. These tubes are terminated by a tube of pipe-clay, to prevent their being burnt. A single well (of gas) heats more than three hundred kettles. The fire thus produced is exceedingly brisk, and the caldrons are rendered useless in a few months. Other bamboos conduct the gas intended for lighting the streets and great rooms or kitchens†."

This connection of inflammable gas with saline springs or salt is not confined to China, but has also been observed in America and Europe. While boring for salt at Rocky Hill, in Ohio, and near Lake Erie, the borer suddenly fell, after they had pierced to a depth of 197 feet. Salt water immediately spouted out, and continued to flow for several hours; after which a considerable quantity of inflammable gas burst forth through the same aperture, and, being ignited by a fire in the vicinity, consumed all within its reach‡.

* Beaufort's Karamania.
† Bibl. Universelle, and Edin. New Phil. Journal, 1830.
‡ Trans. New York Phil. Soc.

It also appears that M. Rœders, inspector of the salt-mines of Gottesgabe, at Reine in the county of Tecklenberg, has for two or three years used an inflammable gas which issues from these mines, not only as a light, but for all the purposes of cookery. He obtains it from the pits that have been abandoned, and conveys it by pipes to his house. From one pit alone a continuous stream of this gas has issued for sixty years. It is supposed to consist of carburetted hydrogen and olefiant gas *.

Inflammable gases are also found to proceed from ground charged with petroleum and naphtha. The inhabitants of Badku, a port on the Caspian Sea, are supplied with no other fuel than that derived from the petroleum and naphtha with which the earth in the neighbourhood is strongly impregnated. About ten miles to the N.E. of this town there are many old temples of Guebres, on each of which there is a jet of inflammable gas, rising from apertures in the earth. The flame is pale and clear, and smells strongly of sulphur. Another and a larger jet issues from the side of a hill. The ground is generally flat, and slopes to the sea. If in the circumference of two miles, holes be made in the earth, gas immediately issues, and inflames when a torch is applied. The inhabitants place hollow canes into the ground, to convey the gas upwards, when it is employed for the purposes of cookery as well as a light †.

Carbonic acid gas is evolved abundantly in coal-pits and volcanic regions. Its occurrence in the Grotto del Carne, of which such overcharged descriptions have been given, is well known. MM. Bischof and Nöggerath notice a pit, on the side of the lake of Laach, in which they found dead birds, squirrels, bats, frogs, toads, and insects, killed by the evolution of carbonic acid gas.

A very copious discharge of carbonic acid gas occurs on the Kyll, nearly opposite Birresborn. The gas rises through fissures of the rock, and traverses a pool of rain-water, resting on it with such violence that the noise is stated to be heard at the distance of 400 yards. Birds are killed when they approach too close, and persons wishing to drink are driven away by the gas, a stratum of which covers the surrounding turf ‡.

In many situations gaseous vapours come to the surface mixed with water or petroleum, with sufficient force to produce 'salses' or mud volcanos. Dr. Daubeny considers those of Maculaba in Sicily as independent of volcanic action, but due to the combustion of the sulphur existing among the rocks. Mud eruptions from the discharge of gaseous vapours and water are known in many other places §.

---

* Journal of Science.   † Edin. Phil. Journal, vol. vi.
‡ Bischof and Nöggerath, Edin. Phil. Journal.
§ Those near Modena have long been celebrated.

Deposits from Springs.

Springs are seldom or ever quite pure, owing to the solvent property of water, which percolating through the earth, always becomes more or less charged with foreign matter. Carbonate, sulphate, and muriate of lime, muriate of soda, and iron, are frequently present in spring waters. Some are more highly charged with these and other substances, such as carbonate of magnesia and even silica, than others, and have hence obtained the name of mineral springs. Many are thermal, as before noticed, and seem not immediately derived from the waters of the atmosphere; as may also be the case with many that are cold, their more elevated temperature having been lost in their passage upwards through colder strata.

Many thermal springs contain silica, though this substance is of exceedingly difficult solution. The siliceous deposits from the Geysers in Iceland are well known. Sir George Mackenzie describes the leaves of birch and willow converted into stone, every fibre being discernible. Grasses, rushes, and peat are in every state of petrifaction. There are also deposits of clay containing iron pyrites,' which decompose and communicate very rich tints to it. The deposits from the Geysers extend to about half a mile in various directions, and their thickness must be more than twelve feet, for that depth is seen in a cleft near the Great Geyser.

The finest exhibition of such deposits as yet noticed occurs in the volcanic district of St. Michael, Azores. Dr. Webster describes the hot springs of Furnas as respectively varying in temperature from 73° to 207° Fahr., and depositing large quantities of clay and siliceous matter, which envelope the grass, leaves, and other vegetable substances that fall within their reach. These they render more or less fossil. The vegetables may be observed in all stages of petrifaction. He found "branches of the ferns which now flourish in the island completely petrified, preserving the same appearance as when vegetating, excepting the colour, which is now ash-grey. Fragments of wood occur, more or less changed; and one entire bed, from three to five feet in depth, is composed of the reeds so common in the island, completely mineralized, the centre of each joint being filled with delicate crystals of sulphur *."

The siliceous deposits are both abundant and various: the most abundant occur in layers from a quarter to half an inch in thickness, accumulated to the depth of a foot and upwards. The strata are nearly always parallel and horizontal, though sometimes slightly undulating. The silex forms stalactites, often two inches in length, in the cavities of the siliceous deposits, and these are

---

* Edin. Phil. Journal, vol. vi. p. 310.

frequently covered with small brilliant quartz crystals. Compact masses of siliceous deposits, broken by various causes, have been re-cemented by silica, and the compound is represented as very beautiful. Some of the elevations of this breccia Dr. Webster considers upwards of thirty feet in height. The general deposit appears to be considerable, and to form low hills. The colours of the clay and siliceous substances are very various, and even brilliant,—white, red, brown, yellow, and purple being the principal tints. Where the acid vapours reach the rocks, they deprive them of their colours. Sulphur is abundant, and the springs occur in a district of lava and trachyte *.

According to James †, the thermal springs of the Washita deposit a very copious sediment, composed of silex, lime, and iron. This shows that hot springs, when propelled through a non-volcanic district, may yet contain silica. The same may be said of some of the springs in India. Dr. Turner found that the thermal springs of Pinnarkoon and Loorgootha, in that country, which produced 24 grains of solid matter in a gallon, contained 21.5 per cent of silica, 19 of chloride of sodium, 19 of sulphate of soda, 17 of carbonate of soda, 5 of pure soda, and 15.5 of water‡. The following is an analysis of the Geyser waters and hot springs of Reikum, Iceland, by Dr. Black. A gallon of each produced:—

	Geyser.	Reikum.
Soda	5.56	3.0
Alumina	2.80	0.29
Silica	31.50	21.83
Muriate of soda	14.42	16.96
Sulphate of soda	8.57	7.53

These analyses do not show the presence of lime, but Sir G. Mackenzie mentions a calcareous deposit from boiling springs (temp. 212°) in the valley of Reikholt, in Iceland, charged with carbonic acid gas. Many thermal and other springs contain this gas, which seems very abundant in volcanic regions. To its power of dissolving lime, when passing through calcareous rocks, those deposits are due, that are so common in some countries, particularly when volcanic, which are known under the general name of Travertino or calcareous tufa. Probably, also, many hot springs may contain carbonic acid gas, which, not meeting with calcareous or magnesian strata, is thrown off when in contact with the atmosphere.

Travertins are of greater geological importance than the siliceous deposits from modern springs, at least so far as their extent

* Edin. Phil. Journal, vol. vi.
† Expedition to the Rocky Mountains.
‡ Elements of Chemistry.

of surface and depth are concerned; though both these have been greatly exaggerated, from the usual mode of comparing such deposits, not with the superficies of the land generally, but with their magnitude relatively to the valleys or plains in which they may occur, and not unfrequently with that of man himself.

The deposit from the fountain of Saint Allire, near Clermont, formed a bridge which was, in 1754, one hundred paces long, eight or nine feet thick at its base, and twenty or twenty-four inches in its upper part *.

Mr. Lyell notices the calcareous deposits from the baths of San Vignone, and states that one stratum, composed of several layers, is fifteen feet thick, and that large masses are cut out of it for architectural purposes †. According to Dr. Gosse, the thermal waters which deposit this travertino are sufficiently hot to boil eggs.

The thermal waters at the baths of San Filippo, not far from the above, have a temperature of 122° Fahr., one spring being about a degree or two higher. They contain silica, sulphate of lime, carbonate of lime, sulphate of magnesia, and sulphur; and, notwithstanding their elevated temperature, *Confervæ* flourish in them. The ground around is formed of travertino deposited by the springs. There are many fissures; one thirty feet deep, and 150 to 200 feet long. In it the water is whitish, and in a state of ebullition, whence its name, Il Bollore. It emits copious discharges of steam and sulphurous vapour. There are other fissures, in which sulphur is sublimed in the same manner as at the Solfatara near Naples, and the produce was sufficient to constitute a branch of industry, now however abandoned. The surfaces of these fissures are penetrated by sulphuric acid. Dr. Gosse observed the siliceous stalagmites mentioned by Professor Santi, and describes them as covering the surface of the travertino to the depth of one-eighth of an inch ‡. Mr. Lyell notices the spheroidal structure of the travertino deposited, and compares it with the magnesian limestone of Sunderland. What the amount of magnesia may be in the San Filippo travertino is not stated, but according to Dr. Gosse it is combined with sulphuric acid; while, if it resembled the magnesian limestone above cited, it should have existed as a carbonate. Moreover sulphate of lime exists in great abundance in these springs, so much so, that before the water is conducted to the places where the well-known medallions are formed, it is allowed to stagnate for the purpose of depositing the sulphate of lime. That the sulphates should be common, would be expected where so much sulphurous vapour is evolved, and it is even stated that sulphur exists in the travertino, though it is principally composed of carbonate of lime.

* Daubuisson, t. i. p. 142.   † Principles of Geology, p. 202.
‡ Gosse, Edin. Phil. Journal, vol. ii.

Deposits of travertino are by no means uncommon from cold springs in the Apennines, particularly near the volcanic region of southern Italy. The celebrated Falls of Terni are, as is well known, artificial, and have been formed by cutting through a previous calcareous deposit, to form a channel for the Velino, which now rushes over a precipice into the Nera beneath. Upon the flat land above, a considerable deposit of lime has taken place; —when, it does not so clearly appear, but probably since the establishment of the present order of things. Notwithstanding the velocity of the water, its cutting powers are trifling, and the upper channel preserves all the appearance of art. The Velino contains much carbonate of lime, which it deposits after the great leap, even in the bed of the Nera, which does not cut it off, but is obstructed to a certain degree by it, as may be seen at a place called the Bridge, over which I crossed the Nera, by taking one or two leaps at the chasms cut by the latter torrent. At this place there must be a constant struggle between the destructive power of the Nera, and the lapidifying power of the Velino. The country around exhibits abundant examples of calcareous deposits from springs charged with carbonate of lime. The usual explanation of this phænomenon seems very probable. It supposes the carbonic acid to be derived from the volcanic regions beneath, (and they appear not far distant on the surface,) which, passing with the water through the calcareous strata, dissolves as much lime as it can take up, giving off the excess of carbonic acid under diminished pressure in the atmosphere, and causing the carbonate of lime to be deposited. The carbonic acid found so abundantly in acidulous springs is ascribed by Von Buch, Brongniart, Boué, Von Hoff, and other geologists, to volcanic or igneous action at various depths beneath the surface. M. Hoffman has further shown that, in certain valleys of elevation, mineral springs are frequent, and cites the Valley of Pyrmont as a good example, where the waters are charged with carbonic acid gas *. In the marshy meadows of the Valley of Istrup (one of elevation), mounds of mud, from fifteen to twenty feet high, and 100 feet in circumference, are produced by currents of carbonic acid gas, and on their surface many small reservoirs of water are kept in a state of ebullition by bubbles of gas of the size of the fist †. After producing other examples of this evolution of carbonic acid gas, either combined with water, or nearly if not altogether free, M. Hoffman observes, that " the

---

* The following are the contents of these waters, according to Bergman, in a wine pint: Carbonic acid, 26 cubic inches; carbonate of magnesia, 10 grains; carbonate of lime, 4.5; sulphate of magnesia, 5.5; sulphate of lime, 8.5; chloride of sodium, 1.5; and oxide of iron, 0.6. —Henry's Elements, and Turner's Elements.

† Hoffman, Journal de Géologie, t. i., and Poggendorf's Annalen, 1829.

country situated on the left bank of the Weser in the direction from Carlshafen to Vlotho, up to the foot of the Teutoburg-Wald, may be compared to a sieve, whose apertures, as yet unclosed, permit the escape of gas, disengaged from volcanic depths by means unknown *."

The travertino of Tivoli, and the famous Lago di Zolfo, near Rome, have been much appealed to by those who ascribe all geological appearances to such causes only as are now in operation; but the former is a mere incrustation, considerable it is true in some situations, if measured by our own magnitude, but insignificant if compared with the country in which it occurs; and the latter is but a pond of water, dignified somewhat strangely by the name of a lake, and containing, according to Sir H. Davy, a saturated solution of carbonic acid, with a very small quantity of sulphuretted hydrogen. The spring is thermal, being about 80° Fahr.; plants thrive in and about it, and they are incased in stone beneath, while they vegetate above, and thus they may become fossil, their most delicate structure preserved, and their ramifications uncompressed.

All the examples hitherto produced of deposits, that can fairly be traced to existing springs, are relatively unimportant; and though they may lead us to understand how great geological deposits may, chemically, have taken place, as the cabinet experiments of the chemist teach us the laws which govern nature on a large scale, they no more could have produced the great limestone or siliceous deposits observed on the earth's surface, than the experiments above alluded to could produce the great chemical phænomena they illustrate, however long continued.

Mr. Lyell has presented us with an account of calcareous deposits in Scotland, which are remarkable, not for their extent, but for the circumstances which attend them. It appears that the Bakie Loch, Forfarshire, has produced a marl used in the agriculture of the country. The following is a section of the beds: 1. Peat, containing trees, one to two feet; 2. Shell-marl, containing in parts tufaceous limestone, provincially termed " *rock-marl,*" one to sixteen feet; 3. Quick-sand, without pebbles, cemented together in some places by carbonate of lime, two feet; 4. Shell-marl of good quality for agriculture, (almost every trace of shell is often obliterated,) one to two feet; 5. Fine sand, without pebbles, resting on transported detritus, at least nine feet. The rock-marl is limited to the vicinity of the springs, irregularly distributed over the lake. The Bakie shell-marl is white, with a yellow tint. The rock-marl has the same yellow tint, and consists almost wholly of carbonate of lime, compact, and even crystalline.

* Hoffman, Journal de Géologie, t. i., and Poggendorf's Annalen, 1829.

Organic remains of the marl. Horns of stags and bulls; wild boar tusks. *Cypris ornata,* Lam. *Limnæa peregra, Valvata fontinalis, Cyclas lacustris, Planorbis contortus, Ancylus lacustris,* all of Lamarck. Mr. Lyell considers this calcareous rock as not immediately due to the springs, but to have been produced through the agency of the testaceous inhabitants of the lake; for though the springs do contain lime, it is in such small quantities, that they could not directly produce the marl. He considers that the testaceous animals obtained the lime either from the water or from the *Charæ* which they fed upon, and that, dying, they left their calcareous exuviæ to form, by accumulation, the shell-marl, which was converted into calcareous rock by the action of the water upon it; the water containing carbonic acid, and forming a solution of carbonate of lime, which might produce a crystalline limestone. Seeds of *Charæ,* or *Gyrogonites,* are converted into carbonate of lime, in which the nut is sometimes found within; but commonly that space is empty, and the integument alone preserved. The *Chara* here found mineralized is the *Chara hispida,* a plant which now abounds in the Bakie Loch, and in the other lakes in Forfarshire. It contains such a proportion of carbonate of lime, as strongly to effervesce with acids when dried.

Mr. Lyell, noticing the deposits of marl in the Loch of Kinnordy, states that it is thickest at that end of the lake where the springs are most common. The shells are the same here as at the Bakie Loch, and are, like them, nearly all young, scarcely one in ten being full-sized. A large skeleton of a stag (*Cervus elaphus*) was dug out of the marl, and was remarkable as being found in a vertical position, the points of the horns being nearly at the surface of the marl, while the feet were about two yards below it. The marl is covered by peat, and in this peat were discovered other skeletons of stags, and (in 1820) the remains of an ancient canoe, hollowed out of the solid trunk of an oak *.

There is something in the formation of these lakes which reminds us strongly of the epoch of the submarine forests and of the lacustrine deposits of East Yorkshire, which will be noticed in the sequel; like them they seem to have succeeded the transport of detritus, and to have been gradually filled up, being surmounted by peat; previous to the formation of which latter production man certainly was an inhabitant of these islands, as his works are entombed in it : the lakes being then, probably, more or less open spaces of water, or else his boat would have been of little service to him.

Naphtha and Asphaltum Springs.

These are distributed over various parts of the world, and cannot be considered as rare. According to Dr. Holland, the

---

* Lyell, Geol. Trans. 2nd series, vol. ii.

petroleum springs of Zante are much in the same state as in the time of Herodotus. They are situated on a small marshy flat, bounded by the sea on one side, and by limestone and bituminous shale-hills on the others. The principal pool is about 50 feet in circumference, and a few feet deep : the sides and bottom of this and the others are thickly covered with petroleum, which by agitation is brought to the surface of the water, and collected: The amount obtained is estimated at 100 barrels annually *.

James states that about 100 miles above Pittsburgh, and near the Alleghany river, there is a spring, on the surface of which float such quantities of petroleum, that a person may collect several gallons in a day. He considers that it may probably be connected with coal strata, as is the case with similar springs in Ohio, Kentucky †.

The pitch lake of Trinidad, estimated at about three miles in circumference, has long been celebrated. According to Dr. Nugent, the asphaltum is sufficiently hard in wet weather to support heavy weights, but during the heats it approaches fluidity. It is intersected with numerous cracks filled with water; and it appears that these cracks sometimes close up again, leaving marks on the surface of the pitch lake. When slightly covered with soil, as it is in some situations, good crops of tropical productions are obtained. From this covering of soil it is difficult to estimate the exact boundaries of the lake ‡

Large quantities of naphtha are obtained on the shores of the Caspian. The inhabitants of the town of Badku, a port on that sea, are supplied with no other fuel than that obtained from the naphtha and petroleum, with which the neighbouring country is highly impregnated. In the island of Wetoy and on the peninsula of Apcheron, this substance is very abundant, supplying immense quantities which are taken away. Thermal springs are found near those of naphtha §.

The naphtha springs at Rangoon, Pegu, appear to be exceedingly abundant. Mr. Coxe estimates their produce at 92,781 tons per annum. In the Indian Islands there are also similar springs. Marsden notices them in Sumatra, at Ipu, and elsewhere.

Coral Reefs and Islands.

In consequence of the numerous situations where these are observable in the Pacific Ocean and Indian Seas, very exaggerated ideas have very generally been entertained of their relative importance. Large masses, supposed to be the work of myriads of polypifers, were considered to have been raised by the labour of these animals from great depths, while immense sheets of coral

* Holland's Travels in the Ionian Isles, Albania, &c.
† Expedition to the Rocky Mountains.
‡ Nugent, Geol. Trans. vol. i.   § Edin. Phil. Journal, vol. v.

rock were supposed to cover the bottom of the seas. During Kotzebue's voyage, M. Chamisso enjoyed opportunities of visiting some remarkable groups of islands, arranged in a circular or oval manner, with openings among them which permitted the passage of a vessel from the outer ocean into the central basin. These islands seemed merely higher portions of a circular or oval ridge of coral reefs of unequal heights. M. Chamisso presented a description of what he considered the stages which the coral reef passed through before it became an island habitable for man. This description has been so often quoted that it must be familiar to most readers.

Subsequently to Kotzebue's voyage, MM. Quoy and Gaimard, who sailed with the expedition of M. Freycinet, paid particular attention to the coral islands and reefs which they had opportunities of examining; and the result of their observations was, that the geological importance of these islands and reefs had been greatly exaggerated. Far from supposing that the polypifers raise masses from great depths, they consider that they merely produce incrustations of a few fathoms in thickness. In those situations where the heat is constantly intense, and where the land is cut into bays, with shallow and quiet water, the saxigenous polypi increase most considerably, incrusting the rocks beneath. The same authors observe, that the species which constantly formed the most extensive banks belong to the genera *Meandrina, Caryophyllia,* and *Astrea,* but especially to the latter; and that these genera are not found at depths exceeding a few fathoms. It is therefore concluded, that unless we are to suppose these animals enjoying the prerogative of inhabiting all depths, under various pressures of water, and different temperatures, they cannot have produced the masses attributed to them. From these and other considerations they infer, that the appearance of coral reefs and islands depends on the inequalities of the mineral masses beneath, the circular character of some being due to the crests of submarine craters *. This conclusion seems far from improbable, for we know that volcanic vents are common in the same seas; and that in the West Indies, and the tropical parts of the Atlantic, where corals are sufficiently numerous, we do not observe these circular groups of islands, where volcanic vents, though existing, are far from attaining the importance of those in the Pacific Ocean or Indian Seas.

MM. Quoy and Gaimard observe, that, neither with the anchor nor the lead, have they ever brought up fragments of Astreæ, alone capable of covering large spaces, except where the water was shallow, about twenty-five or thirty feet in depth, though they

* Quoy et Gaimard, Sur l'Accroissement des Polypes Lithophytes considéré géologiquement, Ann. des Sci. Nat. tom. vi.

found that the branched corals, which do not form solid masses, lived at great depths *. They agree with Forster, that the poly-pifers may form small isles, when masses of land shelter them, by raising their habitations to the level of the sea: thus ex-posing a surface on which sands and other matters are heaped and consolidated: a mode of formation in accordance with what I have observed on the coasts of Jamaica.

With regard to the great depth of water frequently observed close to the coral reefs, the same authors consider, that they may be accounted for on the supposition that the polypifers have erected their dwellings upon the verge of a steep cliff, such as is commonly observed on the sides of mountains and coasts. In support of this opinion they cite the isle of Rota; where corals, resembling those now found in the neighbouring seas, occur on cliffs. There are, however, certain situations where coral reefs run, as it were, in a line with a coast, but separated from it by deep water, which would seem to require a different explanation.

In situations such as those in which these coral isles and reefs abound, where recent, and comparatively recent, volcanic action is so apparent, we should expect to find evidences of the rise of such reefs above the level of the sea; and, accordingly, navigators have presented us with them. MM. Quoy and Gaimard state, that the shores of Coupang and Timor are formed of coral beds, which induced Peron to consider that the whole island was the work of polypifers. But it appears, that, proceeding towards the heights, vertical beds of slate, traversed by quartz, are met with at about five hundred yards from the town; and upon these and other rocks do the coral beds rest, which MM. Quoy and Gaimard estimate as not exceeding twenty-five or thirty feet in thickness. At the Isle of France a similar bed, more than ten feet thick, occurs between two lava-currents; and at Wahou, one of the Sandwich Isles, coral beds extend some little distance into the interior. To this we may add, that round the east coast, and on the northern side of Jamaica, there is an extensive bed that merely fringes the land, about twenty feet thick, which has every appear-ance of a coral bank raised above the waters, and brought within the destructive action of the breakers.

In situations like those in the Pacific, where volcanos and coral reefs are both abundant, we should expect to find some curious combinations of volcanic matter with coral banks, and even alter-nations; even admitting, for the argument, that the principal rock-forming polypifers do not build beneath twenty-five or thirty feet

---

* Sounding off Cape Horn at about 56° S., and in about fifty fathoms of water, they brought up small live branched corals; and sounding in one hundred fathoms on the bank of Laghullas (off the southern point of Africa) they obtained *Reteporæ.*

of water, still with the movements of land which may accompany
volcanic action, such banks may be depressed, and covered by
lava-currents, and again raised and brought to view. The example
adduced in the Isle of France is sufficient to show, that at least
one coral bed may be inclosed between lava-currents.

We cannot conclude this sketch without noticing a singular fact,
observed, as we have been informed, by Mr. Lloyd while engaged
in his survey of the Isthmus of Panama. Seeing some beautiful
polypifers on the coast, he detached specimens of them; and, it
being inconvenient to take them away at the time, he placed
them on some rocks, or other corals, in a sheltered and shallow
pool of water. Returning to remove them a few days afterwards,
it was found that they had secreted stony matter, and fixed themselves firmly to the bottom. Now this property must greatly
assist in the formation of solid coral banks; for if pieces of live
corals be struck off by the breakers, and thrown over into calm
water or holes, they would affix themselves, and add to the solidity
of the mass.

Submarine Forests.

At various points round the shores of Great Britain, and the
northern parts of France, accumulations of wood and plants, which
do not appear to differ from those now existing, but on the contrary to be identical with them, occur at levels beneath those of
high-water, so that the wood and plants thus situated could not
have grown at the present relative levels of sea and land. To
these ligneous and other vegetable remains, which are commonly
seen at the retreat of the tide, a temporary removal of the beach,
or an encroachment of the sea on tracts of land but slightly raised
above the sea, the name of *Submarine Forests* has been given.
To explain this phænome nonvarious hypotheses have been framed;
but probably that which attributes it to a subsidence of the land,
consequent on earthquakes or internal movements of the earth, is
most consonant with known facts and general geological appearances. This explanation was proposed by Mr. Correa de Serra in
1799, and was still further improved by Playfair, who considered
such subsidences as merely forming a part of those depressions
and elevations of the land, which alternately convert it into the
bed of an ocean or into continents and islands.

Correa de Serra describes the submarine forest on the coast of
Lincolnshire as composed of the roots, trunks, branches, and
leaves of trees and shrubs, intermixed with aquatic plants; many
of the roots still standing in the position in which they grew,
while the trunks were laid prostrate. Birch, fir, and oak were
distinguishable, while other trees could not be determined. In
general, the wood was decayed and compressed, but sound pieces

were occasionally found, and employed for œconomical purposes by the people of the country. The subsoil is clay, above which were several inches of compressed leaves, and among them some considered to be those of the *Ilex aquifolium*, as also the roots of *Arundo phragmites*.

These appearances are not confined to the coast, but extend considerable distances into the interior, so that the former merely presents a natural section of that which occupies a large area inland. A well sunk at Sutton afforded the following section.

1. Clay . . . . 16 feet.
2. Substances similar to the submarine forest . 3 to 4 feet.
3. Substances resembling the scouring of a
       ditch-bottom, mixed with shells and silt . 20 feet.
4. Marly clay . . . . 1 foot. -
5.* Chalk rock . . . . 1 foot to 2 feet.
6. Clay . . . . 31 feet.
7. Gravel and water . . . . Not known.

Another boring made inland by Sir Joseph Banks afforded a similar section. This "moor" as Correa de Serra terms it, is considered to extend to Peterborough, more than sixty miles south from Sutton †.

Mr. Phillips presents us with very interesting details respecting some lacustrine deposits in Yorkshire, which are apparently of the age of these submarine forests, and which have become in some places submerged. He remarks that the following may be considered as their general section. 1. Clay, generally of a blue colour and fine texture. 2. Peat, with various roots and plants ; and in large deposits, containing abundance of trees, nuts, horns of deer, bones of oxen, &c. 3. Clay of different colours, with fresh-water *Limnææ*. 4. Peat, as above. 5. Clay, with fresh-water *Cyclades*, &c. and blue phosphate of iron. 6. Shaly curled bituminous clay. 7. Sandy coarse laminated clay, filling hollows in the diluvial formation. Mr. Phillips considers the accumulations of peat along the banks of the Humber and its tributaries as of the same epoch as these deposits, of which, he observes, the most constant beds are Nos. 1, 2, and 5. The species of deer enumerated as found in the peat, are,—the great Irish elk (*Cervus giganteus*), the red deer (*C. elaphus*), and the fallow deer (*C. Dama*). The peat deposit of the marsh-lands is covered by silt and clay, sometimes thirty feet thick, such as is now deposited by the Humber‡. The peat is represented as beneath low-water mark,

* This would seem not to be chalk, properly so called, but merely a chalky substance.
† Correa de Serra, Phil. Trans. 1799.
‡ Phillips, Illustrations of the Geology of Yorkshire, 1829.

therefore the change of the relative level of the land seems as certain here as in the other localities to be hereafter noticed.

Dr. Fleming describes a submarine forest on the shores of the Frith of Tay, extending in detached portions on each side of Flisk beach, three miles to the westward and seven miles to the eastward. It rests on clay of unknown depth. The clay is similar to the carse ground on the opposite side of the Frith, and to the banks in the channel. " The upper portion of this clay has been penetrated by numerous roots, which are now changed into peat, and some of them even into iron pyrites. The surface of this bed is horizontal, and situated nearly on a level with low-water mark. In this respect, however, it varies a little in different places. The peat bed occurs immediately above this clay. It consists of the remains of leaves, stems, and roots of many common plants of the natural orders *Equisetaceæ, Gramineæ,* and *Cyperaceæ,* mixed with roots, leaves, and branches of birch, hazel, and probably also alder. Hazel-nuts destitute of kernel are of frequent occurrence. All these vegetable remains are much depressed or flattened where they occur in a horizontal position, but when vertical, they retain their original rounded form. The peat may be easily separated into thin layers, the surface of each covered with leaves. The lower portion of this peat is of a browner colour than the superior layers; the texture is likewise more compact, and the vegetable remains more obliterated*."

The same author further observes, that stumps of trees, with the roots attached, are observed on the surface of the peat, and no doubt can exist that they are in the positions in which they grew. No alluvial soil stratum was observed above the peat, the surface of which does not occur at a higher level than from four to five feet *below high-water mark.*

Dr. Fleming also describes another submarine forest in the Frith of Forth, at Largo Bay. It rests on a brown clay, into which the roots of the trees have penetrated. The author considers it as lacustrine silt. Over this there is an irregularly distributed covering of sand and fine gravel. The peat is composed of land and fresh-water plants, among which are the remains of birch-, hazel-, and alder-trees; hazel-nuts are also seen. Dr. Fleming traced the root of one tree, apparently an alder, more than six feet from the trunk†.

If we pass from the main land of Scotland to its isles, we shall observe that the same appearances present themselves. Mr. Watt notices a submarine forest in the bay of Skaill, on the west coast of the mainland of Orkney. Stems of small fir-trees, ten feet long and five or six inches in diameter, are found partly imbedded in,

* Trans. Royal Soc. of Edinburgh, vol. ix.
† Journal of Science.

H

and partly resting on, the surface of an accumulation of vegetable matter principally composed of leaves. The stems were still attached to their roots, and the whole was greatly decayed, so as to be easily cut by the spade. Many seeds of the size of a turnip-seed were discovered among the vegetable matter *.

The Rev. C. Smith describes a submarine forest on the coast of Tiree, one of the Hebrides. Beneath a plain of 1500 acres in extent there would appear to be moss-land, similar to that previously noticed, under twelve or sixteen feet of alluvial covering. The moss-land is seen to bound the plain on the east, and the bay in which it appears is open to the whole force of the Atlantic. The general depth of the peat or moss-land amounts to several feet, but at its appearance on the shore it does not exceed four or five inches. This is firm, and adheres strongly to a sandy clay, on which it is based. Besides the remains of trees, which are obvious, there are other and smaller plants, and numerous seeds, which at first looked quite fresh, but afterwards became darker from exposure. " The seeds have the appearance of belonging to some plant of the natural order of *Leguminosæ;* and Mr. Drummond suggests that they may probably be those of *Genista anglica*†:"

According to the same author, submarine forests are by no means uncommon on the shores of Coll. He also cites the Rev. H. Maclean as having noticed similar appearances, not observed by himself in the island of Tiree.

Returning again to the main land, we find similar appearances, described by Mr. Stephenson, on the shores of the flat lands between the Mersey and the Dee, on the coast of Cheshire. Stumps of trees, ramifying in all directions, are stated to appear as if cut off about two feet from the ground. The vegetable matter rests on bluish marl, and is covered by sand‡.

Mr. Horner describes a submarine forest on the coast of the S.W. part of Somersetshire. It is well seen between Stolford and the mouth of the Parret, where the shore is low; a high shingle beach, principally composed of lias (the rock of the vicinity), protects the level land behind from the sea. The vegetable remains present themselves here, as in the other places, as a stratum of peat or decayed leaves, containing the trunks, stems and branches of trees. Among these are twigs, nuts, and a plant, (commonly found entire,) which Mr. Brown considered might be the *Zostera oceanica* of Linnæus. Some of the stems of trees were twenty

---

* Edin. Phil. Journal, vol. iii. p. 100.

† Smith, Edin. New Phil. Journal, 1829.

‡ Edin. Phil. Journal, vol. xviii. Mr. Smith cites the Liverpool Courier of December 1827, to show that after a heavy gale, trunks and roots of trees were found under the sand below high-water mark, which had all the appearance of having grown where then found.

feet long, and the woods were considered to be oak and yew, not generally decayed, but sufficiently hard and tough to be used as timber, and for fuel. Even those trees which were soft when taken out, became hard when dried. The brown vegetable matter was generally a foot or eighteen inches thick, and rested on blue clay *.

From this coast there is an extensive tract of flat land, which extends a considerable distance inland, and from it the hills rise in promontories, islands, and other forms, precisely as they would rise from a level sea. Mr. Horner cites De Luc as stating, that while new channels were digging between the Brue and the Axe, in the vicinity of Bridgewater, a bed of peat was found beneath the surface. This stratum, if it may be so called, has been noticed in other parts of the same flats, and even trees have been reported as found in it; seeming to show that the forest noticed on the shore may be only a section of a large deposit beneath the Bridgewater levels.

A very important addition to our knowledge of submarine forests has been made by Dr. Boase in his description of that in Mount's Bay, Cornwall. The vegetable bed consists of a brown mass, composed of the bark, twigs, and leaves of trees, which appear to be almost entirely hazel. In this there are numerous branches and trunks of trees. The greater part of this wood is hazel mixed with alder, elm, and oak. "About a foot below the surface of this bed, the chief part of the mass is composed of leaves, amongst which hazel-nuts are very abundant. In this layer may also be found filaments of mosses, and portions of the stems and seed-vessels of small plants, many of them evidently belonging to the order of Grasses; together with the fragments of insects, particularly of the elytra and mandibles of the beetle tribe, which still display the most beautiful shining colours when first dug up, but on exposure to the air all these minute objects soon crumble into dust." Beneath this, the vegetable matter becomes closer, and finally earthy and of a lamellar structure. It rests on granitic sand, and this again on clay slate. The vegetable stratum slopes from the interior to the sea at about an angle of two degrees. It is covered by a bed of smoothly polished shingles, about two or three inches in diameter, composed of hornblende rock fragments about sixteen feet thick, which is crowned by a granitic sand about ten feet thick. The vegetable bed, by its rise, appears beneath a marsh inland, having passed under its covering of pebbles and sand †.

M. De la Fruglaye observed that after a heavy gale in 1811, a beach near Morlaix, which previously seemed to consist of sand,

* Horner, Geol. Trans. vol. iii. p. 380, &c.
† Boase, Trans. Geol. Soc. Cornwall.

presented, from the sand being washed away, an appearance of a large mass of vegetable matter and trees united together, and extending along shore for a considerable distance. The leaves were well preserved, but the trunks and branches of trees were rotten. Oak was observed among the wood, and insects with their colours preserved were discovered in the mass. A few days after this event this accumulation of vegetable matter was again covered up by sand*.

Having cited so many examples to show their general similarity, I shall merely notice that I have observed submarine forests on the coasts of Normandy, one to the east of the Vaches Noires cliffs, and the other near St. Honorine, both at the mouths of valleys; and that at the mouth of the Char, coast of Dorset, there are traces of another.

That there has been a change in the relative levels of land and water since these trees and plants vegetated, cannot be doubted, but the manner in which this was produced may admit of a question. From the subsidences sometimes caused by earthquakes, we may presume that Great Britain, with the Shetland Isles, Hebrides, and the north coast of France, have subsided. But if this had taken place suddenly by a violent earthquake, great waves would have been produced, and in that case, the lighter vegetable substances, such as leaves, which constitute so large a proportion of all these deposits, must, one would suppose, have been swept away. Now this is not the case; from which it might be presumed that the relative change of level has been somewhat gradual, though the apparently snapped trees do not quite accord with this supposition, but rather with something sudden, more like a tornado or wave consequent on an earthquake. It may also be supposed that a gradual rise of the sea, which would accumulate protecting banks in front of low land, the breakers propelling them forward as they occupied a higher level, might be the cause. Whatever hypothesis approaches nearest the truth, a change has taken place in the relative levels of sea and land round Great Britain and on the north coast of France since the establishment of climates differing little from, if they were not exactly the same with, those now existing. The absence of marine remains seems to show that the forests were not suddenly overwhelmed by the sea, for had this been the case, some vestiges of its former presence must have appeared. If a sudden rise of land were again to restore the relative levels, the residence of the forests beneath the sea would be very apparent, various marine substances being attached to the trees, which are not uncommonly perforated by the *Pholas.* The above details are perhaps too copious for the plan of this volume; but it seemed important to show that changes

---

* Journ. des Mines, t. xxx.

in the relative levels of the ocean and land had taken place round our own shores at such geologically recent times, more particularly as it will be attempted to prove beneath that at least a partial difference of levels on our southern shores, of quite a contrary kind, has preceded it.

Raised Beaches and Masses of Shells.

At Plymouth and the neighbouring coast there are the remains of a beach, of which the maximum elevation is about thirty feet above high-water mark, sloping gradually to the sea*. The following is a section at the Hoe.

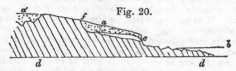

Fig. 20.

Upon the grauwacke limestone beds *d d,* which dip at a considerable angle southwards, rests an accumulation (*c*) of rounded pebbles and sand, with here and there a larger and angular piece of limestone intermixed. The accumulation has every appearance of a sea-beach raised above the present level of the sea *b*, and the shingles and sand are so arranged that the resemblance is quite perfect, more particularly when shells are found in it†. The shingles consist of limestone, slate, red sandstone, reddish porphyry (which occurs, in place, in another part of Plymouth Sound), and of various rocks that form part of the grauwacke series of the neighbourhood. The section annexed is exposed by blasting the rock, the limestone being taken away in great quantities. It will be observed that the beach (*c*) did not extend to *f*, which seems formerly to have been a cliff, in the same manner as the present beach is backed by a low cliff. The beach and part of the limestone hill are covered by a gravel or loose breccia of angular limestone fragments *a a',* which clearly have not received attrition from the action of water upon them. This circumstance seems to

---

* Professor Sedgwick informs me that the Rev. R. Hennah pointed out this beach to him several years since; and Mr. Hennah has noticed it in his account of the Plymouth limestones.

† I was only fortunate enough to see fragments, and these apparently consisted of pieces of Patellæ and small Nerites, the latter with their colours preserved, and resembling those now found on the coast; but many hundreds were found in a cavity of the limestone filled with sand and thrown away by the quarry-men. Beneath the citadel the sand is composed of fragments of shells.

afford us a relative date for the beach, as the reader will recollect that under the head of degradation of land it was observed that the whole of this part of Devon afforded a superficial detritus of the rocks beneath. Now the angular pieces of limestone (*a*) are derived from the hill above, and have slipped by the force of gravity, assisted by meteoric causes, over the beach *c*, as they have also fallen into the cavity *a′*, which being above the old beach *c*, does not contain either pebbles or sand, but is precisely similar to those clefts in the Oreston quarries near Plymouth, where the remains of elephants, rhinoceroses and other animals, occur beneath fragments of the same kind. It therefore seems fair to infer that the beach was raised during the existence of these animals, and previous to that long period of time, during which the action of the atmosphere slowly, though considerably, destroyed the surface of the hills. It seems, moreover, to show a configuration of land in the vicinity not very different from that which now exists. This view is strengthened by a minute examination of the coast from hence towards Tor Bay, along which, similar appearances may here and there be seen; but it must be evident that this will depend upon the quantity of cliff cut back by the sea at its present level, as will be seen by the annexed diagram.

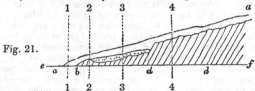

Fig. 21.

Thus, if *a a* represent the angular detritus derived from the slate rock *d d*, which rises into a high hill behind, and *b* an ancient beach now raised above the level of the sea *e f*, and covered by the detritus *a a*, a section made at 1 1 would only show the detritus; another made at 2 2 would only expose detritus or slate, the bottom rising, as is very commonly the case on sea-beaches where rocks project among the shingles, particularly on coasts such as are now under consideration. If the sea cut back the cliff to 3 3, we should have such a section as that at the Hoe; but if the cliff be cut to 4 4, the whole beach is removed, and no traces of it left. This is precisely what happens on this coast, where all the above varieties are observed. Under Mount Edgecumbe, near Plymouth, the rolled shingles are covered by fragments of slate and red sandstone. At Staddon Point, sand is covered by compact red sandstone fragments. Further south, on the eastern side of the Sound, and nearly opposite the Shag Rock, there is the following section, which may or may not be ancient beach covered

with detritus: *c*, the main rock of argillaceous and arenaceous schist; *b*, detritus, some of which approaches a
sandy earth, mixed with small pieces
of slate rarely exceeding the size of a
shilling or sixpence; *a*, detritus, com-
posed of angular pieces of schist and sandstone of the size of an
egg and upwards, mixed with others of smaller dimensions.

Fig. 22.

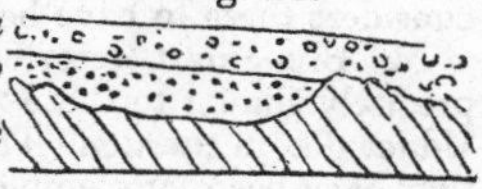

From the apparent date of the elevated Plymouth beach, this
notice might perhaps have been more in its place in the next
section, but it seems so intimately connected with the subject of
the alternate rise and fall of land, that it seemed better to let it
follow the notice of submarine forests. The conclusions from both
phænomena, which should by no means be hastily generalized,
but confined for the present to the places noticed, would seem to
be:—1. A configuration of land not greatly differing from the present, when elephants and rhinoceroses, *perhaps*, existed in this
climate. 2. The beach elevated. 3. A considerable but quiet
destruction of the surface of the hills, covering over the ancient
beach, the general shape of the hills and valleys being not very
different from those we now see. 4. A depression of the land,
submerging woods and forests, and bringing the detritus of epoch
3 into destructive contact with the sea, from which it was in
a great measure previously protected by the usual slopes and
beaches; and, 5. The changes effected since the establishment
of the present relative levels of sea and land.

Captain Vetch describes six or seven terraces or lines of beach
on the Isle of Jura in the Hebrides, which appear to have been successively raised above the present level of the ocean. The lowest
is on a level with high water, the most elevated about forty feet
above it. The terraces or beaches rest partly on the bare rock,
and partly on a thick compound of clay, sand, and angular pieces
of quartz. Their continuity is here and there interrupted by
mountain torrents, or the action of the sea on the supporting
compound. They are well seen at Loch Tarbert. Their aggregate breadth varies "according to the disposition of the ground:
where the slope is precipitous, it may be a hundred yards; where
gentle, as on the north side of the loch, three quarters of a mile
from the shore." These terraces or beaches are formed of round
smooth white pieces of quartz, of the size of cocoa-nuts. They
are precisely similar to those which constitute the present beach
of the Atlantic on this side of the island, and from their forms
they must have been produced by the united action of tides and
waves. Captain Vetch mentions, in confirmation of this opinion,
that a series of caves is to be found on the same level along the
north side of Loch Tarbert, at a considerable height above the sea;
and as he never observed any caverns formed in the quartz rock

of Isla, Jura, and Fair Island, except those on the sea shore, he considers these to have been thus produced*.

M. Brongniart describes a singular accumulation of shells, precisely similar to those which exist in the neighbouring sea, at Uddevalla, in Sweden. Their abundance is very considerable, for they have been long employed on the roads; they are nearly free from any earthy mixture, and though many are broken, there are numbers entire. The largest mass rises among gneiss rocks, to the height of sixty-six metres above the level of the sea. This author, considering that he might find traces of the former residence of the sea upon the fundamental rock, gneiss, searched around with considerable attention, and was rewarded by the discovery of *Balani* still adhering to the rocks on which they grew, now become the summit of a hill. MM. Berzelius, Wöhler, and Ad. Brongniart, were present at this discovery †.

At Nice the sub-fossils of St. Hospice have long attracted attention; they correspond with the present inhabitants of the Mediterranean, and often retain their colours, though they are generally blanched. Of these shells M. Risso has given a long list‡. From personal observation, I have little doubt that the whole has been raised, in comparatively recent times, above the present level of the Mediterranean. Beneath Baussi Raussi, a neighbouring cliff, and from thence to the principal deposit of sub-fossil shells, there is apparent evidence of a raised beach, the pebbles being rounded, and intermixed with sand, in which shells similar to those now existing in the neighbouring sea are discovered. Indeed, between the peninsula of St. Hospice and the cliff above mentioned, the old beach much resembles that near Plymouth, with the exception that the latter has been higher raised§. The elevation near Nice must have taken place after the land had, in a great measure, received its present configuration.

It has been previously observed, that on the west coast of South America a beach was raised during the earthquake of 1822, and there were evidences of former beaches having been so elevated. M. Lesson also observed at Conception, more southerly on the same coast, banks of shells, corresponding with those of the neighbouring sea, now dry and raised above it ‖.

It is almost impossible not to remark in these raised beaches and sea beds, the action of the same forces which have been no-

---

* Vetch, Geol. Trans. second series, vol. i.

† Brongniart, Tableau des Terrains qui composent l'Ecorce du Globe, p. 89.

‡ Hist. Nat. de l'Europe Meridionale.

§ For a more detailed description of these localities, with a view and a section of Baussi Raussi cliff, see my paper in the Geological Transactions, vol. iii. second series.

‖ Brongniart, Tab. des Ter. qui composent l'Ecorce du Globe, p. 92.

ticed under the head of Earthquakes. The land has been liable to rise and fall at various epochs, as will be seen in the sequel ; the intensity of the force, producing these changes, varying materially. It is exceedingly difficult to assign dates to the Plymouth raised beach, to the shells at Uddevalla, and to the other similar appearances above noticed ; but we learn from them, that since the establishment of animal life, such as we now observe it, the relative levels of the sea and land have been liable to change, as they have been previously to this period, and referring to the Temple of Serapis, near Naples, as they have been even since man has erected his temples and other works of art*.

Organic Remains of the Modern Group.

These will necessarily consist of existing animals, but may also include some no longer found in a living state. Man not only greatly modifies the present surface of the land, by destroying tracts of forests, preventing the inundations of low countries, turning torrents, and directing the surface water through innumerable channels to satisfy his own wants and conveniences, but he also drives all animals before him which do not suit his purposes; thus circumscribing the domain of those which are not useful to him, while he covers the country with those that are, and which never could exist in such numbers but for his care and protection. Consequently all terrestrial remains would correspond with the increasing power of man, and therefore a very different suite of such remains would be now entombed, than when his power was more limited. Over the inhabitants of the waters he would exercise little controul, excepting in rivers, small lakes, and around some coasts.

One very material difference would be effected in the quantity of trees and shrubs transported to the sea, more particularly in the temperate and colder regions, where man requires wood, not only for the purposes of various constructions, but also for fuel. We see in the delta of the Mississippi what an abundance of wood

---

* For a detailed account of the geological appearances connected with the celebrated Temple of Serapis, at Puzzuoli, near Naples, consult Lyell's Principles of Geology, vol. i. p. 450—459. The rise and fall of land seem to have been as follows : 1. After the original building of the temple, a sinking of the land, and a covering of the lower part of the columns, so that the boring shell (Lithodomus) only attacked them about twelve feet above their pedestals. The height to which the shells have bored is also about twelve feet ; therefore the columns, without being overthrown, were certainly lowered to the depth of twenty-four feet above their pedestals in water. 2. Elevation of the temple, still standing, above the level of the sea, or nearly so, for the pavement is not flooded to any considerable depth, not more than about one foot.

is now transported there by the river, but which will daily diminish as man converts the forests, whence it is derived, into pastures and corn-fields.

The gigantic animal *Cervus giganteus*, commonly known as the Irish Elk, was once imagined to have existed only at an epoch anterior to man, but it is now considered that he was co-existent with him; although this by no means proves that it did not live upon the earth previous also to him, as seems to have been the case. We have no great certainty when the Mastodons of North America ceased to exist; it is commonly supposed that they became extinct previous to the commencement of the modern group, but of this we have no good proof. The same may be said of some other animals.

The Dodo seems to afford us an example of the extinction of an animal in comparatively recent times; for it is now almost certain that this curious bird existed on the isle of Mauritius, during the voyages of the early navigators to the East Indies. The relative antiquity, therefore, of animals whose remains are only now found entombed, must not be too hastily inferred. The bone of the wolf is that of an extinct animal, as far as the British islands are concerned. In the darkness of ages many animals may have perished, not a tradition of whose existence remains, not only from the advance of man, and the power which civilization affords him, but also from the destruction caused by predaceous animals, though the latter is not so probable as the former*.

* The Rev. R. Hennah possesses some curious fossils in his collection at Plymouth, which appear to be the eggs of the snake enveloped in stalagmite. There are several of them, which have apparently formed a chain, and have given way before pressure, as eggs of snakes would do. They occurred in a cleft of the limestone rock, at the Oreston quarries, near Plymouth. The stalagmite has not entered into the interior of the egg, which presents a spongy structure.

SECTION III.

ERRATIC BLOCK GROUP.

WE must impress upon the geological student the necessity of considering this group as simply one of convenience, formed provisionally for the purpose of presenting certain phænomena to his attention, which in the present state of science could not so easily be done under any other head. The origin of the various transported gravels, sands, blocks of rocks, and other mineral substances scattered over hills, plains, and on the bottoms of valleys, often referred to one epoch, may belong to several. In a word, all that transported matter commonly termed *Diluvium,* requires severe and detailed examination. At the present time, there would appear to be three principal opinions connected with the subject. One, supposing the transport to have been effected at one and the same period; another, that several catastrophes have produced these superficial gravels; while a third would seem to refer them to a long continuance of the same intensity of natural forces as that which we now witness. Perhaps these various opinions may arise from our present inadequate knowledge of the phænomena on which we attempt to reason, and probably also from premature generalizations of local facts. These different opinions, though they cannot each be correct in explanation of all the observed facts, may all be so in part; and it were to be wished that the phænomena here arranged under one head solely, as above stated, for convenience, were examined without the controul of a preconceived theory.

At the close of the last section, a local elevation of land was noticed, of somewhat difficult arrangement in our systems. In order to illustrate the changes which have taken place in the same district, without, however, attempting to consider such appearances as general, I shall continue the description of it. At Oreston quarries, Plymouth, clefts and caverns in limestone rocks have afforded numerous remains of the elephant, rhinoceros, bear, ox, horse, deer, &c. buried, more particularly in the case of clefts, beneath considerable angular masses and smaller fragments of limestone. In one instance which I noticed, the animal remains occurred beneath ninety feet of such accumulations, the bones and teeth being confined to a black clay under the fragments. The remains of bears, rhinoceroses, hyænas, and other animals contained in the celebrated Kent's Hole, near Torquay, belong to the same district. In the superficial gravel of this part of the

country, the remains of animals, of the same kind as those detected in the caverns, have not yet been discovered; but if we
continue our researches eastward, we shall find them in the valleys of Charmouth and Lyme*, where they occur in situations
which would appear anterior to the great weathering, if I may
so express myself, of the circumjacent hills; thus apparently
giving these remains of elephants and rhinoceroses the same relative antiquity as those beneath fragments in the clefts of rocks
near Plymouth, and probably also as those contained in the
caverns at the same place, and in Kent's Hole. Now the raised
beach in Plymouth Sound seems to afford evidence of a configuration of land not widely different, in that place, from the present, and therefore we may perhaps infer the existence of inequalities in the land, or hill and dale, in this district generally,
not widely different from the present. It will be remarked that
the animal remains which seem to imply a warmer climate
existing at that time than at present, occur in low grounds, fissures, and caves. Upon the former they may have lived, and
into the two latter they may have either fallen or been dragged
by beasts of prey. The elephants probably browsing on branches
and herbage, rhinoceroses preferring low grounds, the bears and
hyænas inhabiting caves, and the deer, the ox, and the horse,
ranging through the forest and the plain; all which supposes
land fitted for them, and therefore hill and dale, level plains and
rocky escarpments with open caverns. Consequently valleys were
scooped out previous to the existence of the elephants; and if a
mass of waters acted on the land, destroying these animals, it
must have been influenced in its direction by the previously existing inequalities of surface.

The next question may be, does this district present evidences
of the exertion of a greater intensity of natural force than that
which we now observe? The answer may be, that it does. The
whole district is fractured, or, to use geological terms, so broken
into faults, that the spaces in which, with careful examination, they
may not be detected, are very inconsiderable. Such dislocations
may, or may not, have been contemporaneous with the raised
beach. Perhaps they were previous to it, for there has evidently
been a very considerable dispersion of rock fragments, and this
apparently by water, which would have scattered such a beach as
that noticed at Plymouth. The following section at the Warren
Point, near Dawlish, is not only a good example of a compound
fault, but also of transported gravel upon it.

* The line of coast has been preferred in this description, because the
sections are there more clear and less equivocal.

b b, conglomerates, and *c c*, sand-stones of the red sandstone formation, fractured or broken into faults at *f f*, so that continuous strata are displaced. Upon these fractured strata rests a gravel, *a a*, composed of chalk flints, and green sand chert, mixed with a few pebbles similar to those in the conglomerates *b b*. It has evidently been deposited subsequent to the fracture, for it rests quietly upon it and is unfractured. The chalk and green sand of this district have once

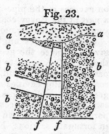

Fig. 23.

covered very considerable spaces, though the latter is now only seen on Haldon Hills ; near this section, it is true, but separated from it by an intervening valley. There are many other dislocations so covered on the same coast, where these appearances can be observed with the greatest ease, particularly at low water.

It might be supposed that these flints and pieces of chert were merely the remains of superincumbent masses of chalk and green sand, which have been destroyed by meteoric agents, the harder parts falling down on the top of the fracture. We can scarcely consider this physically probable, if even possible ; for it supposes the removal of more than 600 feet of sandstone and conglomerate (for not until that height above this section would the green sand and chalk come on), without scarcely leaving any of the pebbles, or large masses of the red conglomerate, while the flints and cherts, which belonged to upper, and consequently first destroyed rocks, remain.

Let us now consider another class of appearances. Over the whole district, where transported gravel occurs, the surface of the rocks, (it being of no importance what they happen to be,) is drilled into cavities and holes, similar to those well known on the chalk of the east of England. The following sections will illustrate this:

a a, gravel, principally of flint and chert, resting in a hollow of the red sandstone *b b*, between Teignmouth and Dawlish, the lines in the gravel following the outline of the cavity.

a a, gravel composed in a great measure of flints, among which are some large rounded pieces of siliceous breccia (the same as that which occurs in blocks on the top of the chalk hills near Sidmouth), resting

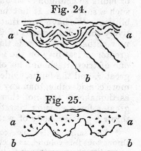

Fig. 24.

Fig. 25.

on cavities in the pipe-clay, near Teign Bridge, which constitutes a part of the Bovey coal formation, and which is not, as has been supposed, contemporaneous with the superficial transported gravel.

Other examples might easily be adduced; but these are here given, because the geological student can very readily observe them*. They seem to point to some general agent, which, in its passage over the land, has produced similar effects on various rocks, forming cavities and depositing fragments, transported from greater or less distances. In addition to this, we have, still in the same country, evidences of a washing of the rock beneath, by which portions of it are mixed with the transported substances, and even, in fortunate sections, have the false appearance of surmounting the transported matter, as the annexed section of the cliff near Dawlish well illustrates.

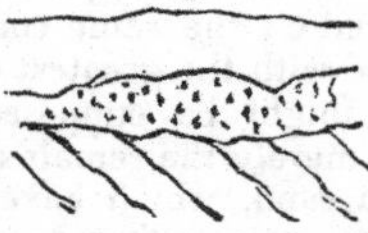

a a, regenerated red sandstone; *b b,* gravel composed of chalk flints, chert, and pebbles derived from the conglomerate interstratified with the red sandstone (*c c*), upon which it rests. To a person unaccustomed to geological investigations it would

Fig. 26.

easily be imagined, from this section alone, that the flints were included in the red sandstone; but the true arrangement is very apparent, even if the stratification of *a a* and *c c* did not show it; for the section is entirely fortuitous, every variety being observable in the vicinity, and this merely selected as an extreme case.

Our limits will not permit greater detail, which would require the necessary maps, but it would go far to support the supposition that a body of waters had passed over this land. The question might now arise, may there be any connection between the mass of water supposed to have passed over the land, and the fractures or faults so common? In answer to this it may be replied, that such a supposition is not inconsistent with possibilities or probabilities. We have seen that during such vibrations and comparatively small dislocations of the earth's surface as those which we now witness, the water is thrown into movement, and breaks with greater or less fury on the land. Still confining our attention to one district, it should be observed, that the dislocations are far greater, and the faults, evidently produced at a single fracture, far more considerable, than any we can conceive possible from modern earthquakes. It is not, therefore, unphilosophical to infer, that a greater force causing vibrations and fractures of the rocks would throw a greater body of water into more violent movement, and

---

* The same motive has governed me in the selection of sections throughout this volume, as it cannot be expected that the student should so readily observe difficult facts as the accomplished geologist.

that the wave or waves bursting upon the land would have an elevation and a destructive sweeping power, proportioned to the disturbing force employed.

The next question that will arise is, are there any other marks of such a deluge passing over the land? To this it may be replied, that the forms of the valleys are gentle and rounded, and such as no complication of meteoric causes, that ingenuity can imagine, seems capable of producing; that numerous valleys occur on the lines of faults; and that the detritus is dispersed in a way that cannot be accounted for by the present action of mere atmospheric waters. I will more particularly remark, that on Great Haldon Hill, about 900 feet above the sea, pieces of rock, which must have been derived from levels not greater than 700 or 800 feet, and even less, occur in the superficial gravel. They are certainly rare, but may be discovered by diligent search. I there found pieces of red quartziferous porphyry, compact red sandstone, and a compact siliceous rock, not uncommon in the grauwacke of the vicinity, where all these rocks occur at lower levels than the summit of Haldon, and where certainly they could not have been carried by rains or rivers, unless the latter be supposed to delight in running up hill.

It may be stated, before we quit this local description, that the faults do not all range in one direction, though east and west are not uncommon; and that as we approach the Weymouth district, this direction predominates. Near Weymouth there is one east and west fault, fifteen miles of which can be traced, but it probably extends further, for it enters the chalk on the east, and therefore cannot be easily observed, while it plunges into the sea on the west. There seems also every probability that these Weymouth faults are connected, as has already been remarked by Prof. Buckland and myself in another place, with the east and west dislocations through the Isle of Weight, and probably also with the east and west upraised, and afterwards denuded country of the Wealds of Sussex. It should also be remarked, that the accumulations of gravel are often most considerable on the eastern sides of the valleys, in the vicinity of Sidmouth and Lyme.

Let us now proceed to consider to what extent these local facts may be more or less general. To begin with England. Lowland valleys, often very considerably broader than those before noticed, and therefore more favourable to the supposition of a moving mass of water, occur very generally; for the surface composed of lowland valleys is very considerably greater than that exhibiting mountain valleys, though both have been modified by rivers and other agents now in operation. Over these valleys, foreign matter, not detritus derived from the weathering of the rocks beneath, is variously distributed. It may sometimes be possible, with the aid of ingenuity, to produce a case of transport by a long continuance of such natural effects as are now seen, but in other situations such explanations seem altogether valueless and unphilosophical.   In like manner also faults covered only by gravel are common, the lines of faults being frequently lines of valley.   I would by no means infer that all faults, only covered by gravel, have been contemporaneous ; on the contrary, it seems only reasonable to conclude, that faults or fractures have accompanied every great convulsion, and that as these have been frequent, so faults may also have been frequent ; yet it is curious that on coasts, ravines, and other great natural sections, there should be a difficulty in finding covered faults, that is, faults covered by substances deposited previous to the gravels we are considering, and that no such difficulty should exist respecting those covered only by gravel, or altogether without superincumbent mineral substances.

Not only are gravels brought from various distances, but even huge blocks, the transport of which by actual causes into their present situations seems physically impossible.   Mr. Conybeare has remarked on the great accumulation of transported gravel in midland England, more particularly at the foot of the inferior oolite escarpments on the borders of Gloucestershire, Northamptonshire, and Warwickshire, and observes that they are composed of such various materials that a nearly complete suite of English geological specimens may there be obtained.   " Portions of the same gravel have been swept onwards through transverse valleys affording openings across the chains of the oolite and chalk hills, as far as the plains surrounding the metropolis ; but the principal mass of diluvial gravel in this latter quarter is derived from the partial destruction of the neighbouring chalk hills, consisting of flints washed out from thence, and subsequently rounded by attrition*."   Mr. Conybeare also notices the occurrence of great blocks among the transported rocks of Bagley Wood, Oxfordshire, as also the presence of flints on the summits of the Bath Downs.   Prof. Buckland mentions that he found, among the transported gravel of Durham, twenty varieties of slate and greenstone, which do not occur, in place, nearer than the lake district of Cumberland.   He also notices a large block of granite at Darlington, composed of the same granite as that of Shap, near Penrith.   Blocks of the same granite occur in the valley of Stokesley, and in the bed of the Tees, near Bernard Castle.   Similar blocks are also found on the elevated plain of Sedgefield, near Durham.   In many of these cases blocks are mixed with rolled pieces of various kinds of greenstone and porphyry, probably derived from Cumberland †.

Prof. Sedgwick notices large transported boulders on parts of the

* Conybeare and Phillips, Outlines of the Geology of England and Wales.
† Buckland, Reliquiæ Diluvianæ.

Derbyshire chain, which overhang the great plain of Cheshire. He also remarks on the boulders accompanying the transported detritus at the base of the Cumberland mountains from Stainmoor to Solway Firth, the plain bordering the hilly region on the north presenting boulders and pebbles that have been transported across the Firth from Dumfrieshire. In the transported rubbish capping a hill near Hayton Castle, about four miles N.E. of Maryport, there are large granitic boulders resembling the rocks of the Criffel. " Among them was one spheroidal mass, the greatest diameter of which was ten feet and a half, and the part which appeared above the ground was more than four feet high." From St. Bees Head to the southern extremity of Cumberland, the coast region is covered by transported detritus, among which are boulders of granite, porphyry, and greenstone, some of large size. In low Furness similar phænomena are observable. Prof. Sedgwick further remarks, that large blocks derived from the green-slate district are found on the granitic hills between Bootle and Eskdale. Millions of large blocks are scattered over the hills on the N.W. boundary of the mountainous region. The syenitic blocks of Carrock-fell can be traced " through the valleys and over the hills of the mid region, to the very foot of the parent rock." Numerous boulders of the Carrock syenite rest on the side of High Pike; the largest, termed " the Golden Rock," being twenty-one feet long, ten feet high, and nine feet wide. Rolled masses of St. John's Vale porphyry abound near Penruddock, and descend the valleys thence into the Eamont. Rounded boulders of Shap granite are numerous on the calcareous hills south of Appleby; some being twelve feet in diameter. Rounded blocks, apparently derived from the green-slate at the head of Kentmere and Long Sleddale, are found on the flat-topped calcareous hills W. of Kendal. Prof. Sedgwick remarks that the blocks of Shap granite, which cannot be confounded with other rocks in the north of England, are not only drifted over the hills near Appleby, but have been scattered over the plain of the new red sandstone; rolled over the great central chain of England into the plains of Yorkshire; imbedded in the transported detritus of the Tees; and even carried to the eastern coast[*].

By comparing these statements with the little district first noticed, we find that the evidences of a transporting power by water are far greater in midland and northern England than in Devon and Dorset, the gravel having been carried far greater distances, and huge blocks added to the transported mass. How far these gravels may be contemporaneous can only be determined by future and exact observation. We shall, therefore, merely confine ourselves to a detail of facts, which must be taken into account in all generalizations on this subject. Between the Thames and the

[*] Sedgwick, Ann. of Phil. 1825.

Tweed, pebbles and even blocks of rock are discovered, of such a mineralogical character, that they are considered as derived from Norway, where similar rocks are known to exist. Mr. Phillips states, that the accumulation, at present termed *diluvium*, in Holderness, on the coast of Yorkshire, is composed of a base of clay, containing fragments of pre-existent rocks, varying in roundness and size. " The rocks from which the fragments appear to have been transported are found, some in Norway, in the Highlands of Scotland, and in the mountains of Cumberland; others, in the north-western and western parts of Yorkshire; and no inconsiderable portion appears to have come from the sea-coast of Durham, and the neighbourhood of Whitby. In proportion to the distance which they have travelled, is the degree of roundness which they have acquired *."

Patches of gravel and sand are stated to occur in the great mass of clay, sometimes amounting to considerable accumulations. In one of these, at Brandesburton, the remains of the fossil elephant were detected.

If, quitting England, we proceed northwards to Scotland, there are evidences of a similar force having acted in that country; and Sir James Hall even considers that a rush over the land has left traces of its course in the shape of furrows, which the transported mineral substances, moving with great velocity, have cut in the solid rocks beneath. From the direction of these marks Sir James Hall infers that the current had a western course in the vicinity of Edinburgh†. Continuing our course still northwards, the evidence of a transport continues; for Dr. Hibbert found fragments of rocks at Papa Stour, Shetland Isles, which must have travelled twelve miles from Hillswick Ness, the latter bearing from the former, N. 47°, E. He also remarks on the large blocks near the mansion of Lunna, on the east of Shetland, named the stones of Stefis, which appear to have been removed a mile or more by a shock from the N.E. The same author mentions many other interesting circumstances: among others, that at Soulam Voe, open to the Northern Ocean, there are boulders about three or four feet high, which do not correspond with any known rock in the country, and were probably derived from the northward‡. It is also probable, from Landt's notice, cited by Dr. Hibbert, that similar phænomena are observable in the Feroe Islands.

The probability therefore, as far as the above facts seem to warrant, is, that a body of water has proceeded from north to south over the British Isles, moving with sufficient velocity to transport fragments of rock from Norway to the Shetland Isles and the

* Phillips, Illustrations of the Geology of Yorkshire.
† Sir James Hall, Trans. Royal Soc. Edinburgh.
‡ Hibbert, Edin. Journal of Science, vol. vii.

eastern coast of England; the course of such body of water having been modified and obstructed among the valleys, hills, and
mountains which it encountered; so that various minor and low
currents having been produced, the distribution of detritus has
been in various directions.

If the supposition of a mass of waters having passed over Britain be founded on probability, the evidences of such a passage or
passages should be found in the neighbouring continent of Europe,
and the general direction of the transported substances should be
the same. Now this is precisely what we do find. In Sweden
and Russia large blocks of rock occur in great numbers, and no
doubt can be entertained that they have been transported southward from the north. In Sweden, the transported materials were
observed by M. Brongniart to run in lines, sometimes inosculating,
but having a general direction north and south*. Similar observations had been previously (1819) made by Count Rasoumovski on
the transported blocks of Russia and Germany, which, having been
unknown to M. Brongniart, render his account of the Swedish
blocks the more valuable. Count Rasoumovski observes, that,
where many blocks are accumulated they form parallel lines, with
a direction from N.E. to S.W. He states that the erratic blocks
are very numerous, and composed of Scandinavian rocks between
St. Petersburgh and Moscow; and remarks, that in some places, especially in Esthonia, the blocks appear and disappear at greater or
less intervals, apparently owing to the form of the land at the time
of their transport; for these masses are discovered where escarpments presented themselves, while, where the land sloped away,
or became more or less horizontal, they disappear; thus seeming
to show that the steep escarpments caught them in their passage
onwards. Count Rasoumovski also remarks that the blocks occur
abundantly on the heights, and but rarely, or thinly scattered,
over the lowlands†. Proceeding south, the course of the waters

---

* Ann. des Sci. Nat. 1828.
† *Ibid.* t. xviii.

Prof. Pusch observes, that the erratic blocks from the Duna to the
Niemen are composed of granite resembling that of Wiborg in Finland;
of another granite, with Labrador felspar from Ingria; of a red quartzose
sandstone from the shores of Lake Onega; and of a transition limestone
from Esthonia and Ingria. In Eastern Prussia, and in that part of Poland
situated between the Vistula and the Niemen, the granitic blocks are abundant: three varieties of granite are the same as those found in Finland,
at Abo and Holsinfors; another coarse-grained granite and a sienite are
also from the north. The hornblende blocks of the same countries are
from southern and central Finland; the quartzose blocks are exactly the
same as the rocks named *Fjall Sandstein*, between Sweden and Norway;
and the porphyry blocks are of the same mineralogical character as the
porphyries of Elfdalen in Sweden. "From Warsaw to the west, towards

seems to have continued in that direction over the low districts of Germany, to the Netherlands, depositing huge blocks in their passage; these blocks proved by their mineralogical composition to have been derived from rocks known to exist in the northern regions.

South of this comparatively low land, various obstructions arise in the shape of mountains; and if the supposition of a mass of waters be correct, it would be thrown out of its original course in various directions, and from lofty mountain ranges, such as the Alps, there would be re-action, and a back wave created, rushing back through the valleys, if the mass of waters had not been previously checked by interposed obstacles, and could not reach the Alps with sufficient force to produce any remarkable effects.

Such a movement as this over part of Europe would, if the supposition of a mass of waters were correct, be observed in other northern regions, for the waters thrown into agitation would cause waves around the centre of disturbance. In America, therefore, we should expect to find marks of such a deluge, the evidences pointing to a northern origin*. Now in the northern regions of that country we do find marks of an aqueous torrent bearing blocks and other detritus before it, the lines of their transport pointing northward, according to Dr. Bigsby, and reminding us of the same appearances observed in Sweden and Germany. The quantity of transported matter covering various large tracts in North America seems quite equal to that scattered over Northern Europe; and as they both point one way, we can scarcely refuse to admit that the course of the disturbance or disturbances was towards the north, the undulations of the waters having been caused by some violent agitation, perhaps beneath the sea in those regions, for it is by no means necessary that it should be above its level.

A convulsion or convulsions of this magnitude, reasoning from the analogy of those minor agitations which we term earthquakes, would be felt over a considerable portion of the globe, and the waters over a large surface would be thrown into agitation. A part of the earth would be greatly disturbed, and we should expect fractures and faults produced in strata where the convulsion was most felt, as similar minor effects are produced at present from the exertion of a less intense force.

Ice would seem to afford a possible explanation of the transport of many masses; for the glaciers which descend the valleys of

Kalisch and Posen, the blocks of the red granite of Finland diminish in number, but those composed of hornblende rocks and gneiss become more abundant, as is also the case with those of porphyry. Few Finland rocks are in general there found, while those of Sweden are common." Pusch, Journal de Géologie, t. ii. p. 253.

* Journal of Science, vol. xviii.

high northern regions are, like those of the Alps, charged with blocks and smaller rock fragments, which have fallen from the heights. A wave, or waves, rushing up such valleys would float off the glaciers, more particularly as northern navigators have shown that they project into the sea. It is considered that the huge masses of ice known as icebergs, are the projecting portions of these boreal glaciers, which having been detached from the parent mass, are borne into more temperate climes, in some cases transporting blocks and smaller fragments of rock. This debris will, as Mr. Lyell has observed, be deposited at the bottom of the seas over which they pass; and therefore, if such bottoms were raised so as to become dry land, blocks might be discovered scattered over various levels of that land, presenting appearances that might be mistaken for the action of diluvian currents. If the present continents bore evident marks of long submergence beneath an ocean immediately previous to their present appearance, and if the blocks were merely scattered here and there, this explanation would by no means be without its weight: but there are too many circumstances tending to other conclusions, to render it probable. The supposition of masses of ice, covered by blocks and smaller rock fragments, borne southwards with violence, though it may account for some appearances, does not, it must be confessed, seem applicable to all, more particularly where blocks can be traced to their sources at comparatively small distances. Supposing a wave, or waves, discharged over Europe and America from the northwards, many phænomena would depend on the time of year at which the catastrophe, or catastrophes, took place; for if in the winter, waters rushing from that quarter would transport a greater quantity of ice, and many superficial blocks and gravels, bound by ice together, might be torn up and carried considerable distances from the possible small specific gravity of the mass; for even in the case of rivers, it has been found that large masses of rocks have occasionally been transported from places, when encased in ice and acted on by the stream. In Sweden and Russia it is more than probable that many blocks would be thus encased during winter, and therefore a flood of waters passing over them would cause them to rise, to float, and to be borne onwards, until the ice melting, the blocks would sink and be finally brought to rest.

Upon the hypothesis of a convulsion in the North, the effects would become less as we receded from the centre of disturbance, and, finally, all traces of them would be lost.

We now arrive at another question,—how far the distribution of blocks from the Alps may have been contemporaneous with the supposed transport of erratic fragments from Scandinavia? To answer this question, without more direct information than we possess, would be difficult; and we should be particularly cautious in applying preconceived theories before we have the requisite data. All that we can safely remark on this subject seems to be, that the blocks in both cases appear to a certain extent superficial and uncovered by deposits which would afford us information respecting their difference of age; and that it is possible a great elevation of the Alps, and distribution of blocks on both sides of the chain, may have been contemporaneous, or nearly so, with a convulsion in the north.

An immense quantity of debris has, at a comparatively recent epoch, been driven from the central chain of the Alps outwards; the consequence, according to M. Elie de Beaumont, of a great elevation in those mountains, extending from the Valais to Austria. MM. von Buch, De Luc, Escher, and Elie de Beaumont, have presented us with a detail of numerous and well observed facts, which all tend to one conclusion; namely, that the great valleys existed previously to the catastrophe which tore blocks and other fragments from the Alps, and scattered them on either side of the chain. M. Elie de Beaumont observes * that the valleys of the Durance, of the Drac, of la Romanche, of the Arc, and of the Isere, present the same appearances as those of the Arve, the Rhone, the Aar, the Reuss, the Limmat, the Rhine, and the valleys which descend into the plains of Bavaria, noticed by different geologists. On the Italian side of the chain, appearances are also similar, and no doubt can exist that the blocks and debris have passed down the respective valleys, where they have left unequivocal marks of their transit. M. Elie de Beaumont has presented us with very detailed accounts of these appearances in the valleys of the Durance, of the Drac, and others, where they are precisely what would have been expected from the passage of a rock-charged mass of waters down the respective channels, the largest fragments having been transported the shortest distances, being most angular, while the smaller and most rounded have been carried the furthest. Thus, in the valley of the Durance, the transported substances become more angular and of greater volume, as we proceed from the great mass of pebbles, called the Crau, to the mountains beyond Gap, whence the debris, judging from its mineralogical characters, have very clearly been derived. Similar phænomena will be observed up the valley of the Drac, which proceeds by another course to the neighbourhood of the same mountains, the two streams of debris not mingling until they join in the Crau †.

From my own observations, I can fully confirm the remarks of various authors respecting the situations of the Alpine blocks, and

---

* Récherches sur les Rev. de la Surface du Globe; Ann. des Sci. Nat., 1829 et 1830.

† Elie de Beaumont, Récherches sur les Rev. du Globe.

their probable derivation from the respective valleys, which they, as it were, appear to face. But I have nowhere observed such striking masses of erratic blocks as those which occur in the vicinity of the lakes of Como and Lecco. They are particularly remarkable on the northern face of the Monte San Primo, a lofty mountain ridge presenting one of its sides to the more open and northern part of the lake of Como, where the latter stretches towards the high Alps; thus presenting a bold front to any shock which should come from the north, leaving open passages to the right and left of it, one down the southern part of the lake of Como, the other down that of Lecco. Not only in front, facing the high Alps, but also round the flanks and shoulders of this mountain, and even behind it, where the eddy-current would have transported them, blocks of granite, gneiss, mica-slate, and others from the central chain, of various sizes, and often accompanied by smaller fragments and gravel, are seen in hundreds, nay thousands, scattered over the dolomite, limestone, and slate of the mountain, and nearly filling up a previously existing valley which faced the north, the direction whence the rock-charged fluid descended. Proceeding down the side valleys, partly occupied by the lower lake of Como, and the lake of Lecco, we find the evidences of such a current in the presence of blocks occurring, as they should do, where direct obstacles were opposed to its course, or in situations where eddies would be produced behind the shoulders of the mountains. One very remarkable instance of such occurrence is behind, or on the southern side of, the Monte San Maurizio, above the town of Como; where numerous blocks are accumulated on the steep flank of the mountain, precisely where a body of water, rushing down the great valley, would produce an eddy at its discharge into the open plains of Italy*. The blocks, though no doubt many have descended from their first positions in consequence of the long-continued action of atmospheric agents, occupy an elevated line, as also on other but lower heights in the vicinity, which opposed more direct obstacles to the debacle: seeming to show that the blocks occurred near the surface of the fluid mass, and were whirled by the eddy, at nearly the same level, against the steep sides of this calcareous mountain, as well as thrown against the more direct obstacle of a range of conglomerate hills.

The following is a section of the Monte San Primo, exhibiting the manner in which the erratic blocks rest on its surface:

* For illustrations of these appearances, see Sections and Views illustrative of Geological Phænomena, plates 31, 32.

Fig. 27.

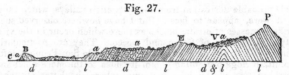

P, Monte San Primo : B, bluff point of Bellaggio rising out of the lake of Como C : *a a a a*, blocks of granite, gneiss, &c., scattered over the surface of the limestone rocks *l l l l*, and the dolomite *d d d*. V, the Commune di Villa, where a previously existing depression or valley is nearly filled with transported matter. E, the Alpi di Pravolta, on the northern side of which is the large granite block figured beneath, remarkable not so much for its size as for its angular character.

Fig. 28.

The accumulation of erratic blocks of the Alps in groups has been particularly remarked by M. De Luc (nephew), who has very carefully examined them round the lake of Geneva and neighbouring country * The levels which the blocks keep on the Jura and other places has been often observed by various authors. Such a common mode of occurrence must, we should suppose, have some common cause, and can scarcely be accidental.

Solutions of the problem of erratic blocks seem not very practicable at present, and our attempts at general explanations can be considered little else than conjectures that may appear more or less probable. The student, therefore, should be careful not to consider such explanations as well ascertained truths, but merely as hypotheses, which future and extensive observations may, or may not, prove correct.

It has been above remarked, that the Alpine erratic blocks frequently occur in groups. To present a general explanation of this phænomenon would, at present, be somewhat difficult ; but it may be asked, as a mere conjecture, whether masses of floating ice charged with blocks and other detritus, rushing down the great valleys into the more open country of lower Switzerland, might

* De Luc, Mém. de la Soc. de Phys. et d'Hist. Nat. de Genève, vol. iii.

not be whirled about by the eddies, and the icy masses be destroyed by collision against each other, so that groups of blocks would afterwards be found beneath the places where the whirlpools had existed. Masses of ice, charged with blocks and pent up for the moment within such basins as might be formed between the Alps and Jura, might also be carried at certain levels against the sides of the opposing mountains, such as the Jura, and be there deposited in groups and in lines of level.

Such passages of bodies of water over land, as have been above noticed, whether contemporaneous or not, could scarcely have failed to destroy the larger portion of the animals previously existing on that land. At the time when the remains of extinct elephants, mastodons, and rhinoceroses, were considered to characterize one set of gravels or transported matter, it was natural to conclude that all such debris were contemporaneous : but as these animals are now found to have existed earlier, if not also later, than was imagined, this supposed guide has failed us; and we gain no very definite ideas relative to the age of the transported matter in which they may occur, further than that they probably come within a certain range of the more recent geological deposits.

The following is a list of those animals which are generally considered as referable to the passage or passages of waters over the land, and which, whether exactly contemporaneous or not, are found in superficial gravels, sands, and clays*.

1. Elephas primigenius, *Blumenbach.* Scattered over various parts of
     Europe. Very common in the northern parts of Asia, where the
     ivory of the fossil tusk, or defence, is so far uninjured as to be
     used for ornamental purposes. Found also on the northern
     coast of the American continent. (Highest transported gravel
     near Lyons. *Beaum.*)
1. Mastodon maximus, *Cuv.* North America. Various authors†.
2. —————— angustidens, *Cuv.* North and South America ; Simorre ;
     Dax; Asti. *Al. Brong.*
3. —————— Andium, *Cuv.* Cordilleras ; Santa Fé de Bogota. *Humboldt.*
4. —————— Humboldtii, *Cuv.* South America. *Humboldt.*

* The student should be careful, if he be so fortunate as to discover any of these remains, to remark, whether they occur in detritus evidently moved from a distance, or in that great mass of weathered fragments which often covers hills and valleys, and which seems principally due to the action of the atmosphere upon them.

† The relative age of the deposit, in which the American mastodons are found, cannot be considered as very satisfactorily ascertained. Some geologists, indeed, suspect that these animals have disappeared more recently than is commonly supposed. It is very desirable that, in this state of uncertainty, some American geologist would thoroughly examine the district in which these remains are principally discovered.

I

5. Mastodon minutus, *Cuv.*   Europe. *Al. Brong.*
6. ———————— tapiroides, *Cuv.*   Europe. *Al. Brong.*
1. Hippopotamus major, *Cuv.*   Walton in Essex; Oxford; Brentford. *Buckl.* Bavaria, *Holl.*
2. ———————— minutus, *Cuv.*   Landes of Bourdeaux. *Al. Brong.*
1. Rhinoceros tichorhinus, *Cuv.*   Very common in Europe.
2. ———————— leptorhinus, *Cuv.*   Common in Europe.
3. ———————— incisivus, *Cuv.*   Germany; Appelsheim. *Al. Brong.*
4. ———————— minutus, *Cuv.*   Moissac. *Al. Brong.* Magdeburg. *Holl.*
1. Tapirus giganteus, *Cuv.* Allan; Vienne in Dauphiné; Chevilly; and other parts of France. *Al. Brong.* Furth, Bavaria; Feldsberg, Austria. *Holl.*
1. Cervus giganteus, *Blum.*   Ireland; Silesia; Banks of the Rhine; Sevran near Paris.
Cervus. Several different species, common in various parts of Europe.
Bos. Remains of, common.
Auroch (fossil), *Cuv.*   Siberia, Gemany, Italy, &c.
Hyæna (fossil), *Cuv.*   Lawford near Rugby; Herzberg and Osterode; Canstadt, near Stutgart; Eichstadt, in Bavaria. *Buckl.* Fourent, near Gray. *Al. Brong.*
Ursus.
Equus. Common in many places.
Trogontherium Cuvieri*, *Fischer.*   Sea coast near Taganrock, Sea of Azof. *Fischer.*
Megalonix*, *Jefferson.* Green Briar, Virginia.
Megotherium*, *Cuv.* Buenos Ayres.

We cannot quit the subject of the large mammalia entombed in superficial gravel, sands and clays, without adverting to the elephant found encased in ice near the embouchure of the river Lena in Siberia. It had been preserved entire, having undergone no decomposition since death; on the contrary, when detached from the ice, it afforded food to various animals, and parts of its skin and hair were collected, and are now preserved with its skeleton in the Museum at St. Petersburgh. Mr. Adams, to whom the scientific public are indebted for the preservation of what remained of the animal, and for the account of its original discovery, relates that Schumachof, a Tungusian chief and owner of the peninsula of Tamset, where the elephant was discovered, first observed a shapeless mass among the ice in 1799; but it was not until 1804 that this mass fell on the sand, and disclosed the ice-preserved elephant, whose tusks were cut off and sold by the Tungusian chief. Two years afterwards Mr. Adams visited the spot, and collected the remains as above stated. According to this observer, the escarpment of ice in which the elephant had been preserved, extended two miles, and rose perpendicularly about

* We are by no means certain of the relative antiquity of these remains; indeed, the list here given requires a thorough revision.

200 or 250 feet. On this ice, which is described as pure and clear, there was a layer of friable earth and moss, about fourteen inches thick*.

M. Cuvier mentions that in 1805 M. Tilesius had received, and had sent to M. Blumenbach, some hair torn from the carcase of a mammoth, or elephant, by a person named Patapof, near the shores of the Icy Sea. He further observes, that some of the hair and skin of this individual was presented to the Jardin du Roi at Paris, by M. Targe, who had received it from his nephew at Moscow†.

Pallas mentions the discovery (in 1770) of an entire rhinoceros with its skin and hair, enveloped in sand on the banks of the Wiluji, which falls into the Lena below Jakoutsk. The animal is described as being very hairy, particularly on the feet. It was an individual of the *Rhinoceros tichorinus,* Cuvier‡.

The causes, whatever they were, which effected the destruction of the elephant at the mouth of the Lena, also destroyed many others; for it now appears, according to M. Hedenstrom, who visited the shores of the Icy Sea, under the direction of the Russian Government, between the Lena and the Colyma, that there are hundreds of elephants, rhinoceroses, oxen, and other animals, in the ice of those regions§. When these discoveries shall have been more fully made known, we may expect to have great light thrown on this subject. It seems probable that there has been a great change of climate on the northern coasts of Asia and America since these animals existed there; for with every allowance for the adaptation of the particular species of elephant, so commonly found fossil, to much colder climates than the existing species now inhabit, (and that they were so adapted seems exceedingly probable, from the woolly hair discovered on the individual encased in ice at the mouth of the Lena,) we must grant them something to live upon, food fitted to their powers of mastication and digestion; and this they could scarcely find, if the climates were such as they now are, permitting the existence of ice disposed as at present, into the crevices of which they have been supposed during their promenades to have accidentally fallen.

Ossiferous Caverns, and Osseous Breccia.

It is to Prof. Buckland that we are indebted for our more intimate acquaintance with the various circumstances under which organic remains are found in caverns; for though bones of bears

---

* From the account of the elephant found in the ice of Siberia, London, 1819 ;—taken from the Mem. of the Imp. Acad. of Science of St. Petersburgh, vol. v.

† Cuvier, Oss. Fossiles, t. i. éd. 1822.   ‡ *Ibid.* t. ii.
§ Journal de Géologie, t. ii. p. 315.

and other animals, occurring in caves, had long attracted attention, more particularly in Germany, it was not until after the discovery of the celebrated Kirkdale cavern, in Yorkshire, that the subject acquired a new interest, and became as much a part of general geological investigation, as the fossil contents of any well established rock had previously been. It is gratifying to observe, that even those who are opposed to the theoretical conclusions that have been deduced from cavern bones, are still willing to pay their tribute of praise to the zeal and activity with which Prof. Buckland conducted his researches.

Prof. Buckland pointed out that the general arrangements in caverns, are: 1. The original sides of the cave, which may or may not be covered with stalagmite. 2. A deposition of animal remains, mixed with mud, silt, rolled stones, or broken fragments; many circumstances sometimes attending this deposition seeming to attest the long-continued residence of certain animals in caverns for successive generations: some, the hyænas for instance, having there dragged their prey, often consisting of parts of the elephant and rhinoceros. 3. The deposition of stalagmite covering up the animal remains, the mud, silt, &c., with a greater or less depth of carbonate of lime; so that to all appearance, in a newly discovered cavern, the bottom is a mere mass of stalagmite, beneath which the organic riches would for ever remain unknown, unless the concealing crust should be fractured by accident, or broken through by the geologist, now aware that animal remains may be found beneath it.

Since the discovery and description of Kirkdale cavern, notices of other ossiferous caves have become so numerous that a mere list of them would be somewhat long; and they multiply so fast, that we may anticipate, at no distant period, a very singular body of evidence on this subject alone. Already has the spirit of inquiry produced singular results in the South of France, where the remains of man are stated to have been discovered in the same mass and in the same cave with those of the extinct rhinoceros and other animals usually found in caverns.

The remains of animals, similar to those contained in caverns, are frequently found in fissures of the rock; in some situations, the whole mass of bones, fragments of rock, and cementing matter being so hard and compact, that it frequently equals, and sometimes exceeds, in durability the rock within which it is inclosed. Of this the *osseous breccias* of Nice and many other places in the Mediterranean are examples.

It becomes daily more necessary to ascertain, as far as may be, the relative ages of these various accumulations of animal remains, investigating the subject with proper attention, and as much as possible without preconceived theory. It also becomes important to examine with attention, in those cases where the mouths of ossiferous caverns are covered up with detritus, whether such detritus be composed of angular fragments of the rock in the vicinity, which might have been gradually accumulated over the external aperture during the long lapse of ages, by causes and effects similar to those in daily operation; or whether it is composed of transported fragments more or less rounded, and which must have travelled from a distance : in the latter case, endeavouring to ascertain whether such transported matter could have been carried to its present situation by actual causes, or whether we must seek a greater intensity of force to account for its presence, physical obstacles opposing its carriage by any other means. If angular fragments, derived from the immediate vicinity, alone cover the cavern's mouth, we have no certainty when it was finally closed; and therefore, even supposing that one set of animals may have been overwhelmed by a rush of waters into the cavern, there is nothing to prevent another race of animals from frequenting the same place, whose bones might become, to a certain degree, mixed with the others, and entombed beneath fragments of rock and stalagmite, from the constant change operating in the interior of caves. Thus the bones of man, and his early rude manufactures, such as unbaked pottery, may become, to a certain extent, mingled, in a mass of stalagmite and rock fragments, with the remains of elephants, rhinoceroses, cavern bears, and hyænas; and the whole might, after the cave became deserted, and the accumulation at the mouth considerable, be covered with a crust of stalagmite : so that upon the discovery of such a cavern, it might be described, if attention had not been paid to the kind of detritus which blocked up the mouth, as being closed externally, and open to a certain height inside, beneath which there was a crust of stalagmite, covering an accumulation of rock fragments and bones, among which those of man were found mingled with those of the elephant and other animals; and it might hence be concluded, that all these remains were of contemporaneous origin, and, consequently, that man existed at the time when the elephants roamed the forests, and hyænas and bears lurked in the caverns of Europe. If the mouths of ossiferous caverns be closed by fragments of rock transported from a distance, such transport being clearly not due to the operation of actual causes, but to the exertion of a greater intensity of force; and if we then find the remains of man entombed with those usually contained in caverns, there would seem little reason to doubt, unless other communications from the surface could be traced, that man was a contemporary with the extinct species of elephants, rhinoceroses, hyænas, and bears, found not only in the caves, but also in masses of transported gravel, and that he existed previous to the catastrophe which overwhelmed him and them. If the co-existence of man and these extinct animals should ever be satisfactorily proved, it would become a curious question, whether his, so found, remains are those of an extinct species; or undistinguishable, like the bones of the horse, from those which now exist. It is a singular circumstance, and one which demands attention, notwithstanding the ingenious remarks that have been made on the subject, that the remains of the monkey tribe should not yet have been discovered among the undisturbed bones and other substances in caves, or in the old transported gravel, or *diluvium* of Prof. Buckland. It has been objected to a remark that man and the monkey tribe were perhaps created about the same period, and were of comparatively modern appearance on the earth's surface, that the countries have not been geologically well examined where the monkey race now exist. This is perfectly true. But is there any reason why monkeys should not have lived in climates and in situations where elephants, rhinoceroses, tigers, and hyænas were common? for the climates and regions in which existing elephants, rhinoceroses, tigers, and hyænas abound, are precisely those where monkeys are now found. To the objection, that if they did then exist, their bones would not be discovered, as their activity would secure them from falling a prey to hyænas and other predaceous animals; it may be opposed, that they must have died like other animals, and that their dead carcases must have fallen to the ground, and that they were quite as likely to have become the food of less nimble creatures, as the birds found in the cavern of Kirkdale.

Kirkdale cavern was discovered by cutting back a quarry, in the summer of 1821, and was visited by Prof. Buckland in December of the same year. Its greatest length is stated at 245 feet, and its height generally to be so inconsiderable, that there are only two or three situations where a man can stand upright. The following is a section *.

Fig. 29.

a a, a a, horizontal beds of limestone, in which the cave is situated; *b*, stalagmite incrusting some of the bones, and formed before the mud was introduced; *c*, stratum of mud containing the bones; *d d*, stalagmite formed since the introduction of the mud, and spreading over its surface; *e*, insulated stalagmite on the mud; *f f*, stalactites depending from the roof.

" The surface of the sediment when the cave was first opened was nearly smooth and level, except in those parts where its regularity had been broken by the accumulation of stalagmite above it, or ruffled by the dripping of water : its substance is an argil-

* From Buckland's Reliquiæ Diluvianæ.

laceous and slightly micaceous loam, composed of such minute particles as would easily be suspended in muddy water, and mixed with much calcareous matter, that seems to have been derived in part from the dripping of the roof, and in part from comminuted bones. . . . . . . At about 100 feet within the cave's mouth the sediment became more coarse and sandy*."

According to Dr. Buckland, the following are the animals, the remains of which were found in the Kirkdale cavern: *Carnivora;* —Hyæna, Tiger, Bear, Wolf, Fox, Weasel. *Pachydermata;*— Elephant, Rhinoceros, Hippopotamus, Horse. *Ruminantia;*—Ox, and three species of Deer. *Rodentia;*—Hare, Rabbit, Water-rat, and Mouse. *Birds;*—Raven, Pigeon, Lark, a small species of Duck, and a bird about the size of a Thrush.

From the mode in which these remains were strewed over the bottom of the cavern when the mud was removed, the great proportion of hyæna teeth over those of other animals, and the manner in which many of the bones were gnawed and fractured, Prof. Buckland inferred that this cavern was the den of hyænas during a succession of years; that they brought in, as prey, the animals whose remains are now mixed with their own; and that this state of things was suddenly terminated by an irruption of muddy water into the cave, which buried the whole in an envelope of mud. The inference of the hyænas having been long resident in the cave, was strengthened by the occurrence of their fæces, precisely as would happen in a den of hyenas at the present day. In addition to which it was further observed, that many bones were rubbed smooth and polished on one side, while the opposite side was not. This, Prof. Buckland considered, was produced by the friction of the animals walking or rubbing themselves upon the bones.

The German caverns of Gailenreuth, Küloch, Bauman, &c. contain an abundance of bones, nearly identical, according to Cuvier, over 200 leagues; by far the greatest proportion being referable to two extinct species of Bear, *Ursus spelæus,* and *U. arctoideus.* The remainder consisted of the extinct Hyæna (the same as at Kirkdale), a Felis, a Glutton, a Wolf, a Fox, and Polecat†. These caverns so far resemble Kirkdale cave, that there is more or less of a stalagmitic crust beneath which the bones are discovered, the stalagmitic matter being frequently transfused through the previously deposited sediment‡. There is, however, one fact con-

* Buckland, Reliquiæ Diluvianæ.

† Sections of some of these caves will be found in Prof. Buckland's Reliquiæ Diluvianæ.

‡ Buckland, Reliquiæ Diluvianæ. According to M. Wagner, the Muggendorf caverns contain the remains of the *Ursus spelæus, Ursus arctoideus* (Cuv.), *Ursus priscus* (Goldf.), *Hyæna spelæa* (Goldf.), *Felis spelæa* (Goldf.), *Canis spelæus* (Goldf.), *Canis minor, Gulo spelæus* (Goldf.), a *Cervus,* and a *Bos.* Wagner, Leonhard and Bronn's Jahrbuch für Geologie, &c. 1830.

nected with these caverns, wherein they differ very considerably
from the Yorkshire cave. In the latter, no rolled pebbles were
observed; while in the former they have been noticed in some
places. Thus, in Bauman's Höhle, pebbles of various sizes are
stated to occur among crushed and pounded bones; leading to the
presumption that the pebbles broke the bones, for the sand and mud of
the same chamber contain them nearly entire. It would therefore
appear that water had rushed into the cave, bringing with it rolled
pebbles of the surrounding country, crushing and distributing the
previously accumulated bones. By reference to Prof. Buckland's section of this cave*, we find the gorge of Bode exposes the
entrance of the cavern, from whence there is a descent into the
chamber where the crushed bones and pebbles occur: so that
the same phænomena may here be explained by two different hypotheses; the one supposing a fracture of strata produced during
a great convulsion permitting the sudden inroad of waters from
above; the other, the gradual cutting of the gorge by the river
Bode, which, so long as it cut across the mouth of the cavern,
would throw rounded pebbles into it, very considerable rushes of
water and pebbles taking place during floods. We thus obtain
little information on the subject. The same remarks apply to the
caves of Rabenstein, and others in Franconia. The Zahnloch
may, perhaps, admit of only one explanation; for it is described as
being on a hill 600 feet above the valley of Muggendorf. The ossiferous mass is stated to be composed of " brown loam, mixed with
numerous pebbles and angular fragments of limestone†."

Be the origin of the pebbles, sand and mud, what it may, it
seems clear that the remains of various animals were enveloped
by them; since which, there has been a long continuance of repose,
permitting, in most cases, the deposit of stalagmite upon the ossiferous mass.

Dr. Buckland informs me that Mr. M⁶Enery fonud rounded pebbles of granite, of the size of an apple, mixed with the bones under
the stalagmite in Kent's Hole, Torquay; and he states that he has
found pebbles of greenstone, completely rounded, in the same
place; and that in some parts of Kent's Hole, particularly the lowest, the bone breccia is full of fragments of grauwacke and slate,
some of them rolled, some angular. The cave itself is situated in
a limestone resting on shale, and the grauwacke and slate are
rocks of the country; but the granite is at some distance, not nearer
than Dartmoor: so that although the situation of the cave is
such as to make it possible, though not perhaps very probable,
that under a variety of combinations the greenstone grauwacke, and slate, may have been conveyed into the cave, by what

* Reliquiæ Diluvianæ, pl. 15.
† *Ibid.* p. 131.

are termed actual causes, the granite pebbles would scarcely seem reconcileable with such an hypothesis.

M. Thirria describes the Grotte d'Echenoz, on the south of Vesoul, near the summit of a high plateau, between the villages of Echinoz, Andelarre, and Chariez (Haute Saone), as formed in the lower system of the Jura limestone, or oolitic group. The upper parts of this cave are very irregular, and in one place (the Grand Clocher) rises so high, that there must be little space remaining between it and the surface of the plateau. The bottom is not far removed from a level, here and there interrupted by stalagmites. These stalagmites are not numerous; but there are some which rise high, and cover a considerable surface. No researches had been attempted in this cavern previous to those of M. Thirria, in August 1827. He broke up the ground " at different points of the four chambers of the cavern, and all afforded bones in greater or less abundance. The researches carried on in the fourth chamber were the most productive, for each blow of the pick-axe brought up a bone. The depth at which the bones were discovered, varied from ten centimetres to a metre : they occurred in the midst of a red clay, mixed with a great number of rounded pebbles with a smooth surface, the size of which often attained that of a man's head. They are all composed of a gray lamellar limestone, resembling that which forms the sides of the cavern and many rocks of the vicinity. Independently of these pebbles, which have evidently been rolled by waters, and could not have penetrated into the cave except through some fissures in its roof no longer visible, pieces of stalagtites and stalagmites are discovered with their angles worn down, showing that they have been moved. The clay deposit, the thickness of which does not appear to exceed one metre thirty centimetres, is nearly everywhere covered by stalagmite a few centimetres deep, and upon this crust, which is mammillated, there rests a bed, from ten to twenty-five centimetres thick, composed of a clay more unctuous but less red than that situated beneath, and frequently blackish from the remains of vegetables, of which it still contains some debris. No rounded pebbles are found above the stalagmitic crust, and they are only seen on the surface when the stalagmite does not exist. Hence it appears evident, that the ossiferous clay containing the rounded pebbles has been carried by the waters and deposited in the cavern, anterior to the formation of the stalagmitic crust, produced by droppings from the roof, before the deposit of the clay bed by which this crust is covered*." M. Thirria further infers, from the resemblance of these pebbles to those of the transported matter (termed diluvium) in the vicinity, that the introduction of

* Thirria, Mém. de la Soc. d'Hist. Nat. de Strasbourg, t. i., where good sections of the cave will be found.

the pebbles and clay mixed with the bones in the Grotte d'Echenoz was contemporaneous with the transport of the diluvium.
The bones were most commonly discovered beneath a certain
thickness of clay; but in many situations they occurred immediately beneath the stalagmitic crust, and sometimes even entirely
in it. " In general the bones constituted a thickness of about eight
to sixteen centimetres in the middle of the clay : they crossed in
various directions, and covered each other with small intermediate spaces, without having preserved their relative position
They have not, however, suffered complete dislocation; for the
dorsal vertebræ were nearly always discovered near the skull and
jaws ; the humerus and cubitus near the pelvis; and the os calcis,
the metatarsal and metacarpal bones or phalanges, near the fe-
murs, the tibias and the cubitus." The bones, examined by
Cuvier, were found to belong to the *Ursus spelæus, Hyæna,
Felis, Deer, Elephant,* and *Boar* ; by far the largest proportion
belonging to the *Ursus spelæus**.

M. Thirria also describes the Grotte de Fouvent, situated at
Fouvent near Champlitte (Haute Saone). This cavern was accidentally discovered by quarrying the rock in such a manner as to
strike into a natural cleft, through which the matters contained
in the cave are supposed to have entered, there being apparently no other aperture. The cave is considered too small for
the habitation of beasts of prey : its upper part is only about two
yards beneath the surface of the plateau; and it was completely filled with bones, a yellow marl, and angular pieces of
the surrounding rock and of those in the vicinity ; the whole
mixed pell-mell, and resembling the detritus termed *diluvium*
covering many plains and valleys in the neighbourhood. A thin
red clay bed covers the bottom of the cave, and a small thickness at the top did not contain animal remains. According to
M. Cuvier, these remains belong to the Elephant, Rhinoceros,
Hyæna, Ursus spelæus, Horse, Ox, and Lion. M. Thirria remarks that this ossiferous mass merely requires a compact cement
to become an osseous breccia.

A very common condition of cavern bones is their being
found mixed with angular fragments of the rock in which the
caverns occur. Banwell Cave, in the Mendip Hills, is a good
example of a large accumulation of the remains of *Ursus, Felis,
Cervus, Bos, Equus,* and other animals; with fragments of carboniferous or mountain limestone, the rock in which the cavern is
formed. The contents of this cave merely require, as M. Thirria
has observed respecting that at Fouvent, a calcareous cementing
matter, to become an osseous breccia, such as is found at Nice
and other places on the shores of the Mediterranean. The osseous

---

* Thirria, Mém. de la Soc. d'Hist. Nat. de Strasbourg, t. i.

breccia of the Chateau Hill at Nice appears indeed to have been
partly a cavern, which has been quarried away by the works constantly carried on there.   The following is a section, fresh when
I observed it in the winter of 1827.

q, quarry ; *a a*, hard brecciated dolomite ; *l l l*, holes bored in the dolomite by some lithodomous shell ;
c, rounded pebbles, composed principally of rock fragments transported
from a distance, cemented by a compact calcareous cement ; *o*, osseous
breccia, cemented by a reddish calcareous cement.

Fig. 30.

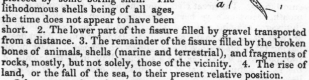

This section seems to point to
the following conclusions :—1. An
open fissure beneath water, the sides
pierced by some boring shell.   The
lithodomous shells being of all ages,
the time does not appear to have been
short. 2. The lower part of the fissure filled by gravel transported
from a distance. 3. The remainder of the fissure filled by the broken
bones of animals, shells (marine and terrestrial), and fragments of
rocks, mostly, but not solely, those of the vicinity. 4. The rise of
land, or the fall of the sea, to their present relative position.

Other osseous breccias are common in the vicinity, some being
at least 500 feet above the level of the present Mediterranean :
the cement reddish, and often vesicular ; the vesicles being lined
with carbonate of lime. A portion at least of this osseous breccia
would seem to have been formed beneath the sea, for it contains marine remains ; and among other things those of a *Caryophyllia* at Vil-
lafranca. Independent of the fissures containing the remains of terrestrial animals, there are others, merely affording marine remains,
which remains do not seem to differ from the actual inhabitants of
the Mediterranean, and the breccia appears to have been contemporaneous with the osseous breccias ; the mineral compound in all
cases taking its character from the rock in which it occurs.

Similar osseous breccias occur at Gibraltar, Cette, Antibes,
Corsica, Sardinia, and various other places on the shores of the
Mediterranean.   The bones appear to belong to three species of
Deer, one at Gibraltar, one at Nice, and one at Pisa ; *Antelope*,
one species (Nice) ; *Felis*, two species, one large, the other small
(Nice) ; *Fox*, one species (Sardinia) ; *Mouse* (Corsica, Sardinia) ;
Lagomys, two species (Sardinia, Corsica) ; *Rabbit*, one species
(Gibraltar, Pisa, Cette) ; and *Lizard*, one species (Sardinia).

M. Brongniart considers that many of the pisiform iron-ores
which occur in the clefts of some rocks, particularly in the Jura,
were of contemporaneous origin with the osseous breccias.   In

support of this opinion, M. Necker de Saussure observes, that at Kropp, in Carniola, clefts of rocks containing iron-ore worked for profitable purposes, contain the remains of the *Ursus spelæus*. It also appears that the remains of mammalia have been discovered under similar circumstances in the district of Wochein*. According to MM. Thirria and Walchner, there are two deposits of pisiform iron-ore in the north-west part of the Jura (Haute Saone) and in the environs of Bâle, one probably derived in a great measure from the partial destruction of the other, which occurs between the oolitic group and the supercretaceous rocks. The most recent deposit sometimes contains the remains of the rhinoceros and bear, and is considered of the same geological date as the osseous breccias †.

There would appear to be much analogy between many ossiferous caverns, the osseous breccias, and some clefts containing iron ore, leading to the presumption that the animal remains contained in them have been introduced under certain general circumstances. The great cleft before noticed, at Oreston near Plymouth, seems to have been quite open when the elephant and rhinoceros remains were introduced into it; the accumulation of angular fragments, many of them very large, and ninety feet deep, having taken place since the remains were deposited; marking no transport from a distance, but a simple falling in of fragments, of the same nature as that of the rock on each side (grauwacke limestone).

Before we conclude this subject, we should notice an ossiferous cavern on the banks of the Meuse, at Chockier, about two leagues from Liege, which exhibits some curious circumstances. Fragments of limestone, of the same kind as that in which the cavern occurs, are mixed with some quartz pebbles, and with bones, mostly broken; the whole being united by a calcareous cement. The bones and teeth occur equally in the solid breccia and mud, which, with three beds of stalagmite, nearly fill the cavern. It is stated that bones were discovered beneath each of these three distinct beds of stalagmite. The remains belong to at least fifteen species of animals,—elephants, rhinoceroses, cavern bears, hyænas, wolf, deer, ox, horse, &c. The most abundant being those of bears, hyænas, and horses‡.

* Ann. des Sci. Nat. Jan. 1829.
† Mém. de la Soc. d'Hist. Nat. de Strasbourg.
‡ Journal de Géologie, t. i. 1830. While this sheet was passing through the press, information was received of osseous breccia in Australia :—see Appendix B.

Section IV.

SUPERCRETACEOUS GROUP.

(Syn. Superior Order, *Conyb.* ; Tertiary Rocks, *Engl. Authors* ; Terrains Tertiares, *Fr. Authors* ; Tertiärgebilde, *Germ. Authors* ; Terrains Izémiens Thalassiques, *Al. Brong.*)

Prior to the labours of MM. Cuvier and Brongniart on the country round Paris, the various rocks comprised within this group were geologically unknown, or were considered as mere superficial gravels, sands, or clays. Subsequent to the publication of their Memoir (1811), it has been found that the geological importance of these rocks is very considerable, and that they occupy a large part of the superficies of the present dry land, entombing a great variety of terrestrial, fresh-water, and marine remains. It was observed that in the vicinity of Paris, and for certain distances around, the organic remains detected in the different beds were not all marine, but that fresh-water shells and terrestrial animals of genera now unknown were not uncommon ; and by prosecuting the discovery, it was found that these remains were deposited in beds, each holding a certain place in a certain series*.

* While these discoveries were proceeding in France, Mr. William Smith,—a name that must always be remembered with respect by the geologists of Britain,—was working on more ancient rocks, and, amid a thousand difficulties, identified strata in various parts of England by means of organic remains. It is true that he did not publish regular works until 1815 ; but it is equally true, and well known, that fossils constituted his mode of tracing equivalent beds long previous to this period.

According to M. Keferstein, Fuchsel (a German geologist) had observed that certain beds between the Hartz and Thuringerwald, and around Rudelstadt, were characterized not only by their mineralogical structure, but by their organic contents, as early as 1762 and 1775. This is proved by two works of Fuchsel, one in 1762, entitled, *Historia Terræ et Maris, ex Historia Thuringiæ per Montium Descriptionem erecta;* the other in 1775, entitled, *Entwurf zu der æltesten Erd-und-Menschengeschichte.* Fuchsel seems to have determined the relative position of the rocks now known, as the Muschelkalk, red or variegated sandstone, the Zechstein, the copper slate, and the Rothe Todte Liegende. His theoretical geology is remarkable, and far superior to that of Werner, which afterwards became so prevalent. " He states that the continents were formerly covered by the sea until after the formation of the muschelkalk : but as certain beds only contained vegetables or terrestrial animals, this sea must have been surrounded by a continent more elevated than it, and which occupied the place of the present ocean. This land has *by degrees*

As might have been expected from their labours and those of Mr. Smith on the older rocks of England, the presence of fossils in particular strata was instantly generalized; and it became a well received theory for a considerable time, that every formation or particular set of beds contained the same organic remains, not to be discovered in those above or beneath. This opinion has gradually given way before facts; and the present theory seems to be, that though certain shells may not be precisely peculiar to certain beds, they are more abundant in them than in others, and that the uniformity of organic contents is greater as we descend in the series of fossiliferous rocks: so that the older the beds, the greater will be the uniformity over considerable spaces; and the newer the series, the less the uniformity. How far this opinion may be correct, can only be determined by an accurate examination of rocks in distant parts of the world; and most probably we shall be indebted to the American geologists for the first great advance on this subject. But while we thus wait for information, it may be remarked, that such an opinion is not inconsistent with that which supposes the world to have once been a heated mass, which has gradually cooled at the surface. These observations have been rendered necessary, as, in the group of rocks under consideration, a great variety of organic remains, in many cases of a different character, is found in deposits not far distant from each other.

During the deposit of the different rocks comprehended within

been swallowed up by the sea. Debacles have often carried masses of vegetables into the sea, which have been covered by marine mud. Similar changes may now take place; *for the earth has always presented phænomena similar to those of the present day.*" Fuchsel may therefore in some measure be considered the first propounder of the theory of actual causes, as indeed is further shown by M. Keferstein in his analysis of the two memoirs above noticed. " He (Fuchsel) found that in the formation of deposits Nature must have followed existing laws; every deposit forms a stratum, and a suite of strata of the same composition constitutes a formation, or an epoch in the history of the world: the currents of the ancient sea may be determined by the direction of the formations. There are many chemical deposits the formation of which remains inexplicable. All the sedimentary deposits have been formed horizontally, and have accommodated themselves to the inferior surface. The inclined beds occur in that position in consequence of earthquakes or oscillations of the ground, catastrophes which have produced a considerable quantity of mud, which distinguishes the deposits which pass from one into the other." (Keferstein, Journal de Géologie, t. ii.)—The above and other observations are mixed with remarks characteristic of an infant science, but such remarks are comparatively few in number. Altogether, Fuchsel seems to have been a very remarkable man; and, as M. Keferstein observes, it was little creditable in Werner, that while he adopted his ideas as to strata and formations, he should have followed them so much less logically.

this group, the various operations of Nature would seem to have proceeded, uninterrupted by a catastrophe so violent, or by any condition so common to a large surface, as to produce a deposit of similar substances, characterized by great depth and by similar organic remains, over Europe; for to this comparatively limited area it would yet seem prudent to confine our generalizations. Under this state of things, springs would deposit the different substances which they are capable of holding in solution : and if the theory of internal heat and of a great decrease of surface temperature be well founded, they would generally be hotter than at present; i. e. the number of thermal springs might be greater ;—so far an important consideration, as perhaps more silex would be dissolved and deposited then, as indeed might be the case with many other substances *. It may be here remarked, that this consideration would have weight throughout the deposits of an older date ; so that the older the class of rocks, the greater would be the probability of an increased number of thermal springs, and consequently, the greater the abundance of the siliceous and some other deposits.

Whether this hypothesis be correct or not, it is geologically certain that the superficial temperature has decreased, and, as Mr. Lyell has observed, shows itself in the rocks under consideration, even when the organic remains they contain are of the same species of animals as those which now exist; for they are found, as is to be seen in Italy, larger than those which live in the neighbouring seas, thereby pointing out their probable growth beneath the influence of a warmer climate.

A difference in climate would also produce other variations visible in the supercretaceous rocks, as also in those which were previously formed. The warmer and more tropical the climate, the greater, perhaps, might be the evaporation and the fall of rain, as also the power of many meteoric agents. Consequently, under this hypothesis, the earlier the deposits, the more they would present evidences of having felt the influence of such climates. Tropical rains bursting upon high mountains like the Alps, even supposing a portion of them not to have been so lofty as at present, would

* The manner in which some solutions of silex are effected seems as yet unexplained. It is well known that the Grasses, Canes, and other plants of the same natural family, have an external coating of silex,—a wise provision of Nature for their protection. But the most remarkable siliceous secretion with which we are acquainted, seems to be that which takes place in the cavities of the Bamboo, and is known by the name of *tabasheer.* Dr. Turnbull Cristie informs me, that the *tabasheer* found in the *green* bamboo of India is perfectly translucent, soft, and moist; but that after its exposure to the atmosphere its moisture evaporates, and it becomes opaque, hard, and of a white or gray colour, such as it appears when brought to Europe.

produce very different effects from those we now witness in the same regions. Torrents of water would be suddenly produced, of which the present inhabitants of those mountains have no conception; and the body of detritus borne down by them would be vastly greater than that carried forward by the present Alpine torrents, though these are by no means inconsiderable. So that the differences produced on land by the greater power of meteoric agents in warm regions should always, supposing this hypothesis correct, be taken into account; particularly when it is apparent from a succession of beds observed in the same district, that the temperature under which the deposits have taken place has gradually diminished.

Let us now inquire how far vegetation could counteract the superior decomposing and transporting power of atmospheric agents in tropical or warm climates. It appears that, all other circumstances being equal, the warmer the climate, the greater the body of vegetation produced in it. The question then is, does vegetation protect land from the destructive agency of the atmosphere? We can scarcely reply, except in the affirmative. Indeed, if we wanted evidences of it, we might find them in the artificial mounds of earth, or barrows, so common in many parts of England, which have been exposed to the action of the atmosphere in this climate for about two thousand years, and yet have not suffered any marked alteration of form, though only covered with a short turf for at least a considerable portion of that time. Now if it be admitted that vegetation, to a certain extent, protects land beneath it, it will follow that the greater the vegetation, the greater the protection; and consequently, that land is always defended from the destructive agency of the atmosphere in proportion to the protection required. Without this provident law of nature, the softer rocks in tropical regions would speedily be washed away, and the soil would be unable to support animal and vegetable life; for though in many tropical countries large tracts of apparently barren wastes suddenly seem to spring into life, and are covered with a brilliant green herbage, as if by enchantment; after two or three days of rain, the roots, which when wetted send up such vigorous shoots, and those of the by-gone annuals, whose seeds now develope green leaves, are matted together in such a manner as to produce considerable resistance to the destructive power of the rains*.

It is by no means intended to infer that the degradation of land is not greater in the tropics generally than in milder climates, but merely to state that there is a relative proportion of vegetable protection in both. Suppose a rainy season, such as is common in the tropics, to fall on England; who would doubt that large tracts of land

---

* In the savannahs of the western world, there is frequently very little vegetation, and the consequent loss of surface is considerable.

would be bared, and that the barrows before noticed would speedily disappear: and that if the rains of the English climate were to fall in the tropics, there would be scarcely such a thing as vegetation in the low lands, the water thus produced being insufficient to support the tropical plants? and though it might tend to degrade the land, it would be so speedily evaporated that little would be effected in that manner. The rains and the vegetation are proportioned to each other: but the destruction of the land still remains in proportion to the quantity of rain and the superior force of many meteoric agents; so that, all other circumstances being the same, the heavier the rains, the greater the destruction of land; and consequently the warmer the climate, the greater the degradation of the hills*.

It must also be borne in mind, that during the epoch in which the supercretaceous rocks were formed, subterraneous forces would probably be not less active than they were previously or have been since. We should expect to find igneous rocks of various kinds intermixed with the aqueous deposits; and, under favourable circumstances, interstratified with them; approaching through a succession of ages so nearly to the character of modern volcanos, more particularly as their exposure to ordinary destructive causes would be gradually less, that it would be exceedingly difficult to say where the modern volcano commenced and the ancient volcano ceased. There is also no reason why the same vent should not have continued to vomit forth various substances for a long succession of ages and during various changes on the earth's surface, as has been previously noticed; so that our endeavours to classify their products may not be very successful. Great movements in the land may have been effected, altering the general levels of various districts; and even ranges of mountains may have been thrown up, producing consequent effects that may have greatly influenced certain deposits.

It has been observed that the supercretaceous rocks present numerous instances of fresh-water deposits, scattered over a considerable surface,—a fact which seems to point to a large continuous body of land; in other words, to the presence of considerable continents or

* In tropical countries the parasitical and creeping plants entwine in every possible direction, so as to render the forests nearly impervious, and the trees possess forms and leaves best calculated to shoot off the heavy rains,—thus affording protection to innumerable creatures which seek shelter, at such seasons, beneath them. The pattering of the tropical rains on such forests is heard at distances which an inhabitant of the temperate regions would little suspect, and is particularly striking to a stranger. The rain, thus broken in its fall, is quickly absorbed by the ground beneath, or thrown into the drainage depressions, where, it must be confessed, the torrents thus produced are sufficiently furious, and cause great destruction.

large islands. And this opinion seems strengthened by finding the remains of large mammiferous animals entombed in the same rocks, which are termed fresh-water, because marine remains are not detected in them, their organic contents being either the exuviæ of animals of which the analogous kinds inhabit lakes or rivers at the present day, or else of animals or vegetables whose analogues are found only on the dry land. It is also inferred that these remains could only have been entombed beneath deposits in rivers or in lakes, whence also they are often named lacustrine rocks. Independent of these lacustrine or fresh-water formations, there are others of a mixed character, wherein the organic remains are terrestrial, fresh-water, and marine; and these are considered as deposited in estuaries, from analogous assemblages of this kind now supposed to be forming in such situations. The rocks containing only marine remains speak for themselves : but it by no means seems to follow, that because a rock may contain terrestrial or fresh-water remains, the origin of the deposit is necessarily an estuary ; for if analogies be always sought in the present state of things, we know that such remains are frequently carried far beyond the mouths of rivers.

It is a common practice to describe the supercretaceous rocks as occurring in basins, such as the London, Paris, Vienna, Swiss, and Italian basins; but this term seems often exceedingly misapplied : for great marine deposits were, one would suppose, no more liable to have been formed in basins formerly than now, when certainly, unless we often term the great bed of the ocean a basin, we should by no means characterize the deposit as taking place in such a cavity. Thus we should ill characterize the delta deposit of the Ganges by terming it basin-shaped. It is a common thing to speak of the London basin, when the supercretaceous rocks which occur in this supposed basin, seem little else than the continuation of a great belt of these rocks which extends through Europe by the north of Germany towards the Black Sea. We also hear of the Isle of Wight basin, as if there had existed a separate cavity or depression in that particular place ; while there is very good reason for supposing, (as has been stated by Prof. Buckland,) that the supercretaceous deposits of London and the Isle of Wight had once been continuous, but that this continuity had been destroyed by the upheaving of the chalk beneath, subsequently to the deposit of these rocks; and that the intervening upraised portion has been removed by denudation, as has happened to much thicker and harder rocks. The same with the Paris basin, which may have easily been connected with those above mentioned, and as easily separated from them, by movements of the earth, and by denudation. It may therefore have happened, that these so called basins were formerly continuous portions of one whole, which various circumstances have disunited, perhaps even during the deposit of the rocks in question; their commencement having been in a sea which washed the older strata, and extended from the west of Europe, between Scandinavia and northern Germany, towards the Black Sea. Outliers of these rocks, similar to those of other deposits, are seen on the hills in the West of England, attesting their elevation, and the denudation which has destroyed the continuity of their mass, and left detached portions like islands fringing a continent. In consequence of the various movements of lands, and the denudation either consequent on them or some other cause, the barriers of many a fresh-water deposit are removed; and though, from analogy, we consider them as formed beneath the waters of lakes, we are totally unable to point out the shores of such pieces of water. The student should be careful to keep this great denudation in mind, not only as applicable to the rocks under consideration, but also to the various changes and deposits that have previously occurred : indeed he may consider that no considerable portion of the earth's surface has ever remained long, geologically speaking, in a state of rest; but that the rise and depression of land, and the removal of a large proportion of it, have been frequent. Even in the rocks now treated of, he will be called upon to consider that there have been an alternate rise and depression of land, to account for an alternation of marine and fresh-water deposits; and this he will perhaps be the more ready to do, as he has already seen that such movements of the land have happened at a more recent period.

Amid so great a variety of deposits, attesting such different modes of formation, it is no easy task to know where to begin in the descending series, or what may be precisely contemporaneous. In this difficulty, perhaps the safer course is to consider those deposits the most modern which contain organic remains bearing the closer resemblance to the animals and vegetables now existing. Now all the terrestrial animals found in caves and superficial gravels, marls, and sands, whatever may be the theory formed to account for their disappearance, must have lived upon lands existing at the period under consideration : and even supposing them in a great measure destroyed by a catastrophe, there is nothing to prevent their having been abundantly entombed during their residence on the earth. For while the extinct bears and hyænas were the inhabitants of caverns, generation succeeding generation in their possession, the great work of nature was proceeding; and the elephants, rhinoceroses, hippopotami and other animals, some of which were dragged into the hyænas' dens, were perishing from old age or accident, and their remains included in the various deposits then forming. The same with land and fresh-water animals, marine remains, and vegetables.

The nearer also, judging from organic remains, that the climates can be considered like those now existing, the greater would appear the probability, that the rocks containing them occupied the higher part of the supercretaceous series. Thus in the tropics we should expect to find, among the most recent of these beds, remains analogous to those now existing in similar regions; while as we approached each pole, we should be prepared to discover organic remains corresponding with the various latitudes. As far as facts have yet gone, this would seem to be the case; for the fossil vegetables found in the more recent strata in the tropics are tropical, while those discovered in contemporaneous deposits in Europe are not so, but more suited to the climate; as for instance, the vegetable remains of Œningen*.

In the supercretaceous rocks of Italy and the South of France, and probably also of other Mediterranean countries, there seems better evidence of the nearer approach of organic life to that now existing, than has yet been pointed out elsewhere, though other evidence is not wanting. Indeed, it may be exceedingly difficult to separate the actual state of animal and vegetable life from that which preceded it in the more recent deposits of Italy, or precisely to say when marine remains similar to those now existing in the Mediterranean were raised to various heights above it.

In the more modern supercretaceous deposits of the Apennines, commonly termed Subapennine rocks, it is well known that there is a mixture of species such as now exist in the Mediterranean, and of those found in warmer climates. The deposit noticed by Mr. Vernon in Yorkshire may not be far removed from this date, as land and fresh-water shells were found precisely similar to those now existing, though mixed with the bones of elephants, &c.†

* Should it eventually be found, that the organic remains discovered in tropical countries are always characteristic of such climates, or of one which may be termed ultra-tropical, it will go far to prove that the axis of the earth has not changed, but that the present equatorial regions have always been under the influence of considerable heat, which, though it may have decreased with that of the surface of the world generally, still produces a far more vigorous vegetation than is to be found in the north or south. Should attentive examination also show that at a certain term in the series of rocks, the nature of the vegetable and animal remains found entombed in the tropics, does not point to a more elevated temperature than a similar term in a general series in Europe, or in any more northern or southern latitude, it would seem to show that the cause of this equal temperature has not been external but internal; for, with any arrangement that may be made in the relative positions of the earth and sun, we cannot conceive one which should produce an equal, or nearly equal temperature over our spheroid; while we might conceive such a state of things possible, if an internal heat be capable of producing an equable surface temperature, independent, in a great degree, of solar heat.

† Phil. Mag. and Annals of Philosophy, 1829—1830.

According to M. Elie de Beaumont, there exists in the valleys of the Isère, Rhone, Soane, and Durance, a large deposit of rolled pebbles and sands, clearly distinguishable from that which accompanies the transported blocks, and more ancient than it. It is not in general distinctly stratified, but seems rather to constitute a deep mass, sometimes several hundred yards thick. The rolled pebbles can all be traced to the Alps, and are unmixed with the fragments of distant rocks. Lignite occurs in it, and apparently bears the marks of slow deposit. At one place, (Vallon de Roize, near Pommiers,) the lignite is covered and supported by rolled pebbles, and is itself inclosed in a fine-grained and earthy bed: the carbonaceous mass is divided into even strata, between which numerous shells of *Planorbes* are discovered. M. Elie de Beaumont remarks, that in places where the parts are slightly agglutinated, the sands, mixed with mica, strongly remind us of those now brought down by the Rhone, the Isère, and the Durance. This sand sometimes becomes marly and schistose, containing fragments of lignite, which often accumulate into sufficient masses to be profitably worked, the lignite being included between strata of clay, marl, or fine sand, alternating with the rolled pebbles. The lignites of St. Didier are composed of the flattened trunks of trees, in which the woody fibre can still be traced. M. Elie de Beaumont considers these lignites as contemporaneous with those in Savoy, at Novalése, Barberaz, Bisses, Motte-Serrolex, and Sonnaz, near Chambery. This deposit of pebbles and sands is traced through the plain of Bresse ; it is observable in the escarpments of the Rhone between the embouchure of the Ain and Lyon, with the same characters as are observable in the department of the Isère. It may be well studied near Lyon, and is seen at the foot of the Jura near Ambronay and Ambrutrix. Near Ajou there is a deposit of bituminous wood, described by M. Héricart de Thury, who notices beneath a mass of rolled pebbles and argillaceous marls: 1. Blue clay; 2. Lignite; 3. A bed of pebbles; 4. Blue clay ; 5. Lignite; 6. Blue clay, containing the branches, trunks, and roots of trees, more or less well preserved; 7. Red and blue clays; 8. A bed of bituminous wood, very thick and compact. In the first bed of lignite there was sometimes an admixture ; pebbles and numerous terrestrial and fluviatile shells were discovered in the mass.

M. Elie de Beaumont traces the deposit in other directions, and considers it may have been one formed in the waters of a shallow lake, which existed subsequent to the elevation of the Alps of Savoy and Dauphiné, but prior to that of the main chain from the Valais into Austria. The various pebbles seem clearly to be derived from the Alps, and the different lignite deposits appear to show that they were not suddenly transported in a mass. It may not therefore be unreasonable to infer that they were

carried forward by the action of rivers from the Alps into the situations where we now find them. The time required for this
would be very considerable; but with the lignite deposit, as a part
cf the mass, we can scarcely refuse it a gradual formation*.

The same author points out that this mass of pebbles should
not be confounded with those collections of Alpine pebbles and
sands which constitute a very considerable deposit on either side
of the Alps, known commonly by the name of Nagelfluhe and
Molasse; and which had not only been previously formed and
consolidated, but also upheaved before the pebbles and sands
under consideration were transported. These observations in
the same district are highly important; for it must rarely happen
that the Nagelfluhe and Molasse, the pebbles and sand now
treated of, and the transported substances of the erratic block
group, can be distinctly seen, as it were, together, under circumstances which mark their difference.

We should expect that, previous to the supposed convulsion at
the erratic block period, such marks of degradation should be
everywhere apparent; and that the occurrence of river-borne
pebbles, sands and clays, would be sufficiently common; and
would, when not removed by subsequent debacles, be often found
beneath deposits formed by such debacles.

The precise age of the celebrated Bovey coal cannot at present
be well determined, but may conveniently find a place here. A
body of water has evidently passed over it, working hollows in the
clay, and leaving a large deposit of transported substances in
some situations. It also appears to have been tranquilly deposited
in a previously existing depression at this spot. The area comprising the surface of the deposit is far more considerable than is
usually given, and it has certainly once occupied a greater elevation, as a mass, than it now does, the upper portion having been
removed by denudation. The principal deposit of lignite occurs
near Bovey Tracy, Devon, at the north-western end of the deposit. The upper part is composed of quartzose sand, probably
derived from the granite country near, of portions of rocks of the
immediate neighbourhood, and of rounded pieces of clay, which
appear portions of the clay that accompanies the Bovey coal
deposit.

Beneath about twenty feet of this *head*, as the workmen term
it, there is an alternation of compressed lignites; shales, or clays;
the whole mass dipping at about 20° to the S.E. or S.S.E. The
lignite is evidently composed of dicotyledonous trees, many of
which are knotted; and among them a curious seed is occasionally
discovered.

* Elie de Beaumont, Récherches sur les Rev. du Globe; Ann. des
Sci. Nat., 1829 et 1830.

Other similar parts of the deposit are worked for profitable purposes: but its most useful product is a clay used in the potteries, in some cases so fine as to constitute what is termed pipe-clay. Large quantities of both these varieties of clay are annually shipped at Teignmouth. Lignite more or less accompanies the clay throughout, occurring, when not in beds, as small detached pieces. Animal remains must be exceedingly rare; for I could not, after diligent search, obtain any traces of them, though I was given to understand some shells had been seen near Teignbridge. This deposit has been considered as part of the transported gravels named Diluvium, as also a representative of the plastic clay. It will have been seen that it existed previous to the great transport of pebbles in this district; and it seems more recent than the plastic clay, as there is good reason to suppose that deposits of that age once covered the chalk and green sand, now so extensively denuded in Devonshire, as will be noticed hereafter. And it does not seem improbable that various undulations of this district had been formed subsequent to the deposit of the plastic clay series; which undulations did not very materially differ in character from those we now see, though they may have been greatly modified since. Now the Bovey coal deposit seems to have taken place in a kind of basin, after a general arrangement of hill and dale in the vicinity; for it is exceedingly conformable to their windings, even seeming to run up some valleys, as at Aller Mills, not far distant from Newton Bushel, where there has evidently been an old valley excavated in red sandstone conglomerate, grauwacke, limestone, and grauwacke slate; and in this the alternate beds of lignite and clay, now worked, have been deposited. The deposit has evidently been at one time more considerable in this valley, and has been denuded; for on Milber Down on the one side of it, and on some hills on the other, there are large accumulations of sands and rolled flints; and although it is possible some part of them may be the remains of the green sand, and even of the plastic clay series, the remainder seems to have formed part of the Bovey coal deposit. The following is a section of these rolled flints and coarse sand, apparently composed of triturated quartz and flints, and possibly also chert, on that part of Milber Down facing Ford.

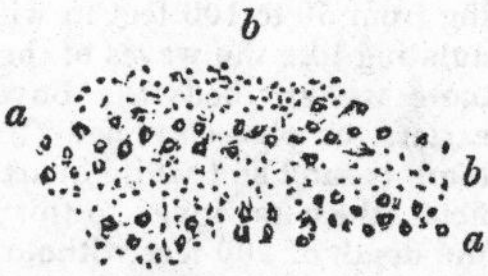

a a, rolled flints; *b b*, coarse sand. The disposition of the two is strongly characteristic of the unequal wash of water, the velocities of which have not been constantly the same over the same spot. A similar mixture of the clay and sand may be seen near Aller Mills. On an inspection of the whole formation, there would,

Fig. 31.

apparently, be little doubt that it was, as before stated, deposited in a pre-existing depression in a variety of rocks. The only question is,—when was this depression formed? For my own part, I should answer,—after the plastic clay on the chalk to the westward had been upheaved. Without, however, the more direct testimony of characteristic organic remains, I should give this answer with much hesitation, it being one for the confirmation or rejection of which future observations are very necessary *. Considering that the relative age of the valleys in this part of England is geologically very important, I have been induced to offer the above notice, as it may lead to further inquiry; though the detail here given somewhat exceeds the limits that should be assigned it.

No doubt, future and delicate observations will detect numerous passages or transitions, in various countries, from a different state of animal and vegetable life to that which now exists, more particularly in marine remains, not so liable to destruction as the inhabitants of dry land. It was long considered that the remains of elephants, rhinoceroses, and mastodons were confined to superficial gravels; but we now know that they are entombed deeper in the series of rocks, and were probably inhabitants of the globe before the *Palæotherium* and some other mammiferous genera became extinct.

It was also once considered that the supercretaceous rocks of England and Paris presented us with all the deposits which were formed between the chalk and the present times; and this theoretical opinion being strongly impressed on the minds of geologists, it was very natural that all supercretaceous or tertiary deposits should be considered as the equivalents of some one or other of those detected in the Paris basin. Such generalizations of local circumstances are common in the history of geology, and are such as would be expected in the progress of any science; for until our knowledge of facts becomes extensive, there is nothing to check such opinions. We must therefore be exceedingly careful not to consider our power of checking such generalizations, as evidence of a clearer sight than those of our predecessors; while, in point of fact, we are merely in possession of a greater mass of

---

* According to Mr. Whiteway and Mr. Kingston, who have possessed the great advantage of continued local observation, the Bovey deposit consists chiefly of five clay beds, and as many of gravel, the latter varying from 50 to 100 feet in width. The clay beds are described as undulating like the waves of the sea; and it is stated that beneath the four more western beds the Bovey coal is found; while below the more eastern or pipe-clay bed (frequently worked to the depth of 80 feet) there is sand and white quartz. Near the S.E. corner of Bovey Heathfield, (the name given to this low district,) the deposit has been bored to the depth of 200 feet without traversing it.—Nat. Hist. of Teignmouth, Tor Quay, Dawlish, &c., by Turton and Kingston.

facts, and are therefore enabled to turn them to a different account. Neither should we be unthankful for these generalizations, for they have promoted inquiry, and have probably contributed, far more than we are often inclined to admit, to that knowledge which we now possess, and which permits us to see that such generalizations are untenable.

The Italian deposits, commonly termed Sub-Apennine from occurring at the lower part of the Apennines, have been appealed to as good examples of a transition or passage from the present state of things to one wherein animals were somewhat different: and this appeal seems well founded; for among the shells discovered in them, there are some which closely correspond with those now existing in the Mediterranean; while there are others whose analogues seem to live in warmer climates, and many are wholly unknown.

In 1829, M. Desnoyers endeavoured to show, 1st, That all tertiary or supercretaceous basins were not contemporaneous, but successively formed and filled. 2nd, That this succession of basins may have resulted from frequent oscillations of the soil, produced during the long series of supercretaceous deposits, by the influence of volcanic agents, then very considerable. 3rd, That this difference in the epoch of the formation of basins may allow us to distinguish many great periods in the supercretaceous or tertiary deposits, some stable, others transitory. 4th, That each of these periods would comprehend deposits formed in the sea, either by the sea waters, or by the rivers, and deposits formed at the same time out of the sea, by lakes, thermal springs, and rivers; both the one and the other offering, according to the basins, every possible variety of sediment. 5th, That the basins of Paris, London, and the Isle of Wight only contain the ancient and middle supercretaceous deposits. 6th, That the last lacustrine rocks of the Seine basin did not therefore terminate the series of these rocks, but that many formations, both marine and fresh-water, have succeeded it in other and more modern basins. 7th, That these more recent formations appear to indicate at least two periods; to which we may add that with which we are contemporaneous. 8th, That all these periods presented, in their deposits and in their fossils, a progressive and insensible passage from one to the other, from the ancient state of nature to the present, from the more ancient supercretaceous basins to the actual basins of our seas. This author also endeavours to establish other opinions, which may however be more questionable; but it seemed necessary to state the above, because there would appear to be much truth in them, and because he was one of the first to point out the probable zoological passage of the ancient supercretaceous deposits to the present state of things, though he was not the first, as he himself remarks, to attribute the variations observed in tertiary or supercretaceous

K

basins to the differences produced by the local action of such causes as we now witness, this having already been done by MM. Prevost, Boué, and other geologists. He also remarks that the continental waters would carry terrestrial and fresh-water shells into the sea, together with the remains of the large mammalia, such as the Elephant, the Rhinoceros, Mastodon, and Hippopotamus, with fluviatile and terrestrial reptiles, which would thus become mixed up with Cetacea and other marine remains *.

The following are some of the facts adduced by M. Desnoyers in support of the opinion, that the large extinct mammalia, such as the fossil species of Elephant, Rhinoceros, &c., are found mixed with the analogues of existing shells on the one hand, and of their having been contemporaries of at least a part of the Palæotherian race on the other. At Montabuzard there is a mixture of the remains of the *Lophiodon* and *Palæotherium*, with the middle-sized Rhinoceros, and the *Mastodon tapiroïdes;* the whole accompanied by terrestrial and fluviatile shells. It is remarked that these more recent supercretaceous rocks of the Loire present a mixture of animals now existing, with those contained in the last marine deposit of the Paris basin.

M. Desnoyers presents a list of fossils, which he considers to be the remains of animals which existed at this epoch. POLYPIFERS: Many species of the genera, *Retepora, Eschara, Flustra, Cellepora, Favosites, Millepora, Theonea, Porita, Alcyonium.* The most common species are the large globular *Favosites* of Guettard (t. iii. pl. 28. fig. 5.), and a polypifer approaching an *Alcyonium.* There are also many other polypiferous genera, such as *Lunulites, Astrea, Caryophyllia,* &c. Species of these genera, more or less similar, are found at Aldborough in Suffolk; in the brown tuff of Carentan; at Rennes; at the Cléons; at Nantes; on the banks of the Layon; near Doué, &c. They are not less abundant in the basin of the Rhone than in those of the Loire. The polypifers occur in various states;—rolled and broken, as on an ancient coast, in Touraine; disposed as a sand, as in a deeper sea, at Doué; in place, and adhering to shells, pebbles, and rocks, on the banks of the Layon (Maine and Loire); and as a solid bed, such as occurs in the ocean, near the Cléons (Loire-Inférieure). ECHINITES: Many large *Scutellæ,* such as the *Scutella subrotunda* (Scilla, tab. 8; Parkinson, vol. iii. pl. 3. fig. 2.), and *Scutella bifora* (Park. vol. iii. pl. 2. fig.6.), are found abundantly in the basins of the Loire, the Gironde, the Rhone, at Malta, and in Sicily; *Clypeaster altus* (Scilla, t. 9. fig. 1 and 2.), *C. marginatus* (Scilla, t. 11. lower fig.), and *C. rosaceus* sometimes accompany them (Reggio in Calabria; Malta; environs of Dax; and Montpellier),

* Desnoyers, Obs. sur un Ensemble de Depôts marins plus recents que les terrains Tertiaires du Bassin de la Seine;—Ann. des Sci. Nat. 1829.

and even seem to replace them (Corsica; Sardinia; Sienna). CIRRIPEDA : *Balanus Tintinnabulum, B. sulcatus, B. Tulipa, B. cylindricus, B. miser, B. pustularis, B. crispatus.* These are common in Italy, and especially in Piedmont, and are for the most part the analogues or varieties of existing species. Some of these species are found in the Loire, where there are also, as in Dauphiné, *B. Delphinus,* and *B. virgatus* (Defrance). Smaller species are found in the tuffs of the Cotentin, which species M. Defrance has named *B. circinatus,* and *B. communis,* and are the same with those named *B. tessellatus,* and *B. crassus,* by Sowerby. *Balani* are abundant in the sands and limestones of Dax, of Beziers, Narbonne and Montpellier; throughout the basin of the Rhone, especially in the environs of Marseille, at Bolène, and Saint-Paul-Trois-Chateaux; in the shelly molasse of Berne and Lucerne ; in the conglomerate of the Leitha, and in the plains of Hungary. M. Desnoyers infers from the habits of modern *Balani,* that the seas containing those enumerated were shallow. Of the CONCHIFERA, the most common species are, *Arca Diluvii, Cyprina Islandicoides, Pectunculus pulvinatus* (numerous varieties): the great *Terebratula perforata,* Defr. (Scilla, t. 16. fig. 6.), considered exceedingly characteristic : the great Oyster with the long spur, of which many species have been made under the names of *Ostrea longirostris, O. crassissima,* and *O. virginica* (Touraine, banks of the Dordogne, the Garonne, and the Lôt; Beziers; Aix; Saint-Paul-Trois-Chateaux; Berne; Bâle; Vienna; Messina): many ribbed species of *Pecten, P. Solarium, P. laticostatus, P. rotundatus, P. benedictus,* Lam., accompanied by small species, *P. lepidolaris, P. striatus,* Lam., *P. gracilis,* Sow. Of the MOLLUSCA the most common are, *Auricula ringens* (very abundant); *Turritella quadriplicata,* Bast., and *T. incrassata,* Sow.; *Pyru laclathrata, P. rusticula; Cyprea Pediculus,* and *C. coccinea; Cerithium margaritaceum, C. papaveraceum,* and *C. granulosum; Rostellaria Pes Pelicani, Crepidula unguiformis, Calyptræa muricata, C. sinensis* (var.), *Conus deperditus,* &c. The above are mixed with terrestrial and fluviatile shells, which sometimes occur irregularly interspersed, while at others they alternate with them. Sharks' teeth and the tritores of fish are common.

MARINE MAMMALIA. Two *Phocæ,* one *Trichecus,* one *Delphinus,* and at least one species of *Lamantin,* described by Cuvier ;—the remains of the latter, common (Doué, Touraine, environs of Rennes and Nantes, Cotentin, near Dax, and some other places in the basin of the Gironde). There are *Cetacea* in the shelly molasse of Dauphiné (Genton), in the Berne molasse (Studer), in the sands of Montpellier (Marcel de Serres).

Without following M. Desnoyers through many cases, of which the relative dates may be questionable, we will proceed to a striking example, where there would appear little doubt as to the occurrence of the remains of the large mammiferous animals buried in more ancient strata than those noticed in the erratic block group. There is a mixture of the remains of *Mastodon* and *Palæotherium* in the basin of the Loire, in the Touraine faluns. According to M. Desnoyers the bones are broken and worn, their substances black and hard, often siliceous, and altogether resembling, in these respects, the marine mammalia which accompany them. The bones are stated to be found in many points of the great faluns to the east of Saint Maure. Some are covered with *Serpulæ* and *Flustræ*, showing that they have remained as bare bones for some time in the sea. The remains are stated to be those of the *Mastodon angustidens, Hippopotamus major ? H. minutus, Rhinoceros minutus,* and also one of the larger species of the *Tapirus giganteus,* of a small *Anthracotherium, Palæotherium magnum,* the Horse, of one of the *Rodentia* of the size of a Hare, and of one or two Deer*.

It has long been known that at Mont de la Molière, near Estavayer, Switzerland, the remains of the Elephant, Rhinoceros, Hog, Hyæna, and Antelope occurred in the molasse of that hill†; and I remember having had the remains of the Mastodon and Castor pointed out to me by Professor Meisner of Berne in 1820, as having been obtained from the lignite of the Swiss molasse‡, so that the probable antiquity of these large mammalia has been for some time remarked.

According to M. Desnoyers, a mixture of bones of the *Lophiodon* and *Palæotherium,* with those of the *Mastodon tapiroides* and middle-sized Rhinoceros, are found, accompanied by terrestrial and fluviatile shells, at Montabuzard.

How far the various deposits, to which the English crag has been added, and which have been referred to one epoch, may really be contemporaneous, it will probably require much time to determine; but at all events the facts stated are important, as they show that the Mastodons, Rhinoceroses, and Hippopotami existed as genera at the same time with the *Lophiodon* and *Palæotherium,* and that the former continued to inhabit certain parts of Europe when many molluscous animals existed, similar or analogous to some of those contemporaneous with ourselves.

Great mammalia are stated to be found in the blue marl of Italy, at Peruggia, Parma, and the Val di Metauro, as also in the sandy deposits of other places of the same country.

The English crag, though often mentioned, is nevertheless not

* Desnoyers, Ann. des Sciences Naturelles, 1829.

† Bourdet de la Nièvre, Soc. Lin. de Paris, 1825.

‡ Professor Meisner had printed a notice of them, with a plate, in a work then in the course of publication at Berne, but of which the exact title has escaped my recollection.

so perfectly known as it should be. It occupies a surface with a
variable outline in Norfolk and Suffolk, as will be seen by Mr.
Taylor's map, and moreover appears to be somewhat changeable
in its character. The same author has given sections of it in his
" Geology of East Norfolk," where it will be seen to rest indiffe-
rently on chalk and London clay. The following is a list of some
of its organic remains, as appears in Mr. Woodward's "British
Organic Remains," including the same author's MS. notes on the
Norfolk crag. POLYPIFER *Turbinolia sepulta.* To this may be
added a great variety in the possession of Mr. Taylor. RADIARIA :
Fibularia Suffolciensis. ANNULATA: *Dentalium costatum.* CIR-
RIPEDA: *Balanus crassus, B. tessellatus, B. balanoides?* (Woodward). CONCHIFERA : *Solen siliqua? Panopæa Faujasii, Mya
arenaria, M. Pullus, M. lata, M. subovata, M. truncata? Mactra
arcuata, Mactra dubia, M. ovalis, M. cuneata, M. magna, M. Listeri? Corbula complanata, C. rotundata, Saxicava rugosa, Petricola laminosa, Tellina obliqua, T. ovata, T. obtusa, T. prætenuis,
Lucina antiquata, L. divaricata, Astarte plana, A. antiquata,
A. obliquata, A. planata, A. oblonga, A. imbricata, A. nitida,
A. bipartita, Venus æqualis, V. rustica, V. lentiformis, V. gibbosa,
V. turgida, Venericardia senilis, Ven. chamæformis, Ven. orbicularis, Ven. scalaris, Cardium Parkinsoni, C. angustatum, C. edulinum, Isocardia Cor? Pectunculus variabilis, Nucula lævigata,
N. Cobboldiæ, N. oblonga, Pecten complanatus, P. sulcatus, P. gracilis, P. striatus, P. obsoletus* (3 var.), *P. Princeps, P. grandis,
P. reconditus, Ostrea Spectrum, Terebratula variabilis.* MOL-
LUSCA: *Chiton octovalvis? Patella æqualis, P. unguis, P. ferru-
ginea,* jun., *Emarginula crassa, E. reticulata, Infundibulum rec-
tum, I. tenerum, Bulla convoluta, B. minuta, Auricula pyramidalis, A. ventricosa, A. buccinea, Paludina subaperta, Natica depressa, N. hemiclausa, N. cirriformis, N. patula, N. glaucinoides*
(var.), *Acteon Noæ, A. striatus, Scalaria frondosa, S. subulata,
S. foliacea, S. minuta, S. similis, S. multicostata, Trochus lævigatus, T. similis, T. concavus* (var.), *Turbo rudis, T. littoreus, Turritella incrassata, Tur. punctata, Tur. striata, Fusus alveolatus,
F. cancellatus, Murex contrarius, M. striatus* (2 var.), *M. rugosus,*
(2 var.), *M. costellifer, M. echinatus, M. Peruvianus, M. tor-
tuosus, M. alveolatus, M. corneus, M. elongatus, M. Pullus, M. bulbiformis, M. lapilliformis, M. gibbosus, M. angulatus, Cassis bica-
tenata, Buccinum granulatum, B. rugosum, B. reticosum, B. tetragonum, B. propinquum, B. labiosum, B. sulcatum* (2 var.), *B. in-
crassatum, B. elongatum, B. elegans, B. Mitrula, B. Dalei,
B. crispatum, B. tenerum, Voluta Lamberti, Ovula Leathsi,
Cypræa coccinelloides, C. retusa, C. avellana.*

It has been stated that the remains of the great mammalia are
mixed with these fossils in the crag, but it does not so clearly
appear that this has been the case. According to Smith the re-

mains of a Mastodon have been there found; and although the bones of Elephants and other animals discovered in the transported rocks above it may, without great care, be easily confounded with the fossils of the crag, there does not appear to be any good reason why such remains should not be discovered in this rock as well as in similar, or nearly similar, strata in other parts of Europe.

The following is, according to Mr. Taylor, a section of the crag strata at Bramerton, near Norwich, whence a large proportion of the organic remains noticed in this rock have been derived. 1. Sand, without organic remains, five feet. 2. Gravel, one foot. 3. Loamy earth, four feet. 4. Red ferruginous sand, containing occasionally hollow ochreous nodules, one foot and a half. 5. Coarse white sand, with a vast number of crag shells, one foot and a half. 6. Gravel, with fragments of shells, one foot and a half. 7. Brown sand, in which is a seam of minute fragments of shells, six inches thick ; fifteen feet. 8. Coarse white sand with crag shells, similar to No. 5.; the *Tellinæ* and *Murices* are the most abundant; three feet and a half. 9. Red sand without organic remains, fifteen feet. 10. Loamy earth, with large stones and crag shells, one foot. 11. Large irregular black flints crowded together, one foot. 12. Chalk, excavated to the level of the river *.

It will be observed from this section that the transporting power of water has been sufficient to carry coarse sand, and even gravel, and that at one time (No. 7.) there has been a drift of broken shells. Mr. Taylor has shown me other sections of the crag strata which present those diagonal lines so frequent in mechanical rocks of all ages, where there have been irregular currents of water. From this circumstance, and from the variations in the component parts of the sections, there would appear reason to believe that the crag strata were deposits from irregular currents of water, varying in their velocities and consequent transporting powers. With regard to the unrolled chalk flints upon which the crag strata rest, they remind us of the apparent dissolution of a portion of the chalk in place, so common over a large part of England and France, previous to the deposit of the supercretaceous rocks.

If we look to the Alps, we find on all sides of that chain beds of various depths of sandstones and conglomerates, forming a whole of very considerable thickness. If we also attentively examine the component parts of the sandstones and conglomerates, we find that the former are generally mere comminuted portions of the latter, and that both have been derived from the Alps. The whole is evidently a detritus of the Alpine rocks, and in it organic remains are by no means common, though they occur in certain situations. Such general appearances would seem to indicate a common origin, and that origin to be the Alps them-

---

* Taylor, Geol. Trans. 2nd series, vol. i.

selves. Rolled and comminuted detritus of the kind found may either be derived by the continued action of what are termed actual causes, or some more violent exertion of forces, which, producing rapid motions in water and greater destruction of the land, should accomplish a far greater quantity of work in a given time.

It is quite evident that in certain parts of the Alps, whatever may be the case in others, these detritus beds rest unconformably on many limestone and other rocks, of which some may be referred to the cretaceous and others to the oolitic series. It also clearly appears that subsequent to their deposit they have been thrown up by some force, which, from the evidence of position of strata, must have proceeded from the interior of the Alps, as the strata are tilted up from it on either side; it thus appearing as if a force had endeavoured to thrust the main body of the Alps higher upwards, and had consequently upheaved the lateral deposits of conglomerates and sandstones with it. The two following sections, one on the north side of the main chain of the Righi near Lucerne, the other on the south side of the same chain near Como, show the disturbed appearance of the conglomerates. Fig. 32 is from the observations of Dr. Lusser; Fig. 33 is a sketch by myself.

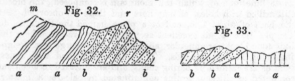

Fig. 32. *m*, Murteberg; *r*, Righi; *a a*, limestone and shales, containing nummulites and other fossils; *b b*, conglomerate of rolled pebbles, composed of pieces of pre-existing Alpine rocks. Fig. 33. *a a*, vertical or nearly vertical beds of gray limestone (containing much silex) covered by the conglomerates and sandstones *b b*, also composed of pre-existing Alpine rocks. There will be little doubt in the mind of the reader, that the conglomerates have been upraised since their deposit, and even have been thrown over at the Righi, if the appearances between that mountain and the Murteberg may not be caused by a fault*. There is also another curious fact, which is, that limestone strata near Como have been upheaved before the deposit of the conglomerate.

If we transport ourselves from Como to the Maritime Alps, we find that these also have been upheaved before the deposit of the

* M. Ebel assured me (while at Zurich in 1829) that this overthrown character was more considerable in other situations in the line of the Righi: it is exceedingly desirable that this should be distinctly determined to be an overthrow of the strata, and not a great longitudinal fault, which might easily accompany a great longitudinal uprise of strata.

rolled fragments, which are clearly derived from the high adjacent country. The rocks upheaved in the vicinity of Nice are compact white limestones, with gypsum, or arenaceous limestones and beds charged with green grains; which latter may, perhaps, be referred to the cretaceous group : but there are other rocks more eastward charged with Nummulites and other fossils, which may belong to some deposits that will be noticed in the sequel.

While on the subject of Nice, it may be as well to notice the supercretaceous rocks of that place generally. After the more regular strata, before noticed, were upheaved, the relative level of the sea and the Maritime Alps must have been very different from what it is at present, for at the height of 1017 feet on the western side of Mont Cao (or Calvo), blocks of the same rock of which the mountain is composed, namely, white compact limestone and dolomite, bear marks of having been pierced by lithodomous shells; and this has been accomplished during a period of comparative tranquillity; for the fragments of rock are angular, and have evidently not been transported from any considerable distance. The same kind of breccia covers the side of the mountain, separating a great mass of rolled pebbles and sandstones from the mass of disturbed white limestone of which the mountain is composed, the blocks still drilled with holes, as may here and there be observed. At the base of Mont Cao, and at a place named the Fontaine du Temple, there is an excellent section of this breccia, where the limestone blocks are angular, and sometimes of large size, weighing many hundred pounds. They are still drilled with holes, such as are formed by boring shells, and are encased in a cement composed of siliceous grains agglutinated by calcareous matter. The whole therefore appears like a state of repose ; but if further proof were requisite, it would be found in certain shells which have much the appearance of *Spondyli*, the lower valves of which are attached to the blocks, not only near the Fontaine du Temple but higher up in the mountain, and the cement has gently covered up the finest of their edges. Now if there had been any considerable motion of the water, more than is common in moderate currents, this could not have happened ; for the fine edges of the shells, and they are very fine, must have been destroyed.

If we proceed to the shores of the present sea, we also find evidences of a tranquil residence of water over the disturbed strata; for beneath the Chateau de Nice we observe an open crack pierced by lithodomous shells, and these shells still remaining in the holes; and we know that this happened before the epoch when such an abundance of rolled Alpine pebbles was carried over this district, because we find part of the cleft filled up by them, burying many of the holes and their inmates. That this residence of the sea was not momentary is shown, as before observed under the head of Osseous Breccia, by the different size of the lithodomous shells and holes, great and small being mixed with each other, affording evidence of a difference of their ages.

In this deposit I have only detected a very large Pecten, found also in Piedmont, the lithodomous and other shells before noticed, the tooth of (perhaps) a Saurian, and some smaller species of Pecten; but no doubt a more extended search would amply repay the geologist.

Near the Fontaine du Temple are some gray marls, resting on the above, which probably constitute the base of the blue or gray marly clay, which next succeeds in the order of superposition. This clay contains a great abundance of marine remains, which have been enumerated by M. Risso *, and of which many are identical with those noticed by Brocchi in the Sub-Apennines. With them vegetable remains are discovered, but these are rare. There is nothing in this deposit which does not mark a continuance of a state of repose. The most delicate shells are well preserved, and all their fine edges are uninjured. Next follows a very different state of things, one in which pebbles of the Alps have been rounded by attrition and conveyed by the force of water over the deposits that have been proceeding so quietly. This force has often torn up the superficies of the clay beds, as must necessarily happen, whether the currents of water thus produced be considered as the currents of rivers or those of the sea; for the force or velocity of water capable of transporting pebbles must necessarily be too great to permit clay or marl to remain at rest; it consequently must cut it up, and leave the surface uneven, producing an irregular mixture of clay, gravel, and sand at the line of junction. Now this is precisely what it has done, as may be seen by the following section, which is not uncommon in the valleys formed in the supercretaceous deposit near Nice, and which only exhibits the unconformable character of the two rocks, it being almost superfluous to adduce examples of the mixture.

Section of the Valley of La Madelaine.   Fig. 34.

c c, bed of the torrent; *a*, blue marly clay; *b b*, beds of rolled Alpine pebbles. This gravel and sand deposit is of very considerable thickness, and dips gently seaward, sloping up to the hills. It spreads out like a fan, the point *a* or centre of the radii being inwards towards the mountains. This form will not help us in determining whether the deposit was successive during a long series of years, by means of a river, or was more sudden, and caused by more violent rushes of water. Be this as it may, it is quite clear that the causes which have operated in this district have not been

* Hist. Nat. de l'Europe Meridionale.

always the same. A period of repose has been succeeded by one of somewhat considerable motion; and if the whole were considered as derived from river detritus, we must suppose that this river was at first by no means rapid, and afterwards acquired considerable velocity; that it continued a quiet river for a considerable period, after which it became a rapid current, no longer transporting mere argillaceous and calcareous particles, but sand and pebbles. The only mode of reconciling these appearances with the river hypothesis seems to be the supposition, that originally, up to and including the period of the clay with its shells, the river was one with a small current, and that the silt was deposited at a distance from the shore;—that the relative levels of sea and land, owing to the elevation of the latter, were somewhat suddenly changed, and that the river-course was lengthened, and the velocity of the current, from the increased declivity of the bed, became sufficient to transport pebbles over the clay*.

Whether we admit this hypothesis, or that of a more sudden rush of waters, a considerable rise of land would seem requisite, as also that the force was exerted between the deposit of the clay and that of the pebbles. If we suppose a sudden rise of land, causing a difference of levels, to the height required, probably a thousand feet and more, the body of waters in the vicinity would be thrown into motion. The waves being in proportion to the disturbing force, and the upraised and fractured land being exposed to all its violence, rounded pebbles would be formed in abundance, and the superficies of the clay washed into inequalities.

It might be considered from a glance at the Maritime Alps, that the clay and the pebbles alternated, and that these alternations merely showed a deposit of one kind at one time, and of another deposit at another; and certainly there are places where they do seem to alternate to a certain extent, particularly at the line of junction. This occurs at Vintimiglia, where the alternating clays contain organic remains; but, nevertheless, the base of the deposit at that place is clay, many hundred feet deep (beneath the Castel d'Appio), and the top is a mass of pebbles. So that, under either hypothesis, we are compelled to admit a great change in the velocities of water passing over the same situation, one from slow to rapid; and it seems difficult to explain this on any other principle than a change, more or less sudden, in the relative levels of the sea and land.

This superposition of gravel, in which the rolled fragments are sometimes by no means small, showing a considerable change in the velocity with which water has passed over the same country,

---

* It should be observed, that in certain situations the marl becomes arenaceous at top, changing into a sand; seeming to show that the transporting power had increased more gradually in some situations than in others.

is not confined to the environs of Nice and Vintimiglia, but is to be noticed in other situations between these places and Genoa, and extends on the other side of the gulf into other parts of Italy. The clay is not always present, the causes that produced it not having acted; but I have here and there observed fragments of rock beneath the mass of sand and pebbles, which, by their angularity, position, and occasional mixture with unbroken fossils, seem to show that they have not participated in the transport of rolled pebbles.

If we enter the body of Italy, and continue towards Florence and Rome, we find a series of sands, marls or clays, which contain many of the organic remains of the Nice rocks, and were probably contemporaneous with them; and we may here also observe a change in the velocity of the water which has deposited the different substances. Thus between Sienna and Florence we shall observe a succession of clay or marl, sand and pebbles, the latter particularly abundant on the approach to Florence, and apparently constituting the upper beds. It would therefore appear that the phænomena noticed near Nice are not altogether local, though they may be modified by local causes, but somewhat general. Indeed the structure of many rocks on the other, or Adriatic side of the Apennines, shows that they merely form a part of some great whole, if we look at their mode of deposit, even independent of organic remains, which are found closely to agree. It would no doubt be easy to state generally certain facts that may be observed in the great gulf of supercretaceous rocks which extends into the northern part of Italy, between the Apennines and Alps, and thus to present an appearance of knowledge, and an intimate acquaintance with the whole mass. The more, however, I have looked into parts of this mass, the more I am convinced that our knowledge of those data, that we ought to possess before we generalize, is imperfect. Certainly the Sub-Apennine marls and sands preserve a general character down the whole range of the Apennines into the Adriatic, and from the abundance and nature of their fossils have attracted considerable attention; but their various connections with other rocks, more particularly with those beneath them, and these again with others, yet requires much attention, judging at least from published documents. If the geologist would make a careful section from Rimini to Foligno, on the road to Rome, over the Apennines, he would find much to reward his labours; or if instead of pursuing the high road, he were to keep the coast from Ancona, and observe the various rocks as they successively plunge into the Adriatic, and thus avail himself of coast sections, he would be rendering good service to science. He would find the white limestone of the main chain contorted and twisted in every direction, and many of those rocks which rest upon it not quite so quietly arranged upon it as, theoretically, they ought to be. He would also observe some curious instances of denudation in the more modern rocks, producing numerous isolated and steep hills, crowned by towns and villages, the picturesque arrangement of which, if he be also an admirer of beautiful scenery, will add not a little to the pleasures of his journey.

In the basin between the Jura and Alps, in Switzerland and thence into Austria, there are immense accumulations of rolled pebbles and sands, known, generally, by the names of Nagelfluh and Molasse, the whole composed of Alpine detritus, and entombing terrestrial, fresh-water, and marine remains. Various artificial divisions have been made in this mass, and parts of it have been considered equivalent to deposits in the Paris basin, *i. e.* of contemporaneous formation with them. M. Studer, who has examined this mass in Switzerland with considerable attention*, agrees with M. Brongniart in referring the molasse to an epoch posterior to the gypseous deposit of the Paris basin. To whatever relative age parts of this mass may be referred, the mineralogical character of the accumulation would seem to show that it was, as a mass, produced by nearly similar causes, such as effected a degradation of, and a transport from, the Alps.

The pebbles are generally of that magnitude which it would require water moving with considerable velocity to transport. We therefore should inquire what current or currents would be able to produce the effects required. If we can obtain a probable explanation of the maximum effects, we may perhaps search for the minor effects in a less intense exertion of the same forces. M. Studer considers that there is evidence of the more recent beds being furthest from the Alps, and nearest to the Jura; this is precisely what would be expected either by the hypothesis of the continued action of meteoric causes, or by that of a series of debacles from the Alps. If rivers have effected the transport of the pebbles, they must, from the size of the pebbles, have had considerable velocity. We should expect the rivers to push forward their detritus into the great basin between the Alps and the Jura: but being once freed from the high mountains and their rocky channels, they would endeavour, as all rivers do when not cut off by rapid tides and currents, to produce deltas, and these might at first cause mixtures of gravel, sand and clay; but the more they were advanced the greater would be their tendency to a horizontal arrangement, and, consequently, the less the velocity of the current, and the smaller the transporting power. Therefore the same river which could once carry large pebbles to the sea would after a lapse of time be unable to do so, unless an elevation of the mountains, whence it flowed, should cause a new system of levels, and the

* Monographie der Molasse, Berne, 1825.

river thus acquire increased velocity, transporting pebbles over the ground, which it formerly only covered with silt or sand. The question then arises, Will the height of the Alps, compared with the distance from these mountains at which large pebbles are found, permit us to consider the transport of these pebbles possible by rivers? In answering this, we must be careful to exclude those superficial gravels scattered over the lower lands, and down the great valley of the Rhine, the transport of which it seems difficult to conceive, except by means of water moving with a greater velocity, and in a greater body, than any river flowing from the Alps could possess. We should only consider those sand- and pebble-beds which constituted the hills on the outskirts of the Alps before they were denuded as we now see them. To do this fairly would require some exceedingly delicate calculations; and we should remember that the warmer the climate the higher the line of perpetual snow, and consequently the greater would be the fall of the running waters. On the river theory, we shall also have to account for the extraordinary equalization of the Alpine pebble-beds, and their general resemblance throughout so long a line of country,—a somewhat difficult task; for if rivers formed the mass, each river would transport its own detritus and push this forward; and though their various deltas might ultimately meet, there would be no stratification common to the whole mass, but one peculiar to each delta. The older or first transported Alpine detritus, marking the commencement of this great degradation of the Alps, rests remarkably even, over considerable spaces, on the rocks beneath them, which is scarcely consistent with their delta or river formation. That these latter now form as much a part of the great transverse valleys as any rock beneath, rising to the height of several thousand feet, is no objection to the river hypothesis; for the causes which upheaved the Alps would upheave these beds with the rest, and they would be traversed by the transverse cracks equally with the lower rocks.

Upon the hypothesis that the pebbles and sands have, in a great measure, been transported from the Alps by debacles, caused by movements in the Alps themselves, which produced corresponding agitation in the seas that bathed their sides, it is not required that these mountains should have been so lofty as seems necessary under the river hypothesis; and the whirling of the waters and currents produced, might equalize the detritus in beds,—not only that detritus which might be broken away during a convulsion, but all that previously formed by the rivers, and on the beaches and deltas, which would give way before the force employed.

While on this subject, let us for a moment consider the Swiss lakes, which occur precisely where they should not, if rivers are to be considered the only excavating forces. The lake of Constance is contained in the rocks under consideration; the lake of

Geneva partly in them and partly out of them in older rocks; the lake of Lucerne the same; and the lake of Neufchatel, with one of its sides bounded by the Jura, and the other by the molasse and nagelfluh. No supposition of river excavation can meet these cases; for the moment the velocity ceases, then will the excavating power cease with it; and we cannot conceive a river cutting out a deep basin bounded on all sides by equal levels, the drainage of which is nearly on a level with the entrance of the river into the basin : but under the hypothesis of a mass of waters thrown into agitation, the difficulty does not appear to be great; for amid the various whirls and great eddies of water inequalities of all kinds must be formed; and although the depressions may appear to us considerable, they are, when compared with the general superficies of land, trifling. If we suppose a body of waters suddenly poured out of the great transverse valleys of the Alps, it would have a tendency to cut up the ground where first discharged upon the low lands, before it had lost its great velocity. I admit that this supposition does not account for all the difficulties; indeed the present remarks are merely made to call attention to the subject, for the lake of Constance is not close to the valley. The position of the lake of Neufchatel is, however, not inconsistent with the idea of a mass of water striking the sides of the Jura. The lake is unequally excavated; and during some soundings which I once made upon it, I found a hill in the middle, but a few fathoms beneath the surface, and with a steep escarpment on one side*. These remarks on the lakes amid the nagelfluh and molasse have been introduced merely to show that other excavating forces than those of rivers would seem necessary to explain some phænomena now observable in this district; and that if such forces have once acted, there does not appear any reason, from the nature of the country generally, that they may not have acted at other times.

In many parts of the mass there would appear evidence of a quiet deposit, as, for instance, the deposits of lignite, such as those of Kæpfnach, which contain the remains of the *Mastodon angustidens*, a Rhinoceros, and a Castor. One of the plants is noticed under the name of *Endogenites bacillaris*. Other lignites occur at Lausanne, Vevay, Ugg, &c., and occur in the lower part of the molasse; *Flabellaria Schlotheimii* being, according to Brongniart, found in that of Lausanne. The remains of the *Palæotherium* have also been discovered in the molasse used for building, near the lake of Zurich. These remains would appear to point to a period when a part of this deposit was forming quietly, and, if fresh-water remains be alone mixed with them, as is stated, by means of fresh water.

* This may be a portion of the more solid rock of the Jura, close to it, which, being harder, better resisted the excavating action than the more easily removed sands and pebbles.

The upper parts of these rocks appear, however, more decidedly to mark the presence of the sea ; for they contain marine remains, such as *Turritella imbricataria,* Lam., *T. Terebra,* Broc., *T. triplicata,* Broc., *T. subangulata,* Broc., *Natica glaucina,* Lam., *Mitra mitræformis,* Broc., *Cancellaria cassidea,* Broc., *Buccinum corrugatum,* Broc., *Cerithium Lima,* Brug., *C. quadrisulcatum,* Lam., *Murex rugosus,* Sow., *M. minax, Pyrula ficoides (Bulla ficoides,* Broc.), *Ostrea virginica,* Lam., *O. edulina,* Sow., *Pecten latissimus,* Broc., *P. medius,* Studer, *Meleagrina margaritacea,* Studer, *Arca antiquata,* Lam., *Cardium edulinum,* Sow., *C. oblongum,* Broc., *C. semigranulatum,* Sow., *C. hians,* Broc., *C. clodiense,* Broc., *C. multicostatum,* Broc., *Tellina tumida,* Broc., *Venus Islandica,* Lam., *Venus rustica,* Sow., *Astarte excavata,* Sow., *Cytherea convexa,* Brong., *Corbula gallica,* Lam., *Panopæa Faujasii, Solen Vagina,* Lam., *S. strigilatus,* Lam. (analogue now living), *S. Legumen,* Linnæus, *Balanus perforatus,* Studer *.

Prof. Sedgwick and Mr. Murchison, in describing the continuation of these rocks on the flanks of the Salzburg and Bavarian Alps, mention great alternating masses of conglomerate, sandstone and marl north of Gmunden ; and still further north, in the higher part of the series, beds of lignite. Detailing the section of the Nesselwang, they state that the lowest supercretaceous or tertiary strata are of great thickness, and are applied vertically against the Alps. The conglomerates are mentioned as extremely abundant, the molasse and marl being entirely subordinate to them. According to these authors, there are three or four distinct lines of lignite, separated from each other by thick sedimentary deposits. Hence they infer that the presence of lignite alone is unimportant, as these occur in very different situations. In a section taken through the hills at the east end of the lake of Constance, the lower part of the supercretaceous or tertiary system is described as composed of green micaceous sandstone, in which beds of conglomerate are subordinate, and it is considered identical with the molasse of Switzerland. The conglomerates alternating with greenish sandstone and variously coloured marls are noticed as forming the upper supercretaceous group, and composing the mass of the mountain ridge extending northwards from Bregenz. Supercretaceous rocks are noticed in the valley of the Inn, containing coal, worked for profitable purposes, thirty-four feet thick, near Haring. The coal is described as accompanied by fetid marls variously indurated. In the coal and overlying beds there are many terrestrial and fluviatile shells, and also in the latter beds numerous impressions of dicotyledonous and other plants. Several marine shells are discovered in these strata. The authors consider that the various sections which they observed prove the com-

---

* Brongniart, Tableau des Terrains qui composent l'Ecorce du Globe.

paratively recent elevation of the neighbouring Alpine chain; and the more recent supercretaceous deposits noticed by them bear the same relation to the neighbouring Alps as the Sub-Alpine rocks in Northern Italy do to the high mountains near them; whence they infer that the northern and western basins of the Danube, and the supercretaceous basin of the Sub-Alpine and Sub-Apennine regions have been left dry at the same period *.

According to Prof. Sedgwick and Mr. Murchison, the supercretaceous rocks of lower Styria consist, in the ascending series of a section from Eibeswald to Radkersburg, 1. Of micaceous sandstones, grits and conglomerates, derived from the slaty rocks on which they now rest at an highly inclined angle. 2. Of shale and sandstone with coal. At Scheineck, where the coal is extensively worked, it contains bones of *Anthracotheria,* and in the shale *Gyrogonites (Chara tuberculata* of the Isle of Wight), flattened stems of arundinaceous plants, *Cypris, Paludinæ,* fish-scales, &c. 3. Of blue marly shale and sand. 4. Of conglomerate, with micaceo-calcareous sand and millstone conglomerate, occupying the whole hilly region of the Sausal. 5. Of coralline limestone and marl. The organic contents of this rock are stated to be, many corals of the genera *Astrea* and *Flustra ; Crustacea ; Balanus crassus, Conus Aldrovandi, Pecten infumatus, Pholas, Fistulana,* &c. The authors refer this rock to the epoch of the Sub-Apennine formations and English crag. 6. Of white and blue marl, calcareous grit, white marlstone, and concretionary white limestone. At Santa Egida, concretionary white limestone, alternating with marls, contains *Pecten pleuronectes, Ostræa bellovicina, Scalaria, Cypræa,* &c. 7. Of calcareous sands and pebble beds, calcareous grits and oolitic limestone. At Radkersburg, where the hills sink into the plains of Hungary, the strata are charged with shells, some being identical with living species (*Mactra carinata* and *Cerithium vulgatum*). The authors consider this group as similar to the more recent rocks of the Vienna basin.

In describing another section Prof. Sedgwick and Mr. Murchison notice that, at the Poppendorf, the marls, sands and conglomerates are crowned by a micaceo-calcareous sand, containing concretionary masses of a perfect oolite, affording a good example, if any were wanting, of the trifling value of mineralogical character in determining rocks far distant from each other †.

Let us now proceed to those parts of the South of France which border the Mediterranean, observing that M. Elie de Beaumont, when remarking on the period at which he considers the Alps to have been thrown up in a direction between Marseille and Zurich, notices numerous situations where the newer supercretaceous strata are

* Sedgwick and Murchison, Proceedings of the Geol. Soc. of London, Dec. 4, 1829.   † *Ibid.* March 5, 1830.

characterised by the remains of Oysters Polypifers, 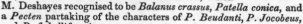 *Patellæ*, the *Balanus crassus* (fig. 35), (which M. Deshayes considers may only be a varietyof *Balanus Tulipa*), *Patella conica*, and other shells. He also identifies these rocks in Provence, Dauphiné, and Switzerland. In the molasse of Pont du Beauvoisin, M. Elie de Beaumont discovered shells, which M. Deshayes recognised to be *Balanus crassus, Patella conica,* and a *Pecten* partaking of the characters of *P. Beudanti, P. Jocobeus,* and *P. flabelliformis*.

Fig. 35.

According to M. Marcel de Serres, the marine supercretaceous rocks of the South of France rest on each other in the following descending order : 1. Sands, generally yellow or white, and more or less argillaceous, calcareous, or siliceous, according to circumstances. These sands abound in the remains of terrestrial and marine mammalia, reptiles, and fish, mixed with the remains of birds, and some wood. Shells are not common, with the exception of *Ostreæ* and *Balani*. 2. Yellow and calcareous marls, of no great thickness, sometimes alternating with stony beds. 3. Beds of limestone, to which the same author has given the name of *calcaire moellon*, usually worked as a building-stone in the South of France. The upper beds generally contain the greater quantity of shells; these and the middle strata also contain the remains of mammalia, fish, crustacea, annulata, and zoophytes. Terrestrial mammalia are very rare, consisting principally of a few bones and isolated teeth, which mostly approach those of the *Palæotherium* and *Lophiodon*. The lower beds contain but few shells. 4. Argillaceous blue marls, well known as the blue Sub-Apennine marls. These marls vary much in their mineralogical character, being more or less calcareous, argillaceous, or sandy, according to circumstances. They have nearly the same colour, passing from a greenish or blueish gray into a blue of greater or less intensity. Their thickness seems to depend on the inequalities of surface, their depth being sometimes very considerable, while at others it is trifling. They contain a large collection of marine remains, principally shells. Terrestrial mammalia and reptiles are exceedingly rare. M. Marcel de Serres only mentions one stag's horn, the bones of a land tortoise, and the vertebræ of a crocodile. Marine mammalia and fish are scarce, as also the remains of zoophytes.

The following is a list of the organic remains, which, according to M. Marcel de Serres, are discovered in the blue marls; which, though long, will be useful in showing the zoological character of

* Elie de Beaumont, Rev. de la Surf. du Globe ;—Ann. des Sci. Nat. 1829 et 1830.

a rock, which seems to preserve the same mineralogical character for a considerable distance *.

LENTICULITES complanata, *Defr.*, Italy, Bordeaux, and C. ;

VAGINELLA depressa, *Bast.*, Bordeaux ; BULLA ampulla, *Lam.*, Italy; B. striata? *Lam.*, Italy, Bord. ; B. hydatis? *Lam.*, Italy; B. truncatula, *Broc.*, Italy, Bord. ; B. lignaria, *Lam.*, Bord., Italy, Paris, England ; TESTACELLA haliotidea, *Draparnaud*, an analogue.

PLANORBIS minutus, *Faujas de St. Fond.*

AURICULA Pisum, *M. de S.*, Italy ; Au. (species resembling Voluta myotis, *Broc.*), Italy ; Au. myosotis, *Draparnaud*, Italy.

TORNATELLA fasciata, *Lam.*, analogous with the existing species, T. allegata, *Desh.*, Paris; T. inflata, *Férussac*, Bord., Paris.

PALUDINA Brardii. AMPULLARIA Faujasii.

MELANOPSIS lævigata, *Lam.* ; M. deperdita, *M. de S.*

MELANIA ventricosa, *Fauj. de St. Fond;* M. pyramidata, *Fauj. de St. Fond.*

NERITA Plutonis, *Bast.*, Bordeaux.

NATICA epiglotina, *Al. Brong.*, Italy and C. ; N. patula, Italy, England ; N. cruentata? *Lam.*, Italy; N. vitellus? *Lam.*, Italy; N. Guilleminii, *Payrandeau*, analogous to the living species, N. Olla, *M. de S.*, Italy ; N. helicina, *Broc.*, Italy.

DELPHINULA solaris (Trochus solaris, *Broc.*), Italy.

TURBO rugosus, *Broc.*, Italy ; T. tuberculatus, *M. de S.*; numerous opercula of the Turbo.

TROCHUS cingulatus, *Broc.*, Italy ; T. striatus, *Broc.*, Italy ; T. magus, *Lam.* ; T. conulus, *Lam.* ; T. Matonii, *Payrandeau* (analogous with the species now living in the Mediterranean); T. Fermonii, *Payrandeau* ; T. zizyphinus, *Lam.*, an analogue; Trochus, resembling T. moniliferus, *Lam.* ; T. patulus, *Broc.*, Bord., Italy; T. agglutinans, *Broc.*, Italy ; T. granulatus †, *M. de S.*

PHASIANELLA pulla, *Payrandeau* (analogous with the existing species) ; Ph. lævis, *M. de S.*

SOLARIUM sulcatum, *Lam.*, Paris; Solarium, very near S. lævigatum, *Lam.*

SCALARIA Textorii, *M. de S.* (Turbo pseudo-scalaris, *Broc.*), Italy and C.; Sc. cancellata (Turbo cancellatus, *Broc.*), Italy ; Sc. lamellosa (Turbo lamellosus, *Broc.*), Italy.

---

* The names which follow those of the authors who have named the species, point out the other localities, or supercretaceous basins as they are termed, in which the same fossil is considered to be found. When the letter C. is appended, it shows that it is also discovered in the calcaire moellon of the South of France.

† This must be carefully distinguished from the *Trochus granulatus* of Sowerby.

TURRITELLA rotifera, *Lam.*, in the marine sands, the calcareous marls, and the calcaire moellon; T. terebralis, *Lam.*, Bord., Italy and C. ; T. terebra, *Lam.* (analogue of the existing species), Italy and C. ; T. turris, *Bast.*, Italy and C. ; T. tricarinata (Turbo tricarinatus, *Broc.*), Italy; T. varricosa (Turbo varricosus, *Broc.*), Italy ; T. cathedralis, *Al. Brong.*, Italy; T. cochleata (Turbo cochleatus, *Broc.*), Italy; T. Archimedis, *Al. Brong.*, Italy; T. serrata (Trochus serratus, *Broc.*), Italy ; T. marginalis (Turbo marginalis, *Broc.*), Italy; T. muricata (Turbo muricatus, *Broc.*), Italy; T. imbricata? *Lam.*, Paris and C.; T. duplicata (Turbo duplicatus, *Broc.*), Italy ; T. perforata? *Lam.*, Paris ; T. acutangula (Turbo acutangulus, *Broc.*), Italy ; T. triplicata (Turbo triplicatus, *Broc.*), Italy and C. ; T. vermicularis, *Al. Brong.*, Italy and C. ; T. fuscata, *Lam.*, (analogous to the species existing in the Mediterranean and Ocean) ; T. Proto, *Bast.*, Bordeaux, Italy and C. ; T. replicata (Turbo replicatus, *Broc.*), Italy; T. quadriplicata, *Bast.*, Bordeaux; T. lata, *M. de S.*; T. corona, *M. de S.*

CERITHIUM marginatum, *Bruguière*, Italy and C.; C. prismaticum, *Al. Brong.*, Italy; C. cinctum, *Bast.* (not C. cinctum, *Bruguière*), Bordeaux; C. cinctum, *Bruguière* (C. lemniscatum, *Al. Brong.*), Italy; C. pictum, *Bast.*, Bord., Italy; C. sulcatum, *Bruguière* (C. plicatum, *Bast.*), Bord., Italy; C. doliolum (Murex doliolum, *Broc.*), Italy; C. plicatum, *Bruguière*, not *Bast.*, Bordeaux; C. papaveraceum, *Bast.*, Bordeaux; C. subgranosum, *Lam.*, Paris; C. tuberculosum? *Lam.*, Paris; C. umbilicatum, *Lam.*, Paris; C. Castellini, *Al. Brong.*, Italy; C. Lima, *Bruguière* (Murex scaber, *Broc.*), Italy; C. mutabile, *Lam.*, Paris; C. bicarinatum, *Lam.*, Paris; C. turbinatum (Murex turbinatus, *Broc.*), Italy; C. vulgatum antiquum, Italy; C. multisulcatum, *Al. Brong.*, Italy; C. calcaratum, *Al. Brong.*, Italy; C. multigranulatum, *M. de S.*; C. alucaster (Murex alucaster, *Broc.*), Italy; C. baccatum, *Al. Brong.*, Italy; C. ampullosum? *Al. Brong.*, Italy.

PLEUROTOMA turricula (Murex turricula, *Broc.*), Italy: P. dimidiata (Murex dimidiata, *Broc.*), Italy; P. muricata, *M. de S.*, Italy; P. auricula (Murex auricula, *Broc.*), Italy; P. textile (Murex textile, *Broc.*), Italy; P. oblonga (Murex oblongus, *Broc.*), Italy; P. contigua (Murex contiguus, *Broc.*), Italy; P. mitræformis (Murex mitræformis, *Broc.*), Italy; P. multinoda, *Lam.*, Paris; P. spiralis, *M. de S.*; P. subulata (Murex subulatus, *Broc.*), Italy; P. Farinensis, *M. de S.*; P. harpula (Murex harpula, *Broc.*), Italy; P. clathrata, *M. de S.*; P. Pannus, *Bast.*, Bordeaux.

FUSUS lignarius, *Payrandeau*, (analogue of the existing species, common in the Mediterranean), Italy; F. subcarinatus, *Al. Brong.*, Italy; F. subulatus (Murex subulatus, *Broc.*), Italy; Fusus, a species between F. Syracusanus of *Lamarck*, and another and undescribed species of the Mediterranean; F. polygonus, *Al. Brong.*,

Italy; F. rugosus, *Lam.*, Paris; F. longirostris (Murex longi-rostris, *Broc.*), Italy; F. uniplicatus, *Lam.*, Paris.

CANCELLARIA clathrata, *Lam.*, Paris.

PYRULA transversalis, *M. de S.;* P. ficoides, *Lam.*, (analogue) of a living species), Italy; P. clathrata, *Lam.*, Italy; P. cla-throïdes, *M. de S.*

RANELLA marginata, *Al. Brong.* (Buccinum marginatum, *Broc.*), Bordeaux, Italy; R. ranina, *Lam.* (an analogue).

MUREX brandaris, *Lam.*, Italy; M. anguliferus, *Lam.* (appa-rently an analogue of the living species), Italy; Murex Motacilla, *Lam.* (an analogue of the living species), Italy; M. craticulatus, *Broc.*, Italy; Murex approaching M. trunculus, Italy; M. inter-medius, *Broc.*, Italy; M. calcitrapoides, *Lam.*, Paris; M. Blainvilii, *Payrandeau* (so like the living species in the Mediterranean, that it cannot be distinguished from it); Murex cornutus, *Lam.* (ap-parently the analogue of the existing species), Italy; M. Haus-tellum, *Lam.* (resembles the living species); M. brevispina, *Lam.* (an analogue of an existing species); M. tenuispina, *Lam.* (an analogue), Bordeaux, Italy; M. crassispina, *Lam.* (analogous to a living species), Italy; M. rarispina, *Lam.* (a complete analogue), Italy; Murex, approaching M. heptagonus of *Brocchi*, Italy; M. tripterus, *Lam.* (var.); M. cristatus, *Broc.*, Italy; M. decus-satus, *Broc.*, Italy; M. transversalis, *M. de S.;* M. rostratus, *Broc.*, Italy; M. oblongus, *Broc.*, Italy.

TURBINELLA infundibulum? *Lam.*, analogous to the existing species.

TRITON corrugatum, *Lam.* (an analogue), Italy; T. pileare, *Lam.* (analogous to a species now living in the Mediterranean), Italy; T. doliare, *Broc.*, Bordeaux, Italy; T. personatum, *M. de S.;* T. intermedium (Murex intermedium, *Broc.*), Italy; T. Chloro-stoma, *Lam.*, (an analogue).

ROSTELLARIA Pes Pelicani (Strombus Pes Pelicani), Italy, Bor-deaux.

STROMBUS pugilis, *Lam.* (a species completely analogous with that now existing in the Mediterranean); S. tuberculiferus, *M. de S.*

CASSIDARIA echinophora (Buccinum. echinophorum, *Broc.*) (an analogue), Italy.

CASSIS Rondeleti, *Bast.*, Bordeaux; C. marginatus, *M. de S.;* C. Diluvii, *M. de S.;* C. striatus, *M. de S.;* C. inflatus, *M. de S.*

DOLIUM, casts of.

NASSA gibba (Buccinum gibbum, *Broc.*), Italy; N. Caronis, *Al. Brong.*, Italy; N. semi-striata, *Al. Brong.*, Italy.

BUCCINUM asperulum, *Broc.*, Italy; B. semi-striatum, *Broc.*, Italy; B. transversale, *M. de S.;* B. corrugatum, *Broc.*, Italy; B. semi-costatum, *Broc.*, Italy; B. Calmeilii, *Payrandeau* (altogether resembling the species so common in the Mediterranean); B. prismaticum, *Broc.*, Italy; B. Lacepedii, *Payrandeau*, C.; Buccinum, apparently approaching B. gemmulatum of *Lamarck*, C.; B. polygonum, *Broc.*, Italy; B. flexuosum, *Broc.*, Italy; B. clathratum, *Lam.*, Italy; B. gibbum, *Broc.*, Italy; B. Miga, *Lam.* (closely approaches the living species); B. angulatum, *Broc.*, Italy; B. reticulatum, *Lam.* (analogous to the existing species), Bordeaux, Italy; B. olivaceum, *Lam.* (apparently an analogue of the living species); B. Turbinellus, *Broc.*, Italy; B. politum, *Bast.*, Bordeaux, Italy; B. mutabile (completely analogous to the living species), Italy; B. crenulatum, *Lam.* (apparently an analogue of the existing species), Italy; B. Carcassonii, *M. de S.;* B. costulatum, *Broc.*, Italy; B. parvulum, *M. de S.;* B. gibbosulum, *Broc.*, Italy; B. pusillum, *M. de S.;* Italy.

TEREBRA duplicata, *Lam.* (an analogue of an existing species, Bordeaux, Italy; T. Vulcani, *Al. Brong.*, Italy; T. pertusa, *Bast.*, Bordeaux, Italy; T. dimidiata (an analogue of the existing species); T. plicaria, *Bast.*, Bordeaux.

MITRA scrobiculata (Voluta scrobiculata, *Broc.*), Italy; M. Brocchii, *M. de S.*, Italy; M. Gervilii, *Payrandeau*, C.; M. pyramidella (Voluta pyramidella, *Broc.*), Italy.

PURPURA Lassaignei, *Bast.*, Bordeaux; P. bicostalis, *Lam.* (analogue of the living species); P. undata, *Lam.* (also an analogue of an existing species).

VOLUTA varricosa, *Broc.*, Italy; V. piscatoria, *Broc.*, Italy; V. citharella, *Al. Brong.*, Italy; V. buccinea, *Broc.*, Italy; V. pyramidella, *Broc.*, Italy; V. tornatilis, *Broc.*, Italy.

RISSOA Cimex, *Bast.*, Bordeaux, Italy; R. cancellata, *Lam.*; R. pusilla (Turbo pusillus, *Broc.*), Italy; R. cochlearella, *Lam.*, Bordeaux, Italy, Paris.

MARGINELLA cypræola (Voluta cypræola, *Broc.*), Italy; M. buccinea (Voluta buccinea, *Broc.*), Italy.

CYPRÆA Amygdalum, *Broc.*, Italy; C. Mus, *Lam.* (analogous to the living species), C.; C. Coccinella, *Bast.*, Bordeaux; C. elongata, *Broc.*, Italy, C.

ANOPLAX inflata, *Al. Brong.*, Italy.

OVULA carnea, *Lam.* (an analogue to the existing species), C.

CONUS betulinoides, *Lam.*, Paris; C. virginalis, *Broc.*, Italy; C. Pyrula, *Broc.*, Italy; C. avellana, *Lam.*, Italy; C. turricula, *Broc.*, Italy; C. Aldrovandi, *Broc.*, Italy; C. Pelagicus, *Broc.*, Italy; C. Mercati, *Bast.*, Bord., Italy; C. canaliculatus, *Broc.*, Italy, and C.; C. deperditus, *Broc.*, Bordeaux, Italy, Paris; C. mediterraneus, *Lam.*, (analogous to the existing species), C.

SIGARETUS costatus (Nerita costata, *Broc.*), Italy, C.; S. striatus, *M. de S.*

PILEOPSIS Paretti, *M. de S.*

CALYPTRÆA lævigata, *Desh.*, Paris; C. muricata (Patella muricata, *Broc.*), Italy, C.

CREPIDULA unguiformis, *Bast.*, Bordeaux, Italy.

PATELLA vulgata? *Lam.*; P. Bonardii, *Payrandeau* (analogous to the existing species), C.; P. Umbella, *Lam.* (also an analogue), C.; P. glabra, *Desh.*, Paris.

FISSURELLA græca, *Desh.* (Patella græca, *Broc.*), Italy, Paris.

EMARGINULA, a species closely approaching E. fissura of *Lamarck*, and E. reticulata of *Sowerby.*

AVICULA, species not determined.

PERNA mytiloides, *Lam.*, Bordeaux, Italy.

LIMA bullata, *Payrandeau* (analogue); L. Breislaki, *Bast.*, Bordeaux, Italy; L. mutica, *Lam.*, Italy; L. nivea (Ostrea nivea, *Renieri, Broc.*), Italy.

PECTEN laticostatus, *Lam.*, Italy and C.; P. benedictus, *Lam.*, Bordeaux, Italy and C.; P. Plica (Ostrea Plica, *Broc.*), Italy; P. scabrellus, *Bast.*, Bord., Italy and C.; P. dubius (O. dubia, *Broc.*,) Italy and C.; P. multiradiatus, Bord., Italy; P. plebeius (O. plebeia, *Broc.*), Italy; P. arcuatus (O. arcuata, *Broc.*), Italy; P. turgidus, *Lam.*, apparently approaches the species found in the American seas; P. lepidolaris, *Lam.*, Italy and C.; P. striatulus, *Lam.*, Italy and C.; P. striatus (O. striata, *Broc.*), Italy; P. inæquicostalis? *Lam.*, Italy; P. Pusio, *Lam.*, Italy and C.; P. scutularis? *Lam.*, Italy and C; P. unicolor, *Lam.*, Italy and C.; P. flabelliformis (O. flabelliformis, *Broc.*), Italy; P. palmatus, *Lam.*, Bordeaux; P. solarium, *Lam.*, Italy; P. terebratulæformis, *M. de S.*, Italy and C.; P. Tournalii, *M. de S.*; P. Phaseolus? *Lam.*, Italy; P. seniensis, *Lam.*, Italy and C.; P. jacobæoides, *M. de S.*, Italy; P. pusioides, *M. de S.*, Italy.

SPONDYLUS gæderopus, *Broc.*, Bord., Italy; S. rastellum, *Lam.*, Italy and C.

HINNITES Brussonii, *M. de S.*, H. Leufroyi, *M. de S.*

PLICATULA, species not determined.

OSTREA canalis, *Lam.*, Paris and C.; O. crassissima, *Lam.*, C.; O. undata, *Lam.*, Bord., Italy and C.; O. virginica, *Al. Brong.*, Italy; O. edulina, *Lam.*, Italy and C.; O. colubrina, *Lam.*, Paris; O. scabrella, *M. de S.*, Italy; O. anomialis, *Lam.*, Italy, Paris; O. flabellula, *Lam.*, Bord., Italy, Paris; O. frondosa, *M. de S.*; O. crenulatoides, *M. de S.*; O. cristata, *Lam.*, apparently analogous to the existing species, Italy; O. corrugata, *Broc.*, Italy.

ANOMIA Ephippium, *Broc.*, analogous to the existing species, Italy; A. costata, *Broc.*, Bord., Italy; A. sulcata, *Broc.*, analogous to the species now living in the Mediterranean, Italy; A. radiata, *Broc.*, Italy; A. cepa, *Lam.*, analogue of the existing species, Italy; A. sinistrorsa, *M. de S.*; A. electrica, *Lam.*, analogue, Italy and C.; A. lens, *Lam.*, closely approaches the living species, Italy and C.; A. Pellis Serpentis, *Broc.*, Italy and C.

PINNA subquadrivalvis, Italy; P. augustana? *Lam.*; P. tetragona, *Broc.*, Italy; P. pectinata, *Lam.*, approaches the living species.

ARCA barbata, *Lam.*, analogue; A. Gaymardi, *Payrandeau,*
apparently analogous to the living species; A. antiquata, *Lam.*,
analogue of the existing species, Italy; A. Diluvii, *Bast.*, Bord.,
Italy and C. ; A. aurita, *Brov.*, Italy; A. biangula, *Lam.*, Italy;
A. lactea, *Lam.*, analogue of the living species, Italy; A. Quoyi,
Payrandeau, analogous to the existing species; A. cardiiformis,
Bast., Bordeaux, Italy; A. Breislaki, *Bast.*, Bord., Italy; A. pec-
tinata, *Broc.*, Italy; A. clathrata, *Bast.*, Bordeaux.

PECTUNCULUS violacescens, *Lam.*, analogous to the living spe-
cies; P. nummarius, (Arca nummaria, *Broc.*), C.; P. pigmæus,
Lam., Paris; P. subconcentricus, *Lam.*, Paris; P. pulvinatus,
Bord., Italy, Paris, England, and C.

NUCULA minuta (Arca minuta, *Broc.*,), Italy; N. pella, *Lam.*,
analogue of the living species, Italy; N. nicobarica, *Lam.*, Italy;
N. rostratus, *Lam.*, (an analogue,) Italy; N. margaritacea, *Lam.*,
analogous to the existing species, Bord., Italy.

MODIOLA discrepans, *Lam.*, C.; M. Semen, analogous to the
existing species; M. subcarinata? *Lam.*; Mytilus edulis, *Bast.*
(*Broc.*), Bord., Italy.

UNIO, species undetermined.

ANODONTA, perhaps many species.

CYPRICARDIA, many species, not determined.

CARDITA ajar, *Lam.* ; C. trapezia, *Lam.*, an analogue ; C. sinuata, *Payrandeau*, an analogue.

CRASSATELLA latissima, *Lam.*

ISOCARDIA Cor, *Lam.*, exactly resembling the living species,
Bord., Italy.

CHAMA intermedia, *Broc.*, approaches Cardita of *Lamarck*, Italy;
C. pectinata, *Broc.*, Italy ; C. gryphoides, *Lam.*, analogous to the
existing species, Bordeaux, Italy.

CARDIUM hians, *Broc.*, Italy; C. punctatum, *Broc.*, Italy ;
C. ciliare, *Broc.*, Italy; C. oblongum? *Broc.*, Italy; C. serratum?
Lam., Italy ; C. rusticum, *Broc.*, Italy ; Cardium approaching
C. tuberculatum, *Lam.*, Italy ; C. rhomboides? *Lam.*, Italy and
C.; C. scobinatum, *Lam.*, C.; C. distans? *Lam.*, Italy; C. lævi-
gatum, analogous to the existing species, Italy; C. edule, *Broc.*
(*Bast.*), an analogue, widely spread, from Antibes to the Pyrenees,
Italy and C.; C. glaucum, *Bruguière*, an analogue; C. fragile,
Broc., Italy; C. striatulum, *Broc.*, Italy; C. planatum, *Broc.*,
Italy ; C. echinatum, *Broc.*, Bord., Italy.

TELLINA stricta, *Broc.*, Italy; T. carinulata, *Desh.*, Paris and
C.; T. zonaria, *Lam.*, Bordeaux; T. tenui-stria, *Desh.*, Italy,
Paris; T. pellucida, *Broc.*, Italy; T. rudis, *Lam.*, Paris; T. sub-
rotunda, *Desh.*, Paris; T. elegans, *Bast.*, Bordeaux; T. depressa,
Lam., analogous to the existing species, Italy; T. elliptica, *Broc.*,
Italy; T. strigosa, *Lam.*, analogous to the existing species, Italy;
T. compressa, Italy and C.; T. pulchella, C.; T. planata, *Lam.*, C.;

T. striatella, *Broc.*, Italy and C.; T. rostralina, *Desh.*; T. nitida, *Lam.*, analogous to the existing species.

Lucina lactea, *Lam.*, analogous to the existing species, C.; L. Scopulorum, *Bast.*, Italy and C.; L. Saxorum, *Desh.*, Paris; L. concentrica, *Lam.*, Paris and C.

Corbis lamellosa, *Lam.*, Paris; C. ventricosa, *M. de S.*

Cyrena, many species not determined.

Cyclas, perhaps many species.

Cyprina islandicoides, *Lam.*, in the marine sands, the calcaire moellon, the blue marls, and in the supercretaceous basins of Bordeaux, Italy, Paris, and England.

Cytherea exoleta, *Lam.*, analogous to the existing species; C. erycinoides, *Bast.*, Bordeaux; C. Lincta, *Lam.*, an analogue, Bordeaux, Italy; C. Chione, *Lam.* (*Broc.*), Italy; C. elegans, *Lam.* (*Desh.*), Paris; C. erycinoides, *Lam.*, Bordeaux, Italy; C. mactroides, *Lam.* ; C. Cypria? (Venus Cypria, *Broc.*), Italy; C. Deshayesiana, *Bast.*, Bordeaux; C. nitidula, *Lam.*, Paris; C. Aphrodite, *M. de S.*, Italy; C. undata, *Bast.*, Bordeaux; C. semisulcata, *Lam.*, Paris; C. incrassata (Venus incrassata, *Broc.*), Italy; C. globulosa?? *Desh.*, Paris.

Venerecardia Jouanetii, *Bast.*, Bordeaux; V. Lauræ, *Al. Brong.*, Italy; V. planicosta, *Lam.*, Paris; V. pinnula, *Bast.*, Bordeaux.

Venus impressa, *M. de S.*; V. angula, *M. de S.*; V. senilis, *Broc.*, Italy; V. Pullastra, *Lam.*, an analogue, Italy; V. Dysera, *Broc.*, Bord., Italy and C.; V. gallina, *Lam.*, an analogue; V. rugosa, *Broc.*, Italy; V. cassinoides, *Bast.*, Bordeaux; V. Pectunculus, *Broc.* Italy; V. radiata, *Broc.*, Italy; V. circinnata, *Broc.*, Italy; V. Lupinus, *Broc.*, Italy.

Donax nitida, *Lam.*, Paris; D. Basterostina, *Desh.*, Paris; D. Fabagella, *Payrandeau*, an analogue.

Mya conglobata, *Broc.*, Italy.

Corbula revoluta, *Bast.*, Bord., Italy. Petricola striata, *Lam.*

Lutraria elliptica? *Lam.*, Italy; L. piperatra, *Lam.* ; L. solenoides? *Lam.*, Italy.

Mactra triangula, *Bast.*, also in the faluns of Touraine; M. crassatella, *Lam.*, England; M. lactea, C.

Solen Vagina, *Lam.*, Italy; S. siliqua? *Lam.*, Italy and C.; S. strigilatus, *Lam.*, Bord., Italy; S. candidus, *Broc.*, Bord., Italy; S. coarctatus, *Broc.*, Italy; all these species of Solen have their existing analogues.

Psammobia Labordei, *Bast.*, Bordeaux; Ps. pulchella, *Lam.*, Italy; Ps. vespertina, *Lam.*, an analogue.

Panopæa Faujasii, *Menard de la Groye*, Bord., Italy and C.

Sanguinolaria, species not determined.

Gastrochæna cuneiformis, *Lam.*, analogous to the existing species, C.

TEREBRATULA ampulla,*M. de S.* (Anomia ampulla, *Broc.*), Italy.
PHOLAS Branderi, *Bast.*, Bordeaux.

CIRRIPEDA. BALANUS tintinnabulum, *Lam.*, also in the marine sands and C.; B. miser? *Lam.*, also in marine sands and C.; B. semiplicatus? *Lam.*, also in marine sands and C.; B. perforatus, *Lam.*, also in marine sands and C.; B. patellaris, *Lam.*, also in marine sands, C., and in Italy; all the above Balani are analogues; B. pustularis, *Lam.*, marine sands, C., Italy; B. crispatus, *Lam.*, marine sands, C., Italy.

ANNULATA. SERPULA quadrangularis? *Lam.;* S. arenaria, *Lam.*, Italy; S. contortuplicata, *Lam.;* S. spirorbis? *Broc.*, analogous to the existing species, Italy; S. spirulæa, *Lam.;* S. ammonoïdes, *Broc.*, Italy; S. annulata? *Lam.;* S. protensa, *Lam.*, analogous to the existing species in the Mediterranean, Italy.

DENTALIUM elephantinum, *Lam.*, apparently analogous to the existing species; D. sexangulum, *Broc.;* D. triquetrum, *Broc.;* D. entalis, *Lam.*, apparently analogous to the existing species, Italy; D. coarctatum, *Lam.*, Italy; D. Tarentinum, *Lam.*, Italy; D. striatum, *Lam.*, Italy, Paris.

CRUSTACEA. PODOPTHALMUS Defrancii, *Desmarest.* (This is the only species noticed by M. Marcel de Serres in the blue marls; but he states that the Atelecylus rugosus, *Desmarest*, is found in the calcaire moellon, near Montpellier. Remains of the genus *Portunus* are also mentioned.)

Species of the echinite family are not stated to occur in the blue marls, but the following are found, according to M. Marcel de Serres, in the calcaire moellon, or calcareous marls.

ECHINUS miliaris, *Lam.*, calc. moel.; E. granularis? *Lam.*, perhaps analogous to the existing species, calc. moel.

SCUTELLA striatula, *M. de S.*, calc. moel.; S. gibercula, *M. de S.*, calc. moel.; Galerites pustulata, *M. de S.*, calcareous marls.

CLYPEASTER altus, *Lam.*, calc. moel., and in Italy; C. marginatus, *Lam.*, cal. moel., and also Bord., Italy; C. politus? *Lam.*, cal. moel., also Italy; Clypeaster, closely approaching C. oviformis of Lamarck, calc. moel.; C. excentricus, *Lam.*, calc. moel.; C. hemisphericus, *Lam.*, calc. moel., also Italy; C. stelliferus, *Lam.*, calc. moel., also Italy; C. gibbosus, *M. de S.*, calc. moel.; C. scutellatus, *M. de S.*, calc. moel.

SPATANGUS canaliferus, *Lam.*, calc. moel.; the specimens of this fossil found highly preserved in the calc. moel. of Barcelona appear quite analogous to the existing species; Sp. lævis? *Deluc*, calc. moel.; Sp. arcuarius, *Lam.*, analogous to the existing species, calc. moel.; Sp. retusus? *Lam.*, calc. moel.

The following section, by M. Marcel de Serres, of the strata of Banyuls, through which the Tech has cut its bed, will remind the geologist of sections to be seen at Nice, and in parts of Italy; 1. (upper bed.) Transported substances, named by the

L

author, *diluvium of the plains*, rolled pebbles of primary rocks, cemented by a brownish red gravelly clay; thickness from one to three yards. 2. Another deposit of transported detritus, named *mountain diluvium* by the author, stated to be distinctly separated from the above, composed of rolled pieces of granite, mica-slate, gneiss, and quartz, cemented by a slightly red clay, more gravelly than the first. The size of the rolled fragments is considerable, the smallest being equal to that of the head; thickness, two to three yards. 3. Yellowish siliceous sands, indurated in parts, the beds thick, varying from four to six yards. Lower portion contains shells and lignites. 4. Argillo-arenaceous marls, blueish gray, and micaceous; sometimes alternating with the upper yellow sands. Shells very abundant; thickness, six to eight yards. 5. Blueish argillaceous and tenacious marls. They contain few shells, and even these become less abundant as the section increases in depth; thickness not known. These marls are supposed to rest upon micaceous clay-slates, from the structure of the Albères chain, at the foot of which these beds of Banyuls dels Aspre are found. Nos. 3. and 4. are stated to contain the remains of mastodons, deer, lamantins, land-tortoises, and sharks, disseminated among the marine shells, but they are represented to be scarce *.

There are many lignite deposits in this part of France, of which the relative ages have not been determined so accurately as could be wished. M. Marcel de Serres, however, shows that some of them are inferior to the calcaire moellon, and probably occur at the lower part of the blue marls. The following is a section at Saint Paulet, about a league and a half from Saint Esprit (order descending) : 1. Yellowish calcareo-siliceous sands, containing the remains of marine shells. 2. Thick beds of the calcaire moellon, containing numerous casts of *Cytherea*, *Venus*, and *Cerithia*. 3. Sands with marine shells resembling No. 1. 4. Alternation of fresh-water limestone (containing *Gyrogonites*), earthy lignite, and sandy marls. 5. Compact limestone, with *Cerithia* or *Potamides* and *Paludinæ*. 6. Thin argillaceous marls, with small oysters. 7. Thin earthy lignite. 8. Argillo-arenaceous marls, with traces of lignite. 9. Compact fresh-water limestone, with *Limnææ* and *Cyrenæ*. 10. Thin yellowish and calcareous marls. 11. Argillaceous blue marls, with traces of more or less fibrous lignite. 12. Argillo-bituminous marls, containing numerous marine and fluviatile shells. These marls, as well as the lignite which succeeds them, contain small pieces of amber. 13. Lignite in beds of two or three yards in thickness, preserving the woody structure, even resembling charcoal: contains amber. 14. Argillo-bituminous marls, with marine and fluviatile shells, the same

* Marcel de Serres, Géognosie des Terrains Tertiaires du Midi de la France. Montpellier, 1829.

as No. 12. 15. Lignite with the same characters as No. 13.—All these beds rest parallel on each other with great regularity, and show that they have been deposited tranquilly and successively.

It has been observed, first I believe by M. Basterot, that there is a great resemblance in the organic contents of the supercretaceous rocks of the South of France, Italy, Hungary and Austria, which would seem to point to circumstances common to them, but not to the supercretaceous basins of the North of France, England, and the Netherlands. Perhaps the remark would more particularly apply to certain parts of the various deposits in each district. It will have been collected from the list of organic remains contained in the blue marls of the South of France, that, though the species enumerated by M. Marcel de Serres are exceedingly abundant, the zoological character of the mass may be said closely to correspond with similar deposits in Italy; a minor quantity of species are common to the Bordeaux district, and the Mediterranean side of France; and a few, and some of these questionable, are referable to species found in the North of France or in England.

Several of the species are analogous with those now existing in the Mediterranean, pointing to some kind of connection between the ancient state of that sea and the present. We therefore seem to arrive at something like a probability, that the blue marls were deposited in a sea, perhaps somewhat similar to the Mediterranean, but presenting more surface than it.

During this state of comparative repose, in which similar mineralogical substances enveloped similar animal remains over a considerable surface, there were some situations in which vegetable matter was more abundantly collected than in others, as might now happen at the embouchures of rivers when the streams possessed no great velocity. The section above given, as observable at Saint-Paulet, is particularly interesting; as it seems to indicate a continuance of the same estuary or delta, in nearly one situation, even after the circumstances which produced the marly or clayey deposit had been somewhat altered, and the velocity of the waters increased from that which transported mere muddy matter, to that which carried forward sands and mud. After the production of the blue marl, circumstances became somewhat altered, and this over a considerable surface,—for the deposit no longer continued the same; sands, showing a greater velocity or transporting power of water, commonly covering these blue marls in the South of France and Italy. There were, however, modifying circumstances; for sheets of calcareous matter, frequently producing limestones, occur mixed with these sands, enveloping terrestrial, fresh-water, or marine remains, as these may come within their influence.

M. Elie de Beaumont notices the following section near the Pertuis de Mirabeau; which, while it shows that the rocks belonging to the cretaceous and oolitic groups of that neighbourhood were disturbed and contorted, previous to the deposit of the supercretaceous rocks which rest upon them, also exhibits the superposition of certain supercretaceous strata of that part of France with which we have been occupied, and which, in the neighbourhood of Aix, present such a curious approach, in their organic contents, to some of the terrestrial inhabitants of the present country.

Fig. 36.

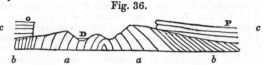

a a, rocks of the oolitic group; b b, rocks of the cretaceous group, containing Ammonites and *Belemnites mucronatus.* D, bed of the Durance at the Pertuis de Mirabeau, on both sides of which rest nearly horizontal beds of supercretaceous rocks (c c) on the upturned edges of the older strata.

On the side P, that of Peyrolles, the supercretaceous rocks constitute a thick fresh-water deposit, "principally composed of gray compact limestone, penetrated by numerous irregular tubular cavities, and of sandstone, analogous to that which near Aix alternates with the variegated marls of the fresh-water series *." On the other side of the Durance, and near the chapel of La Magdelaine (o), the supercretaceous rocks are seen resting on the edges of the older strata, and the following beds are observed in the ascending order. 1. A calcareous sandstone, without shells, in some strata containing calcareous pebbles, and passing into a conglomerate. 2. The above beds, with the remains of marine shells. In these beds M. Elie de Beaumont observed dolomite. 3. A bed containing some limestone pebbles, and a great number of oysters, their hinges elongated, among which are probably the *Ostrea virginica* of the shelly molasse of Piolene and Narbonne; also other shells, among which M. Deshayes recognised *Anomia ephippium, Balanus crassus,* and an undescribed Pecten, resembling the *P. Jacobæus, P. Beudanti,* and *P. flabelliformis.* 4. A considerable thickness of molasse, not very shelly, in one bed of which there are vegetable remains. 5. An oyster-bed, analogous to No. 3, covered by a certain thickness of shelly molasse. 6. A thickness of three yards of a yellow sand, covering an alternation of calcareous sandstone, and a compact blueish gray limestone, with irregular tubular cavities, containing terrestrial and fresh-water shells. M. Elie de Beaumont does not consider this limestone as the same as that noticed on the other side of Durance, but as

* Elie de Beaumont, Rev. de la Surf. du Globe: Ann. des Sci. Nat. 1829 et 1830.

forming the upper part of the supercretaceous series at this place;
while the beds near Peyrolles constitute the lower part of the same
series.

The exact relations of these rocks with the fresh-water deposit
at Aix, remarkable for the insects found entombed in part of it,
do not appear to have been yet well determined. According to
Messrs. Lyell and Murchison the following is a section of the
beds rising above the level of the town of Aix (in the descending
order):—1. White calcareous marls and marlstone, passing gradually into a calcareo-siliceous grit, containing *Cyclas gibbosa*,
Sow.; *Potamides Lamarckii, Bulimus pygmeus*, and an undescribed species of *Cypris;* thickness about 150 feet. 2. Marls,
with plants and shells. 3. Marls, with fish and plants. 4. Bed
with insects, with occasionally *Potamides* and plants. This bed is
described as a brownish green, or light gray calcareous marl, composed of very thin laminæ. 5. Gypsum, with plants. 6. Marls.
7. Gypsum, with fish and plants. 8. Marls, with traces of gypsum. 9. Pink limestone, containing *Potamides, Cyclas gibbosa*,
Sow., and *Cyclas Aquæ Sextiæ*, Sow. This limestone is often highly
contorted, and passes either into a calcareous grit or red sandstone,
and, still lower, into compact calcareous breccia; the whole is
based on a coarse conglomerate. The lower beds dip N.N.E. at
about 25° or 30°. From the section accompanying the memoir
of Messrs. Lyell and Murchison, it would appear that these conglomerates rest, beyond Aix, on red marl, fibrous gypsum and
gray limestone, with *Limnææ* and *Planorbes;* and these again
on the compact limestone, sand, and shale, containing coal at
Fuveau, accompanied by the remains of an *Unio, Melania scalaris*, Sow., *Cyclas concinna*, Sow., *C. cuneata*, Sow., and *Gyrogonites **.

The preservation of the insects is very great, permitting the
determination of genera and species. According to M. Marcel de
Serres, *Arachnides* accompany the insects, properly so called; the
latter, however, being far more abundant than the former, two or
three genera only of *Arachnides* having been determined, while
sixty-two genera of insects have been observed. The most curious
circumstance attending these remains is, that some are considered
identical with those now existing in the country; *Brachycercus
undatus, Acheta campestris, Forficula parallela*, and *Pentotoma
grisea* being, according to M. Marcel de Serres, the more remarkable. It is also worthy of observation, that the greater part of the
insects are of those kinds which generally inhabit arid and dry
places. Although they occur in various positions, they are sometimes spread out, as if by an entomologist for the purpose of dis-
playing their wings. Their colour is generally an uniform tint of

---

* Lyell and Murchison, Edin. New Phil. Journal, 1829.

brown or black.   Some of the fish discovered in the same marls are so small that they do not exceed ten or eleven millimetres in length *.

A large part of the South of France, bounded by the ocean, or rather by the sandy dunes it has thrown up, between the districts of Bordeaux and Bayonne, and extending far into the interior, particularly at the foot of the Pyrenees, is composed of supercretaceous rocks; an exact and detailed account of whose varied relations to each other may still perhaps be considered as wanting, though much has been done respecting them.   This superficies comprises, among other districts, that extensive and monotonous region named the Landes, where the traveller finds little to relieve the sameness which surrounds him, except the peasants stalking over the country mounted on stilts, for the greater convenience of seeing objects afar off.

M. de Basterot has presented us with a very valuable detail of the fossil shells obtained by him from the districts of Bordeaux and Dax, which we have inserted in the Appendix (B.), considering that such lists are of the greatest utility to the geological student; referring him, however, to M. de Basterot's memoir for the detailed description of each shell.   This author remarks, that out of the 330 species of shells noticed by him in the great sandy deposits of the Landes, forty-five only have existing analogues in the neighbouring seas, comprising the Mediterranean; and he further observes, that if the basin of the Gironde be taken as a centre, the shells in similar supercretaceous basins will the more resemble each other as the distances are less.   Thus, out of the 330 species collected in the vicinity of Bordeaux, ninety-one are found in the deposits of Italy, sixty-six in those of the environs of Paris, eighteen in those of Vienna†, and twenty-four in the supercretaceous rocks of England‡.

If reference has been made to M. de Basterot's list, it will have been observed that, though many shells found in this part of France are also discovered at Paris, there is also a very considerable correspondence between them and those of Italy.   It would appear, from the mention of the fresh-water limestone at Saucats, that there was a change of the relative level of sea and land in that situation,

* Marcel de Serres, Géog. des Ter. Tertiaires du Midi de la France, in which some of the insects are figured; as also in the Memoir of Messrs. Lyell and Murchison above noticed, in illustration of the remarks of Curtis on the specimens brought to England.

† M. de Basterot observes that this number will probably become increased as the Vienna basin shall become better known; which we may expect it soon to be, from the labours of M. Parsch.

‡ De Basterot, Description Géologique du Bassin Tertiaire du Sud-Ouest de la France, 1ère partie: Mém. de la Soc. d'Hist. Nat. de Paris, t. 2.

which permitted the envelopment of fresh-water shells in carbonate of lime; and that after this deposit, a change of level was effected, which enabled marine lithodomous shells to bore extensively in the fresh-water rock, and permitted an accumulation of mineral matter and marine shells above it. The analogues of existing species are forty-five; the living species being remarkable for the diversity of their habitats, some being found in the Atlantic and Pacific Oceans, and the Indian and Mediterranean Seas, while not a few inhabit the coasts of the Channel and the Bay of Biscay, to which, from the fall of the land, the Bordeaux and Dax deposits seem naturally to belong. When the ocean covered this part of France, it seems necessary to suppose that the mean temperature of the situation was above that which it now is, in order to suit the animals, many of whose analogues exist in warm climates.

We now proceed to give a short notice of the supercretaceous rocks of the Paris basin, as they long constituted the type to which all deposits of this epoch, wherever found, were referred. However the rocks of this group may be eventually discovered to differ from this type, the labours of MM. Cuvier and Brongniart on the rocks of the Paris basin will not the less retain that place in the annals of Geology, which by common consent has been assigned them. Nor will the zoological discoveries of Cuvier, constituting as they did such a brilliant epoch in the history of geological science, the less claim the gratitude of geologists in succeeding ages.

The following is the classification of the Paris rocks, according to MM. Cuvier and Brongniart (order ascending):

1. First fresh-water formation
{
Plastic clay.
Lignite.
First sandstone.
}

2. First marine formation ...   Calcaire grossier.

3. Second fresh-water formation
{
Siliceous limestone.
Gypsum, with bones of animals.
Fresh-water marls.
}

4. Second marine formation ..
{
Gypseous marine marls.
Upper marine sands and sandstones.
Upper marine marls and limestone.
}

5. Third fresh-water formation
{
Millstone, without shells.
Shelly millstone.
Upper fresh-water marls.
}

Plastic Clay.—So named because it easily receives and preserves the forms given to it, and is used in the potteries. It rests on an unequal surface of chalk beneath, which is hollowed and furrowed in various ways, so as to present hills, valleys, and outstanding knolls, which sometimes have not been covered by the newer and superincumbent rocks; at least, if they have covered

them, the strata which did so have been removed by denudation *.
This clay is variously coloured, being white, gray, yellow, slate-
gray, and red.   It differs considerably in thickness, as might be
expected from the nature of the surface on which it reposes.
Above these beds, to which, strictly speaking, the term "plastic
clay" is alone applicable, there is often another clay, separated
from the former by a bed of sand; the latter clay being black,
sandy, and sometimes containing organic remains.   In it occur
lignites, amber, and shells (both fresh-water and marine). It is
stated, that in this deposit, considered as a mass, the lower parts
do not contain organic remains; that in the central portion the
remains are *commonly* those of fresh-water animals; and that in
the upper part there is a mixture and even an alternation of marine
and fresh-water remains, the latter gradually becoming more
scarce, and the former finally prevailing.—The following is a list
of the organic remains most commonly found in the plastic clay.

Fresh-water Remains.—PLANORBIS *rotundatus*, Al. Brong.;
P. incertus, Defr.; *P. Punctum*, Defr.; *P. Prevostinus*, Defr.

PHYSA *antiqua*, Defr.   LIMNEUS *longiscatus*, Al. Brong.

PALUDINA *virgula*, Defr.; *P. indistincta*, Defr.; *P. unicolor*,
Olivier; *P. Desmarestii*, Prevost; *P. conica*, Prev.; *P. ambigua*,
Prev.

MELANIA *triticea*, Defr.

MELANOPSIS *buccinoidea*, Poiret; *M. costata*, Olivier.

NERITA *globulus*, Defr.; *N. pisiformis*, Defr.; *N. sobrina*, Defr.

CYRENA *antiqua*, Defr.; *C. tellinoides*, Defr.; *C. cuneiformis*,
Sow.

Marine Shells contained in the mixture of the upper part.—CE-
RITHIUM *funatum*, Sow.; *C. melanoides*, Sow.; another *Cerithium*
not determined.

AMPULLARIA *depressa*. Lam.? (var. *minor*); *Ostrea bellovaca*,
Lam.; *O. incerta*, Defr.

Fossil Vegetables.—*Exogenites; Phyllites multinervis; Endogenites echinatus.*

Calcaire grossier.—This, as its name implies, is composed of a
coarse limestone, and is more or less hard, so as to be employed
for architectural purposes.   It alternates with argillaceous beds,
and is remarkable for the constancy of its character throughout a
considerable extent of country.   It is often separated from the
plastic clay beneath by a bed of sand.   The organic remains are
stated to be generally the same in the corresponding beds, presenting rather marked differences when the beds are not identical.
The inferior beds are very sandy, often more sandy than calca-

* A breccia of chalk fragments cemented by clay is found at Meudon,
separating the chalk and plastic clay.

reous, and almost always contain green earth, disseminated either
in powder or grains, which, according to the analysis of M. Berthier, appears to be a silicate of iron. These beds are remarkable
for the abundance of their organic contents. The following is a
list of those fossils which are considered to characterize the different parts of this deposit.

In the lower beds.—MADREPORA, at least three species.

ASTREA, three species at least.

TURBINOLIA *elliptica*, Al. Brong.; *T. crispa*, Lam.; *T. sulcata*,
Lam.

RETEPORITES *digitalia*, Lam.

LUNULITES *radiata*, Lam.; *L. urceolata*, Lam.

FUNGIA *Guettardi.*

NUMMULITES *lævigata; N. scabra; N. numismalis; N. rotun-*
data.

CERITHIUM *giganteum.* LUCINA *lamellosa.*

CARDIUM *porulosum.* VOLUTA *cithara.*

CRASSITELLA *lamellosa.* TURRITELLA *multisulcata.*

OSTREA *flabellula; O. cymbula.*

In the central beds*.—OVULITES *elongata*, Lam.; *O. margari-*
tula, Deroissy.

ALVEOLITES *milium*, Bosc. ORBITOLITES *plana.*

TURRITELLA *imbricata.* TEREBELLUM *convolutum.*

CALYPTRÆA *trochiformis.* CARDITA *avicularia.*

PECTUNCULUS *pulvinatus.*

CITHEREA *nitidula; C. elegans.* MILIOLITES. CERITHIUM?

In the upper beds.—MILIOLITES. AMPULLARIA *spirata.*

CERITHIUM *tuberculatum ; C. mutabile ; C. lapidum ; C. petri-*
colum.

LUCINA *Saxorum.* CARDIUM *Lima.*

CORBULA *anatina? C. striata* †.

Vegetable Remains, according to M. Ad. Brongniart, in the Calc.
Grossier of Paris :—

NAYADÆ—*Caulinites parisiensis.*

EQUISETACEÆ—*Equisetum brachyodon.*

CONIFERÆ—*Pinus Defrancii.* PALMÆ—*Flabellaria parisiensis.*

MONOCOTYLEDONS, OF UNCERTAIN FAMILY—*Culmites nodosus;*
C. ambiguus.

DICOTYLEDONS, OF UNCERTAIN FAMILY—*Exogenites ; Phyllites*
linearis, Ph. nerioides, Ph. mucronata, Ph. remiformis, Ph.
retusa, Ph. spathulata, Ph. lancea ‡.

Siliceous Limestone.—A limestone, sometimes white and soft,

---

* Nearly all the well-known fossils from Grignon are found in these
beds.

† MM. Cuvier and Brongniart, Desc. Géol. des Envir. de Paris. éd.
1822.

‡ Ad. Brongniart, Prod. d'une Hist. des Veg. Fossiles, 1828.

sometimes gray and compact, penetrated by silex, infiltrated in every direction and at all points. It is often cellular, the cells sometimes large and communicating with each other in all directions, the silex lining their sides with mammillary concretions, or with small transparent quartz crystals.

Osseous Gypsum (Fresh-water), and Marine Marls.—The gypseous rocks consist of an alternation of gypsum and calcareous and argillaceous marls. Above this alternation there are thick marl beds, sometimes calcareous, at others argillaceous. In these latter strata are found abundant remains of *Limnææ* and *Planorbes,* and in their lower parts, palms of considerable size are discovered prostrate. The gypseous strata contain the remarkable remains of extinct mammalia and other animals, which the genius of Cuvier may almost be said to have restored to life. Above these beds, which, from the nature of their organic remains, are considered to have been deposited in fresh water, there is a succession of marls, considered as deposited in the sea, because they contain marine remains; the marine and fresh-water systems being separated by calcareous or argillaceous marls, often thick. The upper marl beds contain numerous remains of oysters, considered to have certainly lived in the places where now entombed, more particularly, as M. Defrance discovered them at Roquencourt attached to rounded pieces of marly limestone, which latter are sometimes pierced by Pholades.

Organic Remains in the Gypseous Beds.—MAMMALIA: *Palæotherium magnum*, Cuv.; *P. medium*, Cuv.; *P. crassum*, Cuv.; *P. latum*, Cuv.; *P. curtum*, Cuv.; *P. minus*, Cuv.; *P. minimum*, Cuv.; *Anoplotherium commune*, Cuv.; *A. secundarium*, Cuv.; *A. gracile*, Cuv.; *A. murinum*, Cuv.; *A. obliquum*, Cuv.; *Chæroptamus parisiensis*, Cuv.; *Canis parisiensis*, Cuv.; *Coati; Didelphis parisiensis*, Cuv.; *Sciurus*; &c.

BIRDS. REPTILES: *Crocodile; Trionyx; Emys.* FISH.

Organic Remains of the Fresh-water Marls.—MAMMALIA: *Palæotherium aurelianense*, Cuv. (Orleans); *Lophiodon major*, Cuv. (Soissons, &c.); *L. minor*, Cuv. (Paris); *L. pygmeus*, Cuv. (Paris).

BIRDS. FISH. SHELLS: *Cyclostoma mumia*, Lam.; *Lymnæa longiscata*, Al. Brong.; *L. elongata*, Al. Brong.; *L. acuminata*, Al. Brong.; *L. ovum*, Al. Brong.; *Planorbis lens*, Al. Brong.; *Bulimus pusillus* Brard.

In the Marine Marls (Yellow).—Fish bones; Cytherea? convexa; Cytherea? plana; Spirorbes; Cerithium plicatum.

Yellow Marls separated from the above by Green Marls.—Spears and palates of the Ray; Ampullaria patula? Cerithium plicatum; C. cinctum; Cytherea elegans; C. semisulcata?? Cardium obliquum; Nucula margaritacea.

Calc. Marls, with large Oysters.—Ostrea hippopus; O. Pseudochama; O. longirostris; O. canalis.

Calc. Marls, with small Oysters.—Ostrea cochlearia; O. cyathula; O. spatulata; O. linguatula; Balani; Crabs' feet.

Upper Marine Sands and Sandstones.—This is composed of irregular beds of siliceous sandstone and sand, the lower portion without organic remains that can be supposed to have existed in the places where now found, being broken and very rare. In some situations, where the broken shells are more common, millions of small bodies are discovered, to which M. Lamarck has given the name of Discorbites.

These non-fossiliferous sands are in many places covered by a limestone, sandstone, or calcareo-siliceous rock filled with marine shells, of which the following is a list : Oliva mitreola ; Fusus? approaching F. longævus ; Cerithium cristatum ; C. lamellosum ; C. mutabile? Solarium ; Melania costellata? Melania? another species ; Pectunculus pulvinatus ; Crassatella compressa? Donax retusa? Cytherea nitidula ; C. lævigata ; C. elegans? Corbula rugosa ; Ostrea flabellula.

Upper Fresh-water Formation.—This rock varies very considerably in its mineralogical character, being sometimes composed of white friable and calcareous marls, at others of different siliceous compounds ; among which are the well-known millstones, sometimes without shells, at others charged with Limnei, Planorbes, Potamides, Helices, Gyrogonites (seeds of the Charæ), and silicified wood.

Organic Remains.—ANIMAL. Cyclostoma elegans antiqua ; Potamides Lamarckii ; Planorbis rotundatus ; P. Cornu ; P. Prevostinus ; Limneus corneus ; L. Fabulum ; L. ventricosus ; L. inflatus ; Bulimus pygmeus ; B. Terebra ; Pupa Defrancii ; Helix Lemani ; Helix Demarestina *.

VEGETABLE. Muscites? squamatus ; Chara medicaginula ; C. helicteres ; Nymphæa Arethusæ ; Culmites anomalus ; Carpolithes thalictroides †.

As has been often remarked, there is evidence in the various organic remains entombed in the strata above noticed, that the space comprised within what is commonly termed the Paris basin, has not always been exposed to the influence of the same circumstances since the deposit of the chalk, but that there has been an alternation of three lacustrine or fresh-water deposits with two which are marine ; the former constituting the lower and the upper part of the series. It remains to inquire the probable cause of these variations. By employing the term *basin* for this collection of supercretaceous rocks, we, as before observed, seem to assume that of which we have no great evidence ; the fresh-water deposits may have been, and probably were, effected in basins, but the marine do not require this form. It would seem reasonable to infer that

* Cuvier and Brongniart, Desc. Géol. des Env. de Paris.
† Ad. Brongniart, Prod. d'une Hist. des Veg. Fossiles.

there may have been here, as has been shown to have happened elsewhere, movements in the land, changing its level relatively with the sea. When we regard the mode in which the various deposits are now arrranged, we find that, as a mass, they do not repose horizontally on each other; but that, according to MM. Cuvier and Brongniart, there were various inequalities at different times, commencing with those of the chalk, presenting hills and valleys. In various parts of this unequal soil the lignite and plastic clay were deposited, thus to a certain extent filling up some of the inequalities. Upon this the calcaire grossier was formed, following more or less the inequalities of the surface beneath. To the calcaire grossier succeeded a gypseous deposit, showing an absence of the sea, and the presence of fresh water, of unequal depth. Then followed a large deposit of sand covering up the pre-existing inequalities, in the upper part of which sand are numerous marine remains; the whole presenting a vast plain. A new state of things followed; the sea disappeared, and fresh-water remains became entombed*.

The mechanical and chemical circumstances attending these deposits have also curiously varied. We will not stop to inquire whether the inequalities of the chalk were produced suddenly or slowly, for on this head we possess no very decided evidence; but the deposit of the plastic clay (properly so called) would appear to have been slow, even if the detritus, mechanically suspended, may have resulted from a somewhat violent wash of the inferior rocks. In the sands above this, we have the evidence of a transport by water moving with sufficient velocity to carry sand onwards. This is followed by a deposit, to a certain extent quiet, composed of vegetables and amber derived from them. The nature of the other organic remains mingled with them, at first indicates the presence of fresh-water animals; but finally, some variation in the relative level of the land and sea, apparently occurring gradually rather than suddenly, (for there is no evidence of a rush of waters,) introduces marine animals, which existed at the same time with many fresh-water animals that have gradually become accustomed to live in the same medium with them. This state of things was destined to disappear, and we have a movement of water sufficient to transport sand. This was succeeded by a calcareous deposition, when carbonate of lime, probably in a great measure derived from the ruin of older rocks, was washed away by water, and deposited over a considerable space. It is obvious, from the structure of these rocks, that the materials of which they consist must have been in a state of fine mechanical division, such as to have required no violent rush of waters for their removal: they probably subsided during a period of tranquillity. After the deposit of the calcaire grossier, the production of calcareous rocks,

* Cuvier and Brongniart, Env. de Paris.

remarkable for their cellular structure, took place. The origin of these cells is unknown; but they probably arose from the calcareous matter, during the act of subsidence, enveloping foreign matter more soluble or perishable than itself, which has subsequently been removed by the agency of water. It is remarkable that the cavities are now lined by silex in such a manner as scarcely to admit of any other supposition, than that the silica was deposited within the cells from a liquid in which it had been previously dissolved.

The osseous gypsum presents us with a decidedly new state of things. Singular animals, of which the very genera are now extinct, must have existed somewhere in the district, the remains of which became in some manner entangled in sulphate of lime, considerable deposits of which were then in progress. The question will arise, Whence did such a quantity of sulphate of lime proceed? Certainly it is a new ingredient, at least in any abundance, in this district; and there is no evidence that it was deposited in a sea, as was the case with the carbonate of lime of the calcaire grossier; on the contrary, as it only contains terrestrial and fresh-water remains, it would seem to have been formed through the medium of fresh water. If so, the previous level of the land and sea had been altered, and the springs of the district, if the gypsum was derived from them, must, instead of carbonate of lime, have produced an abundance of sulphate of lime. This state of things changed; the sulphate of lime ceased to be produced or deposited in abundance, the relative level of sea and land again became altered, the result was a formation of marls with marine shells in them; during which, there were at least some places where rolled pebbles were produced, to which oysters became attached, some of the pebbles being pierced by boring shells. These deposits are described as conforming more or less to the surface beneath each, and there is no evidence of any particular movement of water; but to them succeeds a vast quantity of sand, the organic remains in which are broken, and the mass fills up inequalities and forms a plane surface. This appears to show a long continued action of water with a velocity equal to the transport of sand over a considerable space. At the close of this period the causes, whatever they were, that prevented the envelopment of organic remains, ceased, and marine exuviæ became entombed in great abundance. Finally, to crown this curious series, we have a deposit of a very various mineralogical character, containing the remains of such animals and vegetables as are only known to exist on dry land, marshy places, or fresh water. This variety of mineralogical structure is what we should consider probable in a shallow lake, into which springs, holding various substances in solution, entered at various parts. The probability that the water was shallow, at least in part, has been considered probable by MM. Cuvier and Brongniart, from the remains of *Charæ*, so commonly found in this deposit; an opinion exceedingly strengthened by the observations of Mr. Lyell on the Charæ of the Bakie Loch, Scotland. To produce the friable calcareous marls it is not necessary that the waters should be thermal ; but judging from the phænomena of existing springs, this condition would seem requisite for the siliceous deposit ; for we do not know of any such formation now in progress, except in such springs. If the millstone and other siliceous substances were thus produced (and it seems difficult to obtain their formation in any other manner consistent with existing causes), these thermal waters have disappeared, and silex is no longer deposited in this district; seeming to show that very great changes in the solvent powers of water, and in the temperature of springs, may take place in the same district at different epochs. Thus we have a great deposit of carbonate of lime at the epoch of the calcaire grossier; another of sulphate of lime at the period of the osseous marls; and, finally, one of silex at the time of the millstone formation.

Supercretaceous Rocks of England.—Let us now compare the supercretaceous rocks of England with those of the Paris basin. Those of the former country are commonly known by the names of Plastic Clay, London Clay, Bagshot Sands, the Fresh-water formations of the Isle of Wight, and the Crag formerly noticed.

Plastic Clay.—Unlike the deposit to which the same name is applied in the environs of Paris, this rock, though occasionally containing a considerable abundance of clay, employed for various useful purposes, presents us with pebble beds, irregularly alternating with sands and clay; but like the strata of the same name at Paris, they rest upon an unequal surface of chalk beneath. The organic remains also are not principally terrestrial and fresh-water, but for the most part marine, though the others are intermingled with them. These remains are, according to Mr. Conybeare : UNIVALVES—*Infundibulum echinatum ; Murex latus, M. gradatus, M. rugosus, Cerithium funiculatum, C. intermedium, C. melanoides ; Turritella ; Planorbis hemistoma.* BIVALVES— *Ostrea pulchra, O. tener ; Pectunculus Plumstediensis ; Cardium Plumstedianum ; Mya plana ; Cytherea ; Cyclas cuneiformis ; C. deperdita, C. obovata.* In addition to this, traces of lignite and vegetables are observed in several places. The three following sections will convey an idea of this deposit in the neighbourhood of London, according to Prof. Buckland ; and in the Isle of Wight, according to Mr. Webster.

Section near Woolwich (series ascending).—Chalk with flints, above which : 1. Green-sand of the Reading oyster-bed, containing green coated chalk flints, but no organic remains; 1 foot. 2. Light ash-coloured sand, without shells or pebbles; 35 feet. 3. Greenish sand, with flint pebbles; 1 foot.   4. Greenish sand, without shells or pebbles; 8 feet.   5. Iron-shot coarse sand, without shells or pebbles, and containing ochreous concretions disposed in concentric laminæ; 9 feet. 6. Blue and brown clay, striped, full of shells, chiefly Cerithia and Cythereæ; 9 feet. 7. Clay striped with brown and red, and containing a few shells of the above species; 6 feet. 8. Rolled flints, mixed with a little sand, occasionally containing shells like those of Bromley; *e. g.* Ostrea, Cerithium, and Cytherea, disseminated in irregular patches; 12 feet. 9. Alluvium *.

Section at Loam-Pit Hill, three miles S.W. of Woolwich (order ascending).—Chalk with flints, above which: 1. Green-sand, identical with the Reading bed, and in every respect resembling No. 1. at Woolwich; 1 foot. 2. Ash-coloured sand, slightly micaceous, without pebbles or shells; 35 feet. 3. Coarse green-sand, containing pebbles; 5 feet. 4. Thick bed of ferruginous sand, containing flint pebbles; 12 feet. 5. Loam and sand, in its upper part cream-coloured, and containing nodules of friable marl, in its lower part sandy and iron-shot; 4 feet. 6. Three thin beds of clay, of which the upper and lower contain Cythereæ, and the middle, oysters; 3 feet. 7. Brownish clay, containing Cythereæ; 6 feet. 8. Lead-coloured clay, containing impressions of leaves; 2 feet. 9. Yellow sand; 3 feet. 10. Striped loam and plastic clay, containing a few pyritical casts of shells, and some thin leaves of coaly matter; 10 feet. 11. Striped sand, yellow, fine and iron-shot; 10 feet. At a higher level than No. 11. on the same hill, the line of the London clay commences †.

Section of the vertical beds in Alum Bay, Isle of Wight (order ascending).—Above, or rather next to, the chalk: 1. Green, red, and yellow sand; 60 feet. 2. Dark blue clay, containing green earth and nodules of dark limestone, in the latter of which Cythereæ, Turritellæ and other shells are found; 200 feet. 3. A succession of variously coloured sands; 321 feet. 4. Beautifully coloured sands, alternating with pipe-clay, coloured white, yellow, gray, and blackish; 543 feet. In the central parts of these latter deposits are three beds of lignite, and above them, at some distance, five other lignite beds; each one foot thick. 5. Strata of rolled black flint, contained in a yellow sand. 6. Blackish clay, containing much green earth and septaria; analogous to London clay ‡.

It will be observed, from these sections, that the transporting powers of water have not been precisely similar near London and at the Isle of Wight. At the former, there would appear to have been a greater movement than at the latter; the mass of the strata near London containing more pebbles in proportion to its depth than the beds of the Isle of Wight, where there would

---

* Geol. Trans. 1st series, vol. iv.   † *Ibid.*
‡ Webster, Geol. Trans. 1st series, vol. iv.

appear to have been a more calm, as well as a more abundant, deposit. This may perhaps in some measure be accounted for, by supposing the Isle of Wight strata, now thrown into a vertical position, to have been gradually accumulated in a hollow or cavity, more remote from the disturbing power of currents or motions in the water, than in shallower depths. At all events, the transporting power of the waters appears to have been irregular ; their velocities varying in such a manner that pebbles are carried forward at one time, while fine particles of detritus are alone moved at another. In the Isle of Wight beds we also see that circumstances have been favourable to the accumulation of vegetable matter, which is not irregularly disseminated, but occurs in beds ; the circumstances which attended this deposit being continued at irregular intervals, such as might be expected at the mouths of rivers.

London Clay.—This name has been applied to the great argillaceous deposit which underlies the London district. The clay is mostly blueish or blackish, and composed of argillaceous and calcareous matter in variable proportions, the latter rarely attaining a sufficient quantity to constitute marl or imperfect limestone. Layers of calcareous concretions, known by the name of Septaria, are by no means unfrequent ; and it is stated that beds of sandstone are occasionally observed in it.

It has been often remarked, that if the description of the Paris rocks had not preceded that of the country round London and of the Isle of Wight, it never would have been considered that the, so called, Plastic Clay was separated from the London Clay, but rather that they constituted different terms of the same series. It will have been observed that in the above-noticed section at Alum Bay, in the Isle of Wight, there was nothing to warrant such a separation ; neither does there appear to be any good reason why in the London district they should not be regarded as upper and lower portions of a deposit formed under nearly similar general circumstances. The deposit of the London clay would appear to mark a comparatively quiet state of things ; and the clay named Plastic marks a similar state, although it occurs among sands and pebbles. The whole seems merely to show that the velocities of the transporting waters varied, and that they continued for a longer period of little importance during the deposit of the London clay.

This clay varies very considerably in thickness. Thus, one mile east of London it is only seventy-seven feet deep ; at a well in St. James's-street, 235 feet ; at Wimbledon it was not pierced through at 530 feet ; and at High Beech, 700 feet *.

* Conybeare and Phillips's Outlines of the Geology of England and Wales : art. *London Clay.*

Organic Remains.—A Crocodile; a Turtle. Fish. Crustacea, a great variety, few of which have been noticed; among these few, Cancer tuberculatus, *König;* C. Leachii, *Desmarest;* Inachus Lamarckii, *Desm.* Conchifera—Clavagella coronata, *Desh.*, cal. gros. Paris; Fistulana personata, *Lam.*, cal. gros., Paris; Gastrochæna contorta; Pholadomya margaritacea, *Sow.;* Solen affinis, *Sow.;* Panopæa intermedia, *Sow.;* Mya subangulata, *Sow.;* Lutraria oblata, *Sow.;* Crassatella sulcata, *Lam.*, cal. gros., Paris; C. plicata, *Sow.;* C. compressa; Corbula globosa, *Sow.;* C. Pisum, *Sow.;* C. revoluta, *Sow.;* Sanguinolaria Hollowaysii, *Sow.;* S. compressa, *Sow.;* Tellina Branderi, *Sow.;* T. filosa, *Sow.;* T. ambigua, *Sow.;* Lucina mitis, *Sow.;* Astarte rugata, *Sow.;* Cytherea nitidula, *Lam.*, cal. gros., Paris, Bordeaux; Venus incrassata, *Sow.;* V. transversa, *Sow.;* V. elegans, *Sow.;* V. pectinifera, *Sow.;* Venericardia Brongniarti, *Sow.;* Ven. planicosta, *Lam.*, cal. gros., Paris, Ghent; Ven. carinata, *Sow.;* Ven. deltoidea, *Sow.;* Ven. oblonga, *Sow.;* Ven. globosa, *Sow.;* Ven. acuticostata, *Lam.*, cal. gros., Paris; Cardium nitens, *Sow.;* C. semigranulatum, *Sow.*, molasse, Switzerland; C. turgidum, *Sow.;* C. porulosum, *Lam.*, cal. gros., Paris; C. edule, *Brander*, Bordeaux, analogous to the existing species; Cardita margaritacea, *Sow.;* Isocardia sulcata, *Sow.;* Arca duplicata, *Sow.;* A. Branderi, *Sow.;* A. appendiculata, *Sow.;* Pectunculus decussatus, *Sow.;* P. costatus, *Sow.;* P. scalaris, *Sow.;* P. brevirostris, *Sow.;* P. pulvinatus, *Lam.*, cal. gros., Paris, Bordeaux, Turin, Traunstein; Nucula similis, *Sow.;* N. trigona, *Sow.;* N. minima, *Sow.;* N. inflata, *Sow.;* N. amygdaloides, *Sow.;* Axinus angulatus, *Sow.;* Chama squamosa, *Sow.;* Pinna affinis, *Sow.;* P. arcuata, *Sow.;* Avicula media, *Sow.;* Pecten corneus, *Sow.;* P. carinatus, *Sow.;* P. duplicatus, *Sow.;* Ostrea gigantea, *Sow.*, Traunstein; O. flabellula, *Lam.*, cal. gros., Paris, Bordeaux; O. dorsata, *Sow.;* O. cymbula, *Lam.*, cal. gros., Paris, Bordeaux; O. oblonga, *Brander;* Lingula tenuis, *Sow.*, Mollusca.—Patella striata, *Sow.;* Calyptræa trochiformis, *Lam.*, cal. gros., Paris; Infundibulum obliquum, *Sow.;* I. tuberculatum, *Sow.;* I. spinulosum, *Sow.;* Bulla constricta, *Sow.;* B. elliptica, *Sow.;* B. attenuata, *Sow.;* B. filosa, *Sow.;* B. acuminata, *Sow.;* Auricula turgida, *Sow.;* Au. simulata, *Sow.;* Melania sulcata, *Sow.;* M. costata, *Sow.* (Qu. M. costellata, *Brander* and *Lam.*, cal. gros., Paris?); M. minima, *Sow.;* M. truncata, *Sow.;* Paludina lenta, *Sow.;* P. concinna, *Sow.;* Ampullaria ambulacrum, *Sow.;* Am. acuta, *Lam.*, cal. gros., Paris; Am. patula, *Lam.*, cal. gros., Paris; Am. sigaretina, *Lam.*, cal. gros., Paris; Neritina concava, *Sow.;* Nerita globosa, *Sow.;* N. aperta, *Sow.;* Natica Hantoniensis; N. similis, *Sow.;* N. glaucinoides, *Sow.;* N. striata, *Sow.;* Sigaretus canaliculatus, *Sow.;* cal. gros., Paris, Bordeaux; Acteon crenatus, *Sow.;* A. elongatus, *Sow.;* Scalaria acuta, *Sow.;* S. semicostata, *Sow.;*

S. interrupta, *Sow.*; S. undosa, *Sow.*; S. reticulata, *Sow.*; Solarium patulum, *Lam.*, cal. gros., Paris, Bordeaux; Sol. discoideum, *Sow.*; Sol. canaliculatum, *Sow.*; Sol. plicatum, *Lam.*, cal. gros., Paris; Trochus Benettiæ, *Sow.*, Piacenza, Turin, Bordeaux; T. extensus, *Sow.*; T. monilifer, *Lam.*, cal. gros., Paris; Turritella conoidea, *Sow.*; Tur. elongata, *Sow.*; Tur. brevis, *Sow.*; Tur. edita, *Sow.*; Tur. multisulcata, *Lam.*, cal. gros., Paris; Cerithium dubium, *Sow.*; C. Cornucopiæ, *Sow.*; C. giganteum, *Lam.*, cal. gros., Paris; C. pyramidale, *Sow.*; C. geminatum, *Sow.*; C. funatum, *Sow.**; Pleurotoma attenuata, *Sow.*; P. comma, *Sow.*; P. semicolon, *Sow.*; P. colon, *Sow.*; P. exerta, *Sow.*; P. rostrata, *Sow.*; P. acuminata, *Sow.*; P. fusiformis, *Sow.*; P. lævigata, *Sow.*; P. brevirostra, *Sow.*; P. prisca, *Sow.*; Cancellaria quadrata, *Sow.*; C. læviuscula, *Sow.*; C. evulsa, *Sow.*; Fusus deformis, *König*; F. longævus, *Lam.*, cal. gros., Paris; Fusus rugosus, *Lam.*, cal. gros., Paris, Bordeaux F. acuminatus, *Sow.*; F. asper, *Sow.*; F. bulbiformis, *Lam.* (4 var.), cal. gros., Paris; F. ficulneus, *Sow.*; F. errans, *Sow.*; F. regularis, *Sow.*; F. Lima, *Sow.*; F. carinella, *Sow.*; F. conifer, *Sow.*; F. bifasciatus, *Sow.*; F. complanatus, *Sow.*; Pyrula nexilis, *Sow.*; P. Greenwoodii, *Sow.*; P. lævigata, *Lam*, cal. gros., Paris, Traunstein; Murex Bartonensis, *Sow.*; M. fistulosus, *Sow.*; M. interruptus, *Sow.*; M. argutus, *Sow.*; M. tricarinatus, *Lam.*, cal. gros., Paris, Vicentin; M. bispinosus, *Sow.*; M. frondosus, *Lam.*, cal. gros., Paris; M. defossus, *Sow.*; M. Smithii, *Sow.* (2 var.); M. trilineatus, *Sow.*; M. curtus, *Sow.*; M. tuberosus, *Sow.*; M. minax, *Sow.*, Switzerland; M. cristatus, *Sow.*; M. coronatus, *Sow.*; Rostellaria Parkinsoni, *Sow.* (var.); R. lucida, *Sow.*; R. rimosa, *Sow.*; R. macroptera, *Sow.* (2 var.); R. Pes-Pelicani (Strombus Pes-Pelicani, *Linn.*), Piacenza, &c., analogous to the existing species; Cassis striata, *Sow.*; C. carinata, *Lam.*, cal. gros., Paris; Harpa Trimmeri, *Parkinson*; Buccinum junceum, *Sow.*; B. lavatum, *Sow.*; B. desertum, *Sow.*; B. canaliculatum, *Sow.*; B. labiatum, *Sow.*; Mitra scabra, *Sow.*; M. parva, *Sow.*; M. pumila, *Sow.*; Voluta Luctator, *Sow.*; V. spinosa, *Lam.*, cal. gros., Paris; V. suspensa, *Sow.*; V. monstrosa, *Sow.*; V. costata, *Sow.*; V. Magorum, *Sow.*; V. Athleta, *Sow.*; V. depauperata, *Sow.*; V. ambigua, *Sow*; V. nodosa, *Sow.*; V. Lima, *Sow.*; V. geminata, *Sow.*; V. bicorona, *Lam.*, cal. gros., Paris; Volvaria acutiuscula, *Sow.*; Cypræa oviformis, *Sow.*; Terebellum fusiforme, *Sow.*; T. convolutum, *Al. Brong.*, cal. gros., Paris; Ancellaria canalifera, *Lam.*, cal. gros., Paris, Bordeaux; A aveniformis, *Sow.*; A. Turritella, *Sow.*; A. subulata, *Sow.*; Oliva Branderi, *Sow.*; O. Salisburiana, *Sow.*; Conus Dormitor, *Sow.*;

---

* It is remarkable that out of the numerous species of Cerithium found in the calcaire grossier of Paris, the *C. giganteum* should be the only one yet noticed in the London clay.

C. concinnus (2 var.), *Sow.;* C. scabriusculus (2 var.), *Sow.;*
C. lineatus, *Brander;* Nummulites lævigata, *Lam.*, cal. gros.,
Paris, Bordeaux, Traunstein; Num. variolaria, *Sow.;* Num.
elegans, *Sow.;* Nautilus imperialis, *Sow.*, cal. gros., Paris; N.
centralis, *Sow.;* N. ziczac, *Sow.;* N. regalis, *Sow.* *

Vegetable Remains.—The Isle of Sheppy has long been known
as affording a great variety of fruits and seeds; and small portions
and masses of wood are found in the London clay elsewhere, the
argillo-calcareous concretions frequently enveloping species of it.
Some fragments are pierced by a boring shell analogous to the
Teredo navalis, which shows that the wood must have floated in
the sea †.

Bagshot Sands.—These rest on the London clay, and consist,
according to Mr. Warburton, of ochreous meagre sand, foliated
green clay alternating with a green-sand, and alternations of
white, sulphur-yellow, and pinkish foliated marls, containing
abundant grains of green-sand, and fossil shells of the genera
Trochus? Crassatella, Pecten ‡.

Fresh-water Formations, Isle of Wight and Hampshire.—We
are indebted to Mr. Webster for the discovery of these beds, not
long after the labours of MM. Cuvier and Brongniart on the supercretaceous rocks round Paris so strongly excited the attention
of geologists. The fresh-water strata of the Isle of Wight are
divided into two deposits by a rock characterized by the presence
of marine remains, and named the Upper Marine Formation, from
being a supposed equivalent to the sands which intervene between
the two fresh-water deposits of Paris. The lower fresh-water deposit of Binstead near Ryde, consists of a limestone formed of
fragments of fresh-water shells, white shell marl, siliceous limestone and sand; at Headen the equivalent rock is composed of
sandy, calcareous, and argillaceous marls. According to Mr.
Pratt, one tooth of an *Anoplotherium,* and two teeth of a *Palæotherium* have been discovered in the lower and marly beds of the
Binstead quarries; and he further states, that these remains were
" accompanied, not only by several other fragments of bones of
Pachydermata (chiefly in a rolled and injured state), but also by
the jaw of a new species of Ruminant, apparently closely allied
to the genus *Moschus* §."

Prof. Sedgwick observes, that in the upper part of this deposit
there is a mixture of fresh-water and marine species, especially in

---

* Sowerby's Mineral Conchology; Woodward's British Organic Remains; Al. Brongniart, Tableau des Terrains qui composent l'Ecorce
du Globe.
    † Outlines of Geol. Engl. and Wales.
    ‡ Warburton, Geol. Trans., vol. i. 2nd Series.
    § Proceedings of the Geol. Soc. 1831.

Colwell Bay, where a single specimen of rock contained the following genera: *Ostrea, Venus, Cerithium, Planorbis, Lymnæa.* The common fossils in the lower fresh-water deposit would appear to be: *Paludina, Potamides, Melania* (more than one species), *Cyclas* (2 species), *Unio, Planorbis, Lymnæa* (both the last more than one species), *Mya, Melanopsis **.

The Upper Marine Formation, first noticed by Mr. Webster, was called in question by Mr. G. B. Sowerby, who showed that all the shells detected in it were not marine; and he hence inferred that there was no real separation between the fresh-water formations of the Isle of Wight †. Subsequently to Mr. Sowerby's remarks, Prof. Sedgwick has presented us with an account of these strata, in which he remarks that " the lower calcareous beds appear to have been tranquilly deposited in fresh water. But if we ascend to the argillaceous marl which rests immediately upon them, we not only find a complete change in the physical circumstances of the deposit, but a new suite of organic remains; some of which are of a marine origin, others of a doubtful character, and a few are identical with those in the lower beds‡." With regard to the organic remains contained in this rock, Mr. Webster points out a thick oyster-bed in Colwell Bay; and Prof. Sedgwick gives the following list of shells: *Murex* (at least two species), *Buccinum, Ancilla subulata, Voluta,* (resembling *V. spinosa), Rostellaria rimosa,* (two last species rare,) *Murex effossus,* BRANDER, *M. innexus* BRANDER, *Fusus* (fragments), *Natica, Venus, Nucula, Corbula, Corbis ? Mytilus, Cyclas, Potamides, Melanopsis, Nerita,* (2 species, one approaching *N. fluviatilis*), together with other fresh-water shells. These beds would therefore appear to have been deposited, as Prof. Sedgwick observes, in an estuary. But to have produced this estuary, and the circumstances requisite for the presence of marine shells, some physical change, some alteration of the relative levels or of the geographical features of the sea and land, seems necessary, for the previous deposit does not contain marine remains.

Upper Fresh-water Formation.—This, according to Mr. Webster, principally consists of yellowish white marls, in which there are more indurated, and apparently more calcareous portions. The organic remains are either fresh-water, or terrestrial; and therefore the circumstances, whatever they were, which permitted a mixture of marine shells in the beds beneath, no longer existed, and a tranquil deposit in some lake was, probably, the mode in which these beds, about 100 feet thick, were formed.

* Sedgwick, On the Geology of the Isle of Wight: Annals of Philos. 1822.

† G. B. Sowerby, Ann. of Phil. 1821.

‡ Sedgwick, Annals of Phil. 1822.

The fresh-water formation of Hordwell Cliff, Hampshire, was first described by Mr. Webster, in 1821. The cliff is noticed as composed of alternations of clays and marls, some of a fine blueish green colour, in which there were also beds of hard calcareous marls, apparently derived from shells of the genera *Lymnæa* and *Planorbis.* The whole is surmounted by a mass of transported gravel, which covers the various rocks of the vicinity. Mr. Webster observed that these beds seemed the equivalent of the lower fresh-water deposit of the Isle of Wight. Subsequently to these observations of Mr. Webster, Mr. Lyell published a more detailed account of the Hordwell beds; whence it would appear that the upper strata do not show a passage into a marine deposit, as was first supposed, but that all the fossil contents of the beds point to a fresh-water origin, equivalent to the lower fresh-water rocks of the Isle of Wight. The following are the organic remains discovered at Hordwell, according to Mr. Lyell : Tortoise scales (a Tortoise found at Thorness Bay, Isle of Wight); *Gyrogonites*, or seed-vessels of *Charæ* (*C. medicaginula*) ; seed-vessel named *Carpolithes thalictroides*, AD. BRONG.; teeth of crocodile, and scales of fish? *Helix lenta*, BRANDER, abundant; *Melania conica* ; *Melanopsis carinata* ; *M. brevis* ; *Planorbis lens* ; *P. rotundatus* ; *Lymnæa fusiformis* ; *L. longiscata*; *L. columellaris* ; *Potamides* ; *P. margaritaceus* ? *Neritina* ; *Ancylus elegans* ; *Unio Solandri* ; *Mya gregarea* ; *M. plana* ; *M. subangulata*, perhaps the young of *M. Plana* ; *Cyclas* (2 species). Mr. Lyell observes, that though the species are few, the individuals are numerous —a common characteristic of fresh-water deposits*.

Both in the Isle of Wight and on the opposite coast of Hampshire, these fresh-water deposits rest upon a considerable thickness of sand. As a similar sand occurs in the fresh-water rocks of Hordwell, Mr. Lyell considers that there is as much probability of its fresh-water, as of its marine origin. Be this as it may, there must have been a difference in the transporting power of water carrying the sands, from that which permitted the deposit of the marls, which seems to have been very quiet. The sands certainly do not require any considerable velocity of water; still there must have been a difference in the circumstances attending the deposit of the one mass and of the other, though those, which give rise to the mass of sand, partially returned during the formation of the marls.

A very material difference, it will be observed, must have attended the deposit of the supercretaceous rocks in the Parisian and English districts (London and Isle of Wight), as far as respects their mineralogical nature. In the former we have deposits of carbonate of lime (calc. grossier), sulphate of lime (gypseous deposits), and silex (millstones); formations only in part mecha-

* Lyell, Geol. Trans. 2nd series, vol. ii.

nical; while in the latter we have little that may not be considered altogether mechanical, with the exception, perhaps, of the fresh-water marls and the calcareous concretions in the London clay, which latter may have been chemical separations, after deposition, from the argillo-calcareous mass. There is, nevertheless, such an analogy between the organic character of the calcaire grossier of Paris and the London clay, that though not strictly identical, they may have been nearly contemporaneous; so that however the mineralogical character of these deposits may vary, we may suppose them to have been formed at the same or nearly the same epoch, local circumstances and accidents having determined the character of each.

Our limits prevent a proper notice of the labours of Prévost, Boué, Voltz, Parsch, Lill von Lillienbach, Pusch*, and many other geologists, on the rocks of this age in various parts of Europe; but the following section seems so important that it requires a place here.

Prof. Pusch, describing the rocks of Podolia and southern Russia, states, that near Krzeminiec, in Volhynia, (where mountains rise above a plain covered with chalk, flints, and sand,) upper supercretaceous sandstone, occupying a thickness of 396 feet above the river Ikwa and sixty feet beneath it, is composed of: 1. Twenty feet of sand, cemented by a little carbonate of lime, containing many small shells and madrepores, the latter approaching *M. cervicornis.* 2. Forty feet of calcareous sandstone, containing many shells of the genera *Cardium, Venericardia,* and *Arca.* 3. Sixty

---

* Amid a great variety of supercretaceous deposits in Russia and Poland, this author remarks some with an oolitic character, especially near Tiraspol, Latyczew, and Kaluez, on the Dniester, and in the Cecin hills at Czernowitz. The pisolitic structure of some supercretaceous limestones is particularly remarkable in some parts of Poland. The grains are either reniform or rounded, and generally of the size of a pea or a bean, though they here and there become two or three inches in diameter. Good examples of this rock are seen at Rakow. M. Pusch states that repeated observations have convinced him that these concretions are derived from corals, especially *Nulliporæ.* He observes that the large reniform concretions of Rakow are only the *Nullipora byssoides,* Lam., or the *N. racemosa,* Goldf. In some places, particularly at Skotniki, near Busko, a rock of this kind appears as if composed of bullets and cannon-balls.

It should be stated that Prof. Pusch, from a careful comparison of the shells contained in the supercretaceous limestone of Poland with those figured by various authors, considers that the tertiary shells of Poland bear a much greater resemblance to those found at the foot of the Italian Alps and in the Sub-Apennine hills, than those discovered in England or the North of France; moreover, that the species which at first sight do appear identical with those of France and Italy, are found to be varieties of them when examined with attention.

feet of a compact quartzose and porous sandstone, the cavities filled with sand ; contains many *Venericardiæ* ; lowest part most calcareous. 4. Eighty feet of a marly limestone, containing many striated *Modiolæ*, Pectens, and other shells. 5. At sixty feet beneath the surface, a quartzose and slightly calcareous white sandstone, containing numerous *Venericardiæ*, *Trochi*, and *Paludinæ* or *Phasianellæ*. " According to M. Jarocki, while sinking a well in June 1829, the tusk and molar tooth of an elephant were found in the last-mentioned bed (No. 5), which are now preserved in the museum of Krzeminiec. Many other bones were also observed, but they were too firmly fixed in the rock to be extracted *." M. Pusch further remarks, that this rock is the same, both mineralogically and zoologically, with the tertiary sandstone of Szydtow and Chmielnik, in Poland ; and that this fact is analogous to the occurrence of an elephant's molar tooth and tusk in the tertiary sandstone of Rzaka, Wieliczka, which contains *Pecten polonicus*, *Saxicavæ*, and many other marine shells. The reader will also observe that it corresponds with the occurrence of the remains of the great Pachydermata, previously noticed as found mingled with marine exuviæ in other parts of Europe.

It will have been remarked, that throughout this detail of supercretaceous rocks, perhaps too long for a work of this nature, the observations have been confined to certain parts of Europe. Rocks of the same nature no doubt abound in other parts of the world ; indeed we are well assured that very extensive districts are composed of them, as for instance in India ; but our knowledge of them is as yet so imperfect, that we cannot, with safety, compare them with known European deposits. Dr. Buckland, from the information which he obtained from Mr. Crawfurd, who collected an abundance of organic remains on the banks of the Irawadi, considered that supercretaceous rocks probably existed in the kingdom of Ava, containing shells of the genera *Ancillaria, Murex, Cerithium, Oliva, Astarte, Nucula, Erycina, Tellina, Teredo ;* mixed with sharks' teeth and fish scales : these remains are contained in a coarse shelly and sandy limestone. A great abundance of mammiferous and other remains were discovered in the vicinity of some petroleum wells, between Prome and Ava, apparently mixed with much silicified wood in a sandy and gravelly deposit. The bones or teeth of vertebrated animals consist of those of the *Mastodon latidens*, Clift ; *M. elephantoides*, Clift ; *Hippopotamus ; Sus ; Rhinoceros ; Tapir ; Ox ; Deer ? Antelope ; Trionyx ; Emys ;* and *Crocodiles* (2 species)†. Mr. Scott met with beds, probably of the supercretaceous epoch, in the Caribári hills, left bank of the Brahm-putra. The following section (order as-

---

* Pusch, Journal de Géologie, t. 2.

† Buckland and Clift, Geol. Trans. 2nd series, vol. ii.

cending,) was óbserved :—1. Slate clay. 2. Ferruginous concretions and indurated sand. 3. Yellow or green sand. 4. Slate-
clay. 5. Sand and small gravel. Fossil wood is found on the
indurated clay; and in a small isolated hill in the vicinity the
following remains :—Teeth and bones of sharks, fish palates and fin
bones, teeth and bones of crocodiles, remains of quadrupeds,
Ostreæ, Cerithia, Turritellæ, Balani, Patellæ, &c.* These remains have subsequently been examined by Mr. Pentland, who
found that the mammiferous remains were referable to the genus
Anthracotherium, Cuv., to a species allied to the genus *Moschus,*
to a small species of the order *Pachydermata,* and to a carnivorous animal of the genus *Viverra.* The *Anthracotherium* he proposes to name *A. Silistrense* †.

　　These indications are sufficient to show that rocks, probably supercretaceous, exist extensively in India. According to Prof. Vanuxem and Dr. Morton, the supercretaceous or tertiary rocks are
extensively distributed over parts of the United States, occurring in
Nantucket, Long Island, Manhattan Island, the adjacent coasts of
New York and New England; sparingly in New Jersey and Delaware, but extensively in Maryland, and to the southward. The
deposit is stated to be composed of limestone, buhr-stone, sands,
gravels, and clays; and contains the remains of the genera *Ostrea,
Pecten, Arca, Pectunculus, Turritella, Buccinum, Venus, Mactra,
Natica, Tellina, Nucula, Venericardia, Chama, Calyptræa, Fusus,
Panopæa, Serpula, Dentalium, Cerithium, Cardium, Crassatella,
Oliva, Lucina, Corbula, Pyrula, Crepidula, Perna,* &c. Of 150
species of these shells, found in a single locality in St. Mary's
county, Maryland, Mr. Say has described and figured more than
forty as new ‡. That deposits of a similar age are not wanting in
South America seems also certain; but as yet they have not been
examined in sufficient detail to enable us to institute any useful
comparison with rocks of the same antiquity in Europe. Neither
can we, from the same reason, judge of the relative antiquity of
innumerable igneous formations scattered over various parts of
the world. As the science of geology advances, great insight
must be obtained into the superficial appearance of the world at
this period, leading to the most important conclusions; but we
must anticipate very serious obstacles to this advancing knowledge,
arising from hasty generalizations of local facts, and the too common endeavour to force conclusions, more particularly as to the
identity or parallelism of deposits.

　　It is impossible to close this sketch of the supercretaceous rocks

---

* Colebrooke, Geol. Trans. 2nd series, vol. i.
† Pentland, Geol. Trans. 2nd series, vol. ii.
‡ Vanuxem and Morton, Journ. of the Academy of Nat. Sciences of
Philadelphia, vol. vi.

without noticing the important observations of Dr. Boué on those of Gallicia, wherein he establishes the fact, that the celebrated salt deposit of Wieliczka constitutes a portion of the supercretaceous series. Dr. Boué describes this deposit as 2560 yards long, 1066 yards broad, and 281 yards deep. The salt is termed green salt in the upper part of the mine, where it occurs in nodules with gypsum in marl. The salt sometimes contains lignite, bituminous wood, sand, and small broken shells. In the lower part the marl becomes more arenaceous, and there are even beds of sandstone in the salt. Beneath this is a gray sandstone, rather coarse, containing lignite, and impressions of plants, with veins and beds of salt. In the lower part of this stratum an indurated calcareous marl is observed, containing sulphur, salt, and gypsum. Beneath this is an aluminous and marno-argillaceous schist. From the fossils and various other circumstances, Dr. Boué concludes that this great salt deposit forms part of a muriatiferous and supercretaceous clay, subordinate to sandstones (molasse). Most frequently the marly clays are merely muriatiferous ; an abundance of salt, such as at Wieliczka, Bochnia, Parayd in Transylvania, and other places, being more rare *.

Volcanic Action during the Supercretaceous Period.—We have already seen that there was much difficulty in stating at what periods certain products of extinct volcanos had been thrown out. This difficulty is by no means lessened as we descend in the series; for the seat of volcanic action seems to have continued nearly, or very nearly, in the same place for long periods; and the mere circumstance of the interstratification of volcanic matter with aqueous rocks, whose relative age may to a certain extent be known, will not always give that of the igneous rocks so circumstanced, because we cannot be certain that they have not been injected among the aqueous deposits ; and when this may have happened it would be difficult to say. Thus Etna would appear to have been the seat of volcanic action through a long series of ages, commencing with the supercretaceous rocks, on which much of the igneous mass is now based.

In central France, amid the extinct volcanos which there constitute such a remarkable feature in the physical geography of the country, we certainly approach relative dates in some instances. Thus the volcanic mass of the Plomb du Cantal appears to have burst through, to have upset, and to have fractured the fresh-water limestones of the Cantal, which, according to Messrs. Lyell and Murchison, may be equivalent to the fresh-water deposits of the Paris basin, and to those of Hampshire and the Isle of Wight. The following is a list of organic remains obtained by them in the

* Boué, Journal de Géologie, t. i. 1830.

fresh-water rocks of the Cantal :—The rib of an animal resembling that of an *Anoplotherium* or a *Palæotherium;* scales of a tortoise; fish teeth; *Potamides Lamarckii; Limnæa acuminata; L. colu-mellaris; L. fusiformis; L. longiscata; L. inflata; L. cornea; L. Fabulum? L. strigosa? L. palustris antiqua; Bulimus Terebra; B. pygmeus? B. conicus; Planorbis rotundatus; P. cornu; P. ro-tundus; Ancylus elegans.* Plants : *Chara medicaginula,* the seeds (*gyrogonites*), and stems; carbonized wood. It is remarked, that out of this short list there are eight or nine species identical with those found in the upper fresh-water rocks, and five or six with those in the lower fresh-water deposits of the Paris basin *. Here we seem to obtain a relative date for the upburst of the igneous products of the Plomb du Cantal; one posterior to the deposit of the fresh-water rocks of Paris and the Isle of Wight.

With regard to the relative date of the igneous rocks of Au-vergne, it would appear from the labours of MM. Croiset and Jobert, that the Montagne de Perrier, N.W. from the town of Issoire (Puy de Dome), is divided into two stages or terraces, the first about twenty-five yards above the valley of the Allier, the second occupying a height of about 200 yards. The mountain may be considered as based on granite, above which there is a considerable thickness of fresh-water limestone, surmounted by numerous beds of rolled pebbles and sand, of which one in parti-cular is remarkable for the abundant remains of mammalia found in it; the whole crowned by a great mass of volcanic matter.

MM. Croiset and Jobert consider that in this locality and in the neighbouring country there are about thirty beds above the fresh-water limestone, which may be divided into four alternations of alluvial detritus and basaltic deposits. Among the beds there are three which contain organic remains : two belonging to the third of the ancient alluvions, that which succeeded the second epoch of volcanic eruptions; the third fossiliferous deposit being referable to the last epoch of ancient alluvion. The whole of these beds are not seen in the Montagne de Perrier, but are determined from the general structure of the country.

The principal ossiferous bed is about nine or ten feet thick, and can be traced a considerable distance at the foot of the Mon-tagne de Perrier, and in the Vallée de la Couse on the opposite side. The fossil species, according to MM. Croiset and Jobert, are very numerous, consisting of,—Elephant, one species; Masto-don, one or two; Hippopotamus, one; Rhinoceros, one; Tapir, one; Horse, one; Boar, one; Felis, four or five; Hyæna, two; Bear, three; Canis, one; Castor, one; Hare, one; Water-Rat,

---

* Lyell and Murchison, Sur les Dépôts Lacustres Tertiaires du Can-tal, &c. Ann. des Sci. Nat. 1829.

one; Deer, fifteen; and Ox, two. None of these bones are rolled, and no marine remains are found with them. Hence the authors conclude that the remains have not been far removed from the places where the animals existed, and that the lignites found among these beds are the exuviæ of the vegetation upon which many of them subsisted.

MM. Croiset and Jobert notice the following remains in the fresh-water limestone of the country, over which they consider that the first basaltic currents flowed: Anthracotherium? one species; Anaplotherium, two; Canis, one; Marten, one; Lagomys, one; Rat, one; Tortoises, one or two; Crocodile, one; Serpent, one. It should be observed that M. Bertrand-Roux* had some time previously observed the remains of a Palæotherium in a similar rock in the Puy en Velay, and that the fresh-water rocks at Volvic contain birds' bones†.

M. Bertrand de Doue describes the occurrence of bones entombed in and beneath volcanic matter near St. Privat-d'Allier (Velay). After stating that the discovery was due to Dr. Hibbert, who communicated it to him, and that he proceeded to the spot, pointed out, accompanied by M. Deribier, he notices the following descending section :—*a*, third and last flow of basaltic lava; *b*, second flow, four yards thick; *c*, grayish volcanic cinders, two to four decimetres thick; *d*, agglutinated scoriæ and tuff, one or more yards thick, in the upper part of which the bones were discovered; *e*, oldest plateau of basaltic lava; *f*, gneiss. The osseous remains were those of the *Rhinoceros leptorhinus, Hyæna spelæa,* and a large proportion of bones, referable to at least four undetermined species of *Cervi.*

The same author considers, from the fractured character and irregular distribution of the bones over a horizontal and limited space, that this place was the retreat of hyænas, affording them, from the nature of the country, the best shelter they could find. Into this it is considered they dragged their prey, as appears to have been done in the case of Kirkdale. It is observed that the lava-current which passed over the cinders containing these remains has very little altered the bones.

M. Bertrand de Doue does not consider the detrital deposits of the country as produced by transport in a body of waters from a distance, but by a succession of local causes, the substances being all derived from the vicinity. He supposes the distribution of the lateral valleys connected with the Allier (among which is that where the bones were discovered) the same now as when the

* Now M. Bertrand de Doue.

† Croiset and Jobert, Recherches sur les Oss. Foss. du Dept. du Puy de Dome; and Ann. des Sci. Nat. t. xv. 1828.

neighbouring volcanos were in activity; and remarks on the "incertitude in establishing the chronological relations between the epoch when the volcanos of the Velay became extinct, and that in which these animals disappeared from our climates *."

M. Robert, describing the position in which numerous bones have been discovered at Cussac (Haute Loire), mentions that marls, without fossils, rest on the granitic rocks of the country. At Solilhac these marls are surmounted by clayey marls about two or three feet thick, containing plates of mica, grains of quartz, volcanic ashes, basaltic gravel, and impressions of gramineous plants; they also contain the entire skeletons of unknown Deer and Aurochs, with other bones. Above these are beds of volcanic sand two or three yards thick, with small basaltic and granitic pebbles, containing the remains of Ruminants and Pachydermata, the bones being more or less broken. On these rest alluvions of greater solidity, composed of the same volcanic sand, large granitic and basaltic blocks (of which the angles are not rounded), geodes of hydrate of iron, and bones, which appear to have been exposed to the air before they were enveloped. All these substances are cemented by oxide of iron, and beds of ferruginous sands either alternate with, or repose on, the alluvions. M. Robert extracted from these ferruginous beds at Cussac the remains of the *Elephas primigenius*; the *Rhinoceros leptorhinus*; the *Tapir arvernensis*; the Horse, two species; Deer, seven species (to two of which he assigns the names of *Cervus Solilhacus*, and *C. Dama Polignacus*); the *Bos Urus*, and *Bos Velaunus*; and the Antelope. The same author refers the entombment of these remains to a more ancient date than the accumulation of bones at St. Prevat and Perrier, considering it due to some particular cataclysm, which surprised the animals: thus explaining the occurrence of entire skeletons of young and old individuals found mingled at Solilhac; a state of things differing from the accumulations at St. Prevat and Perrier, where the bones seem to have been dragged into their present position by carnivorous animals, whose bones are also mixed with those of their prey†.

* Bertrand de Doue, Edin. Journal. of Sci. vol. ii. new series, 1830.
† Robert, Férussac's Bulletin de Sci. Nat. et de Géologie, Oct. 1830.
The character of the animals thus entombed beneath gravels or sands, and the preservation of the bones, probably protected from attrition by the flesh which covered them, lead us (perhaps hastily) to refer this deposit to about the same age as that of the Upper Val d'Arno, and of the lower accumulation of pebbles and sands, noticed by M. Elie de Beaumont in Provence and Dauphiné; with the difference, that in central France the period required for the arenaceous and pebble deposit was occasionally interrupted by volcanic explosions.
Though not strictly in place, we may here present a sketch of the

Dr. Hibbert considers that the lowest supercretaceous rocks of
the Velay were deposited in fresh-water lakes, entombing the remains of the Palæotherium and Anthracotherium, of terrestrial
and fresh-water shells, and of the vegetation which then existed ;
such deposit being of long continuance, as shown by its depth,
which amounts to 450 feet. This deposit ceased, and the land
became covered with forests and animals ; the forests being of a
marshy growth. The common degradation of land taking place,

Upper Val d'Arno, in order that the student may perceive the connection of the animal remains found in Tuscany and central France.
Though the Val d'Arno has long enjoyed a certain celebrity from the
bones and teeth of various mammalia discovered there, it was not until
within a few years that the deposit has been properly examined.

M. Bertrand-Geslin distinguishes three basins between the source of the
Arno and Florence ; namely, the basins of Casentino, Arrezzo, and Figline ;
the whole valley of the Arno, for that distance, being bounded by a
sandstone named *Macigno*, or by dark-coloured limestone. According to
M. Bertrand-Geslin, the following section (in the descending order) may
be observed between Arrezzo and Incisa :—1. Thick bed of yellow argillaceous sand ; 2. Thick beds of rolled quartzose pebbles, intermixed
with coarse sand ; 3. Fine gray and micaceous sands, many fathoms
thick, containing thin beds of blue sandy marl ; these sands being exceedingly rich in the bones of mammiferous animals in their middle
and lower parts ; 4. Very thick argillaceous blue marl, constituting the
lowest deposit in the basin, and containing many fossils in its upper
part[1].

From his various observations on the Val d'Arno, M. Bertrand-Geslin
concludes, 1. That the rolled pebbles are larger and more abundant in proportion as they approach the mountain chain on the north, whence they
appear to have been derived ; 2. That the coarse sands occupy the central part of the valley, while the finest sands skirt the foot of the mountain range on the south ; 3. That the lower sands and blue marls are
deposited in horizontal beds ; 4. That the bones of mammalia are very
abundant towards the central part of the Val, on the right bank of the
Arno, and rare on the left bank ; 5. That these bones, in good condition,
and sometimes disseminated, are generally deposited in different places,
as if not all at one period ; 6. That the yellow sands contain fluviatile
shells at Monte Carlo ; and 7. That this transported mass contains
neither fragments of marine shells, solid stony beds, nor lignites.

The animals, whose remains are stated to have been discovered in the
Upper Val d'Arno, are the following :—Elephas primigenius, Mastodon
angustidens, Hippopotamus major, Rhinoceros, Tapir, Deer, Horse, Ox,
Hyæna, Felis, Cavern Fox, and Porcupine.

----

[1] Ann. des Sci. Nat. t. xiv. 1828. The reader can scarcely fail to be
struck with the resemblance of this section to those of the South of France,
Nice, &c., where sands and gravel surmount the blue Sub-Apennine
marls, and, in the South of France, contains terrestrial mammalia.

parts of this vegetation were variously entombed, as were also the remains of animals which then existed; such as various species of *Cervi*, some of large size, animals of the *Bos* kind, the *Rhinoceros leptorhinus*, and the *Hyæna spelæa*. Volcanic explosions now took place through various vents, ejecting trachyte and basalt, the latter predominating, piercing the fresh-water deposit in some places, and covering it with lavas in others. Notwithstanding these convulsions, vegetation still flourished in certain situations, and became entombed amid volcanic products, as is seen at Collet, Ronzal, and other places, where vegetable matter contained in black carboniferous clays, "accompanied with ferruginous sands, alternate with rolled masses of trachyte, phonolite, basalt, or volcanic cinders." During the progress of these eruptions, the water-courses became much deranged, lava-currents crossing these channels, damming up the passage, and forming lakes, in which various singular compounds and rock-mixtures were produced. It would appear from the large size and rounded angles of many of the fragments of basalt, that great currents of water had acted upon them in certain situations. After a time this great confusion seems to have ceased, and the large fragments became covered by a deposit of sand and clay, formed into regular strata, as may be observed near Cussac. During this state of things near Cussac, animals of the *Bos* kind, and gigantic stags, became entombed. After this, the district seems to have become the haunt of hyænas, which issuing from their dens in search of food, dragged their prey into their retreats*, in the manner of the Kirkdale hyænas.

In these various localities in central France, the evidence seems generally in favour of the great outburst of volcanos after the deposit of very extensive fresh-water rocks, the volcanic action continuing more or less from that period up to a comparatively recent date.

Quitting central France and proceeding either in the direction of Aix or Montpellier, we find remains of volcanos, which probably were more or less contemporaneous with those of Auvergne. Beaulieu near Aix has been known since the time of De Saussure.

Spain, Italy, and Germany present us with various igneous rocks, which appear referable to the epoch in which the supercretaceous rocks were in the course of formation. As yet, the volcanic rocks of Spain are little known; but those of Germany and Italy, and especially those of the latter, have long engaged the attention of geologists.

The Euganean Hills south of Padua present a mass of trachytic and other volcanic products, which belong to the supercretaceous epoch; as they rest in certain situations on *scaglia*, the equivalent

---

* Hibbert, On the Fossil Remains of the Velay, Edin. Journ. of Sci. vol. iii. 1830.

of chalk. Dr. Daubeny mentions that the trachyte is associated with basalt at Monte Venda. The same author informs us, that at the hill of Belmonte in the Vicentine, a rivulet section exposes five basaltic dykes, which from their mode of occurrence might be mistaken for an interstratification of chalk and basalt. "Dykes of basalt are also frequently seen traversing this formation at Chiampo, Valdagno, and Magre, but without altering the adjacent rock *." An extensive formation of porphyritic augite rock covers the whole district, resting in some places on chalk, in others on older rocks, filling up the pre-existing inequalities in each ; the upper part is amygdaloidal : this is surmounted by various alternations of calcareous beds, with others composed of fragments, basalts, volcanic sand, and scoriform lava ; the aggregate or mixture of volcanic substances containing fossil remains, as well as the calcareous deposits, and often as fully charged with them †. The long celebrated fossil fish from Monte Bolca are derived from the calcareous beds of this deposit. At Ronca there are six alternations of volcanic substances with the calcareous beds, the lowest volcanic product being a cellular basalt.

M. Al. Brongniart presents us with the following list of the shells and zoophytes in these beds of the Vicentine, the locality of each being marked (R. for Ronca; C. G. Castel-Gomberto; V. S. Val-Sangonini; M. M. Monteccio-Maggiore;) :—*Nummulites nummiformis,* Defr., R.; *Bulla Fortisii,* Al. Brong., R. ; *Helix damnata,* Al. Brong., R.; *Turbo Scobina,* Al. Brong., C. G. ; *T. Asmodei,* Al. Brong., R. ; *Monodonta Cerberi,* Al. Brong., V.S. ; *Turritella incisa,* Al. Brong., R. ; *T. asperula,* Al. Brong., R. ; *T. Archimedis,* Al. Brong., R.; *T. imbricataria,* Lam., R.; *Trochus cumulans,* Al. Brong., C. G.; *T. Lucasianus,* Al. Brong., C. G.; *Solarium umbrosum,* Al. Brong., R. ; *Ampullaria Vulcani,* Al. Brong., R.; *A perusta,* Defr., R. ; *A. obesa,* Al. Brong., M. M. and C. G.; *A. depressa,* Lam., R. ; *A. spirata,* Lam., V. S.; *A. cochlearia,* Al. Brong., C. G. ; *Melania costellata,* Lam., (var. *roncana,* Al. Brong.,) R., V. S.; *M. elongata,* Al. Brong., C. G.; *M. Stygii,* Al. Brong., R.; *Nerita conoidea,* Lam., R.; *N. Acherontis,* Al. Brong., R.; *N. Caronis,* C. G.; *Natica cepacea,* Lam., Val de Chiampo; *N. epiglottina,* Lam., R.; *Conus deperditus,* Broc. (var. *roncanus,* Al. Brong.) R.; *C. alsiosus,* Al. Brong., R. ; *Cyprœa Amygdalum,* Broc., R.; *Cyp. inflata,* Lam., R. ; *Terebellum obvolutum,* Al. Brong.; *Voluta subspinosa,* Al. Brong., R.; *V. crenulata,* Lam., V. S.; *V. affinis,* Broc., R. ; *Marginella Phaseolus,* Al. Brong., R.; *M. eburnea,* Lam., R., and V. S.; *Nassa Caronis,* Al. Brong., R.; *Cassis striata,* Sow., R.; *C. Thesei,* Al. Brong., R.; *C. Æneæ,* Al. Brong., R.; *Murex angulosus,* Broc., various parts of the Vicentine ; *M. tricarinatus,* Lam., Vicentine;

* Daubeny, Description of Volcanos.　　† *Ibid.*

Terebra Vulcani, Al. Brong., R.; *Cerithium sulcatum*, Lam. (var. *roncanum*, Al. Brong.), R.; *C. multisulcatum*, Al. Brong., R.; *C. undosum*, Al. Brong., R.; *C. combustum*, Defr., R.; *C. calcaratum*, Al. Brong., R.; *C. bicalcaratum*, Al. Brong., R. &c.; *C. Castellini*, Al. Brong., R.; *C. Maraschini*, Al. Brong., R.; *C. corrugatum*, Al. Brong., R.; *C. saccatum*, Defr., R.; *C. ampullosum*, Al. Brong., C. G.; *C. plicatum*, Lam., R.; *C. lemniscatum*, Al. Brong., R.; *C. Stropus*, Al. Brong., C. G.; *Fusus intortus*, Lam., (var. *roncanus*, Al. Brong.), R.; *F. Noæ*, Lam., R.; *F. subcarinatus*, Lam. (var.), R.; *F. polygonus*, Lam., R.; *F. polygonatus*, Al. Brong., R.; *Pleurotoma clavicularis*, Lam., M. M.; *Pteroceras Radix*, Al. Brong., C. G.; *Strombus Fortisii*, Al. Brong., R.; *Rostellaria corvina*, Al. Brong., R.; *Ros. Pes-carbonis*, Al. Brong., R.; *Hipponix Cornucopiæ*, Defr., R.; *Chama calcarata*, Lam., C. G.; *Spondylus cisalpinus*, Al. Brong., C. G.; *Ostrea*, R.; *Pecten lepidolaris?* Lam., R.; *P. plebeius?* Lam., R.; *Arca Pandoræ*, Al. Brong., C. G.; *Mytilus corrugatus*, Al. Brong., R.; *M. edulis?* Linn., R.; *M. Antiquorum*, Sow., R.; *Lucina Scopulorum*, Al. Brong., R.; *L. gibbosula*, Lam., R.; *Cardita Arduini*, Al. Brong., C. G.; *Cardium asperulum*, Lam., C. G.; *Corbis Aglauræ*, Al. Brong., C. G.; *Cor. lamellosa*, Lam., R.; *Venus? Proserpina*, Al. Brong., R.; *V.? Maura*, Al. Brong., R.; *Venericardia imbricata*, Lam., C. G.; *Ven. Lauræ*, Al. Brong., C. G.; *Mactra? erebea*, Al. Brong., R.; *M.? Sirena*, Al. Brong. R.; *Cypricardia cyclopæa*, Al. Brong., R.; *Psammobia pudica*, Al. Brong., V. S.; *Cassidulus testudinarius*, Al. Brong., R.; *Nucleolites Ovulum?* Lam., R.; *Astrea funesta*, Al. Brong., R.; *Turbinolia appendiculata*, Al. Brong., R.; *T. sinuosa*, Al. Brong., Vicentine *.

It has been concluded, and with great probability, that these rocks were produced by the alternate eruptions of volcanos in the vicinity, and the deposit of calcareous matter in shallow seas. M. Brongniart mentions that parasitical shells and certain corals are seen adhering to fragments of igneous rocks, which shows that these rocks have had abundant time to cool and form the bottom of the sea previous to the deposits above them. And as in some places igneous products and calcareous deposits often alternate, we may infer that a long period elapsed during the formation of the whole.

On the north and south of Rome there is abundant proof of extinct volcanic action. At Viterbo basaltic rocks rest on a compound of pumice and volcanic tuff, in which the bones of mammalia have been discovered; reminding us of Auvergne. Rome itself is founded on rocks of volcanic origin, mixed with others which are aqueous, and mostly of contemporaneous formation.

* Brongniart, Terrains Calcaréo-Trappéens du Vicentine, 1823.

Proceeding hence to Sicily, we find it very difficult to conceive where the volcanic action commenced which now finds a vent at Etna; as volcanic products are found mixed with supercretaceous rock. Dr. Daubeny observes, that the supercretaceous blue marl which occupies a considerable portion of Sicily, contains sulphur, various sulphuric salts, and muriate of soda, all substances sublimed from modern volcanos, and which may have been produced by exhalations from beneath.

Among the variety of volcanic products in the vicinity of the Rhine and neighbouring parts of Germany, are many which seem clearly to belong to the supercretaceous epoch. Among these may be mentioned the Siebengebirge, the Westerwald, the Habichtswald near Cassel, and the Meisner near Eschwege. The Siebengebirge are composed of trachyte, basalt, and volcanic conglomerates, traversed by dykes. The Westerwald is composed of the like substances. Basaltic knolls are scattered over the country between the Westerwald and the Vogelsgebirge. The Kaiserstuhl and the igneous rocks on the north of the lake of Constance would appear to be examples of volcanic rocks which may have been ejected at the supercretaceous epoch.

According to M. Beudant there are five principal volcanic groups in Hungary, referable to the age with which we are now occupied:—1. That in the district of Schemnitz and Kremnitz. 2. That constituting Dregeley mountains, near Gran on the Danube. 3. That of the Matra, in the centre of Hungary. 4. The chain commencing at Tokai, and extending north about twenty-five leagues. 5. That of Vihorlet, connected with the volcanic mountains of Marmorosch, (borders of Transylvania). The whole composed of different varieties of trachytic rocks.

According to Dr. Boué, volcanic rocks of undoubted supercretaceous origin occur in Transylvania. They constitute a range of hills separating Transylvania from Szeckler land, and extending from the hill of Kelemany, north of Remebyel, to the hill Budoshegy, on the north of Vascharhely. They are principally composed of varieties of trachyte, and trachytic comglomerate *.

From the observations of Von Buch and Dr. Daubeny, it appears that Gleichenburg, not far from Gratz, Styria, is composed of trachyte, round which are mantle-shaped strata of volcanic products and supercretaceous beds, alternating with each other.

If we turn from these igneous products on the continent of Europe to our own islands, we find that great igneous eruptions have taken place in the north-eastern parts of Ireland, after the deposit of the chalk, and consequently in the supercretaceous period. The basaltic ranges of the celebrated Giant's Causeway, Fairhead, &c. belong to this eruption, which in its upburst has

* Daubeny's Volcanos.

M 5

torn and rent all which it encountered, entangling enormous masses of chalk, as may be seen at Kenbaan. We find the mass of this erupted igneous rock to be basaltic,—sometimes columnar, at others not; the two varieties being so arranged on the coast between Dunseverie Castle and the Giant's Causeway, that they have the appearance of being interstratified. At Murloch Bay, Fairhead, and Cross Hill, the basalt rests on coal measures; at Knocklead and other places, on chalk *. As an intermixture with supercretaceous rocks has not yet been observed, the relative date of this eruption cannot be well determined. Both the basaltic mass and the rocks on which it rests have been traversed, at a period posterior to the first overflow of the former, by dykes of igneous matter; one of these has produced a singular change in the chalk, which it cuts, together with superincumbent basalt, in the Isle of Raghlin, as will be best explained by the annexed section. *a a a*, trap dykes cutting through chalk *b b*, which it has converted into granular limestone *c c c c*.

Fig. 37.

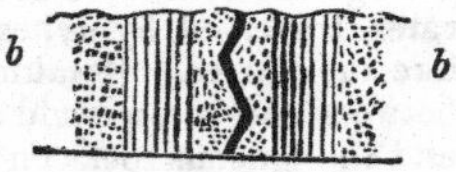

It now only remains to consider those recent observations on the Alps and the vicinity of Maestricht, which seem to point to at least a zoological passage of this group into the next; appearing to show, that from the progress of science, the clear line of separation once supposed to exist between the secondary and tertiary classes, as they are termed, cannot be drawn, but that the zoological character of the upper part of the one and the lower portion of the other would approach each other, as indeed might be expected ; for we cannot conceive a natural destruction of life so general as to cause the complete annihilation of animals, particularly those which are marine, existing at any given time, so that a totally new creation should be necessary. Such a supposition would not appear to accord with what is observable in other rocks, as will be noticed in the sequel. It is not contended that there may not be great specific distinctions in the remains entombed in this and the next group in many parts of Europe, but merely that it does not necessarily follow, because Europe may present us with two classes of rocks, one of which may be named tertiary and the other secondary, from the general nature of their organic contents, that in many parts of the world the whole may not constitute a series in which lines of distinction cannot be drawn. Suppose some violent cause should produce a great debacle which should rush over Europe, the land and fresh-water animals and plants would probably be destroyed; and we will even consider, for the sake of

* Buckland and Conybeare, Geol. Trans. vol. iii. ; and Sections and Views illustrative of Geological Phænomena, pl. 19.

the argument, that the marine inhabitants of our seas perished
also,—does it necessarily follow that the marine, fresh-water, and
terrestrial inhabitants would also be annihilated in Australia?
Should we not rather consider that these would be entombed, if
rocks were there forming, as well after and during the destruction of European life, as previous to it; and that the rocks formed
in those regions, about this supposed period, would by no means
show any alteration in their zoological character. That very great
changes have taken place in the organic character of deposits, in
the same districts, and that somewhat suddenly, does not admit
of a doubt; but it is a subject on which we are, as yet, far from seeing our way clearly. There is always great difficulty in comprehending why the marine remains should be so suddenly changed
in certain deposits, which do not exhibit marks of being the results of violent commotion; for although we can understand why
terrestrial and fresh-water animals should be destroyed by an inroad of the sea, produced by a sudden elevation of a mountain
chain sufficiently near, or any other cause, it is difficult to comprehend why, from that cause alone, the general character of the
marine animals should be changed.

From an examination of portions of the Austrian and Bavarian
Alps, in 1829, Professor Sedgwick and Mr. Murchison concluded
that they had discovered a series of beds intermediate between the
chalk and commonly known supercretaceous rocks, affording as
it were a passage of the, so called, tertiary class into the secondary; yet as they were above the true chalk, being considered as
tertiary. The correctness of this determination is questioned,
more particularly by Dr. Boué, who contends that the disputed
rocks belong to the cretaceous series. According to the former
authors, the valley of Gosau, in the Salzburg Alps, presents a
good example of the correctness of their views. This valley is
described as about 2600 feet above the level of the sea, exhibiting
these newer strata brought suddenly into contact with more ancient rocks on one side. The following is stated to be a section
of them, in the descending order. " 1. Red and green slaty micaceous sandstone, several hundred feet thick (cap of the Horn).
2. Green micaceous gritty sandstone extensively quarried as
whetstone, succeeded by yellowish sandy marls (Ressenberg).
3. A vast shelly series consisting of blue marls alternating with
strong beds of compact limestone and calcareous grit, the upper
beds of which are marked by obscure traces of vegetables; and
the middle and inferior strata by a prodigious quantity of well
preserved organic remains *. The fossils found in the lowest
strata at Gosau bear the impress, according to these authors, of
the cretaceous period; while those of the overlying blue marls ap-

* Proceedings of the Geol. Soc. Nov. 1829.

proach so nearly to many species of the lower supercretaceous or tertiary formations, that they refer the whole deposit to an age intermediate between the chalk and those formations hitherto considered as tertiary *.

Dr. Boué is by no means willing to admit this deposit of Gosau as a tertiary or supercretaceous rock, but as constituting a part of the green sand which extends along the Alps, as will be seen in the next section, from Austria into Savoy †.

It may here be remarked that M. Brongniart long since (1823) considered certain rocks constituting the upper part of the Diablerets (Valais), were referable to the supercretaceous or tertiary series. From the section of this mountain, made by M. Elie de Beaumont, and produced by M. Brongniart, it appears that the strata are singularly contorted; so that the newer beds have been twisted between the older strata in such a manner that the latter not only occur beneath the former, but also above them ‡. The beds considered supercretaceous are described as composed of calcareous sandstone, anthracite, and a black, compact, and carbonaceous limestone, containing *Nummulites*; *Ampullaria* (two species); *Melania costellata*, Lam.; *Cerithium Diaboli*, Al. Brong. (very abundant); *Turbinella? Hemicardium; Cardium ciliare*, Broc.; *Cariophyllia; Madrepora*.

The nummulites found so abundantly in the Alps by no means mark a distinct geological epoch, as they would appear to do in Northern France and in England; for instead of being confined to the supercretaceous group, they pervade the cretaceous, and possibly also some older rocks.

The observations of Dr. Fitton on the Maestricht beds would appear to throw light on these Alpine deposits, as far at least as their zoological character is concerned, it being understood that the celebrated deposit of the Mont St. Pierre contains a mixture, to a certain extent, of the, so called, secondary and tertiary remains; that "it is throughout superior to the white chalk, into which it passes gradually below, but the top bears marks of devastation, and there is no passage from it to the sands above. The siliceous masses which it includes are much more rare than those of the chalk, of greater bulk, and not composed of black flint, but of a stone approaching to chert, and in some cases to chalcedony;

---

* The various labours of Prof. Sedgwick and Mr. Murchison on the Alps will be found in the second part of vol. iii. of the Geol. Transactions, 2nd series, where there are also figures of the fossils discovered by them at Gosau.

† Boué, various memoirs, Edinburgh Phil. Journal, 1831; Journal de Géologie, 1830; and Proceedings of the Geol. Soc. of London, 1830.

‡ Brongniart, Sur les Terrains Calcaréo-Trappéens du Vicentin, p. 47; and Sections and Views illustrative of Geological Phænomena, pl. 38, fig. 5.

and of about fifty species in the author's (Dr. Fitton's) collection, about forty are not found in Mr. Mantell's catalogue of the chalk fossils of Sussex *."

According to M. Dufrénoy, a similar mixture of the organic remains, usually considered as characterising the cretaceous and supercretaceous rocks respectively, is discovered in the chalk series of the Pyrenees. This author observes, that out of 200 species obtained from this deposit, forty are such as are commonly referred to the supercretaceous epoch. These latter, though most abundant in the upper part of the Pyrenean chalk, are nevertheless scattered through its whole height †.

From these data, it would appear that at Maestricht, in the Pyrenees, and in the Alps, there do exist rocks containing organic remains common to the supposed great classes of secondary and tertiary rocks ; therefore it seems established that out no line can, zoologically, be drawn between them. How far other characters may distinguish them, remains to be seen, and probably minute researches in the Alps will eventually afford the necessary information. Such researches are no doubt difficult in these mountains, requiring time, much patience, and favourable circumstances, particularly as to weather: but while they are difficult, they are delightful; for who can roam unmoved among those regions where the best sections are exposed ? Whole mountains are so tossed over and contorted, that the geological student must be exceedingly cautious how he generalizes among the Alps, which require no common care in their examination ; but at the same time he has the satisfaction of knowing that every section, made with proper caution, accompanied by lists of organic remains, extracted from the various rocks with his own hands (not obtained from dealers), and examined by competent persons, possesses the greatest value.

---

* Proceedings of the Geol. Soc. 1830.

† Dufrénoy, Bulletin de la Société Géologique de France, Juin, 1830.

Section V.

CRETACEOUS GROUP.

Syn.—Chalk, (*Craie*, Fr.,—*Kreide*, Germ.,—*Scaglia*, It.,) Chalk Marl, (*Crai Tufau*, Fr.). Upper Green Sand, (*Glauconie Crayeuse*, Fr.,— *Chloritische Kreide*, *Planerkalk*, Germ.). Gault. Lower Green Sand, (*Glauconie Sableuse*, Al. Brong.,—*Grüner Sandstein*, Boué,—Part of German *Quadersandstein*.)

The upper portion of the cretaceous group partakes of a common character throughout a large portion of Western Europe, generally presenting itself under the well-known form of chalk. The upper part of the chalk throughout a large portion of England is characterized by the presence of numerous flints, more or less arranged in parallel lines : seams of this substance not only occur in a line with the flints, but also traverse the beds diagonally. The white chalk, when freed from the flints or siliceous grains mixed with it, is found to be a nearly pure carbonate of lime. According to M. Berthier, the chalk of Meudon, when the sand disseminated in it was separated by washing, contained in 100 parts, carbonate of lime 98, magnesia and a little iron 1, alumine 1. In the lower parts of the English chalk deposit, the flints disappear, becoming gradually more rare in the passage from the upper to the lower parts. From this circumstance, the white chalk has not unfrequently been divided into upper, or chalk with flints, and lower, or chalk without flints. This supposed characteristic is not however available to any great distances ; for at Havre the lower chalk contains an abundance of flint and chert nodules, where it passes into the upper green sand. There is, however, along the line of coast from Cap la Hêvre to the eastward, a considerable accumulation of chalk, in which flints are rare, apparently interposed between the Havre beds and the chalk with numerous flints. In the cliffs of Lyme Regis (Dorset), and Beer (Devon), we observe how little dependence can be placed on minute divisions of rocks, even within the distance of a few miles ; for considerable differences in the development of the cretaceous series will be observed between the two places, as I had formerly occasion to remark *. There are, however, a few beds which are remarkably persistent throughout the district, extending to Weymouth ; they are characterized by the presence of small and irregularly rounded grains of quartz, probably of mechanical origin, occasionally disseminated through the mass in great abundance. These beds are also remarkable for a great variety of organic

* Geol. Trans. 2nd series, vol. ii.

remains. Notwithstanding the very general presence of these beds, they sometimes become almost suddenly replaced by others, wherein the grains of quartz are not seen. Thus at Beer, the Beer stone, worked during centuries for architectural purposes, seems the equivalent of them, though composed of a white rock, principally carbonate of lime, with some argillaceous and siliceous matter. Probably the Beer stone may be the equivalent of the Malm rock of Hants and Surrey, described by Mr. Murchison, and the Merstham firestone noticed by Mr. Webster, and considered as the upper green sand. It may be here observed that the lower part of the chalk, or its passage into the green sand beneath, is extensively used as a building stone in Normandy, and that some of the inferior chalk beds of that country are considerably indurated, even approaching a whitish compact limestone, as may be well seen on the high road, bordering the Seine, between Havre and Rouen. The lower portion of the cretaceous group has, in England more particularly, received various names, though the mass is very commonly known as green sand. These smaller divisions, to the accurate determination of which we are indebted to Dr. Fitton *, should be borne in mind, more particularly in the study of English geology; as by tracing them as far as possible, we may obtain an insight into the causes which have produced them. These divisions are, Upper Green Sand, Gault, and Lower Green Sand; and can be best studied in the south-eastern parts of England †.

The upper green sand generally appears to graduate into the cretaceous mass above, being charged with a large quantity of green grains, which, according to the analysis of M. Berthier, made on those of the equivalent deposit at Havre, contain :—Silex 0·50, protoxide of iron 0·21, alumine 0·07, potash 0·10, water 0·11. The same author found that the green or reddish nodules disseminated through the same rock, also at Havre, contained :—phosphate of lime 0·57, carbonate of lime 0·07, carbonate of magnesia 0·02, silicate of iron and alumine 0·25, water and bituminous matter 0·07. The reader will at once observe the different composition of the nodules and grains. Respecting the former, M. Al. Brongniart observes, that the phosphate of lime sometimes so abounds as nearly to constitute the whole substance ‡.

The gault, or galt, is an argillaceous deposit of a blueish gray colour, frequently composed of clay in the upper, and marls in the

* Fitton, On the beds between the Chalk and Purbeck Limestone: Annals of Philosophy, 1824.

† The student should consult Dr. Fitton's Memoir (above cited); Mr. Murchison's Memoir on North-western Sussex, Geol. Trans. 2nd series, vol. ii.; Mr. Mantell's Geology of Sussex; and Mr. Martin on West Sussex.

‡ Cuvier and Brongniart, Desc. Géol. des Env. de Paris, 1822, p. 13.

lower part, containing disseminated specks of mica; it effervesces strongly with acids.

The lower green sand is formed of sands and sandstones of various degrees of induration, but principally of ferruginous and green colours, the former usually constituting the upper part and the latter most prevalent in the lower portions, which are not unfrequently argillo-arenaceous, particularly at bottom.

Without entering further into the smaller divisions of the cretaceous group, it may be remarked that the whole, taken as a mass, may in England, and over a considerable portion of France and Germany, be considered as cretaceous in its upper part, and arenaceous and argillaceous in its lower part. The divisions established in south-eastern England have been observed by Mr. Lonsdale in Wiltshire. In northern England the arenaceous deposit is scarcely observable, the white chalk resting on red chalk, the latter based on an argillaceous rock, named Speeton clay by Mr. Phillips. In south-western England the chalk rests on a great arenaceous deposit somewhat variable in its composition, sometimes containing thick regular seams of chert, at others being nearly without them; the lower portion being very generally an argillo-arenaceous deposit, characterized by the presence of a great abundance of green particles, and a great variety of organic remains. The central part is formed of yellowish-brown and loosely aggregated sand, in which organic remains are rare; the superior, of a mixture of brownish-yellow and green sands, with and without chert seams, the organic remains being frequently fractured.

In Normandy the sands beneath the chalk assume a great variety of characters. Advancing into the interior of France, amid the sands which emerge from under the chalk, and extend from the coasts of Normandy by Mortagne to the banks of the Loire at Tours, and thence by the vicinities of Auxerre and Troyes to the northward, we soon become sensible of the utility of abandoning the smaller divisions, so valuable in England, and of adopting two great divisions,—Chalk and Green Sand.

This group is extensively distributed over Europe. The chalk and *mulatto*, or green sand of northern Ireland, may be considered as the most western portion yet known. Of the interior of Spain and Portugal so little is yet geologically explored, that we are not aware of the existence of chalk there; except, indeed, the nummulite limestone, observed by Colonel Silvertop in the provinces of Sevilla and Murcia, should be of this age.

According to M. Nilsson, the chalk of Sweden (the continuation of that in Denmark,) is generally incumbent on gneiss, more rarely on rocks of the grauwacke group, and has only been observed resting on beds of the oolitic group at one place near Limhamn, in Scania. In one locality, near Hammar and Kaoseberge, it has a large capping of sand with bituminous wood, which M. Nilsson refers to the cretaceous group. The chalk deposit of Sweden is occasionally of considerable thickness, and abounds in organic remains. The northern portion of the deposit is white or grayish white, more or less abundantly mixed with siliceous substances. The southern portion is stated to present the various modifications from green sand to white chalk *.

According to Professor Pusch the cretaceous rocks occur extensively in Podolia and southern Russia, being a continuation of those of Lemberg and Poland. It occupies the country in the shape of marly chalk between the Bog and the Dniester round Janow, Lubin, Mikolajew, Uniow, and Rohetyn. Concealed beneath the supercretaceous rocks it is prolonged from Halicz to Zalezczyki on the Dniester. On the west of this river it occupies the environs of Tlumacz, Otynia, and other places, to the foot of the Carpathians. On the north of the Dniester it exists beneath the supercretaceous rocks between that river and Brzezan; it extends to Brody and into the plains of Volhynia. " In many places, and especially around Krzeminiec, it is covered by more recent deposits, but its presence is indicated by an abundance of flints and chalk fossils scattered through the sands." The chalk forms considerable eminences round Grodno in Lithuania. According to M. Eichwald, the chalk of the latter country abounds in belemnites, which are wanting in Volhynia, where they are replaced by *Echinites, Terebratulæ, Ostreæ, Placunæ, Inoceramus (Catillus)*, &c. The flints of the two countries contain *Reteporæ, Escharæ, Ananchytes, Encrinites,* &c. †.

According to M. Eichwald, chalk without flints, with shells of the genera *Plagiostoma, Pecten, Ostrea,* &c., rests on argillaceous slate at Ladowa, on the Dniester. At about seven wersts from thence, near Bronnitza, it alternately rests on a coarse sandstone, greywacke, and argillaceous slate ‡. Further south, and in the plains of Moldavia, Podolia, and Bessarabia, it only appears in detached portions, as between Jaroszow and Mohilew on the Dniester, from Raszkow to Jaorlik on the Pruth, near Kolomea, Sniatyn, Sadagora, Seret, Roswan, Illina, and Jassy. " The chalk is found on the south side of the granitic steppe, in the Crimea, and on the borders of the sea of Azof, between the Berda and the Don; it also occurs on the west of the Don, across the south-east and middle of Russia. In the country of the Don Cossacks, in the governments of Worenech, Koursk, and Toula, it here and there appears in hills, and on the banks of the rivers beneath the vegetable soil, and probably constitutes the base of that great and fertile plain.

* Nilsson, Petrificata Suecana Form. Cretaceæ descripta et iconibus illustrata, 1827.

† Journal de Géologie, t. ii. p. 62.

‡ *Ibid.* p. 61.

The marly clay of eastern Gallicia and of Podolia is connected, as in Poland, with gypsum, at Mikulnice, Seret of Podolia, to the east of Trembowla, but more particularly at Zbrycz near Czarnokozienice. The graphic chalk is there more abundant than in the centre of Poland, and more abounds in flints *."

It further appears from the interesting details of M. Pusch, that there is a deposit of lignite upon the upper part of the chalk, reminding us of the lignite sand noticed by M. Nilsson in Sweden, which would thus appear to be similarly situated at various distant points. It seems to be wanting in central Poland, but is found in many situations in eastern Gallicia, and abundantly along the Carpathians, in Pocutia and Bukowine, from Otynia towards Maydan, Lanczyn, Kniazdwor, and mounting the Pruth, from Miszyn to Seret, and near Czorthow and Ulaszkowce, and on the Dniester near Chochim and Mohilew. This lignite deposit is described as a blueish or greenish gray calcareous sandstone, alternating with sand and clay, more or less calcareous, and with laminated marl : it sometimes contains amber, but more frequently pieces of bituminous wood, thin beds of lignite, and trunks of fossil trees. It contains many shells, among which are, *Pectunculus pulvinatus, P. insubricus, Pecten* (smooth species), and more rarely *Nummulites discorbinus, Dentalium eburnium,* and small *Cerithia.* This sandstone is considered distinguishable from the well-known lignite deposits of western and northern Poland by its fossil shells; but it may perhaps admit of a question, how far local circumstances may not have caused a great difference in this respect; and it yet appears doubtful how far the lignitiferous sandstone can be considered as a portion of the cretaceous group.

Prof. Pusch describes the cretaceous rocks as extensively deposited in Poland, and as divisible into marly chalk and white chalk : the marly chalk is a soft calcareous marl, either white or light gray, becoming sandy in some districts (Miechow, Kazimirz); while other beds are coloured green by silicate of iron (Czarkow, Szczerbakow) ; it alternates with more compact white limestone. A shaft sunk through this deposit at Szczerbakow showed that it was 697 English feet thick at that place. M. Pusch considers that certain gypseous deposits of Poland are connected with the marly chalk. The white chalk is described as identical with that of England, containing a much larger proportion of flints than the marly chalk †.

Rocks of the cretaceous group occur in various parts of Germany, in the district of the Hartz, and near Quedlenburg, Paderborn, Dortmund, Münster, and various other places.

* Pusch, Journal de Géologie, t. ii.
† *Ibid.* p. 253.

It has already been noticed in France; but it may be remarked that it rests on the coal measures of Mons and Valenciennes, and that the rocks of the Isle d'Aix, and the embouchure of the Charente, are considered referable to this group. It is well known as contained in some of the valleys of the Jura, and as ranging along a considerable portion of the French side of the Pyrenees. It occurs on both sides of the Alps, and ranges down a large portion of the Apennines.

It occurs extensively in the maritime Alps, containing among its fossils an abundance of *Nummulites,* remains once considered as wholly supercretaceous. Its usual appearance in that district is that of a marno-arenaceous limestone, the arenaceous matter sometimes predominating, and forming a sandstone. Beds of light-coloured limestone charged with green grains, and full of *Belemnites, Ammonites, Nautili,* and *Pectines,* constitute the lower part, and even appear intimately connected with the upper part of a light-coloured limestone deposit, among which crystalline dolomite abounds. The latter rocks are very difficult to classify, and may either belong to the lower part of the cretaceous, or upper part of the oolitic, group. Be the age of these beds what it may, they seem, according to M. Elie de Beaumont, intimately connected with a large proportion of the Alpine nummulitic rocks, the light-coloured limestones of Provence, of Mont Ventoux, of the departments of the Drome, Isère, &c.; the nummulitic rocks being connected with the cretaceous series of Briançonnet (Basses Alpes), of Villard le Lans (Isère), of the mountains of the Grande Chartreuse, of the Mont du Chat, of the high longitudinal valleys of the Jura, of the Perte du Rhone, of Thonne, and of la Montagne des Fis.

Having premised thus much respecting the geographical distribution of the cretaceous group, we will take a slight sketch of the variations in its mineralogical character. Throughout the British Islands, a large part of France, many parts of Germany, in Poland, Sweden, and in various parts of Russia, there would appear to have been certain causes in operation, at a given period, which produced nearly, or very nearly, the same effects. The variation in the lower portion of the deposit seems merely to consist in the absence or presence of a greater or less abundance of clays or sands, substances which we may consider as produced by the destruction of previously existing land, and as deposited from waters which held such detritus in mechanical suspension. The unequal deposit of the two kinds of matter in different situations would be in accordance with such a supposition. But when we turn to the higher part of the group, into which the lower portion graduates, the theory of mere transport appears opposed to the phænomena observed, which seem rather to have been produced by deposit from a chemical solution of carbonate of lime and silex, covering a considerable area *. For the reader will have observed, that white chalk, very frequently containing flints, extends from Russia, by Poland, Sweden, Denmark, Germany, and the British Islands, into France. The great European sheet of chalk and green sand, produced at the cretaceous epoch, has since been so covered up, shattered, upheaved and destroyed by various causes, that we have mere remnants presented to our examination. Still, however, we have enough to show that it overlapped a great variety of pre-existing rocks from the gneiss of Sweden to the Wealden deposits of south-eastern England inclusive.

Thus far no very material difference in the arrangement and mineralogical character of the mass has been observed, of course disregarding small local variations : but arrived at the Alps we meet with rocks, which certainly, from these mineralogical characters alone, would never have been referred to the cretaceous group; yet unless we disregard the evidence of organic remains, they have been formed at the same epoch. Instead of the soft and white chalk, and the abundance of loosely aggregated sands, which constitute so large a proportion of the group in England and northern France, we have compact limestones and sandstones vying in hardness with the oldest rocks, so as, in the earlier days of geology, to been have considered only referable to them. Such is the hard black limestone, (containing an abundance of *Scaphites, Hamites, Turrilites,* and other fossils,) which crowns the summits of the Fis, the Sales, and other mountains of Savoy, that range up to the Buet.

The rocks referable to this group, on the southern side of the Alps, and facing the great Lombardo-Venetian plains, is not so far removed from the mineralogical character of the chalk of western Europe, being often composed of white, greenish, and reddish beds, occasionally very argillaceous. In some parts of the Apennine range, in which a large mass of rocks would seem referable to this epoch, the character is quite cretaceous.

How far the Alpine rocks of this age have been altered since the deposit, consequent on the disturbances they have experienced, or how far their present condition can be attributed to original formation which must always have been influenced by local causes, yet remains a problem to be solved : but it may be remarked

* If we regard present appearances, we find that silex is held in solution by thermal waters, which also, as in the case of those of St. Michaels in the Azores, may contain carbonate of lime. No springs or set of springs that we can imagine are likely to have produced this great deposit of chalk, so uniform over a large surface. But although springs, in our acceptation of the term, could scarcely have caused the effects required, we may, perhaps, look to a greater exertion of the power which now produces thermal waters for a possible explanation of the observed phænomena.

that we can scarcely imagine them to have been exposed to the various circumstances attending great disturbances, without having suffered from such circumstances.

M. Elie de Beaumont endeavours to show that violent disruptions of strata in different situations have preceded the deposit of the cretaceous group; and he infers this from the tranquil position of deposits of this nature on the upheaved beds of more ancient rocks. Thus the chalk and quadersandstein (green sand) of the environs of Dresden, Pirna, and Königstein, extend in horizontal beds over the inclined strata of the Erzgebirge, which, from the parallelism of the chains, M. Elie de Beaumont considers were thrown up at the same time that the Côte d'Or hills were elevated, thus obtaining the date of the disturbance between the deposit of the oolitic group and that of the cretaceous series. The Mont Pilas is also considered as elevated at the same epoch. From these disruptions of strata, M. Elie de Beaumont infers that bodies of water were thrown into motion, carrying detritus with them, and that the cretaceous rocks resulted from this deposit. Supposing this theory probable, we might ask how far it would assist us in explaining the chemical character of the white chalk and flints; and whether the circumstances accompanying considerable disruptions of strata may not have permitted the sea to become charged with carbonate of lime and silex, which were deposited after tranquillity was restored; sands and clays being first thrown down from the liquid which held them, at least partly, in mechanical suspension. The student will be careful to consider this as a mere suggestion,—one, perhaps, somewhat at variance with the organic remains found in the cretaceous group.

M. Partsch describes a series of calcareous and arenaceous rocks containing nummulites in Dalmatia and the neighbouring provinces, which appears to belong to the cretaceous group. These rocks form high mountains, particularly in Croatia. From the direction of the mountain chains, M. Elie de Beaumont infers that these rocks may extend into Livadia and the Morea. Facts can alone determine how far this inference is correct; but in the mean time it may be remarked that rocks of the Dalmatian character seem to prevail extensively in parts of Greece, and even along the coast of Karamania.

From the various memoirs of MM. Keferstein and Boué, Prof. Sedgwick, Mr. Murchison, and M. Lill von Lillienbach, it seems clear that the cretaceous group exists extensively in the Alps of Austria and Bavaria, and in the Carpathians. There may be certain differences of opinion as to where the series commences, or where it ends, but the main fact of the presence of the group would appear to be undisputed : it would also appear that the deposit was in a great measure cretaceous.

After remarking on the stability of the cretaceous rocks of the

Carpathians since their deposit, contrasted with their dislocation in the main chain of the Alps, (a fact subsequently fully confirmed within a certain distance from Vienna by Mr. Murchison,) M. Elie de Beaumont proceeds to observe that " nearly in the prolongation of the Carpathians, to the environs of Dresden, the right and north ern side of the Elbe valley is bordered by a continuation of gra nite and syenite mountains, which extend from Hinterherms, on the frontier of Bohemia, to Weinbohla, about a league and a half east from Meissen, rising suddenly above the plain of quadersand stein and planerkalk (cretaceous rocks). When the contact of the granitic and cretaceous rocks is examined, it is observed that the former cut, and even horizontally cover, the latter in many places; clearly proving that the granitic and syenitic rocks were elevated to the surface since the deposit of the green sand and chalk : and it is not the less remarkable, that the little chain formed of them runs in the direction of the valley of the Elbe, and exactly parallel to that which reigns in the Pyreneo-Apen nine system *.

The most remarkable point is at the quarry of Weinbohla, where, according to M. Weiss, the chalk there worked contains the *Plagiostoma spinosum, Podopsis, Spatangus*, &c. This rock is in horizontal beds; but near the sienite they gradually dip until they plunge beneath it, so that the sienite conformably covers the chalk. A marly and clay bed, partly bituminous, covers the chalk, occurring between it and the granitic rock. M. Klipstein remarking on these appearances, observes, that mount ing the valley of Polenz, from the foot of the Hockstein, the green sand beds on the right, which are generally horizontal, begin gra dually to dip, the angle increasing with their approach to the gra nite, near the latter dipping at 46° or 48° beneath it; and he states that of this fact there can be no doubt. " Coming from Brand, the elevation of the green sand diminishes in such a man ner in the descent of the valley, that a few feet of it are alone vi sible. In a valley extending into the mountains towards the Rothenwald, the chalk with its marls and clays appears between the green sand and granite; and there are places where galleries have been driven through the granite and chalk into the green sand." From these works it would appear that "the chalk with its clays and marls gradually diminishes, so that the granite at first resting on chalk, comes into contact with green sand. The su perposition of the granite is quite evident at some distance from this point, when suddenly there is a change, and the granite cuts

---

* M. Elie de Beaumont cites these curious appearances of the super position of granitic rock, as obtained from the descriptions of Prof. Weiss, inserted in Karsten's Archiv für Mineralogie, &c. t. xvi., and new series, t. i.

the arenaceous beds without at all deranging or altering them : it is even stated, that beneath it commences taking a position under the green sand *." From these appearances M. Klipstein, without denying the plausibility of M. Weiss's conclusions, seems to think, that if the granite had really been forced up after the deposit of the chalk, the beds would have been more deranged: he more-over finds a difficulty in the diminished thickness of the chalk. Looking, however, at the valuable observations of M. Klipstein, one would be inclined to think that they confirmed the views of M. Weiss ; for the granite cutting the green sand, and resting on the chalk, seems like many cases of trap rocks which cut and overflow beds, without at all altering or deranging them. The mere want of marks of violence does not weaken the views of M. Weiss. With regard to the diminished thickness of the chalk, it may admit of various explanations.

Before we terminate this sketch, we should notice certain beds found in the Cotentin (Normandy), which, if they do not show a passage of the chalk into the supercretaceous rocks, exhibit an in-teresting juxtaposition of strata, containing chalk fossils, and those with organic remains of the calcaire grossier, the former contain-ing many fossils found also at Maestricht. The baculite lime-stones, as they are termed, of the Cotentin had often been vi-sited, and more or less noticed; but their real position in the series was not pointed out before they were described by M. Desnoyers in 1825. The baculite limestone is white or yellow, and for the most part compact, varying, however, in its mineralogical cha-racter, being sometimes cretaceous and even arenaceous. It contains the organic remains of the cretaceous group, several being also found at Maestricht; such as *Bacculites vertebralis, The-cidea radians, T. recurvirostra, Terebratula* four or five particu-lar species not named, &c. These beds are surmounted by others, (the whole collectively being of no considerable thickness,) com-posed principally of calcareous matter, not presenting any very considerable difference in appearance, though they do not pre-cisely resemble those beneath. They contain organic remains, such as are found in the calcaire grossier; and M. Desnoyers con-siders that a well defined zoological line can be drawn between the two deposits; observing, however, that at the contact of the lower portion of the one with the upper part of the other, when the rocks were without much coherence, there was sometimes an apparent mixture of the fossils. " But at the same time," ob-serves M. Desnoyers, " it has appeared to me, that independent of this confusion, which may be accidental, the species of the compact chalk, *Trochus* and *Bacculites,* preserving their habitual mode of petrifaction, might have belonged to a previously formed

---

* Journal de Géologie, t. ii. p. 182.

bed, and thus differed from those in the calcaire grossier, *Cerithium, Cornucopiæ, Hipponyx, Clypeaster politus,* &c. filled with *Miliolites,* and with the pisolitic limestone which surrounds them. Small sandstone and quartz pebbles, common in the secondary rocks of the Cotentin, accompany them at Orglandes; the only place where I observed this mixture apparent *.'' The geological student can observe the baculite limestone at Fréville, Cauquigny, Bonneville, Orglandes, Hauteville, and other places in the Cotentin.

M. Desnoyers remarks on the absence of *Turrilites, Gryphæa Columba, G. striata, Ostrea carinata, O. pectinata, Pecten spinosus, Chenendopora, Hallirhoa, Ventriculites, Spongus,* &c., amid a mass of fossils (enumerated in the general list beneath) found in the chalk and green sand.

Organic Remains of the Cretaceous Group.

PLANTÆ.

Confervæ.

1. Confervites fasciculata, *Ad. Brong.* Arnager, Bornholm, *Ad. Brong.* Chalk, Sussex, *Mant.*
2. ———— ægagropiloides, *Ad. Brong.* Arnager, Bornholm, *Ad. Brong.*
   ———— species not determined. Chalk, Sussex, *Mant.*

Algæ.

1. Fucoides Orbignianus, *Ad. Brong.* Isle d'Aix, Rochelle, *Ad. Brong.*
2. ———— strictus, *Ad. Brong.* Isle d'Aix, Rochelle, *Ad. Brong.*
3. ———— tuberculosus, *Ad. Brong.* Isle d'Aix, Rochelle, *Ad. Brong.*
4. ———— difformis, *Ad. Brong.* Bidache, Bayonne, *Ad. Brong.*
5. ———— intricatus, *Ad. Brong.* Bidache, *Ad. Brong.*
6. ———— Lyngbianus, *Ad. Brong.* Arnager, Bornholm, *Ad. Brong.*
7. ———— Brongniarti, *Mant.* Chalk, Sussex, *Mant.*
8. ———— Targioni, *Ad. Brong.* Chalk, Sussex, *Mant.*
   ———— species not determined. Chalk, Gault, Sussex, *Mant.*

Naïades.

1. Zosterites cauliniæfolia, *Ad. Brong.* Isle d'Aix, *Ad. Brong.*
2. ———— lineata, *Ad. Brong.* Isle d'Aix, *Ad. Brong.*
3. ———— Bellovisana, *Ad. Brong.* Isle d'Aix, *Ad. Brong.*
4. ———— elongata, *Ad. Brong.* Isle d'Aix, *Ad. Brong.*

* Desnoyers, Sur la Craie et les Terrains Tertiaires du Cotentin ; Mém. de la Soc. d'Hist. Nat. de Paris, t. ii.

Cycadeæ.

1. Cycadites Nilssonii, *Ad. Brong.*, Chalk, Scania.

Dicotyledonous wood, perforated by some boring shell; Chalk, Sussex, *Mant.;* Green Sand, Lyme Regis, *De la B.*
Cones of Coniferæ, Green Sand, Lyme Regis, *De la B.* Green Sand? Kjoepinge, Scania, *Nils.*
Ferns? Green Sand, Lyme Regis, *De la B.*

ZOOPHYTA.

1. Achilleum glomeratum, *Goldf.* Maestricht, *Goldf.*
2. —————— fungiforme, *Goldf.* Maestricht, *Goldf.*
3. —————— Morchella, *Goldf.* Cretaceous Rocks, Essen, Westphalia, *Sack.*
1. Manon capitatum, *Goldf.* Maestricht, *Goldf.*
2. —————— tubuliferum, *Goldf.* Maestricht, *Goldf.*
3. —————— pulvinarium, *Goldf.* Maestricht, Essen, Westphalia, *Goldf.*
4. —————— Peziza, *Goldf.* Maestricht; Cretaceous Rocks, Essen, Westphalia, *Goldf.*
5. —————— stellatum, *Goldf.* Cretaceous Rocks, Essen, *Goldf.*
1. Scyphia mammillaris, *Goldf.* Essen, Westphalia, *Goldf.*
2. —————— furcata, *Goldf.* Cretaceous Rocks, Essen, *Goldf.*
3. —————— infundibuliformis, *Goldf.* Essen, *Goldf.*
4. —————— foraminosa, *Goldf.* Cretaceous Rocks, Essen, *Goldf.*
5. —————— Sackii, *Goldf.* Essen, Westphalia, *Sack.*
1. Spongia ramosa, *Mant.* Chalk, Sussex, *Mant.;* Chalk? Yorkshire, *Phil.;* Noirmoutier, *Al. Brong.*
2. —————— lobata, *Flem.* Chalk, Sussex, *Mant.*
3. —————— plana, *Phil.* Chalk, Yorkshire, *Phil.*
4. —————— capitata, *Phil.* Chalk, Yorkshire, *Phil.*
5. —————— osculifera, *Phil.* Chalk, Yorkshire, *Phil.*
6. —————— convoluta, *Phil.* Chalk, Yorkshire, *Phil.*
7. —————— marginata, *Phil.* Chalk, Yorkshire, *Phil.*
8. —————— radiciformis, *Phil.* Chalk, Yorkshire, *Phil.*
9. —————— terebrata, *Phil.* Chalk, Yorkshire, *Phil.*
10. —————— lævis, *Phil.* Chalk, Yorkshire, *Phil.*
11. —————— porosa, *Phil.* Chalk, Yorkshire, *Phil.*
12. —————— cribrosa, *Phil.* Chalk, Yorkshire, *Phil.*
1. Spongus Townsendi, *Mant.* Chalk, Sussex, *Mant.*
2. —————— labyrinthicus, *Mant.* Chalk, Sussex, *Mant.*
1. Tragos Hippocastanum, *Goldf.* Maestricht, *Goldf.*
2. —————— deforme, *Goldf.* Cretaceous Rocks, Essen, *Goldf.*
3. —————— rugosum, *Goldf.* Cretaceous Rocks, Essen, Westphalia, *Sack.*

N

4. Tragos pisiforme, *Goldf.* Cretaceous Rocks, Essen, Westpha-
        lia, *Goldf.*
5. ———— stellatum, *Goldf.* Cretaceous Rocks, Essen, *Goldf.*
1. Alcyonium globulosum, *Defr.* Chalk, Sussex, *Mant.; Chalk,
        Beauvais; Meudon; Amiens; Tours; Gien;
        Baculite Limest., Normandy, *Desn.*
2. ————? pyriformis, *Mant.* Chalk, Sussex, *Mant.*
   ———— species not determined. Chalk, Sussex, *Mant.;*
        Upper Green Sand, Warminster, *Lons.*
1. Choanites subrotundus, *Mant.* Chalk, Sussex, *Mant.*
2. ———— Königi, *Mant.* Chalk, Sussex; Warminster, *Mant.*
3. ———— flexuosus, *Mant.* Chalk, Sussex, *Mant.*
1. Ventriculites radiatus, *Mant.* Chalk, Sussex, *Mant.;* Chalk,
        Moen, *Al. Brong.*
2. ———— alcyonoides, *Mant.* Chalk, Sussex; Warminster,
        Mant.
3. ———— Benettiæ, *Mant.* Chalk, Sussex, *Mant.;* Chalk,
        Yorkshire, *Phil.*
1. Siphonia Websteri, *Mant.* Chalk, Sussex, *Mant.*
2. ———— curvicornis, *Goldf.* Chalk, Haldern, Westphalia,
        Goldf.
1. Hallirhoa costata, *Lam.* Green Sand, Normandy, *De la B.;*
        Upper Green Sand, Warminster, *Lons.*
1. Serea pyriformis, *Lam.* Green Sand, Normandy, *Al. Brong.*
1. Gorgonia bacillaris, *Goldf.* Maestricht, *Goldf.*
1. Millepora racemosa, *Goldf.* Maestricht, *Goldf.*
2. ———— Fittoni, *Mant.* Chalk, Sussex, *Mant.*
3. ———— Gilberti, *Mant.* Chalk, Sussex, *Mant.*
4. ———— antiqua? *Defr.* Baculite Limest., Normandy, *Desn.*
   ———— species not determined. Chalk, Meudon, *Al. Brong.*
1. Eschara cyclostoma, *Goldf.* Maestricht, *Goldf.*
2. ———— piriformis, *Goldf.* Maestricht, *Goldf.*
3. ———— stigmatophora, *Goldf.* Maestricht, *Goldf.*
4. ———— sexangularis, *Goldf.* Maestricht, *Goldf.*
5. ———— cancellata, *Goldf.* Maestricht, *Goldf.*
6. ———— arachnoides, *Goldf.* Maestricht, *Goldf.*
7. ———— dichotoma, *Goldf.* Maestricht, *Goldf.*
8. ———— striata, *Goldf.* Maestricht, *Goldf.*
9. ———— filograna, *Goldf.* Maestricht, *Goldf.*
10. ———— disticha, *Goldf.* Meudon, *Goldf.*
1. Cellepora ornata, *Goldf.* Maestricht, *Goldf.*
2. ———— Hippocrepis, *Goldf.* Maestricht, *Goldf.*
3. ———— Velamen, *Goldf.* Maestricht, *Goldf.*
4. ———— dentata, *Goldf.* Maestricht, *Goldf.*
5. ———— crustulenta, *Goldf.* Maestricht, *Goldf.*
6. ———— bipunctata, *Goldf.* Maestricht, *Goldf.*

7. Cellepora escharoides, *Goldf.* Cretaceous Rocks, Essen, Westphalia, *Goldf.*
1. Retepora clathrata, *Goldf.* Maestricht, *Goldf.*
2. ———— lichenoides, *Goldf.* Maestricht, *Goldf.*
3. ———— truncata, *Goldf.* Maestricht, *Goldf.*
4. ———— disticha, *Goldf.* Maestricht, *Goldf.*
5. ———— cancellata, *Goldf.* Maestricht, *Goldf.*
1. Flustra utricularis, *Lam.* Chalk, Sussex, *Mant.*
2. ———— reticulata, *Desm.* Baculite Limestone, Norm., *Desn.*
3. ———— flabelliformis, *Lam.* Baculite Limestone, Normandy,
Desn.
———— species not determined. Chalk, Sussex, *Mant.*
1. Ceriopora micropora, *Goldf.* Maestricht, *Goldf.*
2. ———— crytopora, *Goldf.* Maestricht, *Goldf.*
3. ———— anomalopora, *Goldf.* Maestricht, *Goldf.*
4. ———— dichotoma, *Goldf.* Maestricht, *Goldf.*
5. ———— milleporacea, *Goldf.* Maestricht, *Goldf.*
6. ———— madreporacea, *Goldf.* Maestricht, *Goldf.*
7. ———— tubeporacea, *Goldf.* Maestricht, *Goldf.*
8. ———— verticillata, *Goldf.* Maestricht, *Goldf.*
9. ———— spiralis, *Goldf.* Maestricht, *Goldf.*
10. ———— pustulosa, *Goldf.* Maestricht, *Goldf.*
11. ———— compressa, *Goldf.* Maestricht, *Goldf.*
12. ———— stellata, *Goldf.* Maestricht; Cretaceous rocks,
Essen, *Goldf.*
13. ———— Diadema, *Goldf.* Maestricht, *Goldf.*
14. ———— polymorpha, *Goldf.* Cretaceous Rocks, Essen,
Westphalia, *Goldf.*
15. ———— gracilis, *Goldf.* Cretaceous Rocks, Essen, *Goldf.*
16. ———— spongites, *Goldf.* Cretaceous Rocks, Essen,
Goldf.
17. ———— clavata, *Goldf.* Essen, Westphalia, *Goldf.*
18. ———— trigona, *Goldf.* Cretaceous Rocks, Essen, *Goldf.*
19. ———— Mitra, *Goldf.* Cretaceous Rocks, Essen, *Goldf.*
20. ———— venosa, *Goldf.* Cretaceous Rocks, Essen, *Goldf.*
1. Lunulites cretacea, *Defr.* Maestricht; Tours; Baculite Limestone, Normandy, *Desn.*
1. Orbitolites lenticulata, *Lam.* Chalk, Sussex, *Mant.;* Green
Sand, Perte du Rhone, *Al. Brong.*
1. Lithodendron gibbosum, *Munst.* Green Sand, Bochum, Westphalia, *Munst.*
2. ———— gracile, *Goldf.* Green Sand, Quedlinburg,
Goldf.
1. Caryophyllia centralis, *Mant.* Chalk, Sussex, *Mant.;* Chalk,
Yorkshire, *Phil.;* Baculite Limestone, Nor-
mandy, *Desn.*
2. ———— Conulus, *Phil.* Speeton Clay, Yorkshire, *Phil.*
N 2

 1. Fungia radiata, *Goldf.* Cretaceous Sand, Aix-la-Chapelle, *Goldf.*
 2. ———— cancellata, *Goldf.* Maestricht, *Goldf.*
 3. ———— coronula, *Goldf.* Cretaceous Rocks, Essen, Westphalia, *Goldf.*
 1. Chenendopora fungiformis, *Lam.* Upper Green Sand, Warminster, *Lons.*
 1. Hippalimus fungoides, *Lam.* Upper Green Sand, Warminster, *Lons.*
 1. Diploctenium cordatum, *Goldf.* Maestricht, *Goldf.*
 2. ———————— Pluma, *Goldf.* Maestricht, *Goldf.*
 1. Meandrina reticulata, *Goldf.* Maestricht, *Goldf.*
 1. Astrea flexuosa, *Goldf.* Maestricht, *Goldf.*
 2. —— geometrica, *Goldf.* Maestricht, *Goldf.*
 3. —— clathrata, *Goldf.* Maestricht, *Goldf.*
 4. —— escharoides, *Goldf.* Maestricht, *Goldf.*
 5. —— textilis, *Goldf.* Maestricht, *Goldf.*
 6. —— velamentosa, *Goldf.* Maestricht, *Goldf.*
 7. —— gyrosa, *Goldf.* Maestricht, *Goldf.*
 8. —— elegans, *Goldf.* Maestricht, *Goldf.*
 9. —— angulosa, *Goldf.* Maestricht, *Goldf.*
10. —— geminata, *Goldf.* Maestricht, *Goldf.*
11. —— arachnoides, *Goldf.* Maestricht, *Goldf.*
12. —— Rotula, *Goldf.* Maestricht, *Goldf.*
13. —— macrophalma, *Goldf.* Maestricht, *Goldf.*
14. —— muricata, *Goldf.* Chalk, Meudon, *Goldf.*
 1. Pagrus Proteus, *Defr.* Meudon; Tours; Baculite Limestone, Normandy, *Desn.*
    Polypifers, genera not determined. Green Sand, Grand Chartreuse, *Beaum.*; Green Sand, Maritime Alps, *De la B.*; Lower Green Sand, Isle of Wight, *Sedg.*

Radiaria.

 1. Apiocrinites ellipticus, *Miller.* Chalk, Sussex, *Mant.*; Chalk, Yorkshire, *Phil.*; Chalk, Touraine; Baculite Limestone, Normandy, *Desn.*
 1. ——————— Caput Medusæ, *Miller.* Speeton Clay, Yorkshire, *Phil.*
    Pentacrinites, species not determined. Chalk, Sussex, *Mant.*; Speeton Clay, Yorkshire, *Phil.*
 1. Marsupites Milleri, *Mant.* Chalk, Sussex, *Mant.*
 2. ——————— ornatus, *Miller.* Chalk, Yorkshire, *Phil.*
 1. Glenotremites paradoxus, *Goldf.* Marly Chalk, Speldorf, between Duisberg and Mühlheim, *Goldf.*
 1. Pentagonaster semilunatus, *Park.* Chalk, Sussex, *Mant.*
    Asterias, species not determined. Chalk, Paris; Rouen; *Al. Brong.*; Baculite Limestone, Normandy, *Desn.*

1. Cidaris cretosa, *Mant.* Chalk, Sussex, *Mant.*
2. —— variolaris, *Al. Brong.* Chalk, Sussex, *Mant.;* Green
   Sand, Havre; Green Sand, Perte du Rhone, *Al.*
   Brong.; Cretaceous Rocks, Koesfeld and Essen,
   Westphalia; Cretaceous Rocks, Saxony, *Goldf.*
3. —— corollaris, *Mant.* Chalk, Sussex, *Mant.*
4. —— claviger, *König.* Chalk, Sussex, *Mant.*
5. —— vulgaris, *Lam.* Chalk, Poland, *Al. Brong.*
6. —— regalis, *Goldf.* Maestricht, *Goldf.*
7. —— vesiculosa, *Goldf.* Cretaceous Rocks, Essen, Westphalia, *Goldf.*
8. —— scutiger, *Munst.* Cretaceous Rocks, Kelheim, Bavaria, *Munst.*
9. —— crenularis, *Lam.* Chalk, France.
10. —— granulosa, *Goldf.* Chalk, Aix-la-Chapelle; Maes-
    tricht; Cretaceous Rocks, Essen, Westphalia,
    Goldf.
    —— species not determined. Chalk, Speeton Clay, York-
    shire, *Phil.*
1. Echinus regalis, *Hœninghaus.* Cretaceous Rocks, Essen,
   Westphalia, *Goldf.*
2. —— alutaceus, *Goldf.* Cretaceous Rocks, Essen, *Goldf.*
3. —— granulosus, *Munst.* Cretaceous Sandstone, Kehlheim,
   Bavaria, *Munst.*
4. —— saxatilis, *Park.* Chalk, Sussex, *Mant.*
5. —— Königi, *Mant.* Chalk, Sussex, *Mant.*
6. —— areolatus, *Wahl.* Balsberg, Scania, *Nils.* Green
   Sand, Wilts; Lyme Regis, *König.*
7. —— Benettiæ, *König.* Green Sand, Chute, Wilts, *König.*
   —— species not determined. Green Sand, M. de Fis,
   Al. Brong.; Baculite Limestone, Normandy,
   Desn.; Upper Green Sand, Warminster, *Lons.*
1. Galerites albo-galerus, *Lam.* Chalk, Sussex, *Mant.;* Chalk,
   Yorkshire, *Phil.;* Chalk, Dieppe, *Al. Brong.;*
   Chalk, Quedlinburg and Aix-la-Chapelle, *Goldf.*
   Chalk, Lublin, Poland, *Pusch.* Chalk, Lyme
   Regis, *De la B.*
2. —— vulgaris, *Lam.* Chalk, Sussex, *Mant.;* Chalk,
   Dreux, &c., *Al. Brong.;* Quedlinburg; Aix-la
   Chapelle, *Goldf.* Chalk, Lyme Regis, *De la B.*
3. —— subrotundus, *Mant.* Chalk, Sussex, *Mant.;* Chalk,
   Yorkshire, *Phil.*
4. —— Hawkinsii, *Mant.* Chalk, Sussex, *Mant.*
5. —— abbreviatus, *Lam.* Cretaceous Rocks, Quedlinburg;
   Aix-la-Chapelle, *Goldf.*
6. —— canaliculatus, *Goldf.* Cretaceous Rocks, Büren and
   Brencken, Westphalia, *Goldf.*

7. Galerites Subuculus, *Linnæus.* Cretaceous Rocks, Koesfeld and Essen, Westphalia, *Golaf.*
8. ———- sulcato-radiatus, *Goldf.* Maestricht, *Goldf.*
9. ———? depressus, *Lam.* Green Sand, M. de Fis, *Al. Brong.*
  ——— species not determined. Chalk, Upper Green Sand, Warminster, *Lons.*
Clypeus, species not determined. Upper Green Sand, Warminster, *Lons.*
1. Cassidulus Lapis Cancri, *Park.* Upper Green Sand, Warminster, *Lons.*
1. Clypeaster Leskii, *Goldf.* White Chalk, Maestricht, *Goldf.*
2. ——— fornicatus, *Goldf.* Cretaceous Rocks, Munster, Westphalia, *Goldf.*
3. ——— oviformis, *Lam.* Green Sand, Mans, *Desn.*
1. Echinoneus subglobosus, *Goldf.* Maestricht, *Goldf.*
2. ——— Placenta, *Goldf.* Maestricht, *Goldf.*
3. ——— Lampas, *De la B.* Green Sand, Lyme Regis, *De la B.*
4. ——— peltiformis, *Wahl.* Balsberg, Scania, *Wahl.*
1. Nucleolites Ovulum, *Lam.* Maestricht, *Goldf.*
2. ——— scrobicularis, *Goldf.* Maestricht, *Goldf.*
3. ——— Rotula, *Al. Brong.* Chalk, Rouen; Green Sand, M. de Fis, *Al. Brong.*
4. ——— castanea, *Al. Brong.* Green Sand, M. de Fis, *Al. Brong.*
5. ——— patellaris, *Goldf.* Maestricht, *Goldf.*
6. ——— pyriformis, *Goldf.* White Chalk, Maestricht and Aix-la-Chapelle, *Goldf.*
7. ——— lacunosus, *Goldf.* Cretaceous Rocks, Essen, Westphalia, *Goldf.*
8. ——— cordatus, *Goldf.* Cretaceous Rocks, Essen, *Goldf.*
9. ——— carinatus, *Goldf.* Chalk, Aix-la-Chapelle and Hildesheim; Cretaceous Rocks, Essen, Westphalia, *Goldf.*
10. ——— Lapis Cancri, *Goldf.* Aix-la-Chapelle; Maestricht, *Goldf.*
  ——— species not determined. Baculite Limestone, Normandy; Low. Chalk, Tours; Rouen, *Desn.*
1. Ananchytes scutata, *Lam.* Chalk, Sussex, *Mant.*
2. ——— ovata, *Lam.* Chalk, Sussex, *Mant.*; Chalk, Yorkshire, *Phil.*; Chalk, Moen; Meudon, *Al. Brong.*; Baculite Limestone, Normandy; *Desn.*; Limhamn, Sweden, *Nils.*; Cretaceous Rocks, Coesfeld, Westphalia, *Goldf.*; Chalk, Lublin, Poland, *Pusch.*
3. ——— hemisphærica, *Mant.* Chalk, Sussex, *Mant.*; Chalk, Yorkshire, *Phil.*

4. **Ananchytes** intumescens, Chalk, Yorkshire, *Phil.*
5. ————— pustulosa, *Lam.*   Chalk, Joigny; Paris; Rouen; and Moen, *Al. Brong.;* Chalk, Norwich, *Woodward.*
6. ————— conoidea, *Goldf.*   Cretaceous Rocks, Aubel, Belgium, *Goldf.*
7. ————— striata, *Lam.*   Maestricht; Aix-la-Chapelle; Quedlinburg, *Goldf.*
8. ————— sulcata, *Goldf.*   Chalk, Aix-la-Chapelle, Maestricht, *Goldf.*
9. ————— Corculum, *Goldf.*   Cretaceous Rocks, Coesfeld, Westphalia, *Goldf.*
   ————— species not determined. Chalk, Warminster, *Lons.*
1. **Spatangus** Cor-anguinum, *Lam.* Chalk, Sussex, *Mant.;* Chalk, Yorkshire, *Phil.;* Chalk, Meudon; Joigny; Dieppe; Green Sand, M. de Fis, *Al. Brong.;* Baculite Limestone, Normandy, *Desn.;* Torp, Scania, *Nils.;* Chalk, Dorset and Devon, *De la B.;* Marly Chalk, Paderborn; Bielefeld; Münster; Coesfeld; Aix-la-Chapelle; *Goldf.;* Planerkalk, Saxony, *Munst.;* Chalk, Lublin, Poland, *Pusch.*
2. ————— rostratus, *Mant.* Chalk, Sussex, *Mant.;* Chalk, Joigny, *Al. Brong.*
3. ————— planus, *Mant.* Chalk, Sussex, *Mant.;* Chalk, Yorkshire, *Phil.*
4. ————— retusus, *Park.* Upper Green Sand, Wiltshire, *Lons.*
5. ————— cordiformis, *Mant.* Chalk, Sussex, *Mant.*
6. ————— suborbicularis, *Defr.* Green Sand, Dives, Normandy, *Al. Brong.;* Marly Chalk, Maestricht, *Goldf.*
7. ————— punctatus, *Park.* Upper Green Sand, Warminster, *Lons.*
8. ————— granulosus, *Goldf.* Maestricht, *Goldf.*
9. ————— subglobosus, *Leske.* White Chalk, Quedlinburg, Cretaceous Rocks, Büren, Paderborn, *Goldf.*
10. ————— nodulosus, *Goldf.* Cretaceous Rocks, Essen, Westphalia, *Goldf.*
11. ————— radiatus, *Lam.* Maestricht, *Goldf.*
12. ————— truncatus, *Goldf.* White Chalk, Maestricht, *Goldf.*
13. ————— ornatus, *Cuv.* Chalk, Aix-la-Chapelle, *Goldf.*
14. ————— Bucklandii, *Goldf.* Cretaceous Rocks, Essen, *Goldf.*
15. ————— Bufo, *Al. Brong.* Chalk, Meudon, Hâvre, *Al. Brong.;* Chalk, Sussex, *Mant.**; Baculite Limestone, Normandy, *Desn.;* Chalk, Aix-la-Chapelle; Maestricht, *Goldf.*

---

* *Sp. Prunella* of Mantell, according to Brongniart.

16. Spatangus arcuarius, *Lam.* White Chalk, Maestricht, *Goldf.*
17. ———— Prunella, *Lam.* Marly Chalk, Maestricht, *Goldf.*
18. ———— Amygdala, *Goldf.* Chalk, Aix-la-Chapelle, *Goldf.*
19. ———— gibbus, *Lam.* Cretaceous Rocks, Paderborn, West-
     phalia,, *Goldf.*
20. ———— Cor-testudinarium, *Goldf.* White Chalk, Maestricht
     and Quedlinburg; Cretaceous Rocks, Coesfeld,
     Westphalia, *Goldf.*
21. ———— Buçardium, *Goldf.* Chalk, Aix-la-Chapelle, *Goldf.*
22. ———— lacunosus, *Linnæus.* Chalk, Quedlinburg and Aix-
     la-Chapelle, *Goldf.*
23. ———— Murchisonianus, *Mant.* Upper Green Sand, Sussex,
     Mant.
24. ———— hemisphæricus, *Phil.* Chalk, Yorkshire, *Phil.*
25. ———— argillaceus, *Phil.* Speeton Clay, Yorkshire, *Phil.*
26. ———— lævis, *Defr.* Green Sand, Perte du Rhone, *Al. Brong.*
     ———— species not determined. Gault and Lower Green
     Sand, Sussex, *Mant.;* Green Sand, Grande
     Chartreuse, *Beaum.;* Chalk, Warminster, *Lons.*

ANNULATA.

1. Serpula ampullacea, *Sow.* Chalk, Sussex, *Mant.;* Chalk,
   Norfolk, *Barnes.*
2. ———— Plexus, *Sow.* Chalk, Sussex, *Mant.*
3. ———— Carinella, *Sow.* Green Sand, Blackdown, *Sow.*
4. ———— antiquata, *Sow.* Green Sand, Wilts, *Sow.*
5. ———— rustica, *Sow.* Upper Green Sand, Folkstone, *Goodhall.*
6. ———— articulata, *Sow.* Upper Green Sand, Folkstone, *Sow.*
7. ———— obtusa, *Sow.* Chalk, Norfolk, *Rose.*
8. ———— fluctuata, *Sow.* Chalk, Norfolk, *Barnes.*
9. ————? macropus, *Sow.* Chalk, Norfolk, *Leathes.*
   ———— species not determined. Red Chalk, Speeton Clay,
   Yorkshire, *Phil.;* Chalk, Paris, *Al. Brong.;* Charlottenlund; Kjöpinge, Scania, *Nils.*

CIRRIPEDA.

1. Pollicipes sulcatus, *Sow.* Chalk, Sussex, *Mant.*
2. ———— maximus, *Sow.* Chalk, Norfolk, *Barnes.*

CONCHIFERA.

1. Magas pumilus, *Sow.* Chalk, Meudon, *Al. Brong.;* Maestricht, *Hœn.*
1. Thecidea radians, *Defr.* Chalk, Maestricht, *Fauj. de St.
   Fond;* Baculite Limestone, Normandy, *Desn.*
2. ———— recurvirostra, *Defr.* Maestricht; Baculite Limestone,
   Normandy, *Desn.*

3. Thecidea hieroglyphica, *Defr.* Chalk, Essen, *Hœn.*
1. Terebratula subrotunda, *Sow.* Chalk, Sussex, *Mant.* ; Green Sand, Bochum, *Hœn.*
2. ———— carnea, *Sow.* Chalk, Sussex, *Mant.* ; Chalk, Meudon, *Al. Brong.* ; Green Sand, Bochum, *Hœn.*
3. ———— ovata, *Sow.* Chalk, Lower Green Sand, Sussex, *Mant.* ; Kjöpinge, Scania, *Nils.* ; Green Sand, Bochum, *Hœn.*
4. ———— undata, *Sow.* Chalk, Sussex, *Mant.*
5. ———— elongata, *Sow.* Chalk, Sussex, *Mant.*
6. ———— plicatilis, *Sow.* Chalk, Sussex, *Mant.* ; Chalk, Meudon, Moen ; M. de Fis, *Al. Brong.* ; Green Sand, Grande Chartreuse, *Beaum.* ; Chalk, Gravesend, *Sow.*
7. ———— subplicata, *Mant.* Chalk, Sussex, *Mant.* ; Chalk, Yorkshire, *Phil.* ; Chalk, Maestricht ; Tours ; Beauvais ; Bac. Limestone, Normandy, *Desn.*
8. ———— curvirostris, *Nils.* Kjopinge, Scania, *Nils.*
9. ———— Mantelliana, *Sow.* Chalk, Sussex, *Mant.*
10. ———— Martini, *Mant.* Chalk, Sussex, *Mant.*
11. ———— rostrata, *Sow.* Chalk, Sussex, *Mant.*
12. ———— squamosa, *Mant.* Chalk, Sussex, *Mant.*
13. ———— biplicata, *Sow.* Upper Green Sand, Sussex, *Mant.*, Upper Green Sand, Cambridge, *Sedg.*
14. ———— lata, *Sow.* Lower Green Sand, Sussex, *Mant.* ; Green Sand, Devizes, *Sow.* ; Upper Green Sand, Warminster, *Lons.*
15. ———— subundata, *Sow.* Chalk, Speeton Clay, Yorkshire, *Phil.* ; Chalk, Rouen, *Al. Brong.*
16. ———— pentagonalis, . Chalk, Yorkshire, *Phil.*
17. ———— inconstans, *Sow.* Speeton Clay, Yorkshire, *Phil.*
18. ———— tetraedra, *Sow.* Speeton Clay, Yorkshire, *Phil.*
19. ———— lineolata, *Phil.* Speeton Clay, Yorkshire, *Phil.*
20. ———— Defrancii*, *Al. Brong.* Chalk, Meudon, *Al. Brong.* ; Chalk, Sussex, *Mant.* ; Speeton Clay, Yorkshire, *Phil.* ; Balsberg, Mörby, Sweden, *Nils.* ; Maestricht, *Hœn.*
21. ———— alata, *Lam.* Chalk, Meudon, *Al. Brong.* ; Kjöpinge ; Mörby, Sweden, *Nils.*
22. ———— octoplicata, *Sow.* Chalk, Dieppe, *Al. Brong.* ; Balsberg ; Ignaberga, Sweden, *Nils.* ; Green Sand, Quedlinburg, *Hœn.*
23. ———— Gallina, *Al. Brong.* Green Sand, Perte du Rhone, *Al. Brong.* ; Baculite Limestone, Normandy, *Desn.*

* *T. striatula* of Mantell.
N 5

24. Terebratula ornithocephala, *Sow.*   Green Sand, Perte du
        Rhone, M. de Fis, *Al. Brong.*
25. ——————— pectinata, *Sow.*   Baculite Limestone, Norman-
        dy, *Desn.;* Ignaberga, Scania, *Nils.;* Havre,
        Al. Brong.; Upper Green Sand, Wilts, *Meade;*
        Maestricht, *Hœn.*
26. ——————— recurva, *Defr.*   Maestricht; Baculite Limestone,
        Normandy, *Desn.*
27. ——————— lævigata, *Nils.*   Kjöpinge, Scania, *Nils.*
28. ——————— triangularis, *Wahl.*   Kjöpinge, Scania, *Nils.*
29. ——————— longirostris, *Wahl.*   Balsberg; Kjuge, Sweden,
        Nils.
30. ——————— Lyra, *Sow.*   Upper Green Sand, Warminster, *Lons.*
31. ——————— rhomboidalis, *Nils.*   Kjuge; Mörby, Sweden, *Nils.*
32. ——————— semiglobosa, *Sow.*   Charlottenlund, Sweden,
        Nils.; Chalk, Moen, *Al. Brong.;* Green Sand,
        Bochum, *Hœn.;* Chalk, Yorkshire, *Phil.*
33. ——————— obtusa, *Sow.*   Upper Green Sand, Cambridge,
        Sedg.; Green Sand, Quedlinburg, *Hœn.*
34. ——————— obesa, *Sow.*   Chalk, Warminster, *Lons.;* Chalk,
        Bünde, Kündert, *Hœn.*
35. ——————— dimidiata, *Sow.*   Green Sand, Haldon, *Sow.*
36. ——————— aperturata, *Schlot.*   Chalk, Essen, *Hœn.*
37. ——————— chrysalis, *Schlot.*   Maestricht, *Hœn.*
38. ——————— curvata, *Schlot.*   Green Sand, Quedlinburg, *Hœn.*
39. ——————— dissimilis, *Schlot.*   Green Sand, Bochum; Chalk,
        Speldorf, *Hœn.*
40. ——————— lacunosa, *Schlot.*   Green Sand, Quedlinburg, *Hœn.*
41. ——————— microscopica, *Fauj. de St. F.*   Maestricht.
42. ——————— nucleus, *Defr.*   Green Sand, Bochum; Qued-
        linburg, *Hœn.*
43. ——————— ovoidea, *Sow.*   Green Sand, Bochum, *Hœn.*
44. ——————— peltata,   .   Maestricht, *Hœn.*
45. ——————— semistriata, *Lam.*   Green Sand, Bochum, *Hœn.*
46. ——————— striata, *Sow.*   Green Sand, Bochum, *Hœn.*
47. ——————— varians,   .   Chalk, Essen, *Hœn.*
48. ——————— vermicularis, *Schlot.*   Maestricht, *Hœn.*
49. ——————— vitrea, *Lam.*   Chalk, Essen, *Hœn.*
 1. Crania Parisiensis, *Defr.*   Chalk, Meudon, *Al. Brong.;* Chalk,
        Brighton, *Sow.*
 2. ——— antiqua, *Defr.*   Baculite Limestone, Normandy, *Desn.;*
        Chalk, Schlenacken, *Hœn.*
 3. ——— striata, *Defr.*   Baculite Limestone, Normandy, *Desn.;*
        Balsberg, &c. Sweden, *Nils.*
 4. ——— stellata, *Defr.*   Baculite Limestone, Normandy, *Desn.*
 5. ——— spinulosa, *Nils.*   Kjuge; Mörby, Sweden, *Nils.;* Maes-
        tricht, *Hœn.*

6. Crania tuberculata, *Nils.* Scania, *Nils.*
7. —— Nummulus, *Lam.* Balsberg; Kjuge; Ifö, in Scania, *Nils.*; Schlenacken; Schonen, *Hœn.*
8. —— nodulosa, *Hœn.* Maestricht; Sweden, *Hœn.*
 Orbicula, species not determined. Lower Green Sand Sussex, *Martin*; Speeton Clay, Yorkshire, *Phil.*
1. Hippurites radiosa, *Des M.* Cendrieux, Périgord, *Des M.*
2. —— Cornu Pastoris, *Des M.* Pyles, Périgueux, *Jouannet.*
3. —— striata, *Defr.* Alet, Aude; Manbach, Berne, *Des M.*
4. —— sulcata, *Defr.* Alet, Aude, *Des M.*
5. —— dilatata, *Defr.* Alet, Aude, *Des M.*
6. —— bioculata, *Lam.* Alet, Aude, *Des M.*
7. —— Fistulæ, *Defr.* Alet, Aude, *Des M.*
 —— species not determined. Cretaceous Rocks, South of France, *Beaum.*; Pyrenees, *Dufr.*; Western Alps, *Lill von Lillienbach*; *Murch.*
1. Sphærulites dilatata, *Des M.* Chalk, Royan and Talmont, mouth of the Gironde, *Des M.*
2. —— Bournonii, *Des M.* Royan and Talmont; Vallée de la Couze, Dordogne, *Des M.*
3. —— ingens, *Des M.* Royan and Talmont, *Des M.*
4. —— Hœninghausii, *Des M.* Royan and Talmont; Chalk, Languais, Dordogne, *Des M.*
5. —— foliacea, *Lam.* Isle d'Aix, *Fleurian de Bellevue.*
6. —— Jodamia, *Des M.* Mirambeau, Charente-Inférieure, *Defr.*
7. —— Jouannetii, *Des M.* Vallée de la Couze, Périgord, *Des M.*
8. —— crateriformis, *Des M.* Royan; Languais, Dordogne, *Des M.*
9. —— Moulinii, *Goldf.* Maestricht, *Hœn.*
1. Ostrea vesicularis, *Lam.* Chalk, Sussex, *Mant.*; Chalk, Périgueux, Meudon, *Al. Brong.*; Chalk, Maestricht, *Fauj. de St. F.*; *var.* Baculite Limestone, Normandy, *Desn.*; Kjöpinge; Kjuge, Sweden, *Nils.*
2. —— semiplana, *Mant.* Chalk, Sussex, *Mant.*
3. —— canaliculata, *Sow.* Chalk, Sussex, *Mant.*
4. —— carinata, *Al. Brong.* Upper Green Sand, Sussex. *Mant.*; Green Sand, Normandy, *De la B.*; Green Sand, Grasse, (Dep. of the Var.) *Martin de Martigues*; Green Sand, Bochum; Chalk, Essen, *Hœn.*
5. —— serrata, *Defr.* Chalk, Sweden; Dreux, *Al. Brong.*; Green Sand, Grasse, Var; Maestricht, *Hœn.*
6. —— lateralis, *Nils.* Kjöpinge; Ifö, Scania, *Nils.*; Chalk, Essen, *Hœn.*
7. —— clavata, *Nils.* Mörby, Sweden, *Nils.*
8. —— Hippopodium, *Nils.* Ifö; Carlshamn, Sweden, *Nils.*

9. Ostrea acuminata, *Sow.* Ifö; Kjuge, Scania, *Nils.*
10. —— curvirostris, *Nils.* Ifö; Kjuge, Scania, *Nils.*
11. —— acutirostris, *Nils.* Ifö; Scania, *Nils.*
12. —— flabelliformis, *Nils.* Kjuge, Mörby, Sweden, *Nils.;* Chalk, Essen, *Hœn.*
13. —— pusilla, *Nils.* Kjöpinge, Scania, *Nils.*
14. —— diluviana?* *Lam.* Balsberg; Kjuge; Mörby; Carlshamn, Sweden, *Nils.*
15. —— lunata, *Nils.* Ähus, Yngsjö, Scania, *Nils.*
16. —— parasitica, Green Sand, Bochum, *Hœn.*
17. —— truncata, Green Sand, Griesenbeck, *Hœn.*
1. Hinnites? Dubuissoni, Chalk, Doué, *Hœn.*
1. Exogyra recurvata, *Sow.* Green Sand, Haldon Hill, *Baker.*
2. —— plicata, *Sow.* Green Sand, Haldon Hill, *Baker.*
3. —— digitata, *Sow.* Green Sand, Lyme Regis, *De la B.*
4. —— conica, *Sow.* Green Sand, Sussex; Upper Green Sand, Wilts; Green Sand, Blackdown, *Sow.*
5. —— undata, *Sow.* Green Sand, Blackdown, *Goodhall.*
6. —— haliotoidea, *Sow.* Upper Green Sand, Warminster, *Lons.;* Chalk, Essen, *Hœn.*
1. Gyphæa vesiculosa, *Sow.* Upper Green Sand, Sussex, *Mant.;* Green Sand, Warminster, *Bennet;* Green Sand, Bouches du Rhone, *Hœn.*
2. —— sinuata, *Sow.* Speeton Clay, Yorks., *Phil.;* Green Sand, Grande Chartreuse, *Beaum.;* Lower Green Sand, Isle of Wight, *Sedg.*
3. —— auricularis, *Al. Brong.* Chalk, Perigueux, *Al. Brong.;* Green Sand, Grande Chartreuse, *Beaum.;* Chalk, Kazimirz, Poland, *Pusch;* Green Sand, Apt, Vaucluse, *Hœn.*
4. —— Aquila, *Al. Brong.* Green Sand, Perte du Rhone, *Al. Brong.*
5. —— Columba, *Lam.* Green Sand, Normandy; Green Sand, Maritime Alps, *De la B.;* Green Sand, Northamptonshire, *Sow.;* Chalk, Kazimirz, Poland, *Pusch;* Regenburg; Pirna; Königstein, *Holl;* Chalk, Saumur; Mans, *Hœn.*
6. —— plicata, *Lam.* Green Sand, Boesingfeld, Chalk, Saumur, *Hœn.*
7. —— truncata, *Goldf.* Maestricht, *Hœn.*
—— a small species in the baculite limestone and chalk of other parts of France, *Desn.*
1. Sphæra corrugata, *Sow.* Lower Green Sand, Isle of Wight, *Sedg.*
1. Podopsis lata, Chalk, Sussex, *Mant.*

* M. Brongniart considers that this shell, cited by M. Nilsson as
O. diluviana, may be the *O. serrata* of Defrance.

2. Podopsis obliqua, *Mant.* Chalk, Sussex, *Mant.*
3. —— striata, *Sow.* Chalk, Yorks., *Phil.;* Chalk, Hâvre, *Al.*
    Brong.; Chalk, Essen; Bochum, *Hœn.*
4. —— truncata, *Lam.* Chalk, Normandy, Touraine, *Al.*
    Brong.; Balsberg and other places in Sweden, *Nils.*
5. —— lamellata, *Nils.* Kjuge, Mörby, Sweden, *Nils.*
1. Spondylus? strigilis, *Al. Brong.;* Green Sand, Perte du Rhone,
    Al. Brong.
1. Plicatula, inflata, *Sow.* Chalk, Sussex, *Mant.;* Chalk, Cam-
    bridge, *Sedg.*
2. —— pectinoides, *Sow.* Chalk, Sussex, *Mant.;* Gault,
    Cambridge, *Sedg.*
1. Pecten quinquecostatus, *Sow.* Chalk, Sussex, *Mant.;* Chalk,
    Meudon, *Al. Brong.;* Green Sand, Perte du Rhone,
    Al. Brong.; Baculite Limestone, Normandy, *Desn.;*
    Kjöpinge, and other places in Sweden, *Nils.;* Green
    Sand, Blackdown, *Sow.;* Green Sand, Lyme Regis,
    De la B.; Upper Green Sand, Warminster, *Lons.;*
    Green Sand, Coesfeld, Osterfeld; Chalk, Saumur,
    Hœn.
2. —— Beaveri, *Sow.* Chalk, Sussex, *Mant.*
3. —— triplicatus, *Mant.* Chalk, Sussex, *Mant.*
4. —— orbicularis, *Sow.* Chalk, Gault, Lower Green Sand,
    Sussex, *Mant.;* Kjöpinge, Sweden, *Nils.;* Green
    Sand, Aix la Chapelle, *Hœn.*
5. —— quadricostatus, *Sow.* Lower Green Sand, Sussex, *Mant.;*
    Chalk, Maestricht; Baculite Limestone, Normandy,
    Desn.; Green Sand, Grande Chartreuse, *Beaum.;*
    Green Sand, Haldon, *Baker;* Upper Green Sand,
    Warminster, *Lons.*
6. —— obliquus, *Sow.* Lower Green Sand, Sussex, *Mant.*
7. —— cretosus, *Defr.* Chalk, Meudon, *Al. Brong.;* Chalk,
    Lublin, Poland, *Pusch;* Chalk, Angers; Maes-
    tricht, *Hœn.*
8. —— arachnoides, *Defr.* Chalk, Meudon and Normandy, *Al.*
    Brong.; Chalk, Lublin, Poland, *Pusch.*
9. —— extextus *, *Al. Brong.;* Chalk, Hâvre; Baculite Lime-
    stone, Normandy, *Desn.;* Chalk, Angers, *Hœn.*
10. —— serratus, Balsberg; Kjöpinge, Sweden, *Nils.*
11. —— septemplicatus, *Nils.* Balsberg, Kjuge, Sweden, *Nils.*
12. —— multicostatus, *Nils.* Balsberg, Sweden, *Nils.*
13. —— undulatus, *Nils.* Kjöpinge; Käserberga, Scania, *Nils.*
14. —— subaratus, *Nils.* Balsberg; Kjuge, Sweden, *Nils.*
15. —— pulchellus, *Nils.* Kjöpinge; Balsberg, Sweden, *Nils.*

---

* M. Hœninghaus considers this shell the same with *P. serratus,*
Nilsson.

16. Pecten lineatus, *Nils.* Kjöpinge; Mörby, Sweden, *Nils.*
17. —— arcuatus, *Sow.* Kjöpinge, Sweden, *Nils.;* Green Sand,
      Aix la Chapelle, *Hœn.*
18. —— virgatus, *Nils.* Balsberg; Mörby, *Nils.*
19. —— membranaceus, *Nils.* Kjöpinge, and other places,
      Sweden, *Nils.*
20. —— lævis, *Nils.* Kjöpinge; Yngsjoe, Sweden, *Nils.;* Aix
      la Chapelle, *Hœn.*
21. —— inversus, *Nils.* Kjöpinge, Sweden, *Nils.*
22. —— asper, *Lam.* Upper Green Sand, Warminster, *Lons.;*
      Chalk, Lublin, Poland, *Pusch;* Green Sand, Bo-
      chum; Chalk, Hatteren, *Hœn.*
23. —— asperrimus,   Green Sand, Hardt, *Hœn.*
24. —— gracilis, *Sow.* Green Sand, Aix la Chapelle, *Hœn.*
25. —— gryphæatus, Green Sand, Aix la Chapelle, *Hœn.*
26. —— nitidus, *Sow.* Chalk, Sussex, *Mant.;* Green Sand, Aix
      la Chapelle, *Hœn.*
27. —— regularis, *Schlot.* Maestricht, *Hœn.*
28. —— sulcatus, *Sow.* Green Sand, Hardt; Maestricht, *Hœn.*
29. —— versicostatus,   Green Sand, Aix la Chapelle; Green
      Sand, Minden, *Hœn.*
     —— species not determined;—Chalk, Sussex, *Mant.;* Spee-
      ton Clay, Yorks., *Phil.;* Green Sand, Maritime
      Alps, *De la B.*
1. Lima pectinoides,   Maestricht, *Hœn.*
1. Plagiostoma spinosum*, *Sow.* Chalk, Sussex, *Mant.;* Chalk,
      Meudon, Dieppe, Rouen, Périgueux, Poland,
      Al. Brong.; Kjöpinge, Sweden, *Nils.;* Chalk,
      Dorset and Devon, *De la B.;* Chalk, Wein-
      bohla, Saxony, *Weiss;* Quedlinburg, *Holl;*
      Osterfeld, *Hœn.*
2. ————— Hoperi, *Mant.;* Chalk, Sussex, *Mant.*
3. ————— Brightoniensis, *Mant.;* Chalk, Sussex, *Mant.*
4. ————— elongatum, *Sow.;* Chalk, Sussex, *Mant.*
5. ————— asperum, *Mant.;* Chalk, Sussex, *Mant.*
6. ————— pectinoide, *Sow.*, Green Sand, Perte du Rhone,
      Al. Brong.
7. ————— ovatum, *Nils.* Balsberg and Kjuge, Sweden,
      Nils.
8. ————— semisulcatum, *Nils.* Balsberg and other places,
      Sweden, *Nils.;* Chalk, Künder, Saumur, *Hœn.*
9. ————— Mantelli, *Al. Brong.* Chalk, Dover; Moen,
      Denmark, *Al. Brong.*
10. ————— granulatum, *Nils.* Kjöpinge, Kjuge, Sweden,
      Nils.
11. ————— elegans, *Nils.* Balsberg, Mörby, Sweden, *Nils.*

      * *Pachites spinosa* of Defrance.

12. Plagiostoma pusillum, *Nils.* Balsberg, Kjöpinge, Sweden,*Nils.*
13. ————— turgida, *Lam.* Chalk, Saintes; Green Sand, Osterfeld, *Hœn.*
14. ————— punctata, Maestricht, *Hœn.*
————— species not determined:—Upper Green Sand,
Sussex, *Mant.*
1. Avicula cærulescens, *Nils.* Kjöpinge, Kaseberga, Sweden,
Nils.
————— species not determined. Chalk, Sussex, *Mant.*;
Maestricht? *Hœn.*
1. Inoceramus Cuvieri, *Sow.* Chalk, Sussex, *Mant.*; Chalk,
Yorks., *Phil.*; Chalk, Meudon, *Al. Brong.*;
Balsberg; Ignaberga, Kjuge, Sweden, *Nils.*;
2. ————— Brongniarti, *Sow.* Chalk, Sussex, *Mant.*; Chalk,
Yorks., *Phil.*; Kaserberga, Kjöpinge, Sweden,
Nils.; Chalk, Czarkow, Poland, *Pusch;* Quedlinburg, *Hœn.*
3. ————— Lamarckii, Chalk, Sussex, *Mant.*
4. ————— mytiloides, *Sow.* Chalk, Sussex, *Mant.*; Chalk,
Warminster, *Lons.*; Quedlinburg; Pirna, Königstein, *Holl.*
5. ————— cordiformis, *Sow.* Chalk, Sussex, *Mant.*; Chalk,
Gravesend, *Sow.*
6. ————— latus, *Mant.* Chalk, Sussex, *Mant.*
7. ————— Websteri, *Mant.* Chalk, Sussex, *Mant.*
8. ————— striatus, *Mant.* Chalk, Sussex, *Mant.*
9. ————— undulatus, *Mant.* Chalk, Sussex, *Mant.*
10. ————— involutus, *Sow.* Chalk, Sussex, *Mant.*; Chalk,
Norfolk, *Rose.*
11. ————— tenuis, *Mant.* Chalk, Sussex, *Mant.*
12. ————— Cripsii, *Mant.* Chalk, Sussex, *Mant.*
13. ————— concentricus, *Park.* Gault, Sussex, *Mant.* ;
Green Sand, Perte du Rhone, M. de Fis, *Al.
Brong.*; Chalk, Warminster, *Lons.*; Green
Sand, Quedlinburg, Bochum, and Essen, *Hœn.*
14. ————— sulcatus, *Park.* Gault, Sussex, *Mant.*; Green
Sand, Perte du Rhone, M. de Fis, *Al. Brong.*;
Kjöpinge, Scania, *Nils.*; Green Sand? Nice,
De la B.
15. ————— gryphæoides, *Sow.* Gault, Sussex, *Mant.*; Green
Sand, Lyme Regis, *De la B.*
16. ————— pictus, *Sow.* Chalk, Surrey, *Murch.*
17. ————— rugosus, Quedlinburg, *Hœn.*
————— species not determined. Lower Green Sand,
Sussex, *Martin;* Baculite Limestone, Normandy, *Desn.*
1. Gervillia aviculoides, *Sow.* Lower Green Sand, Sussex, *Mant.*;
Green Sand, Lyme Regis, *De la B.*; Quedlin-

burg, *Holl;* Lower Green Sand? Isle of Wight,
    Sedg.
2. Gervillia solenoides, *Defr.* Lower Green Sand, Sussex, *Mant.;*
    Baculite Limestone, Normandy, *Desn.;* Green
    Sand, Lyme Regis, *De la B.;* Upper Green Sand,
    Warminster, *Lons.;* Maestricht, Marsilly, *Hœn.*
3. —— acuta, *Sow.* Lower Green Sand, Sussex, *Mant.*
1. Crenatula ventricosa, *Sow.* Green Sand, Bochum, *Hœn.*
1. Pinna gracilis, *Phil.* Speeton Clay, Yorks., *Phil.*
2. —— tetragona, *Sow.* Upper Green Sand, Devizes, *Gent.*
3. —— affinis, Chalk, Doué, near Saumur, *Hœn.*
4. —— flabellum, Chalk, Bochum, *Hœn.*
5. —— nobilis, Chalk, Bochum, *Hœn.*
6. —— restituta, Chalk, Valkenburg, *Hœn.*
7. —— subquadrivalvis, Cotentin; Saumur, *Hœn.*
1. Mytiloides labiatus, *Al. Brong.* Chalk near Calne, *Lons.;*
    Aix la Chapelle; Quedlinburg, *Holl.;* Chalk,
    Saumur, *Hœn.*
1. Mytilus lanceolatus, *Sow.* Lower Green Sand, Sussex, *Mant.;*
    Green Sand, Blackdown, *Sow.*
2. —— lævis, *Defr.* Chalk, Bougival, *Al. Brong.*
3. —— edentulus, *Sow.* Green Sand, Blackdown, *Sow.*
4. —— problematicus, Green Sand, Bochum, *Hœn.*
1. Modiola æqualis, *Sow.* Lower Green Sand, Sussex, *Mant.*
2. —— bipartita, *Sow.* Lower Green Sand, Sussex, *Mant.*
1. Pachymya Gigas, *Sow.* Lower Chalk, Lyme Regis, *De la B.*
1. Chama Cornu Arietis, *Nils.* Kjuge; Mörby, Sweden, *Nils.*
2. —— laciniata, *Nils.* Kjuge; Balsberg; Mörby, Sweden,
    Nils.
3. —— haliotoidea, *Sow.* Kjuge; Balsberg; Mörby, Sweden,
    Nils.
4. —— conica, *Sow.* Kjöpinge, Scania, *Nils.*
5. —— recurvata, Chalk, Doué, *Hœn.*
—— species not determined. Chalk, Sussex, *Mant.*
1. Trigonia Dædalea, *Park.* Lower Green Sand, Sussex, *Mant.;*
    Green Sand, Haldon? *Baker*, Lower Green Sand,
    Isle of Wight, *Sedg.*
2. —— aliformis, *Sow.* Lower Green Sand, Sussex, *Mant.;*
    Upper Green Sand? Eddington, *Lons.;* Lower
    Green Sand, Isle of Wight, *Sedg.;* Altenberg,
    Hœn.
3. —— spinosa, *Sow.* Lower Green Sand, Sussex, *Martin;*
    Green Sand, Blackdown, *Steinhauer.*
4. —— rugosa, *Lam.* Green Sand, Perte du Rhone, *Al.
    Brong.*
5. —— scabra, *Lam.* Green Sand, Perte du Rhone, *Al.
    Brong.;* Baculite Limestone? Normandy, *Desn.*
6. —— pumila, *Nils.* Kjöpinge, Scania, *Nils.*

7. Trigonia eccentrica, *Sow.* Green Sand, Blackdown,*Steinhauer.*
8. ———— nodosa, *Sow.* Lower Green Sand, Hythe, Kent, *Sow.*
9. ———— spectabilis, *Sow.* Green Sand, Blackdown, *Goodhall.*
10. ———— arcuata, *Lam.* Aix la Chapelle, *Hœn.*
———— species not determined. Lower Green Sand, Wiltshire, *Lons.*
1. Nucula pectinata, *Mant.* Gault, Sussex, *Mant.*
2. ———— ovata, *Mant.* Gault, Sussex, *Mant.;* Speeton Clay, Yorkshire, *Phil.*
3. ———— impressa, *Sow.* Lower Green Sand, Sussex, *Mant.;* Green Sand, Blackdown, *Sow.*
4. ———— subrecurva, *Phil.* Speeton Clay, Yorkshire, *Phil.*
5. ———— ovata, *Nils.* Kjöpinge, Scania, *Nils.*
6. ———— truncata, *Nils.* Kaseberga, Scania, *Nils.*
7. ———— panda, *Nils.* Kaseberga, Scania, *Nils.*
8. ———— producta, *Nils.* Kaseberga, Scania, *Nils.*
9. ———— antiquata, *Sow.* Green Sand, Blackdown, *Sow.*
10. ———— angulata, *Sow.* Green Sand, Blackdown, *Sow.*
11. ———— undulata, *Sow.* Gault, Folkestone, *Sow.*
1. Pectunculus lens, *Nils.* Balsberg; Kjöpinge, Sweden, *Nils.*
2. ———— sublævis, *Sow.* Green Sand, Blackdown, *Sow.*
1. Arca, carinata, *Sow.* Upper Green Sand, Sussex, *Mant.*
2. ———— exaltata, *Nils.* Carlshamn, Sweden, *Nils.;* Green Sand? Aix la Chapelle, *Hœn.*
3. ———— rhombea, *Nils.* Balsberg, Sweden, *Nils.*
4. ———— clathrata, Chalk, Angers; Saumur, *Hœn.*
5. ———— ovalis, *Nils.* Kjöpinge, Scania, *Nils.*
6. ———— subacuta, Maestricht, *Hœn.*
———— species not determined. Chalk, Gault, Sussex, *Mant.*
1. Cucullæa decussata, *Sow.* Lower Green Sand, Sussex, *Mant.,* Chalk, Rouen, *Al. Brong.*
2. ———— glabra, *Sow.* Green Sand, Blackdown, *Sow.;* Upper Green Sand, Warminster, *Lons.*
3. ———— carinata, *Sow.* Green Sand, Blackdown, *Sow.*
4. ———— fibrosa, *Sow.* Green Sand, Blackdown, *Hill.*
5. ———— costellata, *Sow.* Green Sand, Blackdown, *Sow.*
6. ———— auriculifera, Chalk, Beauvais, *Hœn.*
7. ———— crassatina, Chalk, Beauvais, *Hœn.*
———— species not determined. Chalk, Sussex, *Mant.,* Speeton Clay, Yorkshire, *Phil.*
1. Cardita Esmarkii, *Nils.* Kjöpinge, Scania, *Nils.*
2. ———— Modiolus, *Nils.* Kaseberga, Scania, *Nils.*
3. ———— tuberculata, *Sow.* Upper Green Sand, Devizes, *Gent.*
4. ———— crassa, Chalk, Doué, *Hœn.*
———— species not determined. Upper Green Sand, Sussex, *Mant.*
1. Cardium decussatum, *Sow.* Chalk, Sussex, *Mant.*
2. ———— Hillanum, *Sow.* Green Sand, Blackdown, *Hill.*

3. Cardium proboscideum, *Sow.* Green Sand, Blackdown, *Hill.*
4. —— umbonatum, *Sow.* Green Sand, Blackdown, *Sow.*
5. —— bullatum, *Lam.* Aix la Chapelle, *Hœn.*
Venericardia, species not determined. Chalk, Sussex, *Mant.*
1. Astarte striata, *Sow.* Green Sand, Blackdown, *Sow.;* Upper
       Green Sand, Devizes, *Lons.*
   —— species not determined. Chalk, Sussex, *Mant.;*
       Lower Green Sand, Wilts, *Lons.*
1. Thetis minor, *Sow.* Lower Green Sand, Sussex, *Mant.;*
       Green Sand, Lyme Regis, *De la B.*
2. —— major, *Sow.* Upper Green Sand, Devizes, *Gent;* Green
       Sand, Blackdown, *Hill.*
1. Venus Ringmeriensis, *Mant.* Chalk, Sussex, *Mant.*
2. —— parva, *Sow.* Lower Green Sand, Sussex, *Mant.;* Green
       Sand, Lyme Regis, *De la B.;* Green Sand, Isle of
       Wight, *Sow.*
3. —— angulata, *Sow.* Lower Green Sand, Sussex, *Mant.;*
       Green Sand, Blackdown, *Hill.*
4. —— Faba, *Sow.* Lower Green Sand, Sussex, *Mant.;* Green
       Sand, Blackdown; Green Sand, Isle of Wight, *Sow.*
5. —— ovalis, *Sow.* Lower Green Sand, Sussex, *Mant.*
6. —— lineolata, *Sow.* Green Sand, Blackdown, *Hill;* Green
       Sand, Bochum, *Hœn.*
7. —— plana, *Sow.* Green Sand, Blackdown, *Hill.*
8. —— caperata, *Sow.* Green Sand, Lyme Regis, *De la B.;*
       Green Sand, Blackdown, *Hill.*
1. Lucina sculpta, *Phil.* Speeton Clay, Yorkshire, *Phil.*
1. Tellina æqualis, *Mant.;* Lower Green Sand, Sussex, *Mant.*
2. —— inæqualis, *Sow.* Lower Green Sand, Sussex, *Mant.;*
       Green Sand, Blackdown, *Sow.*
3. —— striatula, *Sow.* Green Sand, Blackdown, *Sow.*
   —— species not determined. Speeton Clay, Yorkshire,
       Phil.
1. Corbula striatula, *Sow.* Lower Green Sand, Sussex, *Mant.*
2. —— punctum, *Phil.* Speeton Clay, Yorkshire, *Phil.*
3. —— gigantea, *Sow.* Green Sand, Blackdown, *Hill.*
4. —— lævigata, *Sow.* Green Sand, Blackdown, *Hill.*
5. —— anatina, *Desh.* Green Sand, Schonen, *Hœn.*
1. Crassitella latissima,   Maestricht, *Hœn.*
1. Lutraria Gurgitis, *Al. Brong.;* Green Sand, Perte du Rhone,
       Al. Brong.; Kjöpinge; Mörby, Sweden, *Nils.*
2. ——? carinifera, *Sow.* Chalk, Lyme Regis, *De la B.*
   —— species not determined. Speeton Clay? Yorkshire,
       Phil.
1. Mya plicata, *Sow.* Green Sand, Osterfeld, *Hœn.;* (var.?)
       Lower Green Sand, Sussex, *Mant.*
2. —— mandibula, *Sow.* Lower Green Sand, Sussex, *Martin;*
       Gault, Isle of Wight, *Fitton.*

3. Mya depressa, *Sow.* Speeton Clay, Yorkshire, *Phil.*
4. —— phaseolina, *Phil.* Speeton Clay, Yorkshire, *Phil.*
5. —— plana, *Sow.* Green Sand, Osterfeld, *Hœn.*
—— ? Chalk, near Calne, *Lons.*
Teredo, species not determined. Maestricht, *Hœn.*
1. Pholas? constricta, *Phil.* Speeton Clay, Yorkshire, *Phil.*
1. Fistulana personata, *Mant.* Chalk, Sussex, *Mant.*
2. —— pyriformis, *Mant.* Gault, Sussex, *Mant.*

MOLLUSCA.

1. Dentalium striatum, *Mant.* Gault, Sussex, *Mant.*
2. —— ellipticum, *Mant.* Gault, Sussex, *Mant.*
3. —— decussatum, *Sow.* Gault, Sussex, *Mant.*
4. —— fissura, Green Sand, Schonen, *Hœn.*
5. —— nitens, Maestricht, *Hœn.*
—— species not determined. Lower Green Sand, Sussex, *Mant.;* Kjöpinge, Sweden, *Nils.*
1. Patella ovalis, *Nils.* Balsberg, Scania, *Nils.*
—— species not determined. Lower Green Sand, Sussex, *Mant.;* Lower Green Sand, Wiltshire, *Lons.*
Pileopsis, species not determined. Lower Green Sand, Sussex, *Mant.*
1. Helix Gentii, *Sow.* Upper Green Sand, Devizes, *Gent.*
1. Auricula incrassata, *Sow.* Chalk, Sussex, *Mant.;* Green Sand, Blackdown, *Hill.*
2. —— obsoleta, *Phil.* Speeton Clay, Yorkshire, *Phil.*
3. —— turgida, *Sow.* Green Sand, Schonen, *Hœn.*
Melania, species not determined. Speeton Clay? Yorkshire, *Phil.*
1. Paludina extensa, *Sow.* Green Sand, Blackdown, *Hill.*
1. Ampullaria canaliculata, Gault, Sussex, *Mant.*
2. —— spirata, Maestricht, *Hœn.*
—— species not determined. Green Sand, M. de Fis, *Al. Brong.*
1. Nerita rugosa, Maestricht, *Hœn.*
1. Natica carena, *Park.* Lower Green Sand, Sussex, *Mant.*
2. —— spirata, Green Sand, Aix la Chapelle, *Hœn.*
—— species not determined. Gault, Sussex, *Mant.;* Lower Green Sand, Wiltshire, *Lons.*
1. Vermetus polygonalis, *Sow.* Lower Green Sand, Hythe, Kent, *Lord Greenock.*
2. —— umbonatus, *Mant.* Chalk, Sussex, *Mant.*
3. —— Sowerbii, *Mant.* Chalk, Sussex, *Mant.;* Speeton Clay, Yorkshire, *Phil.*
4. —— concavus, *Sow.* Lower Green Sand, Sussex, *Mant.;* Upper Green Sand, Wilts, *Lons.*

Vermetus, species not determined.   Lower Green Sand, Isle
    of Wight, *Sedg.*
1. Sigaretus concavus, *Sedg.* Bochum, *Hœn.*
Delphinula, species not determined.   Speeton Clay, Yorkshire,
    Phil.
1. Solarium tabulatum? *Phil.*   Speeton Clay, Yorkshire, *Phil.*
1. Cirrus depressus, *Mant.* Chalk, Sussex, *Mant.*
2. —— perspectivus, *Mant.* Chalk, Sussex, *Mant.*
3. —— granulatus, *Mant.* Chalk, Sussex, *Mant.*
4. —— plicatus, *Sow.* Gault, Sussex, *Mant.*
Pleurotomaria, species not determined.   Maestricht, *Hœn.*
1. Trochus Basteroti, *Al. Brong.* Chalk, Sussex, *Mant.; Kjö-
    pinge, Scania, *Nils.*
2. —— linearis, *Mant.* Chalk, Sussex, *Mant.*
3. —— agglutinans, *Sow.* Chalk? Sussex, *Mant.; Green
    Sand, Aix-la-Chapelle, *Hœn.*
4. —— Rhodani, *Al. Brong.* Upper Green Sand, Sussex,
    *Mant.; Green Sand, Perte du Rhone, *Al.Brong.;*
    Lower Chalk, Lyme Regis, *De la B.; Green
    Sand, Essen; Green Sand, Osterfeld, *Hœn.*
5. —— bicarinatus, *Sow.* Upper Green Sand? Sussex,*Mant.*
6. —— reticulatus? *Sow.* Speeton Clay, Yorkshire, *Phil.*
7. —— Gurgitis, *Al. Brong.* Green Sand, Perte du Rhone,
    *Al. Brong.; Green Sand, Bochum, *Hœn.*
8. ——? Cirroides, *Al.Brong.* Green Sand, Perte du Rhone,
    Al. Brong.
—— species not determined.   Green Sand, M. de Fis,
    Al. Brong.
1. Turbo pulcherrimus, *Bean.* Speeton Clay, Yorkshire, *Phil.*
2. —— sulcatus, *Nils.* Chalk, Kjöpinge, Scania, *Nils.*
3. —— moniliferus, *Sow.* Green Sand, Blackdown, *Sow.*
4. —— carinatus, *Sow.* Green Sand, Coesfeld, *Hœn.*
1. Turritella terebra, *Broc.* Green Sand, Weddersleben, *Hœn.*
2. —— duplicata,   , Maestricht, *Hœn.*
—— species not determined.   Speeton Clay? Yorkshire,
    Phil.
1. Cerithium excavatum, *Al. Brong.* Green Sand, Perte du
    Rhone, *Al. Brong.; Green Sand, Aix-la-Cha-
    pelle, *Hœn.*
—— species not determined.   Green Sand, M. de Fis,
    Al. Brong.
1. Pyrula planulata, *Nils.* Chalk, Kjöpinge, Scania, *Nils.*
2. —— minima, *Hœn.* Green Sand, Aix-la-Chapelle, *Hœn.*
1. Murex quadratus, *Sow.* Green Sand, Blackdown, *Sow.*
2. —— Calcar, *Sow.* Green Sand, Blackdown, *Sow.*
1. Pterocera maxima, *Hœn.* Martigues, *Hœn.*
1. Rostellaria Parkinsoni, *Mant.* Chalk, Lower Green Sand,

Sussex, *Mant.;* Green Sand, Bochum; Coesfeld, *Hœn.*
2. Rostellaria carinata, *Mant.* Gault, Sussex, *Mant.*
3. ————— fissura, *Lam.* Green Sand, Aix-la-Chapelle, *Hœn.*
4. ————— calcarata, *Sow.* Lower Green Sand, Sussex, *Mant.;* Green Sand, Blackdown, *Sow.*
5. ————— composita, *Sow.* Speeton Clay, Yorkshire, *Phil.*
6. ————— anserina, *Nils.* Chalk, Kjöpinge, Scania, *Nils.*
————— species not determined. Lower Green Sand, Isle of Wight, *Sedg.*
1. Strombus papilionatus, Chalk, Maestricht, Aix-la-Chapelle, *Hœn.*
1. Cassis avellana, *Al. Brong.* Chalk, Sussex, *Mant.;* Chalk, Rouen; M. de Fis, *Al. Brong.*
1. Dolium nodosum, *Sow.* Chalk, Sussex, *Mant.*
Eburna, species not determined. Green Sand, Perte du Rhone, *Al. Brong.;* Chalk? Sussex, *Mant.*
1. Voluta ambigua, *Mant.* Chalk, Sussex, *Mant.*
2. ————— Lamberti, *Sow.* Maestricht, *Hœn.*
1. Nummulites lenticulina*, Maestricht; Green Sand, Aix-la-Chapelle, *Hœn.*
2. ————— Faujasii†, Maestricht, *Hœn.*
————— species not determined. Green Sand, Alps of Savoy, Dauphiny, and Provence, *Beaum.;* Maritime Alps, *De la B.;* Chalk, Weinbohla, Saxony, *Klipstein.;* Chalk, Pyrenees, *Dufr.*
1. Lenticulites Comptoni, *Sow.* Chalk, Scania, *Nils.*
2. ————— cristella, *Nils.* Chalk, Charlottenlund, Sweden, *Nils.*
1. Lituolites nautiloidea, *Lam.* Chalk, Paris, *Al. Brong.*
2. ————— difformis, *Lam.* Chalk, Paris, *Al. Brong.*
1. Planularia elliptica, *Nils.* Charlottenlund, Sweden, *Nils.*
2. ————— angusta, *Nils.* Kjöpinge, Scania, *Nils.*
1. Nodosaria sulcata, *Nils.* Chalk and Green Sand, Scania,*Nils.*
2. ————— lævigata, *Nils.* Chalk, Scania, *Nils.*
1. Belemnites mucronatus, *Schlot.* Chalk, Sussex, *Mant.;* Chalk, Yorkshire,*Phil.;* Chalk, Sweden,*Nils.;* Chalk, Meudon, &c., *Al. Brong.;* Baculite limestone, Normandy, *Desn.;* Chalk, Lublin, Poland, *Pusch;* Maestricht, Aix-la-Chapelle, *Hœn.*
2. ————— granulatus, *Defr.* Chalk, Sussex, *Mant.*
3. ————— lanceolatus, *Schlot.* Chalk, Sussex, *Mant.;* Quedlinburg, *Holl.*
4. ————— Listeri, *Mant.* Gault, Sussex, *Mant.;* Red Chalk, Yorkshire, *Phil.*

* *Lycophris lenticularis,* Bast.   † *Lycophris Faujasii.*

5. Belemnites attenuatus,   Gault, Sussex, *Mant.*
6. ———— mamillatus, *Nils.*   Chalk, Scania, *Nils.*
7. ———— Scaniæ, *Blain.*   Chalk, Scania, *Nils.*
8. ———— minimus, *Miller.*   Gault, Folkstone, *Sow.*
———— species not determined.   Speeton Clay, Yorkshire,
       Phil.; Green Sand, Perte du Rhone, *Al. Brong.*
1. Actinocamax verus, *Miller.*   Chalk, Kent, *Miller.*
1. Nautilus elegans, *Sow.*   Chalk, Sussex, *Mant.;* Chalk, Rouen,
       Al. Brong.
2. ———— expansus, *Sow.*   Chalk, Sussex, *Mant.*
3. ———— inæqualis, *Sow.*   Gault, Sussex, *Mant.*
4. ———— obscurus, *Nils.*   Chalk, Scania, *Nils.*
5. ———— simplex, *Sow.*   Rouen, *Al. Brong.;* Green Sand?
       Aix-la-Chapelle, *Hœn.*
6. ———— aperturatus, *Sow.*   Chalk, Maestricht, *Hœn.*
7. ———— pseudo-pompilius? *Sow.*   Maestricht, *Hœn.*
8. ———— undulatus, *Sow.*   Green Sand, Griesenbruch, near
       Bochum, *Hœn.*
———— species not determined.   Lower Green Sand, Sussex,
       Martin; Speeton Clay, Yorkshire, *Phil.;* Green
       Sand, M. de Fis, *Al. Brong.;* Baculite limestone,
       Normandy, *Desn.*
1. Scaphites striatus, *Park.*   Chalk, Sussex, *Mant.;* Chalk,
       Rouen, *Al. Brong.*
2. ———— costatus, *Mant.*   Chalk, Sussex, *Mant.;* Chalk,
       Rouen, *Al. Brong.*
3. ———— obliquus, *Sow.*   Chalk, Rouen; Green Sand, M. de
       Fis, *Al. Brong.*
———— species not determined.   Baculite limestone, Nor-
       mandy, *Desn.*
1. Ammonites varians, *Sow.*   Chalk, Sussex, *Mant.;* Chalk,
       Rouen; M. de Fis, *Al. Brong.;* Baculite lime-
       stone, Normandy, *Desn.;* Chalk and Upper
       Green Sand, Wiltshire, *Lons.;* Green Sand,
       Bochum, *Hœn.*
2. ———— Woollgari, *Mant.*   Chalk, Sussex, *Mant.*
3. ———— navicularis, *Mant.*   Chalk, Sussex, *Mant.*
4. ———— catinus, *Mant.*   Chalk, Sussex, *Mant.*
5. ———— Lewesiensis, *Mant.*   Chalk, Sussex, *Mant.;* Chalk,
       Essen, *Hœn.*
6. ———— peramplus, *Mant.*   Chalk, Sussex, *Mant.*
7. ———— rusticus, *Sow.*   Chalk, Sussex, *Mant.;* Chalk, Lyme
       Regis, *Buckl.;* Green Sand, Bochum, *Hœn.*
8. ———— undatus, *Sow.*   Chalk, Sussex, *Mant.*
9. ———— Mantelli, *Sow.*   Chalk, Sussex, *Mant.;* Hanover,
       Holl.; Green Sand, Bochum; Chalk, Saumur,
       Hœn.

10. Ammonites Rhotomagensis, *Al. Brong.* Chalk, Sussex, *Mant.;*
        Baculite limestone, Normandy, *Desn.;* Rouen,
        Al. Brong.; Chalk, Wilts, *Sow.*
11. ———— cinctus, *Mant.* Chalk, Sussex, *Mant.*
12. ———— falcatus, *Mant.* Chalk, Sussex, *Mant.;* Chalk,
        Rouen, *Al. Brong.*
13. ———— curvatus, *Mant.* Chalk, Sussex, *Mant.*
14. ———— complanatus,   . Chalk, Sussex, *Mant.*
15. ———— rostratus, *Sow.* Chalk, Sussex, *Mant.;* Chalk, Ox-
        fordshire, *Buckl.*
16. ———— tetrammatus, *Sow.* Chalk, Sussex, *Mant.*
17. ———— planulatus, *Sow.* Upper Green Sand, Sussex, *Mant.*
18. ———— Catillus, *Sow.* Upper Green Sand, Sussex, *Mant.*
19. ———— splendens, *Sow.* Gault, Sussex, *Mant.*
20. ———— auritus, *Sow.* Gault, Sussex, *Mant.*
21. ———— planus, *Mant.* Gault, Sussex, *Mant.;* Speeton
        Clay? Yorkshire, *Phil.*
22. ———— lautus, *Mant.* Gault, Sussex, *Mant.*
23. ———— tuberculatus, *Sow.* Gault, Sussex, *Mant.*
24. ———— lævigatus, *Sow.* Gault, Sussex, *Mant.*
25. ———— Goodhalli, *Sow.* Lower Green Sand, Sussex, *Mant.;*
        Green Sand, Blackdown, *Goodhall;* Green Sand,
        Lyme Regis, *De la B.*
26. ———— Lamberti? *Sow.* Speeton Clay, Yorkshire, *Phil.*
27. ———— venustus, *Phil.* Speeton Clay, Yorkshire, *Phil.*
28. ———— concinnus, *Phil.* Speeton Clay, Yorkshire, *Phil.*
29. ———— Rotula, *Sow.* Speeton Clay, Yorkshire, *Phil.*
30. ———— trisulcosus, *Phil.* Speeton Clay, Yorkshire, *Phil.*
31. ———— marginatus, *Phil.* Speeton Clay, Yorkshire, *Phil.*
32. ———— parvus? *Sow.* Speeton Clay, Yorkshire, *Phil.*
33. ———— hystrix, *Phil.* Speeton Clay, Yorkshire, *Phil.*
34. ———— fissicostatus, *Phil.* Speeton Clay, Yorkshire, *Phil.*
35. ———— curvinodus, *Phil.* Speeton Clay, Yorkshire, *Phil.*
36. ———— inflatus, *Sow.* Green Sand, Perte du Rhone; Rouen;
        Hâvre; M. de Fis, *Al. Brong.;* Upper Green
        Sand, Wilts., *Lons.*
37. ———— Deluci, *Al. Brong.* Green Sand, Perte du Rhone;
        M. de Fis, *Al. Brong.*
38. ———— subcristatus, *De Luc.* Green Sand, Perte du Rhone,
        Al. Brong.
39. ———— Beudanti, *Al. Brong.* Green Sand, Perte du Rhone;
        M. de Fis, *Al. Brong.*
40. ———— clavatus, *De Luc.* Green Sand, M. de Fis, *Al.*
        Brong.
41. ———— Selliguinus, *Al. Brong.* Green Sand, M. de Fis,
        Al. Brong.; Chalk, Lublin, Poland, *Pusch;*
        Chalk, Essen, *Hœn.*

42. Ammonites Gentoni, *Defr.* Baculite limestone, Normandy,
        Desn.; Gault, Sussex, *Mant.;* Chalk, Rouen, *Al.*
        Brong.
43. ———— constrictus, *Sow.* Bacculite Limestone, Normandy,
        Desn.; Chalk, Lublin, Poland, *Pusch.*
44. ———— Stobæi, *Nils.* Chalk, Scania, *Nils.*
45. ———— varicosa, *Sow.* Green Sand, Blackdown, *Sow.*
46. ———— Hippocastanum, *Sow.* Chalk with quartz grains,
        Lyme Regis, *De la B.*
47. ———— Benettianus, *Sow.* Gault, Warminster, *Lons.*
48. ———— denarius, *Sow.* Green Sand, Blackdown, *Goodhall.*
49. ———— Nutfieldiensis, *Sow.* Chalk, near Calne, *Lons.*
50. ———— Buchii, *Hœn.* Green Sand, Aix la Chapelle, *Hœn.*
51. ———— ornatus,   Green Sand, Paderborn, *Hœn.*
 1. Turrilites costatus, *Sow.* Chalk, Sussex, *Mant.;* Chalk, Rouen;
        Hâvre, *Al. Brong.;* Chalk, near Calne, *Lons.*
 2. ———— undulatus,   Chalk, Sussex, *Mant.*
 3. ———— tuberculatus, *Montf.* Chalk, Sussex, *Mant.*
 4. ———— Bergeri, *Al. Brong.;* Green Sand, Perte du Rhone;
        M. de Fis, *Al. Brong.*
 5. ————? Babeli, *Al. Brong.* Green Sand, M. de Fis, *Al.*
        Brong.
    ———— species not determined. Green Sand, Maritime
        Alps, *Risso.*
 1. Baculites Faujasii, *Lam.* Chalk, Sussex, *Mant.;* Chalk,
        Norfolk, *Rose;* Maestricht, *Desm.;* Chalk, Swe-
        den, *Nils.;* Bochum; Aix la Chapelle, *Hœn.*
 2. ———— obliquatus, *Sow.* Chalk, Sussex, *Mant.;* Scania, *Nils.*
 3. ———— vertebralis, *Lam.* Chalk, Maestricht, *Fauj. de St.*
        Fond; Baculite Limestone, Normandy, *Desm.*
 4. ———— anceps, *Sow.* Chalk, Scania, *Nils.*
 5. ———— triangularis, *Desm.* Maestricht, *Desm.*
 1. Hamites armatus, *Sow.,* Chalk, Sussex, *Mant.;* Chalk, Ox-
        fordshire, *Buckl.*
 2. ———— plicatilis, *Mant.* Chalk, Sussex, *Mant.* Speeton
        Clay? Yorkshire, *Phil.*
 3. ———— alternatus, *Mant.* Chalk, Sussex, *Mant.;* Speeton
        Clay, Yorkshire, *Phil.*
 4. ———— ellipticus, *Mant.* Chalk, Sussex, *Mant.;* Baculite
        Limestone? Normandy, *Desn.*
 5. ———— attenuatus, *Mant.* Chalk, Gault, Sussex, *Mant.;*
        Speeton Clay, Yorkshire, *Phil.*
 6. ———— maximus, *Sow.* Gault, Sussex, *Mant.;* Speeton
        Clay, Yorkshire, *Phil.*
 7. ———— intermedius, *Sow.* Gault, Sussex, *Mant.;* Speeton
        Clay, Yorkshire, *Phil.;* Green Sand, Aix la Cha-
        pelle, *Hœn.*

8. Hamites tenuis, *Sow.* Gault, Sussex, *Mant.*
9. ———— rotundus, *Sow.* Gault, Sussex, *Mant.;* Speeton Clay, Yorkshire, *Phil.;* Green Sand, Perte du Rhone, *Al. Brong.;* Green Sand, Aix-la-Chapelle, *Hœn.*
10. ———— compressus, *Sow.* Gault, Sussex, *Mant.;* Green Sand, Nice, *Risso.*
11. ———— raricostatus, *Phil.* Speeton Clay, Yorkshire, *Phil.*
12. ———— Beanii, Y. & B. Speeton Clay, Yorkshire, *Phil.*
13. ———— Phillipsii, *Bean.* Speeton Clay, Yorkshire, *Phil.*
14. ———— funatus, *Al. Brong.* Green Sand, Perte du Rhone; M. de Fis, *Al. Brong.*
15. ———— canteriatus, *Al. Brong.* Green Sand, Perte de Rhone, *Al. Brong.*
16. ———— virgulatus, *Al. Brong.* Green Sand, M. de Fis, *Al. Brong.*
17. ———— cylindricus, *Defr.* Baculite Limestone, Normandy, *Desn.*
18. ———— spinulosus, *Sow.* Green Sand, Blackdown, *Miller.*
19. ———— grandis, *Sow.* Lower Green Sand, Kent, *Buckl.*
20. ———— gigas, *Sow.* Lower Green Sand, Hythe, Kent, *G. E. Smith.*

CRUSTACEA.

1. Astacus Leachii, *Mant.* Chalk, Sussex, *Mant.*
2. ———— Sussexiensis, *Mant.* Chalk, Sussex, *Mant.*
3. ———— ornatus, *Phil.* Speeton Clay, Yorkshire, *Phil.*
4. ———— longimanus, *Sow.* Green Sand, Lyme Regis, *De la B.*
   ———— species not determined. Gault, Sussex, *Mant.*
1. Pagurus Faujasii, *Desm.* Chalk? Sussex, *Mant.;* Maestricht.
1. Scyllarus Mantelli, *Desm.* Chalk, Sussex, *Mant.*
   Eryon, species not determined. Chalk, Sussex, *Mant.*
   Arcania, species not determined. Gault, Sussex, *Mant.*
   Etyæa, species not determined. Gault, Sussex, *Mant.*
   Coryster, species not determined. Gault, Sussex, *Mant.*

PISCES.

1. Squalus Mustelus? Chalk, Sussex, *Mant.*
2. ———— Galeus? Chalk, Sussex, *Mant.*
1. Muræna Lewesiensis, *Mant.* Chalk, Sussex, *Mant.*
1. Zeus Lewesiensis, *Mant.* Chalk, Sussex, *Mant.*
1. Salmo? Lewesiensis, *Mant.* Chalk, Sussex, *Mant.*
1. Esox Lewesiensis, *Mant.* Chalk, Sussex, *Mant.*
1. Amia? Lewesiensis, *Mant.* Chalk, Sussex, *Mant.*
   Fish, genera not determined. Speeton Clay, Yorkshire, *Phil.;* Chalk, Paris, *Al. Brong.;* Chalk, Lyme Regis, *De la B.;*

Upper Green Sand, Wilts, *Lons.* Gault, Isle of Wight,
    Fitton.
Fish teeth and palates: common in England and France,
    var. Authors; Bochum; Aix-la-Chapelle, &c. *Hœn.*

REPTILIA.

1. Mososaurus Hoffmanni, Maestricht, *Fauj. de St. Fond;*
        Chalk, Sussex, *Mant.*
1. Crocodile of Meudon, *Cuv.* Chalk, Meudon, *Al. Brong.*
    Reptiles, genera not determined. Speeton Clay, Yorkshire,
        Phil.

From an inspection of the foregoing list it would appear, that
the remains of mammalia have not yet been detected in the cretaceous group; while reptiles, one of them of considerable size, the
Mososaurus Hoffmanni, have been observed in Yorkshire, Sussex,
Maestricht and Meudon. Fish have been observed in Paris, and
in various parts of England. Sharks' teeth and the tritores of
some fish are far from uncommon. Crustacea have been noticed
in Denmark, Yorkshire, Sussex, the Isle of Wight, Dorsetshire,
and Maestricht. Among the polypifers the most abundant would
appear to be different species of the genera *Spongia* and *Alcyonium*
of some authors;—genera, many species of which have been classed
by Goldfuss under the heads of *Achilleum, Manon, Scyphia,* and
Tragos, so that there is much difficulty in presenting a list which
should give the different species under any one arrangement.
Manon pulvinarium, and *M. Peziza,* Goldf., are found at Maestricht, and at Essen in Westphalia; *Spongia ramosa,* Mant., is
discovered in the chalk of Yorkshire, Sussex, and Noirmoutier;
Alcyonium globosum, Defr., in Sussex, Amiens, Beauvais, Meudon,
Tours, Gien, and in the baculite limestone of Normandy; *Hallirhoa
costata,* Lam., in the green sand of Normandy, and the upper
green sand of Wiltshire; *Ceriopora stellata,* Goldf., Maestricht
and Westphalia; *Lunulites cretacea,* Defr., at Maestricht, Tours,
and in the baculite limestone of Normandy; *Orbitulites lenticulata,* Lam., in Sussex, Yorkshire, and at the Perte du Rhone.
According to Goldfuss, numerous polypifers are discovered at Maestricht; consisting of *Achilleum,* 2 species; *Manon,* 4; *Tragos,* 1;
Gorgonia, 1; *Millepora,* 1; *Eschara,* 9; *Cellepora,* 6; *Retepora,* 5;
Ceriopora, 13; *Diploctenium,* 2; *Meandrina,* 1; *Astrea,* 13; to
which should be added, according to M. Desnoyers, *Lunites,* 1.
Among the Radiaria, the *Apiocrinites ellipticus,* Miller, is found
in the chalk of Yorkshire, Sussex, Normandy and Touraine; the
Cidaris variolaris, Al. Brong., in Sussex, and Normandy, at the
Perte du Rhone, in Westphalia, and Saxony; the *C. granulosus,*
Goldf., at Maestricht, Aix-la-Chapelle, and Westphalia; the *Ga-
lerites albo-galerus,* Lam. (Fig. 39.), in Yorkshire, Sussex, Nor-

mandy, Quedlinburg, Aix-la-Chapelle, and Poland; the *G. vulgaris*, Lam., in Sussex and France, at Quedlinburg, and Aix-la-
Chapelle; the *Ananchytes ovata*, in Yorkshire, Sussex, Normandy,
at Meudon, in Westphalia, Poland and Sweden; the *Spatangus*

Fig. 38.   Fig. 39.   Fig. 40.

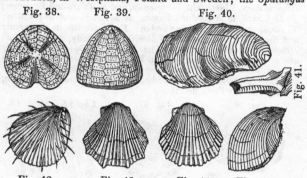

Fig. 42.   Fig. 43.   Fig. 44.   Fig. 45.

Cor-anguinum, Lam. (Fig. 38.), in Yorkshire, Sussex, Dorsetshire,
various parts of France, the Savoy Alps, various parts of Germany,
Poland, and Sweden; *Sp. Bufo*, Al. Brong., Sussex, Normandy,
Maestricht, and Aix-la-Chapelle; the *Sp. Cor-testudinarium*, at
Maestricht and Quedlinburg, and in Westphalia. Among the
shells the most widely distributed would appear to be *Lutraria
Gurgitis*, found at the Perte du Rhone, and in Sweden; *Inoceramus* (or *Catillus*) *Cuvieri* (Figs. 40 and 41.), discovered in the
chalk of Yorkshire, Sussex, Meudon, and Sweden; *Inoceramus*
(or *Catillus*) *Brongniarti*, in the chalk of England, Poland, and
Sweden; *Ino. concentricus*, in Sussex, and in Wiltshire, at the
Perte du Rhone, and in the Savoy Alps; *Ino. sulcatus*, in Sussex,
at the Perte du Rhone, in the Savoy Alps, and in Sweden; *Plagiostoma spinosum* (Fig. 42.), in the chalk of Sussex, Dorsetshire,
Normandy, Meudon, Saxony, Poland, and Sweden; *Gervillia solenoides*, Sussex, Wilts, Dorset, and Normandy; *Pecten quinquecostatus* (Fig. 43.), in Sussex, the West of England, Normandy,
at Meudon, the Perte du Rhone, Sweden, &c.; *P. quadricostatus*
(Fig. 44.), in Sussex, the West of England, Normandy, at Maestricht, and in the Alps of Dauphiné; *P. asper*, Wilts and Poland;
Podopsis truncata (Fig. 45.), in Normandy, Touraine, and Sweden;
Ostrea vesicularis* (Fig. 46.), in Sussex, Normandy and other
places in France, at Maestricht, and in Sweden; *O. carinata*, in
Sussex, Normandy, and the South of France; *Gryphæa vesiculosa*,

* *Grypha globosa, Sowerby.*
O 2

at Périgueux, in the Alps of Dauphiné, and Poland; *G. Columba*
(Fig. 47.), Northamptonshire, Normandy, Maritime Alps, Germany, and Poland; *Terebratula plicatilis,* in Sussex, at Meudon,
and the Alps of Savoy and Dauphiné; *T. subplicata,* in York-
shire, Sussex, Maestricht, Normandy, and at Tours and Beauvais;
T. Defrancii, in Yorkshire, Sussex, at Meudon, and in Sweden;
T. alata, at Meudon and in Sweden; *T. octoplicata,* in Normandy
and Sweden; *T. pectita,* in Wiltshire, Normandy, and Sweden;
Belemnites mucronatus (Fig. 48.), in Yorkshire, Sussex, Normandy
and other parts of France, and in Poland; *Ammonites varians,* in
Sussex, Wiltshire, and the Savoy Alps; *Am. Rhotomagensis,* in
Sussex, Wiltshire, and Normandy; *Hamites rotundus,* Yorkshire,
Sussex, and the Perte du Rhone.

Fig. 46.   Fig. 47.   Fig. 52.   Fig. 48.

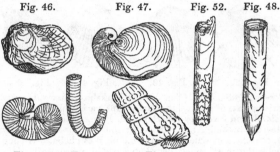

Fig. 49.   Fig. 50.   Fig. 51.

It will be observed that this list is far from large, when we consider the number of species enumerated in the foregoing catalogue,
and that, perhaps, some of those considered identical may be varieties, if not different species. No doubt when we reduce our
view to smaller distances and more minute divisions of the cretaceous group, other species than those above enumerated will be
found occurring under similar circumstances in different situations;
but even then, certain species do not seem to be so constant to
particular beds as has been supposed, though some certainly are
found over considerable distances in similar parts of the group.

The fossil vegetables discovered in the cretaceous group are as
yet found to be principally marine, and much of the fossil wood is
pierced by some boring shell, as if it had long been drifted about.
Hence it has been inferred, that there has been but a slight transport of vegetable matter into the waters where the group was deposited. Very probably this generalization is somewhat too hasty,
but certainly vegetables do appear to be very scarce in the chalk
itself.

It will have been observed, that, among the shells, particular species of the genera *Scaphites, Baculites, Turrilites,* and *Hamites* * have not been observed in many distant places. The student must also have remarked that these genera were not found in any lists of the supercretaceous group; and he will see that in the sequel they have not been discovered in the oolitic group. The remains of a Turrilite have indeed been stated with doubt to have been found in the oolitic series of the North of France. Therefore it would appear, that as far as our information respecting organic remains yet extends, these genera are characteristic of the cretaceous group of Europe; how far they may be more generally so, remains to be ascertained, but if we reason from the analogy of the existing state of things, there is nothing to oppose the inference that the same genera may equally characterise contemporaneous deposits in North America.

Dr. Morton considers that rocks equivalent to the cretaceous group do exist somewhat extensively in North America. He has named it the *Ferruginous Sand Formation* of the United States, and describes it as occupying "a great part of the triangular peninsula of New Jersey, formed by the Atlantic, and the Delaware and Raritan rivers, and extending across the state of Delaware from near Delaware city to the Chesapeake: appearing again near Annapolis, in Maryland; at Lynch's Creek, in South Carolina; at Cockspur Island, in Georgia; and several places in Alabama, Florida, &c." In New Jersey there is a very extensive development of marl. Taken as a mass, the deposit varies considerably in its mineralogical character; most frequently presenting itself in minute friable grains, with a dull bluish or greenish colour, often with a gray tint. The predominant constituent parts of this marl, as it is termed, are described as silex and iron. There are subordinate beds of clay, of siliceous gravel, (the pebbles varyir g in size from coarse sand to one or two inches in diameter,) and calcareous marl. The marl is sometimes yellowish brown and filled with green specks of silicate of iron, and sometimes contains a considerable quantity of mica. The following is a list, according to Dr. Morton, of the organic remains found in this deposit, and described by Mr. Say, Dr. Dekay, and himself†.

Ammonites placenta, Dekay; *A. Delawarensis,* Morton; *A. Vanuxemi,* Morton; *A. Hippocrepis,* Dekay; *Baculites ovatus,* Say; *Scaphites Cuvieri,* Morton; *Belemnites Americanus,* Morton,

---

* To exhibit the forms of these genera the following species have been figured in the opposite page:—*Scaphites obliquus,* Sow. (*Sc. striatus* Mant.), Fig. 49; *Hamites rotundus,* Fig. 50; *Turrilites tuberculatus,* Fig. 51; and *Baculites Faujasii,* Fig. 52.

† Say, American Journal of Science, vols. i. and ii.; Dekay, Annals of the New York Lyceum; and Morton, Journal of the Acad. of Nat. Sciences of Philadelphia, vol. vi.; and American Jour. of Sci. vols. xvii. and xviii.

abundant, (allied to *B. mucronatus*); *B. ambiguus*, Morton, *Turritella; Scalaria annulata,* Morton; *Rostellaria; Natica; Bulla?*
Trochus; Cyprœa (cast); *Terebratula Harlani,* Morton; *T. fragilis,* Morton; *T. Sayi,* Morton; *Gryphœa convexa,* Morton;
G. mutabilis, Morton, (some varieties of this species closely approach *Ostrea vesicularis,* Lam.); *G. Vomer,* Morton; *Exogyra costata,* Say; *Ostrea falcata,* Morton; *O. Crista-Galli; Ostrea,*
two other species; *Anomia Ephippium?* Lam.; *Pecten quinquecostatus,* Sow.; *Pecten,* another species; *Plagiostoma; Cardium;*
Cucullæa vulgaris, Morton; *Cucullæa,* another species; *Mya;*
Trigonia? Tellina; Avicula; Pectunculus; Pinna, resembling
P. tetragona, Sow.; *Venus; Vermetus rotula,* Morton; *Dentalium Serpula; Spatangus Cor anguinum?* Park.; *Sp. stella,* Morton;
Ananchytes cinctus, Morton; *An. fimbriatus,* Morton; *An.? crucifer,* Morton; *Cidaris? Clypeaster.* Crustaceous remains. *Anthophyllum atlanticum,* Morton. *Eschara; Flustra; Retepora,*
resembling *R. clathrata,* Goldf.; *Caryophyllia; Alcyonium; Alveolites.* Teeth and vertebræ of the shark. *Saurodon Leanus,* Say.
Remains of the *Crocodile,* (frequent); of the *Geosaurus*; of the *Mososaurus* (Sandy, Hook and Woodbury, New Jersey); of the *Plesiosaurus*; of a Tortoise; and of some gigantic animal. Lignite
pierced by the *Teredo,* abundant.

It is almost impossible not to be struck, in the foregoing list,
with the great zoological resemblance of this ferruginous sand deposit with the cretaceous rocks of Europe. As has been above
noticed, the genera *Baculites, Scaphites,* and *Turrilites* have
not been discovered out of this series in Europe. The *Pecten quinquecostatus* is a well known and widely distributed cretaceous
fossil. But it is not so much by individual parts as by the general
character of the whole, that Dr. Morton's inference seems in a
great measure established. How far the cretaceous group of the
United States may be separated beneath and above from other
deposits more or less contemporaneous with those in Europe, remains an interesting problem, which it is hoped that Dr. Morton
and other American geologists will endeavour to solve. From
some notices scattered through the memoirs of Dr. Morton and
other authors, it would seem far from improbable that the cretaceous rocks may pass into the supercretaceous group.

Assuming that the American ferruginous sand formation belongs
to the group under consideration, of which there seems great probability, it would appear that the great white carbonate of lime
deposit, or chalk, did not extend there, but that a series of sands,
marls, clays, and gravels constituted the whole group. How far the
marls or clays may be altogether mechanical is perhaps uncertain;
but the gravel would seem to attest the former presence of water,
moving with some velocity, for the pebbles even attain one or two
inches in diameter.

WEALDEN ROCKS.

SYN. Weald Clay, (*Argile Veldienne*, Al. Brong.) Hastings Sands, (*Iron Sand ; Sable Ferrugineux ; Kurzawka* of Poland.) Purbeck Beds, (*Calcaire Lumachelle Purbeckien*, Al. Brong.)

These rocks, characterised in England by the presence of abundant terrestrial and fresh-water remains, occur beneath the lower green sand of the English series. The Weald clay, which constitutes the upper part of the rocks under consideration, does not present a clear line of separation from the marine deposits above it; the lower part of the one and upper portion of the other alternating, according to Mr. Murchison* and Mr. Martin†, in the western part of Sussex;—an important fact, as it shows that the change of circumstances, which permitted the residence of marine animals over a surface previously only covered by fresh-water animals, was not sudden but gradual.

Weald Clay.—According to Dr. Fitton, this clay is composed, in the Isle of Wight, where there are fine sections of it, of slaty clay and limestone, with beds of iron-stone ; the laminæ of the clay, frequently coated with the remains of *Cypris faba*, Desm.‡ Mr. Martin defines the clay of the Weald of Sussex (whence the name) as "a stiff clay, brown on the surface, and blue and slaty beneath, containing concretional iron-stone§." It appears that the iron-stone was once worked, and slags from the ancient furnaces are found in different situations. The thickness of the clay is estimated at 150 or 200 feet in western Sussex. Beneath this there is an alternation of clays and sands, including the limestones full of the *Paludina vivipara*, and known as the Petworth marble.

Hastings Sands.—Mr. Webster, describing this deposit generally, considers that in the upper part a gray calciferous sandstone abounds; that the central portion principally consists of a soft yellow and friable sandstone; and the lower part presents "beds of clay, shale, and ferruginous sandstone, with several layers of iron-stone, and numerous fragments of carbonized vegetables‖." According to Dr. Fitton, the equivalent beds in the Isle of Wight are composed of sands and sandstones, "frequently ferruginous, with numerous alternations of reddish and variegated sandy clays, and concretions of calcareous grit¶."

There are certain local variations, which will be found described

---

* Murchison, Geol. Trans. 2nd series, vol. ii.
† Martin, Geol. Mem. on Western Sussex, 1828.
‡ Fitton, Ann. of Phil. 1824.
§ Martin, Geol. Mem. Western Sussex.
‖ Webster, Geol. Trans. 2nd series, vol. ii.
¶ Fitton, Ann. of Phil. 1824.

in the works treating of particular districts.   The Hastings beds, however, would appear, as a mass, to be principally arenaceous. According to Mr. Mantell, the lower part of the Hastings deposits (the Ashburnham beds) are composed of argillaceous limestone alternating with schistose marls, which are probably connected with the following.

Purbeck Beds.—These are composed of various limestone strata, alternating with marls, many of the former being extensively used for the pavement of London.   Mr. Webster observes, that at Warbarrow Bay, Lulworth Cove, and other places on the coast of Dorsetshire, the upper bed of the Purbeck strata, supporting the Hastings Sands, contains a large proportion of green earth, the calcareous matter being apparently derived from the fragments of a bivalve shell.

Organic Remains of the Wealden Rocks of England.

PLANTÆ.

Calamites, species not determined.   Hastings Sands, Sussex, *Mant.*
1. Sphenopteris Mantelli, *Ad. Brong.*   Hastings Sands, Sussex, *Mant.*
1. Lonchopteris Mantelli, *Ad. Brong.*   Hastings Sands, Sussex, *Mant.*
Lycopodites? ————.   Hastings Sands, Sussex, *Mant.*
1. Clathraria Lyellii, *Mant.*   Hastings Sands, Sussex, *Mant.*
1. Carpolithus Mantelli, *Ad Brong.*   Hastings Sands, Sussex, *Mant.*
Lignite, and undescribed vegetables.   Hastings Sands, Sussex, *Mant.*

CONCHIFERA AND MOLLUSCA.

1. Cardium turgidum? *Sow.*   Weald Clay, Isle of Wight, *Fitton.*
—— species not determined.   Weald Clay, Swanage Bay, *Fitton.*
Pinna? ————.   Weald Clay, Swanage Bay, *Fitton.*
Venus? ————.   Weald Clay, Swanage Bay, *Fitton.*
Ostrea, species not determined.   Weald Clay, Isle of Wight, *Sedg.;* Purbeck Beds, near Weymouth, *Buckl. & De la B.*
1. Cyclas membranacea, *Sow.*   Weald Clay, Hastings Sands, Ashburnham Beds, Sussex, *Mant.;* Weald Clay? Swanage Bay, *Fitton.*
2. —— media, *Sow.*   Weald Clay, Hastings Sands, and Ashburnham Beds, Sussex, *Mant.;* Weald Clay, Isle of Wight, Swanage Bay; Hastings Sands, Isle of Wight; Sussex, *Fitton.*

3. Cyclas cornea, . Hastings Sands? Ashburnham Beds,
    Sussex, *Mant.*
—— species not determined. Weald Clay, Isle of Wight;
    Swanage Bay, *Fitton.*
1. Unio porrectus, *Sow.* Hastings Sands, Sussex, *Mant.*
2. —— compressus, *Sow.* Hastings Sands, Sussex, *Mant.*
3. —— antiquus, *Sow.* Hastings Sands, Ashburnham Beds,
    Sussex, *Mant.*
4. —— aduncus, *Sow.* Hastings Sands, Sussex, *Mant.*
5. —— cordiformis, *Sow.* Hastings Sands, Sussex, *Mant.*
   Succinea? Hastings Sands, Sussex, *Mant.*
1. Paludina vivipara, *Lam.* Weald Clay, Hastings Sands, and
    Ashburnham Beds, Sussex, *Mant. ;* Purbeck Beds,
    Purbeck, *Conyb.*
2. —— elongata, *Sow.* Weald Clay, Hastings Sands, and
    Ashburnham Beds, Sussex, *Mant. ;* Weald Clay,
    Isle of Wight; Swanage Bay, *Fitton.*
3. —— carinifera, *Sow.* Weald Clay, Sussex, *Mant.*
   Potamides, sp. not determined. Weald Clay, Sussex, *Mant.*
1. Melania attenuata, . Weald Clay, Swanage Bay, *Fitton.*
2. —— tricarinata, . Weald Clay, Isle of Wight; Swanage
    Bay, *Fitton.*

PISCES.

Lepisosteus ——. Hastings Sands, Sussex, *Mant.*
Silurus ——. Hastings Sands, Sussex, *Mant.*
Remains of fish, genera not determined. Weald Clay, Ash-
    burnham Beds, Sussex, *Mant. ;* Purbeck Beds,
    Purbeck, *De la B.;* Hastings Sands, Isle of Wight,
    Fitton.

CRUSTACEA.

1. Cypris faba, *Desm.* Weald Clay, Isle of Wight; Swanage
    Bay, &c. *Fitton ;* Weald Clay, Hastings Sands,
    Sussex, *Mant.*

REPTILIA.

1. Crocodilus priscus, Hastings Sands, Sussex, *Mant.*
—— species not stated. Ashburnham Beds, Sussex,
    Mant. ; Purbeck Beds, Purbeck, *Conyb. ;* Weald
    Clay, Swanage Bay, *Fitton.*
   Leptorynchus ——. Hastings Sands, Sussex, *Mant.*
   Iguanodon ——. Hastings Sands, Sussex, *Mant.*
   Megalosaurus ——. Hastings Sands, Ashburnham Beds,
    Sussex, *Mant.*

Reptiles of the genera Trionyx, Emys, Chelonia, Plesiosaurus, and Pterodactylus? Hastings Sands, Sussex, *Mant.*

Tortoise, Purbeck Beds, Purbeck, *Conyb.**

From the above lists it will appear that this deposit of lime-stones, sands, and clays, was formed in water which permitted the existence of shells analogous to those which now live in fresh water. The only shells which do not so live, and are not of questionable genera, are *Ostreæ* and *Cardia*, well known as estuary animals.

It would appear that the *dirt-bed*, first noticed by Mr. Webster in the Isle of Portland, and which has since been observed in the vicinity of Weymouth and elsewhere, commences the phænomena which attest dry land, succeeded by submersion of the same land beneath fresh or estuary waters, in which the whole of the Wealden rocks of south-eastern England were formed; not suddenly, for there are no conglomerates to mark a possible state of violence; but quietly, the shells being tranquilly enveloped by the calcareous, argillaceous, or arenaceous matter which now entombs them. It will be seen that the oolitic group, immediately preceding this state of things, was, judging from the nature of the organic re-mains, formed beneath a sea. Therefore we must suppose a rise of the land, or depression of the sea, to such an amount as to per-mit the sea-formed rocks to become dry land, upon which *Cyca-deoideæ* and dycotyledonous plants of a tropical nature flourished. This land was then depressed; but so tranquilly, that the vegetable soil, mixed with a few pebbles of the subjacent rock, was not washed away; neither were the trees considerably displaced, but they were left much as we have seen other trees in the subma-rine forests which surround Great Britain in various places, and occur on the coasts of France. Like them, also, the trees of the *dirt-bed* are found, some prostrate, others inclined, and others nearly in the position in which they grew; the upright portions being partly included in the limestone strata above. The only difference in the trees in the *dirt-bed*, and those in the submarine forests, would appear to consist in the tropical nature of those in the *dirt-bed*, and the near approach, if not the identity, of the submarine forest vegetation with that now existing in Great Britain and France. There is, therefore, nothing singular in the gradual depression of the land, so quietly as not to cause the removal of the trees and other vegetable matter, as this has since happened in the case of the submarine forests.

Instead of the depression having been effected, in the first in-stance, beneath the waters of the sea, circumstances have so

---

* In this list, the sands, sandstones, and clays, grouped by Mr. Man-tell under the head of Tilgate Beds, are given as Hastings Sands, al-though this arrangement may perhaps clash with one or two local divisions.

existed that it took place beneath fresh water, which gradually acquired sufficient depth to permit a deposit of various mineral substances several hundred feet thick. The circumstances attending this deposit have not been constant. At first calcareous matter was thrown down, with somewhat regular interruptions, which introduced a sufficient quantity of argillaceous matter to produce marl. Although fresh-water and terrestrial animals were now imbedded, there would also appear to have been at least one time when the water near Weymouth and in the Isle of Wight was capable of supporting the life of oysters and cockles, and therefore at least brackish. After this first period, sands were accumulated in great abundance, and in them were entombed a great variety of land and fresh-water Tortoises, Crocodiles, Plesiosauri, Megalosauri, and huge Iguanodons, those monstrous terrestrial reptiles. These must have sported in the waters, or roamed along the banks of this lake or estuary, into which trees and different vegetables were drifted. A clay deposit crowns this succession of rocks, still however not showing any other than a fresh-water origin. How far we may consider the change of the relative level of sea to have produced a constant depression of the land, is uncertain; but be this as it may, the sea was destined again to cover the land and resume its empire, for above the last-noticed clay reposes the whole mass of the cretaceous rocks of south-eastern England, of marine origin. This change, like that which preceded it, was not sudden; there are no marks of violence between the Weald clay and the green sand; on the contrary, there is a passage of one into the other, an alternation of the two at their junction. There is every probability that the sea did not make a furious inroad over the land, but that there was a quiet and gradual change of level, as in the case of the *dirt-bed.* I shall not trace the subsequent changes that have taken place over this spot on the earth's surface, further than to remark, that the sea again disappeared (Isle of Wight), and fresh-water or estuary deposits succeeded.

These conclusions can scarcely be termed hypothetical, for they appear such, however remarkable, as may be considered honest deductions from the phænomena observed.

To form such a deposit as that we have been noticing would be a work of time, and therefore we may infer that equivalent formations were taking place elsewhere, the great operations of nature proceeding in their usual course. The fresh-water character of the deposit can only be considered accidental or local; precisely as formations at the present day, though contemporaneous, may be either marine or lacustrine. Therefore, even supposing various perpendicular movements in the land to have taken place extensively over certain portions of Europe, it does not follow that they should have produced a constant rise of that land above the surface of the sea. On the contrary, we may consider that such movements very frequently caused a mere change in the relative depth beneath the surface-water, and that all deposits in the course of formation, and so circumstanced, partook of the marine character of the surrounding aqueous medium.

M. Thirria describes a considerable superficial deposit of clay with pisiform iron-ore in the department of the Haute Saone, part of which he considers referable to the green sand, and may be equivalent to the Wealden rocks. Above rocks which seem equivalent to the Portland beds of England, there are strata of sand and clay, apparently the denuded remains of a deposit, once more extensive, which has suffered aqueous destruction, the water mixing up portions of the removed strata with the bones of Bears and Rhinoceroses; so that the mass upon reconsolidation much resembles the mineralogical composition of the original beds. The following is a section of beds, which M. Thirria considers as in place, the list of fossils being increased by those which he discovered, also in place, in the department of the Haute Saone : 1. Unctuous green clay; 2. Fine and slightly argillaceous yellow sand; 3. Nodules of yellow limestone contained in greenish clay; 4. Yellow and slightly argillaceous sand; 5. Greenish-yellow and unctuous clay; 6. Greenish clay, with nodules of marly limestone and grains of iron ore; 7. Pisiform iron-ore, contained in an ochreous clay, with *Ammonites binus*, *A. planicostata*, Sow., *A. coronatus*, Schlot., and other species; *Hamites* (new species) ; *Nerinea; Cirrus; Terebratula coarctata*, Sow., and other species ; and *Pentacrinites;* 8. White marl, with nodules of greenish clay and concretions of marly limestone. The whole forming a thickness of about forty feet, and resting on beds considered equivalent to those of Portland*.

The extraordinary mixture of fossils contained in the pisiform iron ore is commented on by M. Thirria, who further remarks that the reniform pieces of ore sometimes contain the empty casts of Jura limestone fossils.

In support of the opinion that some of these pisiform and reniform iron-ore beds were of contemporaneous formation with either the Wealden rocks or green sand and chalk of England, we may cite the observations of Professor Walchner on similar beds near Candern in the Brisgau. He remarks, " that the reniform and pisiform iron-ore deposits in the vicinity of Candern belong to two formations of very different ages ; one of which rests on a compact Jura limestone, apparently corresponding with either the coral rag or Portland stone of the English. It is composed of a mass of sandy clay, containing reniform iron-ore in the lower, and pisiform iron in the upper part; and at the same time spheroids of flint

---

* Thirria, Notice sur le Terrain Jurassique du Departement de la Haute Saone ; Mem. de la Soc. d Hist. Nat. de Strasbourg, tom. i. 1830.

(silex) and jasper. The reniform ores, and the flints which accompany them, contain organic remains: the former of *Astreas* and *Ammonites*, the latter of *Pectines* and *spines of Cidaris*. The whole is covered with the solid beds of conglomerate, more ancient than the molasse, or by the molasse itself. This iron-ore formation may be considered as one of the last of the Jura limestone (oolitic group), and it, without doubt, closely approaches the chalk ; perhaps it may be, like the green sand, intermediate between the Jura limestone and the chalk *."

In further support of this conclusion, Professor Walchner quotes the remarks of MM. Merian and Escher on parts of the Jura, both of whom describe a clay with pisiform or reniform iron-ore, intermediate between the upper beds of the Jura limestone and the molasse (one of the supercretaceous rocks of Switzerland) ; but being sometimes wanting, so that the molasse rests directly on the Jura limestone. M. Merian states that, near Aarau, the ferriferous bed sometimes contains large angular fragments of the limestone on which it rests, as also nodules of flint and jasper; angular fragments of the former containing organic remains, which are the same as those detected in the iron-ore itself. The same author observes, that " the pisiform ore of Aarau is immediately covered by a sandstone and bituminous schist, passing into lignite, which sometimes clearly exhibits a woody texture." The schist, and its accompanying clays, contain an abundance of fossils, among which *Planorbes* and other fresh-water shells could be distinguished.

M. Brongniart notices among the cretaceous rocks of the Isle d'Aix and the embouchure of the Charente, a marl, which he refers to the Wealden clay, containing nodules of amber, pieces of lignite and silicified wood, in which holes, formed by some perforating animal, are replaced by agates†. The latter fact agrees with the presence of pieces of silicified wood, occasionally of large size, found on the green sand of Lyme Regis, where the holes, formed by some perforating animal, are filled with chalcedony or agate. Both examples appearing to show that the wood had drifted, and remained some time in the sea.

According to Professor Pusch there is a ferriferous deposit in Poland, situated between the Jura limestone and the cretaceous rocks, which may be considered as the equivalent to the Weald clay and Iron sand (Hastings Sands) of England. The following is Prof. Pusch's account of these beds, which is too valuable to be abridged: " It fills the valleys (in Poland) of Czarna Przemsa as far as Siewirz, that of Mastonica, that of the Wartha from its origin at Kromolow towards Czenstochau, and of the Liziwarta ; and extending across

* Walchner, Sur les Minerais de Fer pisiforme et réniforme de Candern en Brisgau ; Mém. de la Soc. d'Hist. Nat. de Strasbourg, tom. 1.

† Tab. des Terrains, p. 218.

Higher Silesia to the Oder, running up this river to the country of Ribnyk. It is composed of horizontal beds, often alternating and of little continuity, of a slightly calcareous and schistose clay, either blue or variegated, named *kurzawka;* of a siliceous, quartzose, and compact conglomerate; of a brown ferriferous sandstone; of beds of loose sand, and of thin beds of white or variegated marly limestone. In the country of Kromolow, Poremba, and Siewirzce, this formation contains horizontal beds from six inches to fourteen feet in thickness, of a coarse coaly substance (*moorkohl*), often accompanied with bituminous wood and much pyrites. This combustible is little worked, as the deposit occurs in marshy valleys, but the want of wood may render it useful in the country between Pelica and Czenstochau. From Siewirz, the carbonaceous beds lose themselves on the north. Faint traces of them are found round Czenstochau, Krzepice, and Klobucho; while the unctuous and blue schistose clays are largely developed in these countries, with, as on the top of the carbonaceous deposits, numerous beds of iron-ore, consisting of ranges of spheroidal nodules of compact argillaceous iron-ore, containing numerous ammonites, (especially *Ammonites bifurcatus,*) and bivalves, of the genera *Cardium, Venus, Trigonia, Sanguinolaria,* &c., fossils which partly correspond with those of the Jura limestone. This ferriferous deposit abounds near Panki, near Krzepice, between this point and Wielun, and on the north of Upper Silesia. It furnishes iron for the founderies of Poremba, Miaczow, Panki, Zarki, and various places in Silesia, producing 50 per cent. of iron. A brown ferruginous sandstone, agglutinated by hydrate of iron, covers the blue schistose clays, especially round Kozieglow, Panki, and Prauska*."

The reader will at once perceive the great resemblance of this ferriferous deposit with that above noticed in the Jura; such resemblance being heightened by the occurrence of organic remains, of which ammonites constitute a portion, in the iron-stone nodules of both situations. There would appear to be little difficulty in considering this deposit, with M. Pusch, as the equivalent of the Wealden rocks of England, showing that where local circumstances did not interfere, and the deposit continued to be effected beneath the sea, its zoological character marked a certain connection with the oolitic group; the species of animals existing during the formation of at least a portion of the latter rocks not being suddenly cut off : thus exhibiting a zoological passage of the oolitic into the cretaceous groups, when local circumstances did not interfere, as they have done on the south-east of England. It is remarkable that, notwithstanding the different character of the organic remains, apparently entombed in beds of the same age, which would seem to point out deposits in different waters, iron-

* Pusch, Journal de Géologie, t. ii.

ore should be so common in the Wealden rocks of England, the Jura, and Poland.

When the upper beds of the oolitic series formed dry land, and sustained vegetation in southern England, it seems reasonable to conclude that many parts of the land now constituting Europe were similarly circumstanced; and therefore contemporaneous deposits of various characters may have been produced in different situations; some, by the nature of their organic remains, marking the presence of large lakes, or the embouchures of considerable rivers:—in fact, a state of things, during which there was a mixture of dry land, fresh waters, and sea in this part of the globe. Some cause, with which as yet we are imperfectly acquainted, subsequently produced a great change in the relative levels of sea and land, and the cretaceous rocks (chalk and green sand) became deposited over a very considerable area, one apparently extending over a much larger superficies than that in which the last-formed rocks of the oolitic series were deposited.

SECTION VI.

OOLITIC GROUP.

SYN.—*Oolite formation*, Engl. authors; *Calcaire de Jura, Calcaire Jurassique*, Fr. authors; *Jurakalk*, Germ. authors.

THIS group is, in the southern parts of England, composed of various alternations of clays, sandstones, marls, and limestones; many of the latter being oolitic, whence the name *oolitic series.* At a very early period in the history of English geology, Mr. William Smith affixed names to various portions of this series, many of which are still employed by the geologists of Europe. Several of the divisions and subdivisions are, undoubtedly, very arbitrary, and perhaps separate those things theoretically which nature has united; but their convenience seems proved by their very general adoption. In consequence of three great clay or marl deposits appearing to divide the series in the south of England into three natural groups, Mr. Conybeare has separated it into three systems, as follows, (the Purbeck beds only, for reasons before assigned, being omitted): 1. Upper system, containing, in the descending order, *a.* Portland oolite; *b.* calcareous sand and concretions; *c.* an argillo-calcareous deposit, named Kimmeridge clay. 2. Middle system, *a.* coral rag, and its accompanying oolites; *b.* calcareous sand and grit; *c.* Oxford clay. 3. *a.* Calcareous strata, (sometimes divided by clays or marls,) named cornbrash, forest marble, great or Bath oolite, and inferior oolite; *b.* calcareo-siliceous sands, usually termed sands of the inferior oolite; *c.* an argillo-calcareous deposit named lias.

These three principal divisions, marked by argillaceous deposits, have been traced to various distances, though their subdivisions have not been so readily identified. The extent to which a few fossil shells of each division can be observed, is also deserving of attention.

Mr. Phillips distinguishes this group in Yorkshire into, *a.* Kimmeridge clay; *b.* upper calcareous grit; *c.* coralline oolite; *d.* lower calcareous grit; *e.* Oxford clay; *f.* Kelloway rock (a name given to stony portions of the Oxford clay, near Kelloway Bridge in Wiltshire); *g.* cornbrash limestone; *h.* upper sandstone, shale, and coal; *i.* impure limestone (Bath oolite); *k.* lower sandstone, shale, and coal; *l.* ferruginous beds (inferior oolite); *m.* upper lias shale; *n.* marlstone series; and *o.* lower lias shale. It will be observed that these divisions do not very materially differ from those of the southern parts of England, except in the presence of certain shales and sandstones containing coal, above

and beneath a bed considered equivalent to the Bath oolite.
These carbonaceous beds are stated to have a collective thickness
of 700 feet, the supposed representative of the Bath oolite being
abstracted.

The oolitic series of Normandy also presents a close analogy in
its general, and even in some of its minor divisions, with those of
southern England. Commencing with the vicinity of Havre, and
extending our observations to the Cotentin, we find the following
series: *a.* Kimmeridge clay, in which certain sandstones named
Glos sandstones are subordinate; *b.* limestone and oolitic beds,
referable, from their geological and zoological characters, to the
coral rag; *c.* a ferruginous and calcareous sandstone; *d.* Oxford
clay; *e.* a series of beds, including the well-known Caen stone,
and representing the forest marble and great oolite; *f.* inferior
oolite; *g.* lias*. Mr. Boblaye divides the oolitic series of the
north of France, as follows†: *a.* beds referable to the coral rag,
(the highest of the oolitic series in the district); *b.* a sandy and
ferruginous oolite; *c.* a series of beds representing the cornbrash,
forest marble, and great oolite; *d.* ferruginous limestone, micaceous marls, and sandy limestones, equivalent to the inferior
oolite and its sands; *e.* lias. In Burgundy, M. Elie de Beaumont,
who has remarked on the constancy of the geological facts observable in the oolitic belt of the great geological basin which
contains London and Paris, has found beds which he considers
referable to those of Portland, beneath which is a marly limestone with the *Gryphæa virgula*, a remarkable shell of the Kimmeridge clay, particularly in France. These beds are succeeded
by compact earthy or oolitic limestones, beneath which is gray
marly limestone, supposed equivalent to the Oxford clay. This
is followed, in the descending order, by a series of oolite and other
beds, beneath which there is a limestone remarkable for containing
an abundance of *Entrochi*, and considered equivalent to the inferior
oolite, under which are rocks corresponding with the lias‡.

M. Thirria describing the oolitic series of the department of
the Haute Saone, where it constitutes the north-west limits of the
Jura, notices the following beds (the lias being excluded from the
list according to the views of some of the continental geologists):
—*a.* inferior oolite, composed of various limestones, oolitic, sub-
lamellar, lamellar, and compact, reddish, gray, and yellow; some

---

* De la Beche, Geol. Trans. vol. i. 1822 ; De Caumont, Essai sur la
Topographie Géog. du Calvados, 1828.

† Boblaye, Sur la Form. Jurassique dans le Nord de la France ; Ann.
des Sci. Nat. 1829.

‡ Elie de Beaumont, Note sur l'uniformité qui regne dans la consti-
tution de la Ceinture Jurassique qui comprend Londres et Paris;—Ann.
des Sci. Nat. 1829.

cf the beds being studded with *Entrochi,* or joints of *Crinoidea.*
One bed is remarkable for oolitic hydrate of iron, so abundant as
to be worked for profitable purposes at Calmontiers, Oppenans,
Jussey, and other places; *b.* a yellow marl, considered equivalent
to the Fuller's earth of the English (two yards thick); *c.* great
oolite, composed of oolitic beds, containing among other shells
Ostrea acuminata and *Avicula echinata; d.* limestones with much
red oxide of iron, schistose, suboolitic, or compact, considered
equivalent to forest marble; *e.* marly limestone, gray or yellowish,
full of oolitic grains, supposed equivalent to the cornbrash of England; *f.* schistose blackish gray marls with marly limestone, resting on gray schistose marls containing oolite grains of hydroxide
of iron, worked for profitable purposes in the districts of Orrain
and Saguenay. The whole of this subdivision (*f*) is based on dark
gray and schistose argillaceous limestone, and contains many
fossils, particularly in the ferruginous oolite, among which is
Gryphæa dilatata, a very characteristic shell of the Oxford clay,
to which, and to the Kelloway rock, the whole is referred; *g.* a
series of clay and limestone beds, the latter mostly oolitic; the
upper part containing Corals, and the lower portion numbers of
Nerineæ, the whole considered equivalent to the coral rag; *h.* gray
marls and marly limestone, based on compact gray limestone, the
latter containing abundant remains of *Astarte,* while the other
parts present the *Gryphæa virgula;* these marls are consequently
referred to the Kimmeridge clay; *i.* various limestone beds,
principally of a gray colour, sometimes whitish and yellowish,
at others of a deeper tint, considered equivalent to the Portland
stone *.

M. Dufrénoy, in his remarks on the rocks of this age which
occur in the south-western parts of France, divides the oolitic
group into three distinct systems; admitting, however, at the
same time, that these divisions are not well pronounced, the beds
which apparently correspond with the Oxford and Kimmeridge
clays being replaced by marly limestone. He further observes,
that "the numerous subdivisions noticed by the English geologists
are but very imperfectly seen in the secondary basin under consideration; some, nevertheless, being sufficiently constant." The
lower portion rests on lias, and is composed of micaceous marls,
with *Gryphæa Cymbium, Belemnites,* and other shells, which, as
he observes, may be referred to the sands of the inferior oolite.
There are beds of limestone with oolitic iron, and oolites, considered equivalent to the Bath oolites, the latter only well developed at Mauriac, Aveyron. This lower division is represented
as of considerable thickness. Above this there is a system of

* Thirria, Notice sur le Terrain Jurassique du Département de la
Haute-Saone; Mém. de la Soc. d'Hist. Nat. de Strasbourg, 1830.

marly limestone beds, in some places associated with considerable masses of polypifers and thick beds of irregular and earthy oolite (Marthon, forest of la Braconne, and other places). M. Dufrénoy infers, from the great abundance of the corals, the presence of the oolite and many fossils, that these beds are equivalent to the coral rag and Oxford oolite. Upon this system rests another, composed of marls and marly limestone, abounding in the *Gryphæa virgula,* supporting an oolite (from the environs of Angoulême to the ocean), in which this gryphite is also found. These rocks are referred to the Kimmeridge clay and Portland oolite respectively, and are stated to be surmounted by rocks of the cretaceous group*.

It would thus appear, that throughout a considerable portion of France and England, the causes which have produced the deposit of the oolitic group have not varied materially. Before, however, we attempt any remarks on this apparent uniformity of mineralogical structure over a considerable area, it will be necessary to present a sketch of this deposit in Scotland, Germany, and Sweden.

Our knowledge of the oolitic group of Scotland is more particularly due to Mr. Murchison. The coal deposit of Brora, in Sutherlandshire, has been shown to correspond with the carbonaceous series of Yorkshire, described by Mr. Phillips as occurring between the inferior oolite and cornbrash, and including in its central part a rock considered equivalent to the Bath or great oolite. In the vicinity of Brora there would appear to be various sandstones and shales, containing coal and vegetable impressions. The freestone of Braambury and Hare hills is described as covered by a rubbly limestone, "an aggregate of shells, leaves, stems of plants, lignite, &c." Mr. Murchison considers the organic remains of this bed, and the casts in the freestone, as referable to such as occur in the lower part of the coral rag. At Dunrobin Castle calcareous sandstones are succeeded by beds of "pebbly calciferous grit," covered by shale and limestone with fossils. Other varieties of this oolitic deposit occur on this coast, which consists, in the descending order, of rubbly limestone, white sandstone and shale, shelly limestone, sandstone, shale, and limestone, with plants and coal, considered the same with the Yorkshire carbonaceous deposit.

This oolitic deposit is not confined to the main land of Scotland, but is found in the Hebrides. According to Mr. Murchison, it occurs at Beal near Portree, Sky, the higher part presenting a calcareous agglomerate of fossils, resembling many portions of the English cornbrash and forest marble: it is identical with the shelly limestone of Sutherland, above noticed. At Holm the sandstone rises to a considerable height from beneath the lime-

* Dufrénoy, Annales des Mines, tom. v. 1829.

stone. Impressions of plants are found in the sandstone on the north-east of Holm. Near Tobermory in Mull, sandstone, considered as equivalent to that of the inferior oolite, rests on lias, containing the *Gryphæa incurva.* It also appears that rocks of the oolitic series, including lias, occur in other parts of Mull, the opposite coast of Ross-shire, and in the islands of Rasay and Pabbla, often cut and covered by trap rocks*.

The oolitic group of Germany is not as yet so well known in its details as the same group in England and France, but important additions to our information on this subject may soon be expected from the labours of the German geologists. A dolomitic rock is apparently included in the oolitic series of this country. Von Buch considers much of the German oolite as referable to the coral rag, and the same geologist describes the coral rag as constituting the elevated plateau between the Mein and Switzerland, and as found in the mountains of Streitberg, at Donzdorf in Swabia, at Rathshausen near Bahlingen, and at Mont Randen near Schafhausen. Von Buch observes, that at the latter place there are several beds of polypifers, in which *Cnemidium lamellosum, Cn. striatum,* and *Cn. rimulosum,* are the most characteristic fossils. Beneath these are beds full of Ammonites, such as *A. placatilis, A. triplicatus,* large and very abundant, *A. perarmatus, A. biplex, A. flexuosus, A. bifurcatus,* and *A. canaliculatus.* These coral-rag beds rest on clays and marls, containing the *Gryphæa dilatata* and *Ammonites sublævis* †. The list of organic remains will show that polypifers are abundant on this rock at Streitberg, Muggendorf, &c.

M. Merian has afforded us very valuable details respecting the structure of the Jura near Bâle, and of its continuation into Germany in the same vicinity; whence it appears that the inferior oolite (*Eisen Rogenstein*) and the lias (*Gryphiten Kalk*) constitute clearly marked rocks of the series. The beds which rest on the *Eisen Rogenstein* are divided into older and newer Jura limestone (*Älterer Rogenstein* and *Jüngerer Jurakalk*), the former being considered in a great measure equivalent to the great or Bath oolite, and separated from the latter by beds of clay ‡.

For the superficial distribution of the oolitic group over Germany, the student should consult the geological maps of that country; particularly Hoffmann's map of north-western Germany, and the more general map published by Schropp. The minera-

---

* Murchison, Geol. Trans. 2nd series, vol. ii.

† Von Buch, Recueil de Planches de Pétrifications Remarquables, Berlin, 1831.

‡ Merian, Geognostischer Durchschnitt durch das Jura-Gebirge von Basel bis Kestenholz bey Aarwangen ; Denkschriften der allgemeinen Schweizerischen Gesellschaft für die gesammten Naturwissenschaften, Zurich, 1829.

logical character of the mass does not appear to be very materially different from that above noticed; limestones, sometimes with an oolitic structure, clays, marls, and sandstones, constituting its component parts, and the organic remains hitherto found presenting the same general zoological character with the same group in England and France.

So far, if we except the dolomite in Germany, we have found no great change in the oolitic group, taken as a mass: there is nothing which shows that in the particular parts of Europe above noticed any forces were called violently into action during its deposit. On the contrary, a greater or less degree of repose seems characteristic of it, as also the presence of a large proportion of calcareous matter. The lowest portion, or the lias, preserves certain general characters over a considerable area; and why some geologists have separated it from the oolitic series is not easily understood; for if an apparent passage into the rocks beneath in some situations be the reason, such a reason would hold equally good for not separating it from those above, into which it also passes: if its zoological character be brought forward, there can be little doubt that throughout western Europe this would place it in the group under consideration.

The lias of western Europe may be considered, taken in the mass, as an argillaceous and calcareous deposit, in which sometimes one substance predominates, sometimes the other; sometimes presenting a great abundance of marls or clays, at others of limestones: the latter are however generally most common in the lower portions of the rock. In the Vosges district the lower part of the lias is formed of a sandstone, described by M. Elie de Beaumont as yellow and quartzose, containing mica, a few flattened argillaceous nodules, and small white or black quartz pebbles*. The presence more particularly of the pebbles seems to point to a transport by water. This sandstone extends into the neighbouring parts of Germany, and is one of those to which the name of *Quadersandstein* has been applied. Beneath the oolitic group, which comes into contact with the granitic rocks of central France, M. de Bonnard has described an arenaceous rock, which he has named *Arkose*, and which may represent the arenaceous beds constituting the lowest part of the same rocks in the district of the Vosges. M. Dufrénoy describes an arenaceous deposit corresponding in geological position and external characters with the arkose of M. de Bonnard in the south-western part of France. He also states, that from Châtre, where the coal-measures terminate, to beyond Brives, the separation of the oolitic series and the granitic rocks is marked by the presence of this sandstone, composed of quartz grains and felspathic portions, cemented by

* Elie de Beaumont, Mém. pour servir à une Description Geologique de la France, tom. i.

matter generally marly, but sometimes siliceous; the silex in the latter case becoming sometimes so abundant as to obliterate its character of a sandstone, so that it passes into a jasper. This sandstone seems to pass into the lias limestone, presenting an arenaceous limestone between the two. M. Dufrénoy considers it as the inferior sand of the lias, representing one of the quader-sandsteins of Germany. The same author describes the lias of the south-west of France, particularly at La Salle, near Saint Hyppolyte-du-Gard, as containing a considerable quantity of gypsum. Although sulphate of lime, in the shape of crystals of selenite, is by no means uncommon in the lias marls of other countries, its presence, in that form, does not appear to mark a chemical deposit so much as in the gypsum above noticed. Taken as a whole, the lias seems very persistent in its characters throughout a considerable part of France, England, and Germany, pointing to a somewhat common origin. In the lias of Lyme Regis, Dorset, there would appear evidences of slow deposit in some parts, while in others the animals entombed seem to have been suddenly killed and preserved, so that the animal substances had not time to decay. The ink-bags of fossil Sepiæ, noticed by Prof. Buckland, afford perhaps the best evidence we can adduce of this fact; for had the animal substances which contained ink been exposed but for a short time to decomposition or the attacks of other animals, the ink must have flowed out of the bags. Now the actual forms of this fossil ink are precisely those of the ink-bags found in the Sepiæ and other animals possessing organs of a similar description at the present day; and therefore they appear to have been preserved entire and suddenly in a soft deposit. In the lias of southern England, and many parts of France, the calcareous matter has been more abundant in the lower parts; and limestone beds have been the consequence, interstratified with marl, the latter sometimes schistose. Above the lias we have an arenaceous deposit, into which the marls graduate; and these sandy beds would seem to have been formed over a considerable area, embracing a large portion of France and England, and parts of Scotland and Germany. These are surmounted by limestones, one of which, characterized by the presence of oolitic iron-ore, though not precisely continuous, is remarkable for its occurrence in a similar part of the series, whether it be in the southern parts of England, in the north of France, in the Jura, or in some parts of Germany. Above these beds, termed the Inferior oolite, there is a series which varies much in its mineralogical character, presenting modifications of clays, marls, and limestones; the latter, which are often oolitic, affording beautiful materials for architectural purposes, as is seen in the towns of Bath, Caen, Nancy, and other places. This variety is commonly known by the name of the Bath or Great oolite, while other portions have received the names of Fuller's earth, Bradford clay, Forest marble, and Cornbrash. There can be little doubt that in tracing these supposed minor divisions over parts of Europe, too much attention has been given to them as they exist in southern England and in Normandy, and that conclusions respecting their complete identity elsewhere have been somewhat forced. This is not the case with the next division, one like the lias composed of argillaceous and calcareous matter, known as the Oxford clay, which, with certain modifications, seems to extend through England, and over a considerable portion of France, including the Jura, and probably also into Germany. The next superior rock, termed Coral rag, (from containing in certain situations a great abundance of polypifers,) separating an argillaceous deposit termed Kimmeridge clay from the Oxford clay, seems also to have a wide range, and presents a mixture principally calcareous, and often oolitic, the grains being not unfrequently so large that the rock is named Pisolite. The Kimmeridge clay is also an argillaceous and calcareous mixture, which has a considerable range, particularly over England and France. Its covering, or the beds termed Portland beds, seems very irregularly dispersed, the causes that produced the beds not being so constant as those which formed the clay beneath; it will however have been seen that rocks considered equivalent occur in the south-west of France, and in the Jura.

When we view the oolitic group as a whole, such as it occurs over a considerable part of western Europe, we cannot but be struck with the general uniformity of its structure. The three great argillo-calcareous deposits alternate with as many that are calcareous or arenaceous, but principally the former. When we attempt to apply the operation of such causes as those we daily witness in explanation of this uniformity, we seem to involve ourselves in innumerable difficulties, though to explain certain minor appearances they may be useful. During nearly the whole time, we require the presence and deposition of a large amount of calcareous matter; for even the arenaceous beds, particularly when distributed over a considerable area, contain this substance; as for instance in the sands of the inferior oolite, where the cementing matter is more or less a carbonate of lime. The mere drift of substances into a sea, such as takes place at present, seems quite insufficient for this production of extensive calcareous deposits, setting aside the general uniformity of series, which seems quite at variance with any such mode of formation, unless the transporting powers of, and the matter carried forwards by, rivers, could be so conveniently arranged according to theory, as always to be the same over considerable districts. In a general view of this deposit it would seem better to consider it in connection with the succeeding group. As joined with it, it appears the upper part of one great mass, which has been deposited in various inequalities of surface, the superior portion frequently overlapping the inferior part, so that it rests directly on the older rocks, as is the case in Normandy, where not only the quartz rocks, grauwacke limestones, and grauwacke, appear protruding through the oolitic group, but where various river-courses cut down through the same series to the older rocks enumerated.

As yet we have seen the oolitic group composed of nearly similar mineral substances, and abounding in organic remains. In Poland, however, there would appear, according to Prof. Pusch, to be a change in the general mineral structure, preparing us for other greater changes, which will be noticed in the sequel. M. Pusch describes the lower member of the group under consideration in that country as more or less white and marly. On this rests dolomite, generally of a dazzling whiteness, affording the forms so remarkable in the rocks of this nature, and composing the picturesque country between Oldkusz and Cracow, and near Kromolow, Niegowomie, and other places, rising to the height of 1200 or 1400 feet above the sea. The upper part of the dolomitic limestone from Oldkusz towards Zarki, and especially near Wladowice, contains pisiform iron-ore; it there becomes mixed with a coarse sandstone, and constitutes a problematical agglomerate and red sandstones. The upper portion of the group is formed of gray and oolitic limestones and calcareous agglomerates, and is represented as passing into the beds considered equivalent to the Wealden rocks. The rocks of the oolitic group are seen to rest unconformably on the coal-measures and muschelkalk of Poland, and it is necessary to use some caution not to confound them with the latter rock, when they are in contact, as at Oldkusz and Nowagora. Taken on the large scale, the Polish rocks of this age are stated to have a general direction N.N.W. and S.S.E. From Wielun they plunge beneath the great plain of Poland, here and there appearing in islands through it, and are considered to support it, being met with in sinking through it. The organic remains contained in this deposit are stated to be such as to establish its identity with the oolitic series of other parts of Europe *.

We have now to consider a series of equivalent deposits, with little or no mineralogical resemblance to those noticed above, occurring in the Alps, the Carpathians, and in Italy. Numerous memoirs have been written by different geologists, and some have even considered that certain minor divisions might be established; but it must be confessed, though the evidence is greatly in favour of a considerable development of the oolitic group, with altered mineralogical characters, in the situations above noticed, that the termination of the group either above or beneath is far from possessing that clear and certain character which could be desired.

---

* Pusch, Journal de Geologie, t. ii.

The mineralogical character being so different, recourse has generally been had to organic remains; there are, however, such singular mixtures of these, in the Alps more especially, that the determination of particular deposits is far from certain. Instead of tender, soft marls, clays, sands and light-coloured limestones, we have dark-coloured marbles, masses of crystalline dolomite, gypsum, and schists approaching talcose and micaceous slates. The Alps are also particularly difficult of examination, as from the convulsions by which they have been upraised or otherwise visited, whole mountain masses are thrown over, and the rocks really deposited the latest occur beneath the older strata; and this not in limited spaces, but over considerable distances. These dark-coloured rocks were during the prevalence of the Wernerian theory referred, as was natural, to the transition class, and we are indebted to Dr. Buckland for first pointing out that they were of more recent origin; since that time other geologists have shown the probable relative antiquity of different portions; and among these, M. Elie de Beaumont holds a distinguished place, particularly as respects Savoy, Dauphiné, Provençe, and the Maritime Alps. In a note on the geological position of the fossil plants and *Belemnites* found at Petit Cœur near Moutiers in the Tarentaise, published in 1828 *, this author observes that the system of beds described by M. Brochant in his memoir on the Tarentaise, and which in many places contains considerable masses of granular limestone and micaceous quartz rock, as well as large masses of gypsum, belongs to the oolitic group. He is of this opinion, as he considers that the most ancient secondary rocks of that country, in which no fossil shells have been found that have not been discovered in the lower part of the oolitic series, can be traced to the environs of Digne and Sisteron (Basses Alpes), where they afford a great abundance of those remains supposed to be characteristic of the lias.

In a notice on the geological position of the fossil plants and graphite found at the Col du Chardonnet (Hautes Alpes), M. Elie de Beaumont observes, that as the traveller quits the Bourg d'Oisans (Piedmont) and approaches the continuous range of masses, termed primitive, that extend from the Monte Rosa towards the mountains on the west of Coni, he will perceive that the secondary rocks gradually lose their original character, though certain distinguishing marks may still be seen, thus resembling a half-burnt piece of wood, in which the ligneous fibres may be traced far beyond the part that remains wood †. He has also remarked on the original differences that may have existed between these secondary rocks of the interior of the Alps, and those in the

---

* Annales des Sciences Naturelles, t. xiv. p. 113.

† Ibid. 1828, t. xv. p. 353.

same series of other countries; and thence concludes, that very little importance should be attached to the difference of mineralogical structure observed in the beds above mentioned, and in the lower part of the oolitic group, occurring undisturbed in other parts of Europe, and of which these Alpine rocks appear to him the enlarged prolongation. The vegetables found by M. Elie de Beaumont in the situations above noticed, were examined by M. Ad. Brongniart, and many were found by him to be generally the same with those discovered in the coal-measures. The following is a list of those which he obtained from the Alps, apparently all similarly situated as to geological position:—*Calamites Suckowii,* Ad. Brong., at Pey-Ricard, near Briançon (also in the coal-measures of Newcastle and other places); *C. Cistii,* Ad. Brong., the same locality (also at Wilkesbarre in Pennsylvania); *Lepidodendron,* 2 sp., Pey-Ricard and Pey-Chagnard, near Lamure; *Sigillaria,* the above localities, and La Motte near Lamure; *Stigmaria,* Pey-Chagnard; *Nevropteris gigantea,* Ad. Brong., Servoz, Savoy (also in the coal-measures of Bohemia); *N. tenuifolia,* Ad. Brong., Petit-Cœur, and Col de Balme (also in coal-measures of Liege and Newcastle); *N. flexuosa,* Stern., La Roche Macot, Tarentaise (also coal-measures of Liége and Bath); *N. Soretii,* Ad. Brong., same locality; *N. rotundifolia,* Ad. Brong., La Roche Macot, and Col de Balme (also in the coal mines of Plessis, Calvados); *Odontopteris Brardii,* Ad. Brong., Petit-Cœur (also coal mines of Terrasson, Dordogne); *Od. obtusa,* Ad. Brong., Col de l'Ecuelle, near Chamonix; Petit-Cœur (also at Terrasson); *Pecopteris polymorpha**, Petit-Cœur (also in the coal-measures of St. Etienne, Alais, Litry, Wilkesbarre); *Pe. pteroides,* Ad. Brong., Pey-Chagnard (also in coal-measures at Liége, Mannebach, St. Etienne, and Wilkesbarre); *Pe. arborescens,* Ad. Brong., Val Bonnais, near Lamure; Petit-Cœur (also at Mannebach and Aubin, Aveyron); *Pe. platyrachis,* Ad. Brong., Val Bonnais (also at St. Etienne); *Pe. Beaumontii,* Ad. Brong., Petit-Cœur; this new species is described as resembling the *Pe. nervosa, Pe. bifurcata,* Stern., and *Pe. muricata,* Schlot., found in the coal-measures, and *Pe. tenuis,* found in the oolitic series of Whitby and Bornholm; *Pe. Plukenetii*? Petit-Cœur; Col de l'Ecuelle (also at Alais); *Pe. obtusa,* Ad. Brong., Petit-Cœur (also in coal-measures near Bath); *Asterophyllites equisetiformis,* Tarentaise (also at Alais and Mannebach); *Annularia brevifolia,* Col de Balme (also at Alais and Geislautern †).

These vegetable remains are so far associated with *Belemnites,* that the latter occur both above and beneath them; so that there

---

* This species is common in the coal-measures of France according to M. Ad. Brongniart.

† Ad. Brongniart, Ann. des Sci. Nat. vol. xiv. pp. 129, 130.

can be no doubt as to the *Belemnites* having existed previous to and after the vegetable deposit; and therefore these localities would involve the question of the preference that should be given to the *Belemnites* or to the vegetables, if M. Elie de Beaumont did not appear certain that the same series of beds was continued to Digne and Sisteron, and there contained characteristic lias remains.

M. Necker de Saussure has described a series of beds that compose the upper part of the Buet (Savoy), and which constitute the lowest calcareous deposit of that portion of the Alps, resting, like those above noticed at Petit-Cœur and the Col de Chardonet, on older and non-fossiliferous rocks. The following is a section in the ascending order :—1. Mica slate, which may form part of the protogine rocks of this district. 2. A sandstone, formed of numerous grains of quartz, mixed with a few crystalline grains of felspar, and sometimes with a little talc or chlorite. 3. Red and green argillo-ferruginous schist. This rock is sometimes wanting in the section; but on the east of the Vallée de Vallorsine it alternates with the well-known Vallorsine conglomerate, which is but a similar schist, filled with rounded pebbles of gneiss, mica slate, protogine, &c., among which we neither observe true granite nor limestone ;—an important fact, as is observed by M. Necker, for it appears to show that the Vallorsine granite, which cuts through the gneiss, did not exist before the formation of the conglomerate. 4. A black schist, with impressions of ferns, the vegetable remains being converted into thin talc *. 5. Black or dark blueish-gray limestone, filled with grains of quartz. 6. A black argillaceous schist, containing nodules of Lydian stone. Ammonites are found in this rock, as also in an argillo-talcose schist which alternates with it. 7. A gray calcareous and arenaceous schist, containing *Belemnites*†. The last bed constitutes the summit of the Buet, 10,099 English feet above the sea.

It has been observed by M. Elie de Beaumont, that the calcareous portions of these regions of the Alps are separated from the older and non-fossiliferous rocks by a sandstone more or less coarse, which passes into a conglomerate, seen not only at the Vallée de Vallorsine above noticed, but also at Trient, Ugine,

---

* When crossing and wandering over the Col de Balme in 1819, I picked up specimens of sandstone with impressions of plants upon them; these plants I then considered, from their general character, to be such as are usually found in the coal-measures (Geol. Trans. 2nd series, p. 162.); an opinion which has since been confirmed by M. Ad. Brongniart, though it now appears that they may belong to a more modern deposit.

† Necker, Mém. sur la Vallée de Vallorsine, Mém. de la Soc. de Phys. et d'Hist. Nat. de Genève, 1828.—For a section of the Buet see the same Memoir, and Sections and Views illustrative of Geological Phænomena, pl. 27. fig. 5.

316   *Oolitic Group.*

Allevard, Ferrière, and Petit-Cœur. The same circumstance is observable to the east of the Bourg d'Oisans and Huez, and in other places *. This evidence of the action of water possessing sufficient velocity to transport coarse sands and pebbles should be borne in mind, as, however such sands and pebbles may have been since altered in appearance, it shows that the deposits were not produced quietly; though subsequently, from a change of circumstances, and the establishment of comparative tranquillity, limestones were formed. These appearances are not confined to the Savoy and French Alps, but are seen on the shores of the Lake of Como and of the Gulf of La Spezia. The calcareous beds, of which such fine sections are afforded in the Lakes of Como and Lecco, are separated from the gneiss and mica-slate of the higher Alps, by a conglomerate composed of rounded pieces of quartz, red porphyry, and other rocks, associated with sandstone beds†. The limestone series incumbent on the conglomerate is in some situations strangely mixed with dolomite more or less crystalline, as will be noticed in the sequel. Taken as a mass, the limestones occupy a thickness of many thousand feet, and are more or less gray. They are siliceous, and contain seams of chert in the upper part (near Como), become slaty, with apparently little siliceous matter in their central parts, and are finally compact and more thickly bedded in their lowest situations. Ammonites greatly resembling *A. Bucklandi* and *A. heterophyllus* are discovered in it, as are also *Turritellæ*, and other shells. Anthracite is here and there found. I have little doubt that the oolitic group is represented by at least a part of this calcareous mass; but how much, and what other equivalents there may be, my present information will not permit me to hazard an opinion. The general circumstances are however so similar, that it does not seem unreasonable to conclude that the causes, whatever they were, which produced the Vallorsine conglomerates and the sandstone associated with them in that part of the Alps, were contemporaneous with those which formed the conglomerates and associated sandstones of the lakes of Como and Lugano.

To present a detail of the various observations on those Alpine rocks which are considered as referable to the oolitic group, would far exceed our limits ; the student will consult with advantage the various labours of Studer, Boué, Sedgwick, Murchison, Lill von Lillienbach, Lusser, and others. There may be occasionally some difference of opinion among authors, as to where the series may commence, or where it may end; but the main fact, the existence of the group itself, seems established beyond all doubt. When we

---

* Elie de Beaumont, Ann. des Sci. Nat. t. xv. p. 354.

† For a map, sections, and a description of this district, see Sections and Views illustrative of Geological Phænomena, pl. 31, 32.

consider the disturbed nature of the country to be examined, and the difficulty of attaining certain situations perfectly necessary to a right understanding of the subject, except under very favourable circumstances, we should be more surprised that so much has been accomplished in so short a time, than at finding discordant opinions on certain minor points.

Mr. Murchison observes that, accompanied by M. Lill von Lillienbach, he found in the dark-coloured limestone and shale, at the gorge of the Mertelbach, below Crispel (Austrian Alps),— *Ammonites* 2 species (one approaching *A. Conybeari*), *Pecten* 3 species, small *Gryphæa, Mya, Perna* 2 species, *Ostrea,* Corallines, &c. This group is referred to the lias. An overlying red encrinite limestone contains several species of *Ammonites,* and some *Belemnites.* According to Professor Sedgwick and Mr. Murchison, most of the salt-mines of the Austrian Alps are contained in the oolitic group (Halstadt, Aussee, &c.). The upper part of the oolitic series of this part of the Alps contains semi-crystalline, brecciated, compact, and dolomitic limestones *.

I cannot conclude this sketch of the oolitic group, without adverting to certain limestones of La Spezia which may be referable to it. On the west side of the celebrated Gulf of La Spezia, there is a range of mountains extending along the coast nearly to Levanto, their breadth augmenting as they advance N.W. The sections of these mountains expose the following rocks, easily observed up any of the cross valleys. The annexed wood-cut exhibits a section over Coregna.

Coregna. Fig. 53.

S.   M.

a   *b c d e f*   *g*

S. Gulf of La Spezia. M. Mediterranean. *a.* Limestone series:— Upper beds compact and gray, varying in intensity of tint; more or less traversed by calcareous spar; here and there interstratified with schistose beds, and even argillaceous slate. The beds most commonly thick. The limestone with light-brown veins, so long known by the name of Porto Venere marble, forms part of these beds. *b.* Dolomite:—varying in appearance; not unfrequently crystalline; when most so, nearly white; in some places beds may be distinguished, in others stratification cannot be traced. *c.* Numerous thin beds of dark-gray limestone. *d.* The same kind of beds alternating with light-brown schist, containing an abundance of small nodules of iron pyrites, *Belemnites, Orthoceratites,* and

* Proceedings of the Geological Society, 1831. Phil. Mag. and Annals, vol. ix. 1831.

Ammonites, enumerated beneath. The limestones which alternate with the schist become occasionally light-coloured as they approach the next rock, from which however they are separated by a repetition of the dark-coloured limestone and brown schist. *e.* Brown shale which does not effervesce with acids. *f.* Variegated beds:—greenish blue and argillo-calcareous rocks; more or less schistose, the calcareous matter being often in very small quantity. *g.* Brown sandstone;—principally siliceous, though some of it does contain calcareous matter. It is sometimes micaceous, and occurs either in thick, thin, or schistose beds. It has sometimes been called grauwacke, and it is one of the *macignos* of the Italians.

The organic remains from Coregna were first discovered by M. Guidoni, of Massa; a few indications only of the presence of such bodies in the limestone under consideration having been noticed by M. Cordier some years previously. The strata being perpendicular, the weather acts on the edges of the shale beds, in which the remains are found, and they are thus brought to light. At my request Mr. Sowerby examined the remains that I brought from thence, and he considers that out of fifteen different species of *Ammonites*, one seemed the same with the *A. erugatus*, Phil., discovered in the lias of Yorkshire, while two resembled *A. Listeri** and *A. biformis*, shells discovered in the coal-measures of the same part of England. The remainder he considers undescribed. From the great scarcity of organic remains of these limestones in Italy, I have inserted Mr. Sowerby's descriptions of the various species, together with figures, considering that they may be of service in the examination of other parts of Italy, as well as Greece, and various countries eastward.

Fig. 54.   Fig. 55.   Fig. 56.   Fig. 57.

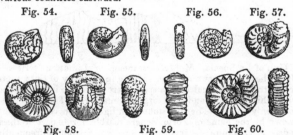

Fig. 58.   Fig. 59.   Fig. 60.

Fig. 54. *Ammonites cylindricus.* Inner whorls perfectly concealed; sides slightly concave about their centres, flat towards the margin; surface smooth; aperture oblong, deeply indented by the preceding whorl; the front square, which distinguishes it from *A. heterophyllus*, Sow.

Fig. 55. *A. Stella.* A small portion of the inner whorls exposed;

* This shell is also discovered, according to M. Hœninghaus, in the coal-measures at Werden.

the sides rather convex, largely umbilicated; of the inner whorls, plain; of the outer, two thirds covered by large convex rays; aperture elongated, its front elliptical, its inner angles truncated.

Fig. 56. *A. Phillipsii.* Inner whorls almost wholly exposed; whorls slowly increasing, about four, their sides flat, irregularly and obscurely undulated; aperture four-sided, rather longer than wide, the sides nearly straight. The cast is contracted at distant intervals by the periodical thickening of the edge of the aperture. Named in honour of Mr. Phillips *.

Figs. 57 and 59. *A. biformis.* Inner whorls partly visible; whorls three or four, rapidly increasing, crossed by many prominent sharp ribs; each rib suddenly becomes obscure, and spreads into two as it passes over the broad convex front; aperture transversely oblong, twice as wide as long, slightly arched.

Upon the inner whorls, which have the front plain, the ribs are contracted into round tubercles. The extremities of the longer ribs almost form spines. This species is found in the coal-measure near Leeds.

Fig. 58. *A. Listeri.* See Min. Con. tab. 501. Also discovered in the coal-measures of Yorkshire.

Fig. 60. *A. Coregnensis.* Inner whorls much exposed; whorls, three or four, crossed by many straight, prominent, sharp ribs, which bend forward, and suddenly terminate upon the nearly plain front; aperture transversely obovate.

This shell is intermediate between *A. biformis* and *A. planicostata,* Sow.: it is, however, nearer the former, as it has tubercles upon the inner whorls, where *A. planicostata* is quite smooth.

Fig. 61.   Fig. 62.   Fig. 63.

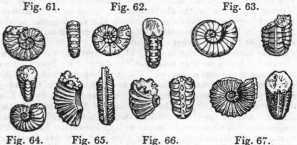

Fig. 64.   Fig. 65.   Fig. 66.   Fig. 67.

Fig. 61. *A. Guidoni.* Inner whorls much exposed; whorls few, their sides flat and crossed by distant flattened ribs; each rib split, the posterior branch most prominent, and raised into a low tubercle before it passes over the narrow convex margin. Named in honour of Sig. Guidoni, the discoverer of these remains at Coregna.

Fig. 62. *A. articulatus.* Inner whorls nearly exposed; whorls few, each divided by eight or ten furrows into as many imbricating joints;

* Author of Illustrations of the Geology of Yorkshire.

the anterior edge of each joint elevated, and crossed by the edges of the
septa.

Fig. 63. *A. discretus.* Inner whorls partly exposed in a large umbi-
licus ; globose ; whorls three or four, crossed by many prominent ribs,
which split as they cross over the convex front ; keel sharp, entire ;
aperture transversely oval, slightly arched.

Fig. 64. *A. ventricosus.* Inner whorls slightly exposed ; whorls about
three ; half the fourth whorl much inflated ; sides ornamented with arched
ribs, that are often flattened and united in pairs as they pass over the
front, which in the last whorl has a furrow along it ; aperture circular,
large.

Fig. 65. *A. comptus.* Inner whorls almost wholly exposed, rapidly
increasing in size ; sides flat ; whorls crossed by very numerous, sharp,
straight radii, which terminate in obscure spines near the narrow concave front ; aperture oblong, narrowest towards the front.

Fig. 66. *A. catenatus.* Inner whorls much exposed ; whorls rapidly
increasing, crossed by strong curved ribs, which enlarge as they approach
the margin ; front ornamented with a chain of hollow squares ; apertures
rather square, notched by the preceding whorl ; the hollow squares
around the margin united by two of their angles to the extremities of
corresponding radii.

Fig. 67. *A. trapezoidalis.* Inner whorls exposed ; whorls three or
four, rapidly increasing in size, crossed by many prominent nearly equal
ribs reaching to the narrow front ; aperture trapezoidal, indented by the
preceding whorl ; the acute angle truncated by the front.

The above figures are all of the natural size of the *Ammonites.*
The remains of *Orthoceratites,* which abundantly accompany the
Ammonites, resemble the *O. Steinhaueri,* found in the coal-mea-
sures of Yorkshire ; they also approach the *O. elongatus* of the
Dorsetshire lias. The remains of *Belemnites* consist only of their
alveoles, and are somewhat common.

As far therefore as the evidence of the *Ammonites* and *Orthoceratites* extends, we may refer the limestone of La Spezia either
to the lias or the coal-measures. There will be observed a curious
correspondence in the organic character of the rocks of the Savoy
and French Alps above noticed, and considered as lias by M. Elie
de Beaumont, with that of the limestones of La Spezia. In the
former, coal-measure plants are found with *Belemnites ;* in the latter
coal-measure *Ammonites* also occur with *Belemnites.* The organic
character of the oolitic group in the Alps is far from being well ascertained, and the undescribed organic remains found in the same
series of the South of France are exceedingly numerous, so that
it may be possible to discover some of the La Spezia *Ammonites* in
both situations ; and the organic remains of the south-east of France,
the Alps, and La Spezia, may hereafter mutually assist in determining the relative ages of the rocks in which they are discovered*.

* It should be observed, that M. Passini states he has discovered red
ammonitiferous limestones in the midst of sandstones in Tuscany, which

The dolomite found among the limestones of La Spezia rises so perpendicularly, that it might be considered as a dyke elevating the strata; while at the same time it has the appearance of an included bed, or series of beds. It preserves a very constant position, and extends in a line across the mountains of La Castellana, Coregna, Santa Croce, Parodi, and Bergamo, towards Pignone. M. Laugier, at the request of M. Cordier, very obligingly made for me an analysis of some crystalline dolomite of La Castellana. One hundred parts were found to contain,—carbonate of lime. 55·36; carbonate of magnesia, 41·30; peroxide of iron and alumine, 2; silex, 0·50; loss, 0·84.

These limestones occur on the other or eastern side of the Gulf of La Spezia, and dolomitic rocks are also found among them. The mode on which they repose on the older rocks is particularly instructive, and is well seen at Capo Corvo, of which the annexed wood-cut is a section, laid bare by the sea.

Fig. 68.

G.   M.

a   b c d   e f g h i k l m   n

G. Gulf of La Spezia. M. Embouchure of the Magra. *a.* Gray compact limestones mixed with schist. *b.* Thick beds of gray compact limestone. *c.* Schist with mica. *d.* Thick beds of hard conglomerate, containing pieces of quartz, varying from the size of a pea to that of a walnut, and even larger, agglutinated by a siliceous cement. Two or three beds of coarse sands are associated with this. *e.* The same, mixed with chlorite schist, often in the same bed. The quartzose beds contain veins of specular iron-ore. *f.* Brown micaceous and schistose beds, with a small proportion of limestone. *g.* A mixture of brown and white crystalline limestone. *h.* Compact chlorite. *i.* White saccharine limestone. *k.* Brown micaceous beds. *l.* White saccharine limestone, rendered schistose by mica. *m.* Brown semi-crystalline limestone, mixed with white. *n.* Micaceous schist, curving round to the eastward.

The crystalline limestones and micaceous schist of this section would seem to form part of the system of rocks, which in the neighbouring mountains of Massa Carrara, now again known by the name of Alpi Apuani, furnishes the long celebrated Carrara marbles. The limestones appear the same as those on the western side of the Gulf of Spezia; but instead, like them, of resting upon a mass of sandstone, they repose upon a conglomerate, seen, between

he considers may be referred to the same age as the limestones of La Spezia. Journal de Geologie, t. ii. p. 98.

the mouth of the Magra and Ameglia, to become far more developed than at the Capo Corvo section, where it is in some manner squeezed between the crystalline limestones and the compact
gray limestones.   Amid this greater development, which appears
to mark an unconformable superposition, a conglomerate will be
observed (particularly on the shore of the Magra), closely resembling that commonly known as the Vallorsine conglomerate, and
noticed above.

I cannot avoid connecting this conglomerate, and that of the
Lake of Como, with the conglomerates and sandstones of the Vallorsine and other parts of the Western Alps, and referring them
to the same epoch of formation;—one in which water, with a
certain velocity, ground down portions of pre-existing rocks, and
which was succeeded by a state of things when a great abundance of carbonate of lime was deposited.   This deposit appears
to have been extensive, not only in the Alps, but in Italy; and in
both situations, where it occurs close to the rocks of an older
date, such as protogine, gneiss, micaceous slates, and associated
saccharine marble, and talcose rocks of that age, it seems to be
separated from them by strata which mark a mechanical origin.
As we may suppose great inequalities to have existed during this
deposit, and others immediately preceding it, we may perhaps in
this way account for the almost close contact of the gray compact
limestones with the saccharine limestone and other associated rocks
at Capo Corvo, while on the western side of the gulf they rest on
arenaceous rocks of considerable thickness, which again repose on
gray siliceo-calcareous schists and sandstones, that extend over a
considerable part of Liguria.   How far these beds, which separate
the limestones of the Alps, Liguria, and Tuscany, may be equivalent to the sandstone found beneath the lias in Southern Germany
and various parts of France, may perhaps be now difficult to
determine, but there is a certain general resemblance which
seems to point to that conclusion.

Supposing that these Italian and Alpine limestones do represent
the oolitic series of Western Europe, (and it seems very possible
that they may do so,) it remains to account for the very great
abundance of organic remains in the one, and their very great
scarcity in the other.   It has often struck geologists, that some
deposits may have taken place in shallow seas, and others in deep
water.   This mode of viewing the subject has, if I mistake not,
induced M. Elie de Beaumont to consider that the oolitic series of
the Western Alps was deposited in a deep sea, at the same time
that the same series was in the course of formation in shallow seas
in other places.   This observation may be extended into Italy and
Greece, where the absence or very great scarcity of organic remains
at this epoch seems to afford it support.   That great inequalities
existed at all periods on the earth's surface it seems fair to infer, as

well beneath the sea as on land. It would be unphilosophical to conclude that marine animals were ever more capable of supporting very considerable differences of pressure than at the present day. Now we know that certain kinds of marine animals, particularly some *Mollusca* and *Conchifera*, are only found on coasts where they can find support beneath a moderate pressure of water, while others, such as the *Nautilidæ*, are so provided with floating apparatus, that they are discovered in parts of the ocean where there may be considerable depth. We have only to consider that in those parts of Western Europe where organic remains are abundant, shallow seas existed, while the same ocean was deep over that point of the globe's surface where we find Italy and Greece, and an explanation would seem to be afforded, not only of the abundance of shells in one place, and their scarcity in another, but also of the kind of shells found; for as yet camerated shells only, such as *Belemnites*, *Orthoceratites* and *Ammonites*, have been discovered in these rocks of central Italy; in other words, animals capable of swimming in deep seas. Organic remains are not only scarce in the limestones in Italy, but also in the sandstones or macignos, which occur in great thickness above and beneath them; the organic remains yet noticed in these sandstones being *Fucoides*, marine plants which may be easily drifted great distances, as the *gulf weed* now is. The differences of depth and consequent pressure may also in some measure account for the different mineralogical structure of the rocks composing the oolitic group in different situations. Still, however, the question of whence all this great mass of carbonate of lime was derived remains unanswered. To attempt to account for it by means of springs at all resembling those we now see, seems quite unphilosophical; and to consider it entirely due to animals which have separated lime from the water, leaving their shells produced through millions of ages to be gradually converted into limestone, appears also a cause inadequate to the effect required, though it cannot be denied that the mass of many limestones is nearly made up of organic remains. With every allowance for calcareous deposits formed by springs, and organic bodies, there remains a mass of limestone to be accounted for, distributed generally over a very large surface, which requires a very general production, or rather deposit, of carbonate of lime, contemporaneously, or nearly so, over a great area.

Organic Remains of the Oolitic Group.

PLANTÆ.

Algæ.

1. Fucoides furcatus, *Ad. Brong.* Stonesfield slate, *Ad. Brong.*
2. ———— Stockii, *Ad. Brong.* Solenhofen, *Ad. Brong.*
3. ———— encelioides, *Ad. Brong.* Solenhofen, *Ad. Brong.*

Equisetaceæ.

1. Equisetum columnare, *Ad. Brong.* Lower carbonaceous series, Yorkshire, *Phil.*; Brora, *Murch.*

Filices.

1. Pachypteris lanceolata, *Ad. Brong.* Coal, shale, &c. between inferior and great oolite, Yorkshire, *Phil.*
2. ————— ovata, *Ad. Brong.* Coal, shale, &c. between inferior and great oolite, Yorkshire, *Phil.*
1. Pecopteris Reglei, *Ad. Brong.* Forest marble, Mamers, *Desn.*
2. ————— Desnoyersii, *Ad. Brong.* Forest marble, Mamers, *Desn.*
3. ————— polypodioides, *Ad. Brong.* Coal, shale, &c. between cornbrash and great oolite, Yorkshire, *Phil.*
4. ————— denticulata, *Ad. Brong.* Coal, shale, &c. between cornbrash and great oolite, Yorkshire, *Phil.*
5. ————— Phillipsii, *Ad. Brong.* Coal, &c. of the oolitic series, Yorkshire, *Ad. Brong.*
6. ,————— Whitbiensis, *Ad. Brong.* Coal, shale, &c. between cornbrash and great oolite, Yorkshire, *Phil.*
1. Sphænopteris hymenophylloides, *Ad. Brong.* Stonesfield slate, *Buckl.*; Coal, shale, &c. between great and inferior oolite, Yorkshire, *Phil.*
2. ————— ? macrophylla, *Ad. Brong.* Stonesfield slate, *Buckl.*
3. ————— Williamsonis, *Ad. Brong.* Coal, &c. of the oolitic series, Yorkshire, *Ad. Brong.*
4. ————— crenulata, *Ad. Brong.* Coal, &c. of the oolitic series, Yorkshire, *Ad. Brong.*
5. ————— denticulata, *Ad. Brong.* Coal, &c. of the oolitic series, Yorkshire, *Ad. Brong.*
1. Tæniopteris latifolia, *Ad. Brong.* Coal, shale, &c. between cornbrash and great oolite, Yorkshire, *Phil.*
2. ————— vittata, *Ad. Brong.* Coal, shale, &c. between cornbrash and great oolite, Yorkshire, *Phil.*

Cycadeæ.

1. Pterophyllum Williamsonis. Coal, shale, &c. between cornbrash and great oolite, Yorkshire, *Phil.*
1. Zamia pectinata, *Ad. Brong.* Stonesfield slate, *Buckl.*
2. ——— patens, *Ad. Brong.* Stonesfield slate, *Ad. Brong.*
3. ——— longifolia, *Ad. Brong.* Coal, shale, &c. between cornbrash and great oolite, Yorkshire, *Phil.*
4. ——— pennæformis, *Ad. Brong.* Coal, shale, &c. between great and inferior oolite, Yorkshire, *Phil.*
5. ——— elegans, *Ad. Brong.* Coal, shale, &c. between great and inferior oolite, Yorkshire, *Phil.*

6. Zamia Goldiæi, *Ad. Brong.* Coal, &c. of the oolitic series, Yorkshire, *Ad. Brong.*
7. —— acuta, *Ad. Brong.* Coal, &c. of the oolitic series, Yorkshire, *Ad. Brong.*
8. —— lævis, *Ad. Brong.* Coal, &c. of the oolitic series, Yorkshire, *Ad. Brong.*
9. —— Youngii, *Ad. Brong.* Coal, shale, &c. between great and inferior oolite, Yorkshire, *Phil.*
10. —— Feneonis, *Ad. Brong.* Coal, &c. of the oolitic series, Yorkshire, *Ad. Brong.*
11. —— Mantelli, *Ad. Brong.* Coal, shale, &c. between great and inferior oolite, Yorkshire, *Phil.*
1. Zamites Bechii, *Ad. Brong.* Forest marble, Mamers, *Desn.;* Lias, Lyme Regis, *De la B.*
2. —— Bucklandii, *Ad. Brong.* Forest marble, Mamers, *Desn.;* Lias, Lyme Regis, *De la B.*
3. —— Lagotis, *Ad. Brong.* Forest marble, Mamers, *Desn.*
4. —— hastata, *Ad. Brong.* Forest marble, Mamers, *Desn.*

Coniferæ.

1. Thuytes divaricata, *Sternb.* Stonesfield slate, *Buckl.*
2. —— expansa, *Sternb.* Stonesfield slate, *Buckl.*
3. —— acutifolia, *Ad. Brong.* Stonesfield slate, *Buckl.*
4. —— cupressiformis, *Sternb.* Stonesfield slate, *Buckl.*
1. Taxites podocarpoides, *Ad. Brong.* Stonesfield slate, *Buckl.*

Lilia.

1. Bucklandia squamosa, *Ad. Brong.* Stonesfield, *Buckl.*

Class uncertain.

1. Mamillaria Desnoyersii, *Ad. Brong.* Mamers, *Desn.*
Many undescribed vegetables, Lias, Lyme Regis, *De la B.*

Zoophyta.

1. Achilleum dubium, *Goldf.* Solenhofen, *Goldf.*
2. —— cheirotonum, *Goldf.* Oolitic rocks, Baireuth, *Munst.*
3. —— muricatum, *Goldf.* Streitberg, *Munst.*
4. —— tuberosum, *Munst.* Hattheim, *Munst.*
5. —— cancellatum, *Munst.* Hattheim, *Munst.*
6. —— costatum, *Munst.* Streitberg, *Munst.*
1. Manon Peziza, *Goldf.* Streitberg; Hattheim; Giengen; Regensberg, *Goldf.*
2. —— marginatum, *Munst.* Streitberg; Muggendorf, *Munst.*
3. —— impressum, *Munst.* Muggendorf, *Munst.*
1. Scyphia cylindrica, *Goldf.* Muggendorf, *Munst.*
2. —— elegans, *Goldf.* Thurnau; Baireuth, *Goldf.*
3. —— calopora, *Goldf.* Thurnau; Baireuth, *Goldf.*

4. Scyphia pertusa, *Goldf.* Streitberg; Baireuth, *Goldf.*
5. ——— texturata, *Goldf.* Giengen, Wirtemberg, *Goldf.*
6. ——— texata, *Goldf.* Legerberg, Switzerland; Streitberg, *Goldf.*
7. ——— polyommata, *Goldf.* Baireuth & Switzerland, *Goldf.*
8. ——— clathrata, *Goldf.* Streitberg; Baireuth, *Goldf.*
9. ——— milleporata, *Goldf.* Baireuth, *Goldf.*
10. ——— parallela, *Goldf.* Streitberg, *Munst.*
11. ——— psilopora, *Goldf.* Muggendorf, *Goldf.*
12. ——— obliqua, *Goldf.* Muggendorf, *Munst.*
13. ——— rugosa, *Goldf.* Streitberg, *Munst.*
14. ——— articulata, *Goldf* Muggendorf, *Goldf.*
15. ——— pyriformis, *Goldf.* Streitberg, *Munst.*
16. ——— radiciformis, *Goldf.* Streitberg, *Goldf.*
17. ——— punctata, *Goldf.* Streitberg, *Munst.*
18. ——— reticulata, *Goldf.* Streitberg, *Goldf.*
19. ——— dictyota, *Goldf.* Streitberg, *Munst.*
20. ——— procumbens, *Goldf.* Baireuth, *Goldf.*
21. ——— paradoxa, *Munst.* Streitberg & Amberg, *Munst.*
22. ——— empleura, *Munst.* Streitberg, *Munst.*
23. ——— striata, *Munst.* Streitberg & Muggendorf, *Munst.*
24. ——— Buchii, *Munst.* Streitberg, *Munst.*
25. ——— Munsteri, *Goldf.* Regensburg; Streitberg, *Goldf.*
26. ——— propinqua, *Munst.* Streitberg; Muggendorf, *Munst.*
27. ——— cancellata, *Munst.* Streitberg; Muggendorf, *Munst.*
28. ——— decorata, *Munst.* Muggendorf, *Munst.*
29. ——— Humboldtii. *Munst.* Muggendorf, *Munst.*
30. ——— Sternbergii, *Munst.* Streitberg, *Munst.*
31. ——— Schlotheimii, *Munst.* Thurnau; Streitberg, *Munst.*
32. ——— Schweiggeri, *Goldf.* Baireuth, *Goldf.*
33. ——— secunda, *Munst.* Heiligenstadt; Streitberg, *Munst.*
34. ——— verrucosa, *Goldf.* Streitberg & Wurgau, *Goldf.*
35. ——— Bronnii, *Munst.* Wirtemberg & Baireuth, *Munst.*
36. ——— milleporacea, *Munst.* Thurnau; Aufseess; Streitberg, *Munst.*
37. ——— pertusa, *Goldf.* Streitberg & Amberg, *Goldf.*
38. ——— intermedia, *Munst.* Hattheim; Streitberg, *Munst.*
39. ——— Neesii, *Goldf.* Streitberg, *Goldf.*
1. Tragos pezizoides, *Goldf.* Muggendorf, *Goldf.*
2. ——— Palella, *Goldf.* Wirtemberg & Switzerland; Rabenstein; Heiligenstadt, *Goldf.*
3. ——— sphærioides, *Goldf.* Sigmaringen, Wirtemberg, *Goldf.*
4. ——— tuberosum *, *Goldf.* Inferior Oolite, Rabenstein; Streitberg, *Munst.*
5. ——— acetabulum, *Goldf.* Streitberg; Randen, *Goldf.*

* *Limnorea lamellosa* of Lamouroux according to M. Goldfuss.

6. Tragos radiatum, *Munst.* Streitberg, *Munst.*
7. —— rugosum, *Munst.* Streitberg, *Munst.*
8. —— reticulatum, *Munst.* Streitberg, *Munst.*
9. —— verrucosum, *Munst.* Streitberg, *Munst.*
1. Spongia floriceps, *Phil.* Coral Oolite, Yorkshire, *Phil.*
2. —— clavaroides, *Lam.* Great Oolite, Wiltshire, *Lons.*
——, species not determined. Lower Calcareous Grit, Yorkshire, *Phil.*; Inferior Oolite, Middle and South of England, *Conyb.*; Forest Marble, Wiltshire, *Lons.*
Alcyonium, species not determined. Forest Marble, Normandy, *De Cau.*; Great Oolite? Wilts, *Lons.*
1. Cnemidium lamellosum, *Goldf.* Randen, Switzerland, *Goldf.*
2. —— stellatum, *Goldf.* Randen, Switzerland, *Goldf.*
3. —— striato-punctatum, *Goldf.* Randen, *Goldf.*
4. —— rimulosum, *Goldf.* Randen, *Goldf.*
5. —— mammillare, *Goldf.* Streitberg, *Goldf.*
6. —— Rotula, *Goldf.* Thurnau, *Goldf.*
7. —— granulosum, *Munst.* Streitberg, *Munst.*
8. —— astrophorum, *Munst.* Hattheim; Regenberg, *Munst.*
9. —— capitatum, *Munst.* Amberg, *Munst.*
1. Limnorea mammillaris *, *Lam.* Forest Marble, Normandy, *De Cau.*
1. Siphonia pyriformis, *Goldf.* Streitberg, *Goldf.*
1. Myrmecium hemisphæricum, *Goldf.* Thurnau, *Goldf.*
1. Gorgonia dubia, *Goldf* Glücksbrunn, Thuringia, *Goldf.*
1. Millepora dumetosa, *Lam.* Forest Marble, Normandy, *De Cau.*
2. —— corymbosa, *Lam.* Forest Marble, Normandy, *De Cau.*
3. —— conifera, *Lam.* Forest Marble, Normandy, *De Cau.*
4. —— pyriformis, *Lam.* Forest Marble, Normandy, *De Cau.*
5. —— macrocaule, *Lam.* Forest Marble, Normandy, *De Cau.*
6. —— straminea, *Phil.* Great Oolite and Cornbrash, Yorkshire, *Phil.*
——, species not determined. Cornbrash and Forest Marble, North of France, *Bobl.*; Forest Marble, Mamers, Normandy, *Desn.*; Forest Marble and Great Oolite, Wiltshire, *Lons.*
Madrepora, species not determined. Bradford Clay, North of France, *Bobl.*; Coral Rag, Normandy, *De Cau.*; Portland Stone, Wiltshire, *Conyb.*; Inferior

---

* Is this *Limnorea mammillosa*, Lam.? If it be, it is the *Cnemidium tuberosum* of Goldfuss.

Oolite, Mid. and South of England, *Conyb.;*
Mauriac Beds, S. of France, *Desfr.*
Eschara, species not determined. Forest Marble, Normandy,
De Cau.
1. Cellepora orbiculata, *Goldf.* Streitberg, *Munst. ;* Oxford Clav,
Haute Saone, *Thir.*
2. ———— echinata, *Goldf.* Inferior Oolite, Haute Saone,
Thir.
————, species not determined. Inferior Oolite, Midland
and Southern England, *Conyb.*
Retepora?————. Great Oolite, Yorkshire, *Phil.*
Flustra, species not determined. Great Oolite, Wiltshire,
Lons.
1. Ceriopora radiciformis, *Goldf.* Thurnau, Baireuth, *Goldf.*
2. ———— striata, *Goldf.* Streitberg; Thurnau, *Munst.*
3. ———— angulosa, *Goldf.* Thurnau, *Munst.*
4. ———— alata, *Goldf.* Thurnau, *Munst.*
5. ———— crispa, *Goldf.* Thurnau, *Munst.*
6. ———— favosa, *Goldf.* Streitberg; Thurnau, *Munst.*
7. ———— radiata, *Goldf.* Thurnau, *Munst.*
8. ———— compressa, *Munst.* Thurnau, *Munst.*
9. ———— orbiculata,   Inferior Oolite, Haute Saone,
Thir.
1. Agaricia rotata, *Goldf.* Randenberg, Switzerland, *Goldf.*
2. ———— crassa, *Goldf.* Randen, Switzerland, *Goldf.*
3. ———— granulata, *Munst.* Bâle; Hattheim, *Munst.*
1. Lithodendron elegans, *Munst.* Wirtemberg, *Munst.*
2. ———— compressum, *Munst.* Heidenhein, Wirtemberg,
Munst.
1. Caryophyllia cylindrica, *Phil.* Coralline Oolite, Yorks., *Phil.*
2. ———— truncata, *Lam.* Forest Marble, Normandy, *De
Cau.*
3. ———— Brebissonii, *Lam.* Forest Marble, Normandy,
De Cau.
4. ———— convexa, *Phil.* Inferior Oolite, Yorkshire, *Phil.*
5. ———— like C. cespitosa, *Ellis.* Coral Oolite, Yorks.,*Phil.;*
Great Oolite, Mid. and S. of England, *Conyb.*
6. ———— like C. flexuosa, *Ellis.* Coral Oolite, Yorkshire,
Phil. ; Great Oolite, Midland and Southern
England, *Conyb.*
7. ———— approaching C. Carduus, *Park.* Coral Rag, Great
Oolite, Middle and South of England, *Conyb.*
————, species not determined. Inferior Oolite, North of
France, *Bobl. ;* Rochelle Beds, *Dufr. ;* Forest
Marble, Mamers, Normandy, *Desn. ;* Forest
Marble, Bradford Clay, and Great Oolite,Wilt-
shire, *Lons.*

1. Anthophyllum turbinatum, *Munst.* Hattheim; Heidenheim, *Munst.*
2. ———————— obconicum, *Munst.* Hattheim; Heidenheim, *Munst.*
1. Fungia orbiculites, *Lam.* Forest Marble, Normandy, *De Cau.;* Cornbrash, Wiltshire, *Lons.*
    ——— species not determined. Inferior Oolite, Midland and Southern England, *Conyb.*
1. Cyclolites elliptica, *Lam.* Inferior Oolite, Midland and Southern England, *Conyb.*
    ——— species not determined. Bradford Clay, Midland and Southern England, *Conyb.*
1. Turbinolia dispar, *Phil.* Coral Oolite, Yorkshire, *Phil.*
    ——— species not determined. Inferior Oolite and Lias, North of France, *Bobl.*
1. Turbinolopsis ochracea, *Lam.* Forest Marble, Normandy, *De Cau.*
1. Cyathophyllum Tintinnabulum, *Goldf.* Banz; Staffelstein; Bamberg, *Goldf.*
2. ———————— Mactra, *Goldf.* Banz; Bamberg, *Goldf.*
1. Meandrina Sœmmeringii, *Munst.* Hattheim; Heidenheim, *Munst.*
2. ——— astroides, *Goldf.* Coral Rag, Haute Saone, *Thir.;* Giengèn, *Goldf.*
3. ——— tenella, *Goldf.* Giengen, *Goldf.*
    ———, species not determined. Inferior Oolite and Coral Oolite, Yorks., *Phil.;* Inferior Oolite? Mid. and Southern England, *Conyb.;* Kimmeridge Clay, Haute Saone, *Thir.;* Great Oolite, Wilts. *Lons.*
1. Astrea microconos, *Goldf.* Biberbach, near Muggendorf, *Goldf.*
2. ——— limbata, *Goldf.* Giengen, *Goldf.*
3. ——— concinna, *Goldf.* Giengen, *Goldf.*
4. ——— pentagonalis, *Munst.* Hattheim; Heidenheim, *Munst.*
5. ——— gracilis, *Munst.* Boll, Wirtemberg, *Munst.*
6. ——— explanata, *Munst.* Wirtemberg, *Munst.*
7. ——— tubulosa, *Goldf.* Wirtemberg, *Goldf.;* Coral Rag, Haute Saone, *Thir.*
8. ——— oculata, *Goldf.* Giengen, *Goldf.;* Coral Rag, Haute Saone, *Thir.*
9. ——— alveolata, *Goldf.* Heidenheim, Wirtemberg, *Goldf.*
10. ——— helianthoides, *Goldf.* Heidenheim; Giengen, *Goldf.;* Inferior Oolite, Coral Rag, Haute Saone, *Thir.*
11. ——— confluens, *Goldf.* Heidenheim; Giengen, *Goldf.;* Coral Rag, Haute Saone, *Thir.*
12. ——— caryophylloides, *Goldf.* Giengen, *Goldf.;* Coral Rag, Haute Saone, *Thir.*

13. Astrea cristata, *Goldf.* Giengen; Heidenheim, *Goldf.*
14. ———— sexradiata, *Goldf.* Giengen, *Goldf.*
15. ———— favosioides, *Smith.* Coral Oolite, Yorkshire, *Phil.;*
    Coral Rag and Great Oolite, Midland and Southern
    England, *Conyb.*
16. ———— inæqualis, *Phil.* Coral Oolite, Yorkshire, *Phil.*
17. ———— micastron, *Phil.* Coral Oolite, Yorkshire, *Phil.*
18. ———— arachnoides, *Flem.* Coral Oolite, Yorkshire, *Phil.*
19. ———— tubulifera, *Phil.* Coral Oolite, Yorkshire, *Phil.*
20. ———— resembling A. siderea. Inferior Oolite, Midland and
    Southern England, *Conyb.*
    ————, species not determined. Coral Rag, Normandy, nu-
    merous, *DeCau.;* Great Oolite, Midland and South-
    ern England, *Conyb.;* Lias, Hebrides, *Murch.;* Great
    Oolite, Wiltshire, *Lons.*
1. Aulopora compressa, *Goldf.* Rabenstein; Grafenberg, *Munst.*
1. Entalophora cellarioides, *Lam.* Forest Marble, Normandy,
    De Cau.
Favosites, species not determined. Forest Marble, Mamers,
    Normandy, *Desn.*
1. Spiropora tetragona, *Lam.* Forest Marble, Normandy, *De
    Cau.*
2. ———— cæspitosa, *Lam.* Forest Marble, Normandy, *De
    Cau.;* Great Oolite, Wiltshire, *Lons.*
3. ———— elegans, *Lam:* Forest Marble, Normandy, *De Cau.*
4. ———— intricata, *Lam.* Forest Marble, Normandy, *De Cau.*
1. Eunomia radiata, *Lam.* Forest Marble, Normandy, *De Cau.;*
    Great Oolite, Wiltshire, *Lons.*
1. Crysaora damæcornis, *Lam.* Forest Marble, Normandy, *De
    Cau.;* Great Oolite, Wiltshire, *Lons.*
2. ———— spinosa, *Lam.* Forest Marble, Normandy, *De Cau.*
1. Theonoa chlathrata, *Lam.* Forest Marble, Normandy,*DeCau.;*
    Great-Oolite, Wiltshire, *Lons.*
1. Idmonea triquetra, *Lam.* Forest Marble, Normandy, *DeCau.;*
    Great Oolite, Wiltshire, *Lons.*
1. Alecto dichotoma, *Lam.* Great Oolite, Wiltshire, *Lons.;*
    Forest Marble, Normandy, *De Cau.*
    ————, species not determined. Inferior Oolite, Middle and
    South of England, *Conyb.*
1. Berenicea diluviana, *Lam.* Great Oolite, Wiltshire, *Lons.;*
    Forest Marble, Normandy, *De Cau.*
    ————, species not determined. Great Oolite, Haute Saone,
    Thir.; Forest Marble, Wiltshire, *Lons.*
1. Tere bellaria ramosissima, *Lam.* Forest Marble and Great
    Oolite, Somerset, *Lons.;* Forest Marble, Nor-
    mandy, *De Cau.*
2. ———— antilope, *Lam.* For. Marble, Normandy, *De Cau.*

1. Cellaria Smithii, *Phil.* Cornbrash, Yorkshire, *Phil.*
1. Thamnasteria Lamourouxii, *Le Sauvage.* Coral Rag, Normandy, *De Cau.*
1. Explanaria mesenterina, *Lam.* Inferior Oolite, Middle and South of England, *Conyb.*
——————, species not determined. Great Oolite, Wilts., *Lons.*
Polypifers, genera not determined. Lias (rare), Lyme Regis, *De la B.;* Lias (rare), Yorkshire, *Phil.;* Lias (rare), Normandy, *De Cau.;* Coral Rag (numerous), North of France, *Bobl.;* Coral Rag (abundant), Burgundy, *Beaum.;* Coral Rag (abundant), South of France, *Dufr.*

RADIARIA.

1. Cidaris florigemma, *Phil.* Coral Oolite, Yorkshire, *Phil.*
2. ———— intermedia, *Park.* Coral Oolite, Yorkshire, *Phil.*
3. ———— monilipora, *Y. & B.* Coral Oolite, Yorkshire, *Phil.*
4. ———— vagans, *Phil.* Calcareous Grit, Cornbrash, and Great Oolite, Yorkshire, *Phil.*
?5. ———— papillata, *Park.* Coral Rag, Midland and Southern England, *Conyb.*
?6. ———— diadema, *Park.* Coral Rag, Midland and Southern England, *Conyb.*
?7. ———— subangularis, *Park.* Inferior Oolite, Midland and Southern England, *Conyb.*
8. ———— ornata,   . Bradford Clay, North of France, *Bobl.*
9. ———— globata, *Schlot.* Coral Rag, North of France, *Bobl.*
10. ———— maxima, *Munst.* Baireuth, *Munst.*
11. ———— Blumenbachii, *Munst.* Thurnau, Muggendorf, Pretzfeld and Theta, *Goldf.*
12. ———— nobilis, *Munst.* Baireuth, *Munst.*
13. ———— elegans, *Munst.* Baireuth, *Munst.;* Kelloway Rock, Haute Saone, *Thir.*
14. ———— marginata, *Goldf.* Regensburg, Heidenheim, *Goldf.*
15. ———— coronata, *Goldf.* Streitberg, Thurnau, Staffelstein, Heidenheim, Randen, *Goldf.*
16. ———— propinqua, *Munst.* Streitberg, *Munst.*
17. ———— glandifera, *Goldf.* Altdorf, Bavaria; Wirtemberg; Randen, *Goldf.*
18. ———— Schmidelii, *Munst.* Dischingen, Switzerland, *Munst.*
19. ———— subangularis, *Goldf.* Thurnau, Muggendorf, *Goldf.*
20. ———— variolaris, *Al. Brong.* Streitberg, Regensburg, Heidenheim, *Goldf.*
————, species not determined. Inf. Oolite, Yorkshire, *Phil.;* Lias, Lyme Regis, *De la B.;* Cornbrash, Bradford Clay, Great Oolite, Inferior Oolite and Lias, Midland and Southern England, *Conyb.;* Coral Rag,

Forest Marble, Normandy, *De Cau.;* Forest Marble, Great Oolite, Wiltshire, *Lons.*

Cidaris, spines of. Great Oolite and Lias, Yorkshire, *Phil.;*
Lias, Mid. and South of England, *Conyb.;* Oolite
beds, Lower System, South of France, *Bobl.;*
Coral Rag, Normandy, *Desn.;* Coral Rag, Haute
Saone, *Thir.*

1. Echinus germinans, *Phil.* Coral Oolite, Calcareous Grit, and
Great Oolite, Yorkshire, *Phil.*

2. ———— lineatus, *Goldf.* Regensburg, Bâle, *Goldf.*

3. ———— excavatus, *Leske.* Regensburg, *Goldf.*

4. ———— nodulosus, *Munst.* Baireuth, *Munst.*

5. ———— hieroglyphicus, *Goldf.* Regensburg: Thurnau, *Goldf.*

6. ———— sulcatus, *Goldf.* Thurnau; Streitberg; Muggendorf;
Heidenheim, *Goldf.*

———, species not determined. Coral Rag, North of France,
Bobl.

1. Galerites depressus, *Lam.* Wirtemberg; Bavaria, *Goldf.;*
Coral Oolite, Calcareous Grit, Cornbrash, York-
shire, *Phil.;* Oxford Clay, Normandy, *Desn.;*
Oxford Clay, Haute Saone, *Thir.*

2. ———— speciosus, *Munst.* Heidenheim, Wirtemberg, *Munst.*

3. ———— Patella,   . Oxford Clay, Normandy, *Desn.*

1. Clypeaster pentagonalis, *Phil.* Calcareous Grit, Yorks. *Phil.*

———, species not determined:—Coral Rag, Normandy,
De Cau. Kimmeridge clay, Haute Saone, *Thir.*

1. Nucleolites scutata,   . Oxford Clay, Normandy, *Desn.*
Oxford Clay, Haute Saone, *Thir.*

2. ———— columbaria, Cornbrash, Forest Marble, North of
France, *Bobl.*

3. ———— granulosus, *Munst.* Amberg; Streitberg; Würgau,
Munst.

4. ———— semiglobus, *Munst.* Pappenheim; Monheim; Bavaria, *Munst.*

5. ———— excentricus, *Munst.* Kehlheim, Bavaria, *Munst.*

6. ———— canaliculatus, *Munst.* Blaubeuren, Wirtemberg,
Munst.

———, species not determined:—Oxford Clay, North of
France, *Bobl.*

1. Ananchytes bicordata. Oxford Clay, Normandy, *Desn.*

1. Spatangus ovalis, *Park.* Coral Oolite, Calcareous Grit, Kelloway Rock, Yorkshire, *Phil.*

2. ———— intermedius, *Munst.* Blaubeuren, Wirtemberg,
Munst.

3. ———— carinatus, *Goldf.* Baireuth, Wirtemberg, *Goldf.*

4. ———— capistratus, *Goldf.* Baireuth, *Goldf.;* Oxford Clay,
Haute Saone, *Thir.*

Spatangus, species not determined :—Cornbrash, Forest Marble, North of France, *Bobl.*
1. Clypeus sinuatus, *Park.* Coral Oolite, Yorkshire, *Phil.*;
Coral Rag, Cornbrash, Great Oolite, Inferior Oolite, Mid. and Southern England, *Conyb.*; Forest
Marble, Normandy, *De Cau.*
2. ———— emarginatus, *Phil.* Coralline Oolite, Yorkshire, *Phil.*
3. ———— clunicularis, *Smith.* Coral Oolite, Cornbrash, Yorkshire, *Phil.*; Coral Rag, Cornbrash, Great Oolite,
Inferior Oolite, Midland and Southern England,
Conyb.; Forest Marble, Normandy, *De Cau.*;
Coral Rag, Weymouth, *Sedg.*
4. ———— dimidiatus, *Phil.* Coral Oolite, Yorkshire, *Phil.*
5. ———— semisulcatus, *Phil.* Coralline Oolite, Yorkshire, *Phil.*
6. ———— orbicularis, *Phil.* Cornbrash, Yorkshire, *Phil.*
————, species not determined;—Cornbrash, Great Oolite,
Wiltshire, *Lons.*
Echinites, genera not determined. Inferior Oolite, Normandy, *De Cau.*
————, spines of. Coral Rag, Burgundy, *Beaum.*; Coral Rag,
North of France, *Bobl.*; Forest Marble, Mamers,
Desn.; Mauriac beds, South of France, *Dufr.*
1. Ophiura Milleri, *Phil.* Lias, Yorkshire, *Phil.*; Inferior
Oolite sands, Bridport, *De la B.*
1. Encrinites echinatus, *Schlot.* Oxford Clay, Haute Saone,
Thir.
2. ———— mespiliformis, *Schlot.* Kimmeridge Clay, Haute
Saone, *Thir.*
1. Eugeniacrinites caryophyllatus, *Goldf.* Baireuth; Wirtemberg;
Switzerland, *Goldf.*
2. ———— mutans, *Goldf.* Streitberg; Muggendorf, *Goldf.*
1. Apiocrinites rotundus, *Miller.* Forest Marble, Normandy,
DeCau.; Bradford Clay, Great Oolite, Mid. and
S. England, *Conyb.*; Forest Marble, *Buckl.*;
Great Oolite, Alsace, *Al. Brong.*; Forest
Marble, Normandy, *De Cau.*; Forest Marble,
Wiltshire; Great Oolite, Somerset, *Lons.*
2. ———— Prattii, *Gray.* Great Oolite, Somerset, *Lons.*
1. Pentacrinites Caput Medusæ, *Miller.* Cornbrash, Coral Oolite,
and Lias, Yorks., *Phil.*; Inf. Oolite, and Lias,
Mid. and S. England, *Conyb.*; Lias, Alsace,
Gundershofen, Figeac, *Al. Brong.*
2. ———— subangularis, *Miller.* Inferior Oolite and Lias,
Middle and South of England, *Conyb.*
3. ———— briareus, *Miller.* Lias, Midland and Southern
England, *Conyb.*; Lias, Yorkshire, *Phil.*

4. Pentacrinites basaltiformis, *Miller.* Lias, Midland and South-
      ern England, *Conyb.;* Lias, Alsace, *Voltz.*
5. —————— tuberculatus, *Miller.* Lias, Midland and South-
      ern England, *Conyb.;* Lias, Alsace, *Voltz.*
6. —————— subteres, *Goldf.* (Var.) Oxford Clay, Haute
      Saone, *Thir.*
7. —————— Jurensis, *Munst.* Coral Rag, Haute Saone, *Thir.*
   ——————, species not determined. Forest Marble, Nor-
      mandy, *De Cau.* Bradford Clay, North of
      France, *Bobl.;* Cornbrash, Forest Marble,
      Great Oolite, Mid. and South of England,
      Conyb.; Inferior Oolite, Wotton-under-Edge,
      Forest Marble, Great Oolite, Somerset, *Lons.*
   Crinoidea, genera not determined. Coral Rag and Inferior
      Oolite, North of France, *Bobl.* Mauriac Beds,
      South of France, *Dufr.;* Calcareous Grit, York-
      shire, *Phil.;* Coral Rag, Haute Saone, *Thir.*

ANNULATA.

1. Serpula squamosa, *Bean.* Coral Oolite, Yorkshire, *Phil.*
2. —————— lacerata, *Phil.* Calcareous Grit, and Great Oolite,
      Yorkshire, *Phil.*
3. —————— intestinalis, *Phil.* Oxford Clay, and Cornbrash,
      Yorkshire, *Phil.*
4. —————— deplexa, *Bean.* Inferior Oolite, Yorkshire, *Phil.*
5. —————— capitata, *Phil.* Lias, Yorkshire, *Phil.*
6. —————— triquetra. Inf. Oolite, Mid. and S. England, *Conyb.*
7. —————— quadrangularis. Oxford Clay, Normandy, *Desn.*
8. —————— sulcata, *Sow.* Calcareous Grit, Oxford, *Sow.*
9. —————— tricarinata, *Sow.* Calcareous Grit, Oxford; Coral
      Rag, Steeple Ashton, Wilts, *Sow.;* Oxford Clay,
      Haute Saone, *Thir.*
10. —————— triangulata, *Sow.* Bradford Clay or Great Oolite,
      Bradford, *Sow.*
11. —————— runcinata, *Sow.* Coral Rag, Oxford, *Sow.*
   ——————, species undetermined. Coral Rag, Oxford Clay,
      Cornbrash, Forest Marble, Bradford Clay, Great
      Oolite, Mid. and South of England, *Conyb.;* Ox-
      ford Clay, Inferior Oolite, Haute Saone, *Thir.;*
      Cornbrash, Forest Marble, Bradford Clay, Great
      Oolite, Fuller's Earth, Wiltshire, *Lons.*

CONCHIFERA.

1. Spirifer Walcotii, *Sow.* Lias, Yorkshire, *Phil.;* Lias, Bath,
      Lyme Regis, *De la B.;* Lias, Normandy, *De Cau.;*
      Lias, South of France, *Dufr.;* Lias, Western
      Islands, Scotland, *Murch.*

1. Delthyris* verrucosa, *Von Buch.* Lias, Bahlingen, Wirtemberg, *Von Buch.*
2. ———— rostrata, *Schlot.* Lias, Wirtemberg, *Von Buch.*
1. Terebratula intermedia, *Sow.* Coral Oolite, and Great Oolite, Yorks., *Phil.;* Cornbrash, Mid. and S. England; Inferior Oolite, Dundry, *Conyb.*
2. ———— globata, *Sow.* Coral. Oolite? Great Oolite, Yorkshire, *Phil.;* Forest Marble, Normandy, *De Cau.;* Oolite, Env. of Bath, *Sow.;* Fuller's Earth, Env. of Bath, Great Oolite, Haute Saone, *Thir.*
3. ———— ornithocephala, *Sow.* Coralline Oolite, and Kelloway Rock, Yorkshire, *Phil.;* Kelloway Rock, Cornbrash, Lias? Mid. and South of England; Inferior Oolite, Dundry, *Conyb.;* Oxford Clay and Lias, Normandy, *De Cau.;* Inferior Oolite, Uzer, South of France, *Dufr.;* Kimmeridge Clay, Great Oolite, Haute Saone, *Thir.;* Inferior Oolite, Wiltshire, *Lons.;* Soleure, Buxwiller, *Hœn.*
4. ———— ovata, *Sow.* Coralline Oolite? Yorkshire, *Phil.;* Inferior Oolite, Mid. and South of England, *Conyb.;* Coral Rag, Haute Saone, *Thir.*
5. ———— obsoleta, *Sow.* Coralline Oolite? Inferior Oolite, Yorkshire, *Phil.;* Cornbrash, Bradford Clay, Great Oolite, and Inferior Oolite, Mid. and South of England, *Conyb.;* Great Oolite, Normandy, *De Cau.;* Lias and Inferior Oolite, South of France, *Dufr.;* Forest Marble, Wiltshire, *Lons.*
6. ———— socialis, *Phil.* Calcareous Grit, and Kelloway Rock, Yorkshire, *Phil.*
7. ———— ovoides, *Sow.* Cornbrash? Yorkshire, *Phil.;* Inferior Oolite, Normandy, *De Cau.;* Rubbly Limestone, &c., Braambury Hill, Brora, *Murch.*
8. ———— digona, *Sow.* Cornbrash, Yorks., *Phil.;* Cornbrash and Bradford Clay, Mid. and S. England; Inferior Oolite, Dundry, *Conyb.;* Forest Marble, Normandy, *De Cau.;* Bradford Clay and Coral Rag? North of France, *Bobl.;* Forest Marble, Bradford Clay, Great Oolite, Wilts, *Lons.*
9. ———— spinosa, *Townsend and Smith.* Great Oolite, Yorkshire, *Phil.*
10. ———— trilineata, *Y. & B.* Inferior Oolite and Lias, Yorkshire, *Phil.*
11. ———— bidens, *Phil.* Inferior Oolite and Lias, Yorks., *Phil.*

---

* The genus *Delthyris*, Dalman, is the same with the genus *Spirifer*, Sowerby; both names have been retained above for the purpose of more easy reference.

12. Terebratula punctata, *Sow.* Lias, Yorkshire, *Phil.* ; Inferior Oolite, Mid. and South of England, *Conyb.* ; Lias, Western Islands, Scotland, *Murch.*
13. ———— resupinata, *Sow.* Lias, Yorkshire, *Phil.* ; Inferior Oolite, Mid. and South of England, *Conyb.*
14. ———— acuta, *Sow.* Lias, Yorkshire, *Phil.* ; Inferior Oolite and Lias, Mid. and South of England, *Conyb.* ; Lias, Normandy, *DeCau.;* Fuller's Earth, Frome, *Lons.* ; Lias, Wirtemberg, *Von Buch.*
15. ———— triplicata, *Phil.* Lias, Yorkshire, *Phil.* ; Lias, Wirtemberg, *Von Buch.*
16. ———— tetraëdra, *Sow.* Lias, Yorkshire, *Phil.* ; Inferior Oolite, Mid. and South of England, *Conyb.* ; Lias, South of France, *Dufr.* ; Forest Marble? Mauriac, South of France, *Dufr.* ; Lias and Micaceous Sandstone, Western Islands, Scotland, *Murch.* ; Echterdingen, Buxweiller, *Hœn.*
17. ———— subrotunda, *Sow.* Cornbrash and Inferior Oolite, Mid. and South of England, *Conyb.* ; Cornbrash and Forest Marble, North of France, *Bobl.* ; Forest Marble? Mauriac, South of France, *Dufr.*; Inferior Oolite, Env. of Bath, *Lons.*
18. ———— obovata, *Sow.* Cornbrash, Mid. and South of England, *Conyb.* ; Inferior Oolite, Env. of Bath, *Lons.*
19. ———— reticulata, *Smith.* Bradford Clay, Mid. and South of England, *Conyb.* ; Forest Marble, Normandy, *De Cau.*
20. ———— carnea, *Sow.* Inferior Oolite, Dundry, *Conyb.* ; Coral Rag, Great Oolite, Haute Saone, *Thir.*
21. ———— semigloba, *Sow.* Inferior Oolite, Dundry, *Conyb.*
22. ———— media, *Sow.* Inferior Oolite, Dundry, *Conyb.* ; Inferior Oolite, Great Oolite, and Bradford Clay, North of France, *Bobl.* ; Dunrobin Oolite, Scotland, *Murch.* ; Fuller's Earth, Inferior Oolite, Env. of Bath, *Lons.*
23. ———— crumena, *Sow.* Inf. Oolite and Lias? Mid. and S. England, *Conyb.* ; Echterdingen, *Hœn.*
24. ———— lateralis, *Sow.* Fuller's Earth, Mid. and South of England, *Conyb.*
25. ———— concinna, *Sow.* Fuller's Earth, Mid. and South of England, *Conyb.*; Inferior Oolite, Normandy, *De C.;* Forest Marble? Mauriac, South of France, *Dufr.* ; Fuller's Earth, Frome; Inferior Oolite, Env. of Bath, *Lons.*
26. ———— biplicata, *Sow.* Oxford Clay, Forest Marble, Great Oolite, and Inferior Oolite, Normandy, *De C.* ; Soleure, *Hœn.*

27. Terebratula tetrandra, . Forest Marble, Normandy, *DeC.*
28. ———— coarctata, *Sow.* Forest Marble, Normandy, *De C.;* Bradford Clay, North of France, *Bobl.;* Bradford Clay, Bath, *Loscombe.*
29. ———— plicatella, *Sow.* Forest Marble, Normandy, *De C.*
30. ———— serrata, *Sow.* Forest Marble, Normandy, *De C.;* Lias, Lyme Regis, *De la B.*
31. ———— truncata, *Sow.* Forest Marble, Normandy, *De C.*
32. ———— lata, *Sow.* Inferior Oolite, Normandy, *De C.*
33. ———— dimidiata, *Sow.* Inferior Oolite, Normandy, *De C.*
34. ———— bullata, *Sow.* Inferior Oolite, Normandy, *De C.;* Inferior Oolite, Bridport, Dorset, *Sow.;* Cornbrash, Wiltshire; Fuller's Earth, Env. of Bath, *Lons.*
35. ———— sphæroïdalis, *Sow.* Inferior Oolite, Normandy, *De C.;* Inferior Oolite, Dundry, *Braikenridge.*
36. ———— emarginata, *Sow.* Inferior Oolite, Normandy, *De C.;* Inferior Oolite, Env. of Bath, *Lons.*
37. ———— quadrifida, . Lias, Normandy, *De C.*
38. ———— numismalis, *Lam.* Lias, Norm., *De C.;* Lias, Bahlingen; Gonningen, *Von Buch.*
39. ———— perovalis, *Sow.* Inf. Oolite, Dundry, *Braikenridge;* Forest Marb.? Mauriac, and Kim. Clay, Cahors, S. of Fr.; Rochelle Limestone, *Dufr.*
40. ———— maxillata, *Sow.* Inf. Oolite, Env. of Bath, *Sow.;* Forest Marb., Wilts., *Lons.*
41. ———— flabellula, *Sow.* Great Oolite, Ancliff, near Bradford, Wilts., *Cookson.*
42. ———— furcata, *Sow.* Great Oolite, Ancliff, *Cookson.*
43. ———— orbicularis, *Sow.* Lias, Bath, *Sow.*
44. ———— hemisphærica, *Sow.* Great Oolite, Ancliff, *Cookson.*
45. ———— inconstans, *Sow.* Shelly Limestone and Calc. Grit, Portgower, &c. N. of Scotl., and Shell Limestone, Beal, Isle of Skye, *Murch.;* Coral Rag, Weymouth, *Sedg.*
46. ———— perovalis, *Sow.* Oxford Clay, Kell. Rock, Haute Saone, *Thir.*
47. ———— bisuffarcinata, *Schlot.* Thurnau, *Hœn.*
48. ———— loricata, *Schlot.* Baireuth, *Hœn.*
49. ———— pectunculus, *Schlot.* Thurnau, *Hœn.*
50. ———— rostrata, *Schlot.* Soleure, *Hœn.*
51. ———— spinosa, *Lam.* Baireuth, *Hœn.*
52. ———— substriata, *Schlot.* Thurnau, *Hœn.*
53. ———— vulgaris, *Schlot.* Porta Westphalica, *Hœn.*
54. ———— Defrancii, *Al. Brong.* Amberg, *Hœn.*
55. ———— Hœninghausii, *Blain.* Baireuth, *Hœn.*
56. ———— sexangula, *Defr.* Muggendorf, *Hœn.*

Q

57. Terebratula rimosa, *Von Buch.* Lias, Bahlingen, Wurtemberg,
       Von Buch.
1. Orbicula reflexa, *Sow.* Lias, Yorks., *Phil.*
2. ———? radiata, *Phil.* Coral. Oolite, Yorks., *Phil.*
3. ——— granulata, *Sow.* Great Oolite, Ancliff, Wilts.,
       Cookson.
   ———, sp. not determined. Inferior Oolite, Yorks., *Phil.*
1. Lingula Beanii, *Phil.* Inferior Oolite, Yorks., *Phil.*
1. Ostrea gregarea, *Sow.* Coral Rag, Yorks., Wilts., &c. ;- Calc.
       Grit and Great Oolite? Yorks., *Phil.;* Coral Rag,
       Mid. and S. of Eng.; Inf. Oolite, Dundry, *Conyb.;*
       Coral Rag and Oxford Clay, Norm., *De C.;* Ox-
       ford Clay and Coral Rag, N. of Fr., *Bobl.;* Kim.
       Clay, Havre, *Phil.;* Coral Rag, Weymouth, *Sedg.*
2. ——— solitaria, *Sow.* Coral Rag and Inf. Oolite, Yorks.,
       Oxon, &c. *Phil.;* Kim. Clay, Haute Saone, *Thir.;*
       Coral Rag, Weymouth, *Sedg.*
3. ——— duriuscula, *Bean.* Coralline Oolite, Yorks., *Phil.*
4. ——— inæqualis, *Phil.* Oxford Clay, Yorks., *Phil.*
5. ——— undosa, *Bean.* Kell. Rock, Yorks., *Phil.*
6. ——— archetypa, *Phil.* Kell. Rock, Yorks., *Phil.*
7. ——— Marshii, *Sow.* Kell. Rock, Cornb., and Great Oolite,
       Yorks., *Phil.;* Cornb. and Fuller's E., Mid. and S.
       of Eng., *Conyb.;* Oxford Clay, Forest Marb., and
       Inf. Oolite, Norm., *De C.*
8. ——— sulcifera, *Phil.* Great Oolite, Yorks., *Phil.;* Inf.
       Oolite, Haute Saone, *Thir.;* Cornb., Wilts., *Lons:;*
       Coral Rag, Weymouth, *Sedg.*
9. ——— deltoidea, *Sow. and Smith.* Kim. Clay, Yorks., *Phil.;*
       Oxford Clay, N. of. Fr. *Bobl.;* Kim. Clay, S. and
       Mid. of England, *Conyb.;* Shell Limestone and
       Calc. Grit? Portgower, &c. Scotl., *Murch.;* Kim.
       Clay, Havre, *Phil.;* Sandst., Limest., and Shale,
       Inverbrora, Scotl., *Murch.;* upper part of Coral
       Rag, Weymouth, *Sedg.*
10. ——— expansa, *Sow.* Portland Stone, *Conyb.*
11. ——— Crista Galli, *Smith.* Coral Rag, Forest Marb., Brad.
       Clay, and Great Oolite, Mid. and S. of Eng., *Conyb.;*
       Great Oolite, Norm., *De C.*
12. ——— palmetta, *Sow.* Oxford Clay, Mid. and S. of Eng.,
       Conyb.; Oxf. Clay and For. Marb., Norm., *De C.*
13. ——— acuminata, *Sow.* Bradford Clay and Inf. Oolite,
       Mid. and S. of Eng., *Conyb.;* Great Oolite and
       Brad. Clay, N. of Fr., *Bobl.;* Great Oolite, Haute
       Saone, *Thir.;* Fuller's E., Inf. Oolite, Env. of
       Bath, *Lons.*
14. ——— rugosa, *Sow.* Inf. Oolite, Mid. and S. of Eng., *Conyb.*

15. Ostrea minima, *Desl.* Coral Rag and Oxford Clay, Norm., *De C.*
16. —— plicatilis. Oxford Clay, Norm., *De C.*
17. —— carinata, *Lam.* Oxford Clay, Norm., *De C.*
18. —— costata, *Sow.* Brad. Clay, N. of Fr., *Bobl.;* Great Oolite, Ancliff, near Bath, *Cookson.*
19. —— pectinata. Oxford Clay, N. of Fr., *Bobl.*
20. —— pennaria. Oxford Clay, N. of Fr., *Bobl.*
21. —— flabelloides, *Lam.* Oxford Clay, N. of Fr., *Bobl.*
22. —— læviuscula, *Sow.* Lias, Eng., *Sow.*
23. —— obscura, *Sow.* Great Oolite, Ancliff, Wilts., *Cookson.*
24. —— Meadii, *Sow.* Inf. Oolite, Env. of Bath, *Lons.*
    ——, species not determined. Many in the For. Marb. and Brad. Clay, Wilts., *Lons.*
1. Exogyra digitata? *Sow.* Kell. Rock., Mid. and S. Eng., *Conyb.*
    ——, species not determined. Kim. Clay, Haute Saone, *Thir.* For. Marb.? Wilts., *Lons.*
1. Gryphæa chamæformis, *Phil.* Calc. Grit, Yorks.; and Oolite, Sutherland, *Phil.*
2. —— bullata, *Sow.* Coral. Oolite? Calc. Grit? *Phil.;* Oxford Clay, Lincolnshire, *Sow.;* Oolite of Braambury Hill, Brora, *Murch.*
3. —— inhærens, *Phil.* Calc. Grit, Yorks., *Phil.*
4. —— dilatata, *Sow.* Kell. Rock, Yorks., *Phil.;* Oxford Clay, Mid. and S. of Eng., *Conyb.;* Oxford Clay and Lias, Norm., *De C.;* Oxford Clay, N. of Fr., *Bobl.;* Oxford Clay, Burgundy, *Beaum.;* Great Arenaceous Formation, Western Islands, Scotl. *Murch.;* Oxford Clay, Haute Saone, *Thir.;* Lower part of Coral Rag, Weymouth, *Sedg.;* Oxford Clay, Beggingen, Schafhausen, *Von Buch.*
5. —— incurva, *Sow.* Lias, Yorks., *Phil.;* Lias, Mid. and S. Eng. *Conyb.;* Lias, Norm., *De C.;* Lias and Inf. Oolite, N. of Fr., *Bobl.;* Lias, S. of Fr., *Dufr.;* Lias, Metz, Salins, Amberg, *Al. Brong.;* Lias, Western Islands, Scotl.; Lias, Ross and Cromarty, Scotl., *Murch.;* Göppingen, Bahlingen, *Hœn.*
6. —— nana, *Sow.* Kim. Clay, Oxford, *Sow.;* Shale and Grit, Dunrobin Reefs, Scotl., *Murch.;* Lias and Oxford Clay? N. of Fr., *Bobl.*
7. —— Maccullochii, *Sow.;* Lias, Western Islands, Scotl., *Murch.;* Lias, Yorks., *Phil.;* Oxford Clay, Norm., *De C.;* Lias, S. of Fr., *Dufr.;* Lias, Env. of Bath, *Lons.*

Q 2

8. Gryphæa depressa, *Phil.* Lias, Yorks., *Phil.*
9. ———— obliquata, *Sow.* Lias, Mid. and S. Eng., *Conyb.;*
　　　Lias, S. of Fr., *Dufr.;* Lias, Western Islands,
　　　Scotl., *Murch.;* Lias, Env. of Bath, *Lons.*
10. ———— cymbium, *Lam.* Inf. Oolite, N. of Fr., *Bobl.;*
　　　Lias, S. of France; Inf. Oolite, Villefranche,
　　　S. of France, *Dufr.;* Inf. Oolite, Haute Saone,
　　　Thir.; Lias, Bahlingen, *Von Buch.*
11. ———— lituola, *Lam.* Brad. Clay, Cornb., and For. Marb.,
　　　N. of Fr., *Bobl.*
12. ———— gigantea, *Sow.* Lias, S. of Fr., *Dufr.;* Lias, Ross
　　　and Cromarty, Scotl.; Great Arenaceous Forma-
　　　tion, Western Islands, Scotl., *Murch.*
13. ———— minuta, *Sow.* Great Oolite, Ancliff, Wilts.,
　　　Cookson.
14. ———— virgula, *Defrance.* Kim. Clay, Havre, *Al. Brong.;*
　　　Kim. Clay, Burgundy, *Beaum.;* Kim. Clay, S.
　　　of Fr., *Dufr.;* Kim. Clay, Weymouth, *Buckl.*
　　　& De la B.; Kim. Clay, Haute Saone, *Thir.*
1. Plicatula spinosa, *Sow.* Lias, Yorks., *Phil.*; Lias, Mid. and
　　　S. Eng., *Conyb.;* Lias, Norm., *De C.;* Inf. Oo-
　　　lite, N. of Fr., *Bobl.;* Great Arenaceous Forma-
　　　tion, Western Islands, Scotl., *Murch.;* Lias, Gun-
　　　dershoffen, *Voltz.*
1. Pecten abjectus, *Phil.* Coral Rag, Yorks. and Oxon; Calc.
　　　Grit, Great Oolite, and Inf. Oolite, Yorks., *Phil.*
2. ———— inæquicostatus, *Phil.* Coralline Oolite, Yorks.; Calc.
　　　Grit, Oxon, *Phil.*
3. ———— cancellatus, *Bean.* Coralline Oolite, Yorks.; Oolite,
　　　Sutherland? *Phil.*
4. ———— demissus, *Phil.* Coralline Oolite, Kell. Rock, Corn-
　　　brash, and Great Oolite, Yorks., *Phil.*
5. ———— lens, *Sow.* Coralline Oolite, Kell. Rock, Great Oolite,
　　　Inf. Oolite, and Lias, Yorks., *Phil.;* Coral Rag,
　　　Mid. and S. Eng.; Inf. Oolite, Dundry, *Conyb.;*
　　　Coral Rag and Oxford Clay, Norm., *De C.;*
　　　Cornb. and For. Marb., N. of Fr., *Bobl.;* Inf.
　　　Oolite, Alsace, and Stranen near Luxembourg, *Al.*
　　　Brong.; Sandst., Limest., and Shale, Inverbrora,
　　　Scotl., *Murch.;* Inf. Oolite, Haute Saone, *Thir.*
6. ———— viminalis, *Sow.* Coral Rag, Yorks., Oxon, and Wilts.,
　　　Phil.
7. ———— vagans, *Sow.* Coral Rag, Yorks. and Oxon; Calc.
　　　Grit, Yorks., *Phil.;* For. Marb., Norm., *De C.;*
　　　Sandst. and Rubbly Limest., Braambury Hill,
　　　Brora, *Murch.;* For. Marb., Wilts., *Lons.*
8. ———— fibrosus, *Sow.* Kell. Rock and Cornbrash, Yorks.,

Phil.; Coral Rag, Kell. Rock, Cornb., For. Marb.,
Brad. Clay, and Inf. Oolite, Mid. and S. Eng.,
Conyb.; Coral Rag, Norm.? *De C.;* Cornb. and
For. Marb., N. of Fr., *Bobl.;* For. Marb.? Mauriac,
S. of Fr., *Dufr.;* Rubbly Limestone, &c., Braam-
bury Hill, Brora, *Murch.;* For. Marb., Wilts., *Lons.;*
Soleure, *Hœn.*

9. Pecten virguliferus, *Phil.;* Inferior Oolite, Yorks., *Phil.*
10. ———— sublævis, *Y. & B.* Lias, Yorks., *Phil.*
11. ———— æquivalvis, *Sow.* Lias, Yorks., *Phil.;* Inf. Oolite, Mid.
and S. Eng., *Conyb.;* Lias, Norm., *De C.;* Lias,
S. of Fr., *Dufr.;* Lias, Western Islands, Scotl.,
Murch.; Inf. Oolite, Env. of Bath, *Lons.*
12. ———— lamellosus, *Sow.* Portland Stone, *Conyb.*
13. ———— arcuatus, *Sow.* Coral Rag, Mid. and S. Eng., *Conyb.;*
Portland Beds, Kim. Clay, Haute Saone, *Thir.*
14. ———— similis, *Sow.* Coral Rag, Mid. and S. Eng., *Conyb.;*
Coral Rag, Norm.? *De C.;* Great Oolite, Haute
Saone, *Thir.*
15. ———— laminatus, *Sow.* Cornb., Mid. and S. Eng., *Conyb.*
16. ———— barbatus, *Sow.* Inf. Oolite, Dundry, *Conyb.;* Lias,
Norm., *De C.;* Inf. Oolite, Lias, Env. of Bath,
Lons.
17. ———— vimineus, *Sow.* Oxford Clay, For. Marb., and Inf.
Oolite, Norm., *De C.;* Forest Marble, Malton, *Sow.;*
Rubbly Limestone, &c., Braambury Hill, Brora,
Murch.; Coral Rag, Haute Saone, *Thir.*
18. ———— corneus, *Sow.* For. Marb., Great Oolite, and Inf.
Oolite, Norm., *De C.*
19. ———— obscurus, *Sow.* For. Marb.? Mauriac, S. of Fr.,
Dufr.
20. ———— annulatus, *Sow.* Cornb., Felmersham, *Marsh.*
21. ———— concinnus, . Namen, near Minden, *Hœn.*
22. ———— marginatus, . Wasseralfingen, *Hœn.*
1. Plagiostoma læviusculum, *Sow.* Coralline Oolite, Yorks.;
Coral Rag and Calcareous Grit, Oxon, *Phil.;*
Coral Rag, Marthon, S. of Fr., *Dufr.*
2. ———————— rigidum, *Sow.* Coralline Oolite, Yorks.; Coral
Rag, Oxon, *Phil.;* Inf. Oolite, Dundry, *Co-
nyb.;* Coral Rag, N. of Fr., *Bobl.;* Coral Rag,
Haute Saone, *Thir.*
3. ———————— rusticum, *Sow.* Coralline Oolite, Yorks.; Calc.
Grit., Oxon, *Phil.*
4. ———————— duplicatum, *Sow.* Coralline Oolite, Oxford Clay,
and Kell. Rock, Yorks., *Phil.;* Inf. Oolite,
Norm., *De C.;* Dunrobin Oolite, Scotl.,
Murch.; Lias, Env. of Bath, *Lons.*

5. Plagiostoma rigidulum, *Phil.* Cornbrash, Yorks., *Phil.*
6. ————— interstinctum, *Phil.* Cornb. and Great Oolite, Yorks., *Phil.*
7. ————— cardiiforme, *Sow.* Great Oolite, Yorks., *Phil.;* Cornb. and For. Marb., N. of France, *Bobl.*
8. ————— giganteum, *Sow.* Inf. Oolite and Lias, Yorks., *Phil.;* Inf. Oolite, Dundry? Lias, Mid. and S. Eng., *Conyb.;* Lias, Norm., *De C.;* Lias, N. of Fr., *Bobl.;* Lias, Western Islands, Scotl., *Murch.;* Inf. Oolite, Haute Saone, *Thir.;* Bahlingen, *Hœn.*
9. ————— obscurum, *Sow.* Kell. Rock, Mid and S. Eng., *Conyb.*
10. ————— pectinoïdes, *Sow.* Lias, Yorks., *Phil.;* Shale and Grit, Reefs at Dunrobin, Scotl., *Murch.*
11. ————— punctatum, *Sow.* Inf. Oolite, Dundry. Lias, Mid. and S. England, *Conyb.;* For. Marb. and Inf. Oolite, Norm., *De C.;* Lias, N. of France, *Bobl.;* Lias, S. of France, *Dufr.;* Lias, Western Islands, Scotl., *Murch.*
12. ————— sulcatum. Lias, S. of France, *Dufr.*
13. ————— ovale, *Sow.* For. Marb.? Mauriac, S. of France, *Dufr.*
14. ————— Hermanni, *Voltz.* Lias, Alsace, *Voltz;* Lias, Env. of Bath, *Lons.;* Lias, Lyme Regis, *De La B.*
15. ————— obliquatum, *Sow.* Sandstone and Limestone, Braambury Hill, Brora. Sandst., Limest., and Shale, Inverbrora, Scotl., *Murch.*
16. ————— acuticosta, *Sow.* Sandst., Limest., and Shale, Inverbrora, Scotl., *Murch.*
17. ————— concentricum, *Sow.* Lias, Ross and Cromarty, Scotl., *Murch.*
————— , sp. not determined. Bradford Clay and Great Oolite, Mid. and S. Eng., *Conyb.;* Lias, Gundershofen, *Voltz.*
1. Posidonia Bronni, *Goldf.* Lias, Ubstadt, near Bruchsal, *Hœn.*
1. Lima rudis, *Sow.* Coralline Oolite, Calc. Grit, Kell. Rock, and Great Oolite, Yorks., *Phil.;* Coral Rag, Mid. and S. Eng., *Conyb.;* Coral Rag, N. of Fr., *Bobl.;* Rubbly Limestone, &c., Braambury Hill, Brora, *Murch.*
2. —— proboscidea, *Sow.* Inf. Oolite? Yorks., *Phil.;* Inf. Oolite, Dundry, *Conyb.;* Oxford Clay, For. Marb., and Inf. Oolite, Norm., *De C.;* Inf. Oolite, Haute Saone, *Thir.;* Soleure, Bâle, *Hœn.;* Coral Rag, Weymouth, *Sedg.*
3. —— gibbosa, *Sow.* Cornb. and Inf. Oolite, Mid. and S. Eng., *Conyb.;* Great Oolite and Inf. Oolite, Norm., *De C.*

4. Lima antiqua, *Sow.* Lias, Mid. and S. Eng., *Conyb.*; Lias, S. of France, *Dufr.*; Inf. Oolite, Haute Saone, *Thir.*
———, sp. not determined. Great Oolite, Wilts., *Lons.*
1. Avicula expansa, *Phil.* Coralline Oolite, Oxford Clay? Kell. Rock and Great Oolite, Yorks., *Phil.*
2. ——— ovalis, *Phil.* Coralline Oolite and Calc. Grit, Yorks., *Phil.*
3. ——— elegantissima, *Bean.* Coralline Oolite, Yorks., *Phil.*
4. ——— tonsipluma, *Y. & B.* Coralline Oolite, Yorks., *Phil.*
5. ——— Braamburiensis, *Sow.* Sandstone, Braambury Hill, Brora, *Murch.*; Kell. Rock, Great Oolite, and Inf. Oolite, Yorks., *Phil.*
6. ——— inæquivalvis, *Sow.* Inf. Oolite and Lias, Yorks., *Phil.*; Great Oolite and Inf. Oolite, Norm., *De C.*; Lias, S. of Fr., *Dufr.*; Great Arenaceous Formation, Western Islands; and Shell Limest. and Grit, Portgower, Scotland, *Murch.*; Lias, Lyme Regis, *De la B.*; Bahlinghen, *Hœn.*; Lias, Gandershofen, *Voltz*; Full. E., Inf. Oolite, and Lias, Env. of Bath, *Lons.*
7. ——— echinata, *Sow.* Lias? Yorks., *Phil.*; Cornb., Mid. and S. Eng., *Conyb.*; For. Marb., Norm., *De C.*; Brad. Clay, Cornb., and For. Marb., N. of Fr., *Bobl.*; Great Oolite, Haute Saone, *Thir.*; Full. E., Env. of Bath, *Lons.*
8. ——— cygnipes, *Y. & B.* Lias, Yorks., *Phil.*; Lias, Western Islands, Scotl., *Murch.*
9. ——— costata, *Sow.* Cornb. and Brad. Clay, Mid. and S. Eng., Inf. Oolite, Dundry, *Conyb.*; For. Marb., Norm., *De C.*
10. ——— lanceolata, *Sow.*; Lias, Lyme Regis, *De la B.*
11. ——— ovata, *Sow.* Stonesfield Slate, *Sow.*
1. Inoceramus dubius, *Sow.* Lias, Yorks., *Phil.*
1. Gervillia aviculoides, *Sow.* Coralline Oolite, Yorks., Calcareous Grit, Oxfordshire, *Phil.*; Oxford Clay, Mid. and S. Eng., Inf. Oolite, Dundry Hill, *Conyb.*; Oxford Clay, Norm., *De la B.*; Sandst., Limest., and Shale, Inverbrora, Scotl., *Murch.*; Lias, Gundershofen, *Voltz*; Coral Rag, Weymouth, *Sedg.*
2. ——— acuta, *Sow.* Great Oolite, Yorks., *Phil.*
3. ——— lata, *Phil.* Inf. Oolite, Yorks., *Phil.*
4. ——— pernoides, *Desl.* Oxford Clay, For. Marb., Great Oolite, and Inf. Oolite, Norm., *De C.*; Gundershofen, *Hœn.*
5. ——— siliqua, *Desl.* Oxf. Clay and For. Marb., Norm., *De C.*
6. ——— monotis, *Desl.* For. Marb., Norm., *De C.*
7. ——— costellata, *Desl.* For. Marb., Norm., *De C.*

344 *Organic Remains of the Oolitic Group.*

Gervillia, species not stated. Coral Rag, Norm., *De C.; Kim.*
    Clay and Inf. Oolite, Haute Saone, *Thir.*
1. Perna quadrata, *Sow.* Coralline Oolite, Kell. Rock, and Great
    Oolite, Yorks., *Phil.;* Cornb., Bulwick, *Sow.*
2. —— mytiloides, *Lam.* Lias, Gundershofen, *Voltz.*
3. —— isogonoides, . Wurtemberg, *Hœn.*
——, sp. not determined. Oxford Clay, Yorks., *Phil.*
1. Crenatula ventricosa, *Sow.* Lias, Yorks., *Phil.*
————, sp. not determined. Portland Stone, *Conyb.*
1. Trigonellites antiquatus, *Phil.* Coral. Oolite, Yorks., *Phil.*
2. ———— politus, *Phil.* Oxford Clay, Yorks., *Phil.*
1. Pinna lanceolata, *Sow.* Coralline Oolite and Calcareous Grit,
    Yorks., *Phil.;* Inf. Oolite, Dundry, *Conyb.;* Lias,
    Norm., *De C.;* Oxford Clay, N. of Fr., *Bobl.;* Coral
    Rag, Weymouth, *Sedg.*
2. —— mitis, *Phil.* Oxford Clay and Kell. Rock? Yorks., *Phil.*
3. —— cuneata, *Bean.* Cornbrash and Great Oolite, Yorks.,
    Phil.
4. —— folium, *Y. & B.* Lias, Yorks., *Phil.*
5. —— pinnigena. Coral Rag, For. Marb., and Inf. Oolite,
    Norm., *De C.*
6. —— granulata, *Sow.* Kim. Clay, Weymouth, *Sedg.;* Kim.
    Clay, Cahors, S. of Fr., *Dufr.;* Lias, Skye, *Murch.*
——, sp. not determined. Inf. Oolite, Env. of Bath, *Lons.*
1. Mytilus cuneatus, *Phil.* Inf. Oolite, Yorks., *Phil.*
2. ———— amplus. Great Oolite, Norm., *De C.*
3. ———— pectinatus, *Sow.* Kim. Clay, Weymouth, *Sedgwick;*
    Rochelle Limestone, *Dufr.*
4. ———— sublævis, *Sow.* Cornb., Eng., *Sow.*
5. ———— solenoïdes. Kim. Clay, Cahors, S. of Fr., *Dufr.*
——, sp. not determined. Coral Rag and Inf. Oolite, Mid.
    and S. Eng., *Conyb.;* Coral Rag, Norm., *De C.;*
    Portland beds, Haute Saone, *Thir.*
1. Modiola imbricata, *Sow.* Coralline Oolite? and Great Oolite,
    Yorks., *Phil.;* Cornb., Mid. and S. Eng., *Conyb.;*
    Cornb., Wilts., *Lons.*
2. ———— ungulata, *Y. & B.* Coralline Oolite, Great Oolite,
    and Inf. Oolite, Yorks., *Phil.*
3. ———— bipartita, *Sow.* Calc. Grit, Yorks., *Phil.;* Sandstone
    and Limestone, Braambury Hill, Brora, *Murch.*
4. ———— cuneata, *Sow.* Oxford Clay, Kell. Rock? and Cornb.,
    Yorks., *Phil.;* Inf. Oolite, Mid. and S. Eng.,
    Conyb.; Lias, Norm., *De C.;* Lias, Western
    Islands, Scotl.; Sandst., Limest., and Shale, In-
    verbrora, Scotl., *Murch.*
5. ———— pulchra, *Phil.* Kell. Rock, Yorks., *Phil.;* Oolite,
    Sutherland.

6. Modiola plicata, *Sow.* Inf. Oolite, Yorks., *Phil.* ; Cornb.,
   Mid. and S. Eng., Inf. Oolite, Dundry, *Conyb.* ;
   Portland Beds, Haute Saone, *Thir.* ; Full. E., Somerset, *Lons.*
7. ———— aspera, *Sow.* Inf. Oolite, Yorks., *Phil.* ; Cornb.
   Mid. and S. Eng., *Conyb.*
8. ———— scalprum, *Sow.* Lias, Yorks., *Phil.* ; Lias, S. of Fr.,
   Dufr.
9. ———— Hillana, *Sow.* Lias, Yorks., *Phil.* ; Lias, Mid. and
   S. Eng., *Conyb.* ; Full. E. ? Env. of Bath, *Lons.*
10. ———— lævis, *Sow.* Lias, Mid. and S. Eng., *Conyb.*
11. ———— depressa, *Sow.* Lias, Mid. and S. Eng., *Conyb.*
12. ———— minima, *Sow.* Lias, Mid. and S. Eng., *Conyb.*
13. ———— subcarinata, *Lam.* Oxford Clay, Norm., *De C.*
14. ———— elegans, *Sow.* For. Marb., Norm., *De C.*
15. ———— tulipea, *Lam.* Oxford Clay, N. of Fr., *Bobl.*
16. ———— pallida, *Sow.* Shale and Grit, Dunrobin Reefs, &c.,
   Scotl., *Murch.*
17. ———— gibbosa, *Sow.* Inf. Oolite, Env. of Bath, *Lons.*
18. ———— livida, *Goldf.* Chaufour, *Hœninghaus.*
19. ———— ventricosa, *Goldf.* Soleure, *Hœn.*
———, sp. not determined. Lias, Gundershofen, *Voltz* ;
   Lias, Bath, *Lons.*
Lithodomus, sp. not determined. Inf. Oolite, N. of Fr.,
   Bobl. ; Inf. Oolite, Env. of Bath, *Lons.*
1. Chama mima or Gryphæa mima, *Phil.* Coral. Oolite and
   Calc. Grit., Yorks., *Phil.*
2. ———— crassa, *Sow.* Bradford Clay, Mid. and S. Eng.,
   Conyb.
———, sp. not determined. For. Marb., Cornb., and Brad.
   Clay, Wilts., *Lons.*
1. Unio peregrinus, *Phil.* Cornb., Yorks., *Phil.*
2. ——— abductus, *Phil.* Inferior Oolite and Lias, Yorks., *Phil.*
3. ——— concinnus, *Sow.* Lias, Yorks., *Phil.* ; Inf. Oolite, Mid.
   and S. Eng., *Conyb.* ; Inf. Oolite, Lias, Env. of Bath,
   Lons.
4. ——— crassiusculus, *Sow.* Lias, Yorks., *Phil.*
5. ——— Listeri, *Sow.* Lias, Yorks., *Phil.* ; Inf. Oolite, Mid.
   and S. Eng., *Conyb.*
6. ——— acutus, *Sow.* Cornb., Mid. and S. Eng., *Conyb.*
7. ——— crassissimus, *Sow.* Lias, Mid. and S. Eng., *Conyb.* ;
   Lias, Norm., *De C.* ; For. Marb. ? Mauriac, and Inf.
   Oolite, Uzer, S. of Fr., *Dufr.*
1. Trigonia costata, *Sow.* Coralline Oolite, Great Oolite, and
   Inf. Oolite, Yorks., *Phil.* ; Cornb., For. Marb.,
   and Brad. Clay, Mid. and S. Engl., Inf. Oolite,
   Dundry, *Conyb.* ; Oxford Clay, For. Marb., and

346   *Organic Remains of the Oolitic Group.*

Inf. Oolite, Norm., *De C.* ; Oxford Clay, N. of Fr.,
Bobl. ; Kim. Clay and Inf. Oolite, Haute Saone,
Thir. ; Lias, Gundershofen, *Voltz* ; Inf. Oolite,
Env. of Bath, *Lons.* ; Coral Rag, Weymouth, *Sedg.*
2. Trigonia clavellata, *Sow.* Coralline Oolite, Kell. Rock, and
Cornb., Yorks., *Phil.* ; Portland Stone and Cornb.,
Mid. and S. Engl., Inf. Oolite, Dundry, *Conyb.* ;
Oxford Clay, Norm., *De la B.* ; Oxford Clay, N.
of Fr., *Bobl.* ; Kim. Clay? Angoulême, *Dufr.* ;
Sandst., Shale, &c., Inverbrora, Scotl., *Murch.* ;
Coral Rag and Inf. Oolite, Haute Saone, *Thir.* ;
Coral Rag, Weymouth, *Sedg.*
3. ———— conjungens, *Phil.* Great Oolite, Yorks., *Phil.*
4. ———— striata, *Sow.* Inferior Oolite, Yorks., *Phil.* ; Inf.
Oolite, Dundry, *Conyb.* ; Inf. Oolite, Norm., *De*
C. ; Lias, S. of Fr., *Dufr.*
5. ———— angulata, *Sow.* Inf. Oolite, Yorks., *Phil.* ; Inf.
Oolite, near Frome, *Sow.*
6. ———— literata, *Y. & B.* Lias, Yorks., *Phil.*
7. ———— gibbosa, *Sow.* Portland Stone, *Conyb.* ; Forest Marb.,
Norm., *De C.*
8. ———— duplicata, *Sow.* Inf. Oolite, Mid. and S. Eng.,
Conyb. ; For. Marb., Norm., *De C.*
9. ———— elongata, *Sow.* Oxford Clay, Norm., *De C.* ; Oxford
Clay, Eng., *Sow.* ; Great Oolite, Alsace, *Voltz* ;
Cornb., Wilts., *Lons.*
10. ———— imbricata, *Sow.* Great Oolite, Ancliff, Wilts., *Cookson.*
11. ———— cuspidata, *Sow.* Great Oolite, Ancliff, *Cookson* ;
var. Coral Rag, Haute Saone, *Thir.*
12. ———— pullus, *Sow.* Great Oolite, Ancliff, *Cookson.*
13. ———— navis, *Lam.* Lias, Gundershofen, *Voltz.*
————, sp. not determined. Coral Rag, Mid. and S. Eng.,
Conyb. ; Coral Rag, Norm., *De C.*
1. Nucula elliptica, *Phil.* Oxford Clay, Yorks., *Phil.*
2. ———— nuda, *Y. & B.* Oxford Clay, Yorks., *Phil.*
3. ———— variabilis, *Sow.* Great Oolite and Inf. Oolite, Yorks.,
Phil. ; Great Oolite, Ancliff, near Bath, *Cookson.*
4. ———— lachryma, *Sow.* Great Oolite and Inf. Oolite, Yorks.,
Phil.
5. ———— axiniformis, *Phil.* Inferior Oolite, Yorks., *Phil.*
6. ———— ovum, *Sow.* Lias, Yorks., *Phil.*
7. ———— pectinata, *Sow.* Oxford Clay, Norm., *De C.* ; Brad.
Clay, Wilts., *Lons.*
8. ———— clariformis. Lias, S. of Fr., *Dufr.*
9. ———— mucronata *Sow.* Great Oolite, Ancliff, Wilts., *Cookson.*

Nucula, species not determined. Coralline Oolite, Yorks.,
   Phil.; Inf. Oolite, Dundry; Lias, Mid. and S. Eng.,
   Conyb.
1. Pectunculus minimus, *Sow.* Great Oolite, Ancliff, Wilts.,
   Cookson.
2. ———— oblongus, *Sow.* Great Oolite, Ancliff, Wilts.,
   Cookson.
1. Arca quadrisculata, *Sow.* Coralline Oolite, Yorks., *Phil.*
2. —— æmula, *Phil.* Coralline Oolite, Yorks., *Phil.*
3. —— pulchra, *Sow.* Great Oolite, Ancliff, Wilts., *Cookson;* Rochelle Limestone, *Dufr.*
4. —— trigonella,   Wasseralfingen, *Hœn.*
5. —— elongata,   Wasseralfingen, *Hœn.*
6. —— rostrata,   Wasseralfingen, *Hœn.*
——, sp. not determined. Lias, Mid. and S. Eng., *Conyb.;*
   Brad. Clay, Wilts.; Full. E., Inf. Oolite, Env. of
   Bath, *Lons.*
1. Cucullæa oblonga, *Sow.* Coralline Oolite, Yorks., *Phil.;* Inf.
   Olite, Dundry, *Conyb.*
2. ———— contracta, *Phil.* Coralline Oolite, Yorks., *Phil.*
3. ———— triangularis, *Phil.* Coralline Oolite, Yorks., *Phil.*
4. ———— pectinata, *Phil.* Coralline Oolite, Yorks., *Phil.*
5. ———— elongata, *Sow.* Coralline Oolite? and Great Oolite.
   Yorks., *Phil.;* Rochelle limestone, *Dufr.*
6. ———— concinna, *Phil.* Oxford Clay and Kell. Rock?
   Yorks., *Phil.*
7. ———— imperialis, *Bean.* Great Oolite, Yorks., *Phil.*
8. ———— cylindrica, *Phil.* Great Oolite, Yorks., *Phil.*
9. ———— cancellata, *Phil.* Great Oolite, Yorks., *Phil.*
10. ———— reticulata, *Bean.* Inf. Oolite, Yorks., *Phil.*
11. ———— decussata, *Sow.* Inf. Oolite, Norm., *De C.*
12. ———— minuta, *Sow.* Great Oolite, Ancliff, Wilts., *Cookson.*
13. ———— rudis, *Sow.* Great Oolite, Ancliff, Wilts., *Cookson.*
——, sp. not determined. Oxford Clay, Haute Saone,
   Thir. Lias, Yorks., *Phil.;* Lias, Mid. and S.
   Eng., *Conyb.*
1. Hippopodium ponderosum, *Sow.* Coralline Oolite and Lias,
   Yorks., *Phil.;* Lias, Mid. and S. Eng.,
   Conyb.
1. Isocardia rhomboidalis, *Phil.* Coralline Oolite, Yorks., *Phil.*
2. ———— tumida, *Phil.* Calc. Grit., Yorks., *Phil.*
3. ———— minima, *Sow.* Cornb. and Great Oolite? Yorks.,
   Phil.; Cornb., Wilts., *Lons.*
4. ———— concentrica, *Sow.* Great Oolite and Inf. Oolite,
   Yorks., *Phil.;* Oxford Clay, Norm., *De C.;*

Cornb., Northamptonshire, *Sow.;* Full. E., Somerset, *Lons.*
5. Isocardia angulata, *Phil.* Great Oolite? Yorks., *Phil.*
6. ———— rostrata, *Sow.* Inf. Oolite, Yorks., *Phil.*
7. ———— striata, *D'Orb.* Kim. Clay, Portland Beds, Haute
Saone, *Thir.*
————, sp. not determined. For. Marb., Norm., *De C.*
1. Cardita similis, *Sow.* Coralline Oolite, Great Oolite, and Inf.
Oolite, Yorks., *Phil.;* Inf. Oolite, Dundry, *Conyb.*
2. ———— lunulata, *Sow.* Inf. Oolite, Dundry, *Conyb.;* Inf.
Oolite, Norm., *De C.*
3. ———— striata. Lias, Norm.? *De C.*
————, species not determined. Portland Stone, *Conyb.*
1. Cardium lobatum, *Phil.* Coralline Oolite, Yorks., *Phil.*
2. ———— dissimile, *Sow.* Kell Rock, Yorks., *Phil.;* Portland
Stone, Portland, *Sow.;* Rocks of the Oolite series,
Braambury Hill, Brora, *Murch.*
3. ———— citrinoideum, *Phil.* Cornb., Yorks., *Phil.*
4. ———— cognatum, *Phil.* Great Oolite, Yorks., *Phil.*
5. ———— acutangulum, *Phil.* Great Oolite and Inf. Oolite,
Yorks., *Phil.*
6. ———— semiglabrum, *Phil.* Great Oolite, Yorks., *Phil.*
7. ———— incertum, *Phil.* Inf. Oolite, Yorks., *Phil.*
8. ———— striatulum, *Sow.* Sandst., Limest. and Shale, Inverbrora, Scotland, *Murch.;* Inferior Oolite, Yorks.,
Phil.
9. ———— gibberulum, *Phil.* Inf Oolite, Yorks., *Phil.;* Cornb.?
Wilts., *Lons.*
10. ———— truncatum, *Sow.* Lias, Yorks., *Phil.;* Sandst.,
Limest. &c., Inverbrora, *Murch.*
11. ———— multicostatum, *Bean.* Lias, Yorks., *Phil.*
1. Astarte cuneata, *Sow.* Portland Stone, S. Eng.; Inf. Oolite?
Dundry, *Conyb.*
2. ———— excavata, *Sow.* Inf. Oolite, Dundry, *Conyb.;* Inf.
Oolite, Norm., *De C.*
3. ———— lurida, *Sow.* Inf. Oolite, Dundry, *Conyb.*
4. ———— ovata, *Sow.* Inf. Oolite, Dundry, *Conyb.*
5. ———— planata, *Sow.* Inf. Oolite, Norm., *De C.;* Bradf.
Clay, N. of Fr., *Bobl.*
6. ———— rugata, *Sow.* Inf. Oolite, Norm., *De C.*
7. ———— imbricata, *Sow.* Inf. Oolite, Norm., *De C.*
8. ———— trigonalis, *Sow.* Inf. Oolite, Dundry.
9. ———— orbicularis, *Sow.* Great Oolite, Ancliff, Wilts.,
Cookson.
10. ———— pumila, *Sow.* Great Oolite, Ancliff, Wilts., *Cookson;* Rochelle Limestone, *Dufr.*
11. ———— elegans, *Sow.* Rochelle Limestone, *Dufr.;* Shell

Limest. and Calc. Grit., Portgower, &c. ; Sandst.,
Limest. and Shale, Inverbrora, Scotl., *Murch.*
12. Astarte Voltzii,   Fullon, near Vesoul, *Hœn.*
——, sp. not determined. Lias, Mid. and S. Eng. *Conyb.;*
Coral Rag and Kim. Clay, Haute Saone, *Thir.;*
Cornb., Wilts., *Lons.*
Venus, sp. not determined.   Coral. Oolite, Calc. Grit and
Lias, Yorks., *Phil.;* Portland Stone, *Smith;* Coral
Rag, Norm., *De C.;* Sandst., Shale, &c., Inverbrora, Scotl., *Murch.*
1. Cytherea dolabra, *Phil.* Great Oolite, Yorks., *Phil.*
2. —— trigonellaris, *Voltz.* Lias, Gundershofen, *Voltz.*
3. —— lucinea, *Voltz.* Lias, Gundershofen, *Voltz.*
4. —— cornea, *Voltz.* Lias, Gundershofen, *Voltz.*
——, sp. not determined. Coralline Oolite, Yorks., *Phil.;*
Lias, N. of Fr., *Bobl.*
1. Pullastra recondita, *Phil.* Great Oolite, Yorks., *Phil.*
2. —— oblita, *Phil.* Inferior Oolite, Yorks., *Phil.*
——, sp. not determined. Lias, Yorks., *Phil.*
1. Donacites Alduini, *Al. Brong.* Inf. Oolite? N. of Fr., *Bobl.;*
Kim. Clay, Havre and the Jura, *Al. Brong.*
1. Corbis lævis, *Sow.* Coralline Oolite? Kell. Rock? Yorks.,
Phil.
2. —— ovalis, *Phil.* Kell. Rock, Yorks., *Phil.*
1. Tellina ampliata, *Phil.* Coralline Oolite, Yorks., *Phil.*
1. Psammobia lævigata, *Phil.* Coralline Oolite, Great Oolite,
and Inf. Oolite, Yorks., *Phil.*
1. Lucina crassa, *Sow.* Sandstone and Rubbly Limestone, Braambury Hill, Brora; Great Arenaceous Formation,
Western Islands, Scotl., *Murch.;* Calc. Grit,
Yorks., *Phil.;* Lincolnshire, *Sow.*
2. —— lyrata, *Phil.* Kell. Rock, Yorks., *Phil.*
3. —— despecta, *Phil.* Great Oolite, Yorks., *Phil.*
——, species not stated. Coral Rag and For. Marb., Norm.,
De Cau.; Inf. Oolite, Yorks., *Phil.;* Shale, &c.,
Inverbrora, Scotl., *Murch.*
1. Sanguinolaria undulata, *Sow.* Sandst., Limest., and Shale,
Inverbrora, Scotl., *Murch.;* Calc. Grit, Ox-
ford Clay, and Cornbrash, Yorks., *Phil.*
2. —— elegans, *Phil.* Lias, Yorks., *Phil.*
——, sp. not determined. Lias, Ross and Cromarty,
Scotl., *Murch.;* Lias, Yorks., *Phil.*
1. Corbula curtansata, *Phil.* Coralline Oolite and Kell. Rock,
Yorks., *Phil.*
2. —— depressa, *Phil.* Great Oolite, Yorks., *Phil.*
3. ——? cardioides, *Phil.* Lias, Yorks., *Phil.*
4. —— obscura, *Sow.* Brora, *Murch.*

Corbula, sp. not determined. For. Marb., Wilts., *Lons.*
1. Crassina ovata, *Smith.* Coralline Oolite, Yorks., *Phil.*
2. ———— elegans, *Sow.* Coralline Oolite and Inf. Oolite, Yorks., *Phil.*
3. ———— aliena, *Phil.* Coralline Oolite, Yorks., *Phil.*
4. ———— extensa, *Phil.* Coralline Oolite, Yorks., *Phil.*
5. ———— carinata, *Phil.* Calc. Grit, Oxford Clay, and Kell. Rock, Yorks., *Phil.*
6. ———— lurida, *Sow.* Oxford Clay, Yorks., *Phil.*
7. ———— minima, *Phil.* Great Oolite, Inf. Oolite, and Lias, Yorks., *Phil.;* Kim. Clay and Coral Rag, Haute Saone, *Thir.*
1. Mactra gibbosa, For. Marb., Norm., *De C.*
1. Amphidesma decurtatum, *Phil.* Cornb. and Great Oolite, Yorks., *Phil.;* Kim. Clay? and Great Oolite? Haute Saone, *Thir.*
2. ————— recurvum, *Phil.* Coralline Oolite? and Kell. Rock, Yorks., *Phil.;* Kim. Clay, Havre, *Phil.*
3. ————— securiforme, *Phil.* Cornb., Inf. Oolite, Yorks., *Phil.;* Kim. Clay, Havre, *Phil.*
4. ————— donaciforme, *Phil.* Lias, Yorks., *Phil.*
5. ————— rotundatum, *Phil.* Lias, Yorks., *Phil.*
1. Lutraria Jurassi, *Brong.* For. Marb., Ligny, Meuse, *Brong.*
1. Gastrochæna tortuosa, *Sow.* Inf. Oolite, Yorks., *Phil.*
1. Mya literata, *Sow.* Coralline Oolite, Calc. Grit., Oxford Clay, Kelloway Rock, Cornb., Inf. Oolite, and Lias, Yorks., *Phil.;* Shale, Sandstone, and Limestone, Inverbrora, Scotl., *Murch.*
2. ——— depressa, *Sow.* Oxford Clay? Yorks., *Phil.;* Kim. Clay? Angoulême, *Dufr.;* Kim. Clay, Havre, *Phil.;* Shale, Limestone, and Sandstone, Inverbrora, Scotl., *Murch.*
3. ——— calceiformis, *Phil.* Kell. Rock, Great Oolite, and Inf. Oolite, Yorks., *Phil.*
4. ——— dilata, *Phil.* Inferior Oolite, Yorks., *Phil.*
5. ——— æquata, *Phil.* Inferior Oolite, Yorks., *Phil.*
6. ——— V scripta, *Sow.* Inf. Oolite, Dundry, *Conyb.;* Great Oolite, Alsace, *Brong.;* Micaceous Sandstone, Western Islands, Scotl., *Murch.*
7. ——— mandibulata, *Sow.* Kim. Clay? Env. of Angoulême, *Dufr.*
8. ——— angulifera, *Sow.* Great Oolite, Haute Saone, *Thir.;* Lias, Alsace, *Voltz;* Fuller's Earth, Environs of Bath, *Lons.*
1. Pholadomya Murchisoni, *Sow.* Sandstone, Limestone and Shale, Inverbrora, Scotl., *Murch.;* Coralline Oolite? and Cornbrash, Yorks., *Phil.;* Inf. Oolite, Normandy, *De Cau.*

2. Pholadomya simplex, *Phil.*   Calc. Grit, Yorks., *Phil.*
3. —————— deltoidea, *Sow.*   Calc. Grit, Yorks., *Phil.*; Kell. Rock and Cornbrash, Midl. and S. Engl., *Conyb.*
4. —————— obsoleta, *Phil.*   Oxford Clay and Kell. Rock, Yorks., *Phil.*
5. —————— ovalis, *Sow.* Cornbrash, Yorks., *Phil.*; Portland Stone, *Conyb.*; Oxford Clay, Normandy, *De C.*; Kim. Clay? Angoulême. Rochelle Limestone, *Dufr.*
6. —————— acuticostata, *Sow.*   Great Oolite, Yorks., *Phil.*; Kim. Clay, Cahors, S. of Fr., *Dufr.*; Kim. Clay? Angoulême, *Dufr.*; Kim. Clay, Haute Saone, *Thir.*
7. —————— nana, *Phil.*   Great Oolite, Yorks., *Phil.*
8. —————— producta, *Sow.*   Great Oolite? Yorks., *Phil.*; Cornb. and Inf. Oolite, Midl. and S. Eng., *Conyb;* Cornb., Wilts., *Lons.*
9. —————— obliquata, *Phil.*   Great Oolite, Inf. Oolite, and Lias, Yorks., *Phil.*
10. —————— fidicula, *Sow.*   Inf. Oolite, Yorks., *Phil.*
11. —————— lyrata, *Sow.* Cornb., Mid. and S. of Eng.; Inf. Oolite, Dundry, *Conyb.*; Lias, Norm., *De C.*; Cornb., Wilts.; Full. E., Env. of Bath, *Lons.*; Soleure, *Hœn.* Inf. Oolite, Haute Saone, *Thir.*
12. —————— obtusa, *Sow.*   Inf. Oolite, Dundry, *Conyb.*
13. —————— ambigua, *Sow.*   Inf. Oolite, Dundry, *Conyb.*; Oxford Clay, Norm., *De C.*; Lias, S. of Fr., *Dufr.*; Lias, Alsace, *Voltz;* Lias, Bath, *Lons.*; Lias, Soleure; Porta Westphalica, *Hœn.*; Lias, Bahlingen, *Von Buch.*
14. —————— æqualis, *Sow.*   Inf. Oolite, Norm., *De C.*
15. —————— gibbosa, *Sow.* Lias, Norm., *De C.*; Soleure, *Hœn.*
16. —————— Proteii, *Brong.*   Rochelle Limest., *Dufr.*; Kim. Clay, Havre and the Jura, *Brong.*; Portland Beds and Kim. Clay, Haute Saone, *Thir.*
—————— , species not determined.   Oxford Clay, Haute Saone, *Thir.*
1. Panopæa intermedia, *Sow.*   Inf. Oolite, Dundry, *Conyb.*
2. —————— gibbosa, *Sow.*   Great Oolite? Yorks., *Phil.*; Inf. Oolite, Dundry, *Conyb.*
1. Pholas recondita, *Phil.*   Coralline Oolite, Yorks., *Phil.*
2. ——————? compressa, *Sow.*   Kim. Clay, Oxford, *G. E. Smith.*

Mollusca.

1. Dentalium giganteum, *Phil.*   Lias, Yorks., *Phil.*
2. —————— cylindricum, *Sow.*   Lias, Mid. and S. Eng., *Conyb.*

Dentalium, sp. not determined.   Calc. Grit, Yorks., *Phil.*
1. Patella latissima, *Sow.*   Oxford Clay, Yorks., *Phil.;* Oxford
             Clay, Mid. and S. of Eng., *Conyb.*
2. ——— rugosa, *Sow.*   For. Marb., Mid. and S. of Eng., *Conyb.;*
             For. Marb., Norm., *De C.*
3. ——— lævis, *Sow.*   Lias, Mid. and S. of Eng., *Conyb.*
4. ——— lata, *Sow.*   Stonesfield Slate, *Sow.*
5. ——— ancyloïdes, *Sow.*   Great Oolite, Ancliff, Wilts., *Cook-*
             son.
6. ——— nana, *Sow.*   Great Oolite, Ancliff, Wilts.,*Cookson.*
7. ——— discoides, *Schlot.*   Lias, Gundershofen, *Voltz.*
1. Emarginula scalaris, *Sow.*   Great Oolite, Ancliff, Wilts.,
             Cookson.
1. Peleolus plicatus, *Sow.*   Great Oolite, Wilts., *Lons.*
Ancilla, species not stated.   Great Oolite and For. Marb.,
             Norm., *De C.*
1. Bulla elongata, *Phil.*   Coral. Oolite, Yorks., *Phil.*
1. Helicina polita, *Sow.*   Inf. Oolite, Cropredy, *Conyb.*
2. ——— compressa, *Sow.*   Lias, Mid. and S. of Eng., *Conyb.*
3. ——— expansa, *Sow.*   Lias, Mid. and S. of Eng., *Conyb.*
4. ——— solarioides, *Sow.*   Lias, Mid. and S. of Eng., *Conyb.*
1. Auricula Sedgvici, *Phil.*   Inf. Oolite, Yorks., *Phil.*
1. Planorbis euomphalus, *Sow.*   Inf. Oolite, Mid. and S. of Eng.,
             Conyb.
1. Melania Heddingtonensis, *Sow.*   Coral. Oolite, Cornb., Great
             Oolite and Inf. Oolite, Yorks., *Phil.;* Coral Rag,
             Mid. and S. of Eng.; Inf. Oolite, Dundry,*Conyb.;*
             Coral Rag and Inf. Oolite, Norm., *De C.;* Rub-
             bly Limest., &c., Braambury Hill, Brora, *Murch.;*
             Kim. Clay, Havre, *Phil.;* Inf. Oolite? Haute
             Saone, *Thir.;* Coral Rag, Weymouth, *Sedg.*
2. ——— striata, *Sow.*   Coral. Oolite and Great Oolite? Yorks.,
             Phil.; Coral Rag and Lias, Mid. and S. of Eng.,
             Conyb.; Coral Rag, N. of Fr., *Bobl.;* Kim. Clay,
             Havre, *Phil.;* Coral Rag, Weymouth, *Sedg.*
3. ——— vittata, *Phil.*   Cornb., Yorks., *Phil.*
4. ——— lineata, *Sow.*   Inf. Oolite, Yorks., *Phil.;* Inf. Oolite,
             Dundry, *Conyb.;* Inf. Oolite, Norm., *De C.*
  ———, sp. not determined.   Great Oolite, Mid. and S. of
             Eng., *Conyb.*
Paludina, species not determined.   Portland Beds, Haute
             Saone, *Thir.*
Ampullaria, sp. not determined.   Coral Rag, Cornb. and Inf.
             Oolite, Mid. and S. Eng., *Conyb.;* Coral Rag,
             Norm., *De C.;* Brad. Clay, N. of Fr., *Bobl.*
1. Nerita costata, *Sow.*   Inf. Ooolite, Yorks., *Phil.;* Great
             Oolite, Ancliff, Wilts., *Cookson.*

2. Nerita sinuosa, *Sow.* Portland Stone, *Conyb.*
3. —— lævigata, *Sow.* Inf. Oolite, Dundry, *Conyb.;* Shell Limestone and Calc. Grit, Portgower, &c., Scotland, *Murch.*
4. —— minuta, *Sow.* Great Oolite, Ancliff, Wilts., *Cookson.*
1. Natica arguta, *Smith.* Coral. Oolite, Yorks., *Phil.*
2. —— nodulata, *Y. & B.* Coral. Oolite, Yorks., *Phil.*
3. —— cincta, *Phil.* Coral. Oolite, Yorks., *Phil.*
4. —— adducta, *Phil.* Great Oolite and Inf. Oolite, Yorks., *Phil.*
5. —— tumidula, *Bean.* Inf. Oolite, Yorks., *Phil.*
——, sp. not determined. Lias, Yorks., *Phil.*
Tornatilla, sp. not determined. Lias, Mid. and S. Eng., *Conyb.*
1. Vermetus compressus, *Y. & B.* Coral. Oolite, Inf. Oolite, Yorks., *Phil.*
2. —— Nodus, *Phil.* Cornb., Great Oolite, Yorks., *Phil.*
——, sp. not determined. Cornbrash, Wilts., *Lons.*
Dephinula, sp. not determined. Coral. Oolite and Great Oolite, Yorks., *Phil.*
1. Solarium calix, *Bean.* Inf. Oolite, Yorks., *Phil.*
2. —— conoideum, *Sow.* Portland Stone, *Conyb.*
1. Cirrus cingulatus, *Phil.* Calc. Grit, Yorks., *Phil.*
2. —— depressus, *Phil.* Kell. Rock, Yorks., *Phil.*
3. —— nodosus, *Sow.* Inf. Oolite, Dundry, *Conyb*
4. —— Leachii, *Sow.* Inf. Oolite, Dundry, *Conyb.*
5. —— carinatus, *Sow.* Inf. Oolite, Wilts., *Lons.*
——, species undetermined. Lias, N. of Fr., *Bobl.;* Oxford Clay, Haute Saone, *Thir.*
1. Trochus granulatus, *Sow.* Coral. Oolite, Calc. Grit, Cornb., and Inf. Oolite, Yorks., *Phil.;* Inf. Oolite, Dundry, *Conyb.;* Inf. Oolite, Norm., *De C.*
2. ——? tornatus, *Phil.* Coral. Oolite, Yorks., *Phil.*
3. —— bicarinatus, *Sow.* Calc. Grit, Yorks., *Phil.;* Coral Rag, Mid. and S. Eng., Inf. Oolite, Dundry, *Conyb.;* Inf. Oolite, Norm., *De C.*
4. —— guttatus, *Phil.* Kell. Rock, Yorks., *Phil.*
5. —— monilitectus, *Phil.* Great Oolite, Yorks., *Phil.*
6. —— bisertus, *Phil.* Inf. Oolite, Yorks., *Phil.*
7. —— pyramidatus, *Bean.* Inf. Oolite, Yorks., *Phil.*
8. —— anglicus, *Sow.* Lias, Yorks., *Phil.;* Lias, Mid. and S. Eng., *Conyb.;* Inf. Oolite, Haute Saone, *Thir.*
9. —— similis, *Sow.* Inf. Oolite, Dundry, Lias, Mid. and S. Eng., *Conyb.*
10. —— concavus, *Sow.* Inf. Oolite, Mid. and S. Eng., *Conyb.;* Inf. Oolite, Norm.
11. —— dimidiatus, *Sow.* Inf. Oolite, Mid. and S. Eng., *Conyb.*

12. Trochus duplicatus, *Sow.* Inf. Oolite, Mid. and S. Eng.,
        Conyb.; Inf. Oolite, Haute Saone, *Thir.;* Lias,
        Gundershofen, *Voltz.*
13. ——— elongatus, *Sow.* Inf. Oolite, Dundry, *Conyb.;* For.
        Marb. and Inf. Oolite, Norm., *De C.;* Inf. Oolite,
        Wilts., *Lons.*
14. ——— punctatus, *Sow.* Inf. Oolite, Dundry, *Conyb.;* Inf.
        Oolite, Norm., *De C.*
15. ——— abbreviatus, *Sow.* Inf. Oolite, Dundry, *Conyb.;* Inf.
        Oolite, Norm., *De C.*
16. ——— fasciatus, *Sow.* Inf. Oolite, Dundry, *Conyb.;* Inf.
        Oolite, Norm., *De C.*
17. ——— sulcatus, *Sow.* Inf. Oolite, Dundry, *Conyb.;* Inf.
        Oolite, Norm., *De C.*
18. ——— ornatus, *Sow.* Inf. Oolite, Dundry, *Conyb.;* Inf.
        Oolite, Norm., *De C.;* Lias, N. of Fr., *Bobl.*
19. ——— imbricatus, *Sow.* Lias, Mid. and S. Eng., *Conyb.;*
        Inf. Oolite, Norm., *De C.;* Lias, S. of Fr., *Dufr.;*
        Soleure, *Hœn.*
20. ——— Gibsii, *Sow.* Oxford Clay, Norm., *De C.*
21. ——— reticulatus, *Sow.* Inf. Oolite, Norm., *De C.;* Coral
        Rag, Weymouth, *Sedg.*
——, sp. not determined.   Portland Stone and Bradford
        Clay, Mid. and S. Eng., *Conyb.;* Coral Rag, Norm.,
        De C.; Oxford Clay, Coral Rag, and Great Oolite,
        Haute Saone, *Thir.*
1. Rissoa lævis, *Sow.* Great Oolite, Ancliff, Wilts., *Cookson.*
2. ——— acuta, *Sow.* Great Oolite, Ancliff, *Cookson.*
3. ——— obliquata, *Sow.* Great Oolite, Ancliff, *Cookson.*
4. ——— duplicata, *Sow.* Great Oolite, Ancliff, *Cookson.*
1. Turbo muricatus, *Sow.* Coral. Oolite, Great Oolite, and Inf.
        Oolite, Yorks., *Phil.;* Coral Rag, Mid. and S. Eng.
        Conyb.; Coral Rag, Weymouth, *Sedg.*
2. ——— funiculatus, *Phil.* Coral. Oolite, Yorks., *Phil.*
3. ——— sulcostomus, *Phil.* Kell. Rock, Yorks., *Phil.*
4. ——— unicarinatus, *Bean.* Inf. Oolite, Yorks., *Phil.*
5. ——— lævigatus, *Phil.* Inf. Oolite, Yorks., *Phil.*
6. ——— undulatus, *Phil.* Lias, Yorks., *Phil.*
7. ——— ornatus, *Sow.* Inf. Oolite, Mid. and S. Eng., *Conyb.;*
        Inf. Oolite, Norm., *De C.;* Lias, Gundershofen, *Voltz.*
8. ——— rotundatus, *Sow.* Inf. Oolite, Norm., *De C.*
9. ——— obtusus, *Sow.* Great Oolite, Ancliff, *Cookson.*
——, sp. not determined. Cornb. and Great Oolite, Norm., *De C.*
1. Phasianella cincta, *Phil.* Great Oolite, Yorks., *Phil.*
2. ——— angulosa, *Sow.* Porta Westphalica, *Hœn.*
1. Turritella muricata, *Sow.* Coral. Oolite, Calc. Grit., Kell.
        Rock, and Inf. Oolite, Yorks., *Phil.;* Rochelle

Limestone, *Dufr.;* Shell Limestone and Grit,
Portgower, &c., Scotland, *Murch.*
2. Turritella cingenda, *Sow.* Coral. Oolite? Great Oolite, and
Inf. Oolite, Yorks., *Phil.*
3. ———— quadrivittata, *Phil.* Inf. Oolite, Yorks., *Phil.*
4. ———— concava, *Sow.* Portland Stone, Tisbury, *Benett.*
5. ———— echinata, *Von Buch.* Banz, Langheim, *Von Buch.*
———, sp. not determined. Portland Stone, Coral Rag?
Cornb., For. Marb., and Brad. Clay, Mid. and S.
Eng., *Conyb.;* Brad. Clay, N. of Fr., *Bobl.;* Portland beds and Coral Rag, Haute Saone, *Thir.;*
Lias, Bath, *Lons.*
1. Nerinea tuberculata, *Blain.* Bailly, near Auxerre, *Hœn.*
———, sp. not determined. Coral Rag and For. Marb.,
Norm., *De C.;* Brad. Clay, N. of Fr., *Bobl.;*
Coral Rag, Inf. Oolite, Haute Saone, *Thir.*
1. Cerithium intermedium (var.). Böhlhorst, near Minden, *Hœn.*
2. ———— muricatum,   Mühlhausen, Bas Rhin, *Hœn.*
———, sp. not determined. Lias, Gundershofen, *Voltz.*
1. Murex Haccanensis, *Phil.* Coral. Oolite, Yorks., *Phil.*
2. ———— rostellariformis, *Von Buch.* Coral Rag, Randen,
Schafhausen, *Von Buch.*
1. Rostellaria bispinosa, *Phil.* Calc. Grit? and Kell. Rock,
Yorks., *Phil.*
2. ———— trifida, *Bean.* Oxford Clay, Yorks., *Phil.*
3. ———— Parkinsonii, *Sow.* Inf. Oolite, Norm., *De C.*
4. ———— composita, *Sow.* Sandst., Limest., and Shale, Inverbrora, Scotl., *Murch.;* Great? and Inf. Oolite,
Yorks., *Phil.;* Oxford Clay, Weymouth, *Sow.;*
Kim. Clay, Havre, *Phil.*
———, sp. not determined. Lias, Yorks., *Phil.;* Oxford
Clay, Kell. Rock, Cornb., Forest Marb., and Inf.
Oolite, Mid. and S. Eng., *Conyb.;* Oxford Clay,
Norm., *De C.*
1. Pteroceras Oceani, *Al. Brong.* Kim. Clay, Havre, the Jura,
Al. Brong.; Portland Beds, Kim. Clay? Haute
Saone, *Thir.*
2. ———— Ponti, *Al. Brong.* Kim. Clay, Havre and the Jura,
Al. Brong.; Kim. Clay, Haute Saone, *Thir.*
3. ———— Pelagi, *Al. Brong.* Kim. Clay, Havre and the
Jura, *Al. Brong.*
1. Actæon retusus, *Phil.* Calc. Grit, Yorks., *Phil.*
2. ———— glaber, *Bean.* Great Oolite and Inf. Oolite, Yorks.,
Phil.
3. ———— humeralis, *Phil.* Inf. Oolite, Yorks., *Phil.*
4. ———— cuspidatus, *Sow.* Great Oolite, Ancliff, Wilts., *Cookson.*

5. Actæon acutus, *Sow.* Great Oolite, Ancliff, Wilts., *Cookson.*
——, sp. not determined. Lias, Yorks., *Phil.*
1. Buccinum unilineatum, *Sow.* Great Oolite, Ancliffe, Wilts., *Cookson.*
——, sp. not determined. Shale, Sandst., and Limest., Inverbrora, Scotl., *Murch.*
1. Myoconcha crassa, *Sow.* Inf. Oolite, Norm., *De C. ;* Inf. Oolite, Dundry, *Sow.*
1. Terebra melanoïdes, *Phil.* Coral. Oolite, Yorks., *Phil.*
2. ——? granulata, *Phil.* Coral. Oolite and Cornb., Yorks., *Phil.*
3. —— vetusta, *Phil.* Great Oolite and Inf. Oolite, Yorks., *Phil.*
4. —— sulcata,   Coral Rag, N. of Fr., *Bobl.*
1. Belemnites sulcatus, *Mill.* Coral. Oolite? Calc. Grit, Oxford Clay, and Kell. Rock, Yorks., *Phil. ;* Shale, Sandst., and Limest., Inverbrora, Scotl., *Murch. ;* Lias, S. of France, *Dufr.*
2. —— fusiformis, *Mill.* Coral. Oolite? Yorks., *Phil.*
3. —— gracilis, *Phil.* Oxford Clay, Yorks., *Phil.*
4. —— abbreviatus, *Mill.* Great Oolite, Yorks., *Phil. ;* Lias, Ross and Cromarty, Scotland, and Micaceous Sandstone, Western Islands, Scotland, *Murch.*
5. —— elongatus, *Miller.* Lias, Yorks., *Phil. ;* Lias, Ross and Cromarty, Scot., *Murch.*
6. —— trisulcatus, *Blain.* Inferior Oolite, N. of Fr., *Bobl.*
7. —— compressus, *Blain.* Fuller's E., N. of Fr., *Bobl. ;* Inf. Oolite, Yorks., *Sow. ;* Lias, Gundershofen, *Voltz.*
8. —— dilatatus,   Fuller's E., N. of Fr., *Bobl.*
9. —— apicicurvatus, *Bl.* Lias, S. of Fr., *Dufr. ;* Lias, Alais, *Al. Brong.*
10. —— pistilliformis, *Blain.* Lias, S. of Fr., *Dufr. ;* Lias, Gundershofen, *Voltz.*
11. —— brevis, *Blain.* Lias, Alais, *Brong.*
12. —— longissimus, *Miller.* Lias, Bath, *Lons.*
13. —— canaliculatus, *Schlot.* Oxford Clay and Inf. Oolite, Haute Saone, *Thir.*
14. —— ellipticus, *Miller.* Inf. Oolite, Haute Saone, *Thir.*
15. —— longus, *Voltz.* Great Oolite, Haute Saone, *Thir.*
16. —— ferruginosus, *Voltz.* (Var.) Oxford Clay, Haute Saone, *Thir.*
17. —— aduncatus, *Miller.* Lias, Bath, *Lons.*
18. —— subclavatus, *Voltz.* Lias, Gudershofen ; Lias, Boll, *Voltz.*

19. Belemnites tenuis, *Stahl.*   Lias, Gundershofen, *Voltz.*
20. ————— subdepressus, *Voltz.* Lias, Gundershofen, *Voltz.*
21. ————— subaduncatus, *Voltz.* Lias, Gundershofen, *Voltz.*
22. ————— digitalis, *Biguet.* Lias, Gundershofen, *Voltz.*
23. ————— breviformis, *Voltz.* Lias, Gundershofen, *Voltz.*
24. ————— ventroplanus, *Voltz.* Lias, Béfort, Haut Rhin, *Voltz.*
25. ————— paxillosus, *Schlot.* Lias, Béfort; Lias, Boll, *Voltz.*
26. ————— longisulcatus, *Voltz.* Lias, Wurtemberg, *Voltz.*
27. ————— trifidus, *Voltz.* Lias, Gundershofen, *Voltz.*
28. ————— comprimatus, *Voltz.* Lias, Bahlingen, *Von Buch.*
————— sp., not determined. Kim. Clay and Inf. Oolite, Yorks., *Phil.* Kim. Clay, Coral Rag, Oxford Clay, Kell. Rock, Stonesfield Slate, Bradford Clay, and Inf. Oolite, Mid. and S. Eng., *Conyb.;* Oxford Clay, For. Marb., Great Oolite, Inf. Oolite, and Lias, Norm., *De C.;* Lias, N. of Fr., *Bobl.*
1. Orthoceras elongatum, *De la B.* Lias, Lyme Regis, *De la B.*
1. Nautilus hexagonus, *Sow.* Kell. Rock? Yorks., *Phil.;* Calc. Grit, Oxford, *Sow.*
2. ——— lineatus, *Sow.* Inf. Oolite and Lias, Yorks., *Phil.;* Inf. Oolite, Dundry, *Conyb.;* Inf. Oolite? Haute Saone, *Thir.;* Lias, Bath, *Lons.*
3. ——— astacoides, *Y. & B.* Lias, Yorks., *Phil.*
4. ——— annularis, *Phil.* Lias, Yorks., *Phil.*
5. ——— obesus, *Sow.* Inf. Oolite, Mid. and S. Eng., *Conyb.;* Inf. Oolite, Norm., *De C.*
6. ——— sinuatus, *Sow.* Inf. Oolite, Mid. and S. Eng., *Conyb.;* Oxford Clay, Norm.? *De la B.*
7. ——— intermedius, *Sow.* Lias, Mid. and S. Eng., *Conyb.*
8. ——— striatus, *Sow.* Lias, Mid. and S. Eng., *Conyb.;* Lias, Alsace, *Brong.*
9. ——— truncatus, *Sow.* Lias, Mid. and S. Eng., *Conyb.;* For. Marb. and Lias, Norm., *De Cau.*
10. ——— angulosus, *D'Orbigny.* Portland Stone, Isle d'Aix, *Brong.*
———, species not stated. Great Oolite, Yorks., *Phil.;* Kim. Clay, Coral Rag, Oxford Clay, Kell. Rock, and Stonesfield Slate, Mid. and S. Eng., *Conyb.;* Coral Rag, Norm., *De Cau.;* Fuller's Earth, N. of Fr., *Bobl.*
1. Ammonites perarmatus, *Sow.* Coral. Oolite, Calc. Grit, and Kel. Rock, Yorks., *Phil.;* Oolitic Rocks, Braambury Hill, Brora, *Murch.;* Coral Rag,

Wilts., *Lons.;* Coral Rag, Randen, *Von Buch.*

2. Ammonites plicomphalus, *Sow.* Kim. Clay? Yorks., *Phil.;* Oxford Clay, Norm., *De C.*

3. ———— triplicatus, *Sow.* Coral Oolite, Yorks., *Phil.;* Inf. Oolite, Norm., *De C.;* Coral Rag, Randen, *Von Buch.*

4. ———— plicatilis, *Sow.* Coral Oolite and Kell. Rock, Yorks., *Phil.;* Coral Rag, Mid. and S. Eng., *Conyb.;* Oxford Clay and Kell. Rock, Haute Saone, *Thir.;* Coral Rag, Randen, *Von Buch.*

5. ———— Williamsoni, *Phil.* Coral Oolite, Yorks., *Phil.*

6. ———— Sutherlandiæ, *Sow.* Sandstone, Braambury Hill, Brora, *Murch.;* Coral Oolite, and Calc. Grit, Yorks., *Phil.*

7. ———— sublævis, *Sow.* Coral. Oolite and Kel. Rock, Yorks., *Phil.;* Full. E., env. of Bath, *Lons.;* Oxford Clay, Beggingen, Schafhausen, *Von Buch;* Kell. Rock, Mid. and S. Eng., *Conyb.;* Oxford Clay, Norm., *De la B.*

8. ———— lenticularis, *Phil.* Coral. Oolite? Kell. Rock, and Lias, Yorks., *Phil.*

9. ———— vertebralis and cordatus, *Sow.* Coral. Oolite, Calc. Grit, and Oxford Clay, Yorks., *Phil.;* Coral Rag, Mid. and S. Eng., *Conyb.;* Oolite of Braambury Hill, Brora, *Murch.;* Kim. Clay and Oxford Clay, Haute Saone, *Thir.;* Coral Rag, Wilts., *Lons.*

10. ———— instabilis, *Phil.* Calc. Grit., Yorks., *Phil.*

11. ———— solaris, *Phil.* Calc. Grit., Yorks., *Phil.*

12. ———— oculatus, *Phil.* Oxford Clay, Yorks., *Phil.*

13. ———— Vernoni, *Bean.* Oxford Clay, Yorks., *Phil.*

14. ———— athleta, *Phil.* Oxford Clay and Kell. Rock, Yorks., *Phil.*

15. ———— Kœnigi, *Sow.* Kell. Rock, Yorks., *Phil.;* Kell. Rock, Mid. and S. Eng., *Conyb.;* Micaceous Sandst., Western Islands, Scot., *Murch.;* Solenhofen, *Hœn.*

16. ———— bifrons, *Phil.* Kell. Rock, Yorks., *Phil.*

17. ———— Gowerianus, *Sow.* Shale, Sandst. and Limest., Inverbrora, Scotl., *Murch.;* Kell. Rock, Yorks., *Phil.*

18. ———— Calloviensis, *Sow.* Kell. Rock, Yorks., *Phil.;* Kell. Rock, Mid. and S. Eng., *Conyb.*

19. ———— Duncani, *Sow.* Kell. Rock, Yorks., *Phil.;* Oxford Clay, Mid. and S. Eng., *Conyb.;* Oxford Clay, Norm., *De Cau.;* Oxford Clay, Haute Saone, *Thir.*

20. Ammonites gemmatus, *Phil.* Kell. Rock, Yorks., *Phil.*
21. ———— Herveyi, *Sow.* Kell. Rock? and Cornb., Yorks., *Phil.;* Inf. Oolite, Mid. and S. Eng., *Conyb.*
22. ———— flexicostatus, *Phil.* Kell. Rock, Yorks., *Phil.*
23. ———— funiferus, *Phil.* Kell. Rock, Yorks., *Phil.*
24. ———— terebratus, *Phil.* Cornb., Yorks., *Phil.*
25. ———— Blagdeni, *Sow.* Great Oolite, Yorks., *Phil. ;* Inf. Oolite, Dundry, *Conyb. ;* Inf. Oolite, Norm., *De Cau.*
26. ———— striatulus, *Sow.* Inf. Oolite and Lias, Yorks., *Phil.*
27. ———— heterophyllus, *Sow.* Lias, Yorks., *Phil.* Lias, Midland and Southern England, *Conyb. ;* Grafenberg, *Hœn.*
28. ———— subcarinatus, *Y. & B.* Lias, Yorks., *Phil.*
29. ———— Henleii, *Sow.* Lias, Yorks., *Phil. ;* Lias, Mid. and S. Engl., *Conyb.*
30. ———— heterogeneus, *Y. & B.* Lias, Yorks., *Phil.*
31. ———— crassus, *Y. & B.* Lias, Yorks., *Phil.*
32. ———— communis, *Sow.* Lias, Yorks., *Phil. ;* Lias, Mid. and S. Eng., *Conyb. ;* Lias, Western Islands, Scot., *Murch.;* Soleure, *Hœn. ;* Lias, Wurtemberg, *Zieten.*
33. ———— angulatus, *Sow.* Lias, Yorks., *Phil. ;* Lias, Mid. and S. Eng., *Conyb.*
34. ———— annulatus, *Sow.* Lias, Yorks., *Phil. ;* Inferior Oolite and Lias, Mid. and S. Eng., *Conyb. ;* Oxford Clay, For. Marb., and Inf. Oolite, Norm., *De C. ;* Inf. Oolite, Uzer, S. of Fr. ; Rochelle Limestone, *Dufr. ;* Inf. Oolite and Lias, Montdor, Lyon, *Al. Brong.;* Coral Rag, Inf. Oolite, Wilts., *Lons.;* Coburg, *Holl;* Inf. Oolite, Gamelshausen, Wurtemberg, *Zieten.*
35. ———— fibulatus, *Sow.* Lias, Yorks., *Phil.*
36. ———— subarmatus, *Sow.* Lias, Yorks., *Phil.*
37. ———— maculatus, *Y. & B.* Lias, Yorks., *Phil.*
38. ———— gagateus, *Y. & B.* Lias, Yorks., *Phil.*
39. ———— planicostatus, *Sow.* Lias, Yorks., *Phil. ;* Lias, Mid. and S. Eng., *Conyb.;* Lias, Bath, *Lons. ;* Kahlefeld, Hartz, Amberg, Altdorf, *Holl;* Lias, Bahlingen, *Von Buch.*
40. ———— balteatus, *Phil.* Lias, Yorks., *Phil.*
41. ———— arcigerens, *Phil.* Lias, Yorks., *Phil.*
42. ———— brevispina, *Sow.* Lias, Western Islands, Scot., *Murch. ;* Lias, Yorks., *Phil.*
43. ———— Jamesoni, *Sow.* Lias, Western Islands, Scot., *Murch.;* Lias, Yorks., *Phil.*
44. ———— erugatus, *Bean.* Lias, Yorks., *Phil.*

45. Ammonites fimbriatus, *Sow.** Lias, Yorks., *Phil.*; Lias, Mid. and S. Eng., *Conyb.*; Lias, Norm., *De C.*; Lias, Wurtemberg, *Zieten* ; Lias, Mende, Lozere, Banz.; Randen, *Von Buch.*
46. ——————— nitidus, *Y. & B.* Lias, Yorks., *Phil.*
47. ——————— anguliferus, *Phil.* Lias, Yorks., *Phil.*
48. ——————— crenularis, *Phil.* Lias, Yorks., *Phil.*
49. ——————— Clevelandicus, *Y. & B.* Lias, Yorks., *Phil.*
50. ——————— Turneri, *Sow.* Lias, Yorks., *Phil.*; Lias, South of France, *Dufr.*; Lias, Wurtemberg, *Zieten.*
51. ——————— geometricus, *Phil.* Lias, Yorks., *Phil.*
52. ——————— vittatus, *Y. & B.* Lias, Yorks., *Phil.*
53. ——————— sigmifer, *Phil.* Lias, Yorks., *Phil.*; Inf. Oolite, Haute Saone, *Thir.*; Lias, Wurtemberg, *Voltz.*
54. ——————— Hawskerensis, *Y. & B.* Lias, Yorks., *Phil.*
55. ——————— Conybeari, *Sow.* Lias, Yorks., *Phil.*; Lias, Mid. and S. Eng., *Conyb.*; Lias, Gundershofen and Buxweiller, *Al. Brong.*; Lias, Western Islands, Scot., *Murch.*
56. ——————— Bucklandi, *Sow.* Lias, Yorks., *Phil.*; Lias, Mid. and S. Eng., *Conyb.*; Lias, Norm., *De Cau.*
57. ——————— obtusus, *Sow.* Lias, Yorks., *Phil.*; Lias, Mid. and S. Eng., *Conyb.*
58. ——————— Walcotii, *Sow.* Lias, Yorks., *Phil.*; Inf. Oolite and Lias, Mid. and S. Eng., *Conyb.*; Lias, Norm., *De C.* ; Lias, S. of Fr., *Dufr.*; Lias, Béfort, Haut Rhin ; Lias, Boll,*Voltz.* ; Achelberg, *Hœn.*
59. ——————— ovatus, *Y. & B.* Lias, Yorks., *Phil.*
60. ——————— Mulgravius, *Y. & B.* Lias, Yorks., *Phil.*
61. ——————— exaratus, *Y. & B.* Lias, Yorks., *Phil.*
62. ——————— Lythensis, *Y. & B.* Lias, Yorks., *Phil.*
63. ——————— concavus, *Sow.* Lias? Yorks., *Phil.*; Inf. Oolite, Mid. and S. Eng., *Conyb.*; Lias, Norm., *De C.*; Coburg, *Holl.*
64. ——————— elegans, *Sow.* Lias? Yorks., *Phil.*; Inf. Oolite, Dundry, *Conyb.* ; Lias, Norm., *De C.*; Inf. Oolite, Uzer, S. of Fr., *Dufr.*
65. ——————— discus, *Sow.* Inf. Oolite, Dundry, Cornb., Mid. and S. Eng., *Conyb.*; Inf. Oolite, Norm., *De C.*; Cornb., Wilts., *Lons.*
66. ——————— Banksii, *Sow.* Inf. Oolite, Dundry, *Conyb.*
67. ——————— Braikenridgii, *Sow.* Inferior Oolite, Dundry,

---

* According to Von Buch, this ammonite is the same with *A. lineatus,* and *A. hircinus* of Schlotheim.

Conyb. Inf. Oolite, Norm., *De C.;* Porta
Westphalica, *Hœn.*
68. Ammonites Brocchii, *Sow.* Inf. Oolite, Dundry, *Conyb.;*
Inf. Oolite, Haute Saone, *Thir.*
69. ———— Sowerbii, *Miller.* Inf. Oolite, Dundry, *Conyb.*
70. ———— falcifer, *Sow.* Inf. Oolite, Dundry, *Conyb.;*
Lias, Norm., *De C.;* Lias, S. of Fr., *Dufr.;*
Lias, Wurtemberg, *Zieten.*
71. ———— Brownii, *Sow.* Inferior Oolite, Dundry, *Conyb.*
72. ———— læviusculus, *Sow.* Inf. Oolite, Dundry, *Braik-
enridge ;* Inf. Oolite, Norm., *De C.*
73. ———— acutus, *Sow.* Oxford Clay, Inf. Oolite, Norm.,
De C.; Lias, Western Islands, Scotl., *Murch.;*
Inf. Oolite, Haute Saone, *Thir.*
74. ———— contractus, *Sow.* Inf. Oolite, Dundry, *Sow.;* Inf.
Oolite, Norm., *De C.*
75. ———— giganteus, *Sow.* Portland Stone, Coral Rag, and
Lias, Mid. and S. Eng., *Conyb.;* Portland
Stone, Isle d'Aix, *Brong.;* (var.) Inf. Oolite,
Haute Saone, *Thir.*
76. ———— Lamberti, *Sow.* Portl. Stone, *Conyb.;* Rochelle
Limest., *Dufr.;* Coburg, Heinberg, Bamberg,
Holl.
77. ———— Nutfieldiensis, *Sow.* Portland Stone, *Conyb.*
78. ———— excavatus, *Sow.* Coral Rag, Mid. and S. Engl.,
Conyb.; Oxford Clay, Norm., *De la B.;* Lias,
Norm., *De C.;* Altorf, *Holl.*
79. ———— splendens, *Sow.* Coral Rag, Mid. and S. Eng.,
Conyb.
80. ———— armatus, *Sow.* Oxford Clay and Lias, Mid. and
S. Eng., *Conyb.;* Oxford Clay, Norm., *De
la B.;* Oxford Clay, Haute Saone, *Thir.;*
Lias, Bath, *Lons.*
81. ———— modiolaris, *Sow.* Fuller's Earth? Mid. and S.
Eng., *Conyb.*
82. ———— jugosus, *Sow.* Inf. Oolite, Mid. and S. Eng.,
Conyb.
83. ———— Stokesii, *Sow.** Inf. Oolite, Mid. and S. Eng.,
Conyb.; Lias, Norm., *De C.;* Lias, S. of Fr.
Dufr.; Inf. Oolite, Haute Saone, *Thir.;* Lias,
Wurtemberg, *Zieten.*
84. ———— Strangwaysii, *Sow.* Inf. Oolite, Mid. and S.
Eng., *Conyb.;* Lias, Norm., *De C.*
85. ———— falcatus, *Sow.* Inf. Oolite, Mid. and S. Engl.,
Conyb.

* *A. Amaltheus.*
R

86. Ammonites Brookii *Sow.*   Inf. Oolite and Lias, Mid. and
S. Engl., *Conyb.*
87. —————— Bechii, *Sow.*   Inf. Oolite and Lias, Mid. and
S. Eng., *Conyb.;* Lias, Lyme Regis, *De la B.;*
Coburg, *Holl.*
88. —————— stellaris, *Sow.* Lias, Mid. and S. Eng., *Conyb.;*
Lias, Norm., *De C.*
89. —————— Greenovii, *Sow.* Lias, Mid. and S. Engl.,
Conyb.; Lias, Lyme Regis, *De la B.*
90. —————— Loscombi, *Sow.* Lias, Mid. and S. Engl.,
Conyb.; Lias, Lyme Regis, *De la B.*
91. —————— Birchii, *Sow.* Lias, Mid. and S. Engl., *Conyb.;*
Lias, Lyme Regis, *De la B.*
92. —————— omphaloides, *Sow.* Oxford Clay, Norm., *De
la B.;* Gt. Arenaceous Formation, Western
Islands, Scotl., *Murch.*
93. —————— quadratus.   Inf. Oolite, Norm., *De C.*
94. —————— Gervillii, *Sow.*   Inf. Oolite, Norm., *De C.*
95. —————— Brongniartii, *Sow.*   Inf. Oolite, Norm., *De C.*
96. —————— biplex, *Sow.*   Inf. Oolite, Norm., *De C.;* Lias,
Ross and Cromarty, Scotl., *Murch.;* Oxford
Clay, Haute Saone, *Thir.;* Solenhofen, *Hœn.*
97. —————— rotundus, *Sow.*   Inf. Oolite, Norm., *De C.;*
Kim. Clay, Purbeck, *Sow.*
98. —————— complanatus.   Inf. Oolite, Norm., *De C.*
99. —————— decipiens.   Lias, Norm., *De C.*
100. —————— Deslongchampi.   Inferior Oolite, N. of Fr.,
Bobl.
101. —————— vulgaris.   Bradford Clay, N. of Fr., *Bobl.*
102. —————— coronatus.   Oxford Clay? N. of Fr., *Bobl.*
103. —————— Humphresianus, *Sow.*   Lias, S. of Fr., *Dufr.*
Inf. Oolite, Sherborne, *Sow.*
104. —————— Parkinsoni, *Sow.*   Lias, Bath, *Sow.*
105. —————— Gulielmii, *Sow.*   Oxford Clay, S. Engl., *Sow.*
106. —————— Davæi, *Sow.*   Lias, Lyme Regis, *De la B.*
107. —————— planorbis, *Sow.*   Lias, Watchet, Somerset, *Sow.*
108. —————— Johnstonii, *Sow.* Lias, Watchet, Somerset, *Sow.;*
Lias, Bath, *Lons.*
109. —————— corrugatus, *Sow.*   Inf. Oolite, Dundry, *Braikenridge.*
110. —————— rotiformis, *Sow.* Lias, Yeovil, *Sow.;* Lias, Bath,
Lons.
111. —————— multicostatus, *Sow.*   Lias, Bath, *Sow.*
112. —————— lævigatus, *Sow.*   Lias, Lyme Regis, *De la B.*
113. —————— latæcostata, *Sow.*   Lias, Lyme Regis, *Murch.*
114. —————— Murchisonæ, *Sow.*   Micaceous Sandst., Holm

Cliff, Western Islands, Scotl., *Murch.;* Inf.
Oolite, Allington near Bridport, *Murch.*
115. Ammonites serpentinus, *Schlot.* Inf. Oolite, Haute Saone,
Thir.; Lias, Gundershofen, *Voltz.*
116. ———— cristatus, *Sow.* Oxford Clay, Haute Saone, *Thir.*
117. ———— interruptus, *Schlot.* Oxford Clay, Haute Saone,
Thir.; Thirnau, *Holl.*
118. ———— fonticula, *Menke.* Oxford Clay, Haute Saone,
Thir.
119. ———— opalinus, *Reinecke.* Lias, Gundershofen, *Voltz.*
120. ———— latina, *Sow.* Coral Rag, Wilts., *Lons.*
121. ———— ammonius, *Schlot.* Lias, Gundershofen, *Voltz;*
Altdorf, *Holl.*
122. ———— comptus, *Reinecke.* Lias, Gundershofen, *Voltz.*
123. ———— planulatus, *De Haan.* Baireuth, *Holl.*
124. ———— Knorrianus, *De Haan.* Boll, Wurtemberg, *Holl.*
125. ———— Reineckii, *Holl.* Coburg, *Holl.*
126. ———— pustulatus, *De Haan.* Coburg; Thurnau, *Holl.*
127. ———— granulatus, *Brug.* Coburg, *Holl.*
128. ———— bifurcatus, *Brug.* Coburg; Baireuth, *Holl;*
Coral Rag, Germany, *Von Buch.*
129. ———— trifurcatus, *De Haan.* Coburg, *Holl.*
130. ———— macrocephalus, *Schlot.* Arau; Coburg, *Holl.*
131. ———— gracilis, *Munst.* Donzdorf, Wurtemberg, *Zieten.*
132. ———— Planula, *Heyl.* Donzdorf, *Holl.*
133. ———— Fonticola*, *Mencke.* Ferruginous Beds, Thurnau; Langheim; *Von Buch;* Gamelshausen,
Zieten; Oxford Clay, Haute Saone, *Thir.*
134. ———— scutatus, *Von Buch.* Lias, Banz, near Bamberg, *Von Buch.*
135. ———— canaliculatus, *Munst.* Coral Rag, Germany,
Von Buch.
136. ———— flexuosus, *Munst.* Coral Rag, Streitberg, near
Erlangen; Donzdorf, Swabia; Rathhausen,
near Bahlingen; summit of Mont Randen,
near Schafhausen, *Von Buch.*
137. ———— crenatus, *Rein.* Coral Rag, Germany, *Von Buch.*
138. ———— subfurcatus, *Schlot.* Lias, Göppengen, *Zieten.*
139. ———— costulatus, *Schlot.* Lias, Wasseralfingen, *Zieten.*
140. ———— striolaris, *Rein.* Eybach, Wurtemberg, *Zieten.*
141. ———— mæandrus, *Rein.* Inf. Oolite, Gamelshausen,
Zieten.
142. ———— abruptus, *Stahl.* Eybach, *Zieten.*
143. ———— sublævis, *Munst.* Donzdorf, *Zieten.*

* According to Von Buch this ammonite is figured as *A. Lunula* by
M. Zieten.

144. Ammonites punctatus, *Stahl.*   Inf. Oolite, Gamelshausen,
        Zieten.
145. ————— undulatus, *Stahl.* Lias, Gamelshausen, *Zieten.*
146. ————— complanatus, *Rein.* Inf. Oolite, Gamelshausen,
        Zieten.
147. ————— hecticus, *Rein.*   Inf. Oolite, Gamelshausen,
        Zieten.
148. ————— refractus, *Rein.*   Inf. Oolite, Gamelshausen,
        Zieten.
149. ————— annularis, *Rein.*   Inf. Oolite, Gamelshausen,
        Zieten.
150. ————— discus, *Rein.* Ganslosen ; Gruibingen, *Zieten.*
151. ————— Pollux, *Rein.* Inf. Oolite,Gamelshausen, *Zieten.*
152. ————— Bollensis, *Zieten.* Lias, Wurtemberg, *Zieten.*
153. ————— æquistriatus, *Munst.* Lias, Wurtemberg, *Zieten.*
154. ————— inæqualis, *Merian.* Bâle, *Merian.*
155. ————— tenuistriatus, *Munst.* Solenhofen, *Hœn.*
156. ————— dubius, *Schlot.* Lias, Gamelshausen, *Zieten.*
157. ————— Kridion, *Heyl.* Lias, Stutgard, *Zieten.*
158. ————— Jason, *Rein.* Lias, Gamelshausen, *Zieten.*
159. ————— alternans, *Von Buch.* Coral Rag, Muggendorf,
        Gailenreuth, &c. *Von Buch.*
  1. Trigonellites lamellosus,   . Discovered with the Ammonites
        discus, *Rein. ;* (Pseudo-ammonites, *Rüppell,*)
        Solenhofen, *Hœn.*
  1. Onychoteuthis angusta, *Munst.* Solenhofen, *Hœn.*
  1. Loligo priscus, *Rüppell.* Solenhofen, *Rüppell.*
  2. ——— antiqua, *Munst.* Solenhofen, *Hœn.*
  1. Sepia hastiformis, *Rüppel.* Solenhofen, *Rüppell.*
       ——, remains of, with ink-bags preserved, Lias, Lyme Regis,
        Buckl.
     Rhyncolites, or Sepia beaks, Lias, Lyme Regis, *De la B. ;*
        Lias, near Bristol, *Miller.*

CRUSTACEA.

  1. Pagurus mysticus, *Holl.* Solenhofen, *Holl.*
  1. Eryon Cuvieri, *Desm.* Solenhofen ; Erchstadt, Pappenheim,
        Holl.
  2. ——— Schlotheimii, *Holl.* Solenhofen, *Holl.*
  1. Scyllarus dubius, *Holl.* Solenhofen, *Holl.*
  1. Palæmon spinipes, *Desm.* Pappenheim, Solenhofen, *Holl.*
  2. ——— longimanatus,   Solenhofen, *Holl.*
  3. ——— Walchii, *Holl.* Pappenheim, *Holl.*
  1. Astacus modestiformis, *Holl.* Solenhofen, *Holl.*
  2. ——— minutus, *Holl.* Solenhofen, *Holl.*
  3. ——— rostratus, *Phil.* Kelloway Rock and Coral. Oolite,
        Yorks., *Phil.*

Astacus, species not determined. Oxford Clay and Lias, Yorks., *Phil.*

Crustacea, not yet determined. Lias, Midl. and S. Engl. *Conyb.*; Lyme Regis, *De la B.*; Forest Marble, Normandy, *De Cau.*; Stonesfield Slate, *Conyb.*; Bradford Clay, North of France, *Bobl.*

INSECTA.

Insects of the Libellula family, with others. Solenhofen, *Munst., Murch.*

Elytra of coleopterous insects, *Leach.* Stonesfield Slate, *Buckl.*

PISCES.

1. **Dapedium politum,** *De la B.* Lias, Lyme Regis, *De la B.*; Lias, and Oxford Clay of Normandy, *De Cau.*
1. **Clupea sprattiformis,** *Blain.* Solenhofen, *Holl.*

Fish, species not yet determined. Several in the Lias, Lyme Regis, *De la B.*; Barrow, Leicestershire, *Conyb.*

Ichthyodorulites, *Buckl. & De la B.* Different kinds. Lias, Lyme Regis, and elsewhere in Southern and Midland England, *Conyb. & De la B.*; Kimmeridge Clay, near Oxford, *Buckl.*; Stonesfield Slate, *Buckl.* In the Great Oolite, Normandy, *De Cau.*

Fish palates and teeth. Lias, Lyme Regis, and Somersetshire, &c. *Conyb.*; Stonesfield Slate, *Buckl.*; Great Oolite, Normandy, *De Cau.*; Cornbrash and Forest Marble, North of France, *Bobl.*; Coral. Oolite, Oxford Clay, Yorks., *Phil.*

REPTILIA.

1. **Pterodactylus macronyx,** *Buckl.* Lias, Lyme Regis, *Buckl.*
2. ———————— longirostris, *Cuv.* Aichstadt, *Collini.*
3. ———————— brevirostris, *Cuv.* Aichstadt, *Cuv.*
4. ———————— grandis, *Cuv.* Solenhofen, *Holl.*
5. ———————— crassirostris, *Goldf.* Solenhofen, *Goldf.*
6. ———————— medius, *Munst.* Monheim, *Schnitzlein.*
7. ———————— Munsteri, *Goldf.* Monheim, *Goldf.*
   ——————————, species not known. Stonesfield Slate, *Buckl.*
1. **Crocodilus Bollensis,** *Jäg.* Lias, Boll in Wurtemberg, *Jäg.*
2. ———————— priscus, *Soemmering.* Monheim, *Soemmering.*
3. **Gavial,** short-snouted. Kim. Clay, Havre, *Al. Brong.*
4. ———— long-snouted. Kim. Clay, Havre, *Al. Brong.*
5. **Crocodile of Mans,** *Cuv.* Great Oolite, *Brong.*
   ——————— remains, species not determined. Lias, Yorks., *Phil.*; Lias? Lyme Regis, *De la B.*; Cornbrash, Engl. *Conyb.*; Stonesfield Slate, *Buckl.*; Coral. Oolite, Yorks., *Phil.*

1. Tileosaurus, *Geoffroy St. Hilaire.* Great Oolite, Caen, *De Cau.*
1. Megalosaurus Bucklandi. Stonesfield Slate, *Buckl.*
————, species not known. Great Oolite, Normandy, *De Cau.*
1. Geosaurus Bollensis, *Jäg.* Lias, Boll, *Jäg.*
2. Geosaurus, *Cuv.* Monheim, *Soemmering.*
1. Lacerta Neptunia, *Goldf.* Monheim, *Goldf.*
1. Plesiosaurus dolichodeirus, *Conyb.* Lias, Lyme Regis, &c.
2. ———— recentior, *Conyb.* Kim. Clay, Engl. *Conyb.;* Kim. Clay, Honfleur, *Brong.*
3. ———— carinatus, *Cuv.* Great Oolite, Boulogne, *Brong.*
4. ———— pentagonus, *Cuv.* Great Oolite, Ballon and Chaufour, *Brong.*
5. ————? trigonus, *Cuv.* Great Oolite, Calvados, *Brong.*
6. ———— macrocephalus, *Conyb.* Lias, Lyme Regis, *De la B.*
————, species not determined. Oxford Clay, Stenay, *Bobl.;* Oxford Clay, Calvados, *De la B.;* Lias, N. of Ireland, *Bryce.*
1. Ichthyosaurus communis, *De la B.;* Lias, Lyme Regis, &c. Engl., *Conyb.,* &c.; Lias, Boll, Wurtemberg, *Jäg.*
2. ———— platyodon, *De la B.* Lias, Lyme Regis, &c. Engl., *Conyb.* &c.; Lias, Boll, *Jäg.*
3. ———— tenuirostris, *De la B.* Lias, Lyme Regis, &c., *Conyb.,* &c.; Lias, Boll, *Jäg.*
4. ———— intermedius, *Conyb.* Lias, Lyme Regis, &c. *Conyb.,* &c.; Lias, Boll, *Jäg.*
————, species not determined. Lias and Inferior Oolite, Normandy, *De Cau.;* Lias, Yorks., *Phil.;* Oxford Clay, England, *Conyb.;* Oxford Clay, Normandy, *De la B.;* Great Oolite, Reugny, *Brong.;* Coral. Oolite, Yorks., *Phil.;* Calc. Grit, Midl. Engl. *Conyb.;* Kim. Clay, Oxford, *Buckl.;* Kim. Clay, Weymouth, *De la B.;* Kim. Clay, Honfleur, *Brong.*
Saurian bones occur in the Kelloway Rock and Bath Oolite, Yorks., *Phil.;* in the Portland Stone, *Buckl. & De la B.*
Tortoise. Stonesfield Slate, *Buckl.;* Lias? Engl. *Conyb.*

MAMMALIA.

1. Didelphis Bucklandi, *Broderip.* Stonesfield Slate, *Buckl.*

It would appear from the foregoing lists, that our knowledge of the vegetable remains is too limited to enable us to form any general conclusions respecting them. Mammalia have only been

found in one locality, Stonesfield, where probably there are the
remains of more than one species of Didelphis. Pterodactyles
have been discovered at Solenhofen, where there would appear to
be many species; and at Lyme Regis, where there is another species. The remains of this strange genus probably also occur at
Stonesfield. Crocodiles seem to have existed during the whole
deposit of the oolitic group, and have been discovered in England,
France, and Germany. The Megalosaurus is found in Oxfordshire, in Normandy, and near Besançon. The Tileosaurus is discovered near Caen, Normandy. The Geosaurus has as yet been
noticed only in the lias of Wurtemberg, and in the Monheim beds *.
Ichthyosauri and Plesiosauri would appear to have been somewhat
widely distributed. Neither Pterodactyles, Crocodiles, Ichthyosauri, nor Plesiosauri, have yet been observed in the South of France.
Respecting Tortoises, Turtles, and Fish, we do not possess information which can lead to any useful conclusions. Insects have
been detected in the oolite of Stonesfield, and at Solenhofen. Polypifers occur in considerable abundance in particular places, and,
as it would appear, principally in the beds that have been named
Coral Rag, and in the upper part of the great oolite, which has
thus obtained in Normandy the name of *Calcaire à Polypiers.*
It has been imagined that the coral rag is a constant rock in the
oolitic series; which is supposing that during the depositions of
this group there was a time when the whole bottom of an extensive
sea was covered with an universal coral reef, and that the same
polypifers could exist under different pressures of water;—supposi-
tions which are at variance with the habits of existing polypifers.
Although it may be considered that in certain favourable situations
corals might produce sheets of calcareous matter, entombing a
variety of substances, there is no evidence that such are formed at
great depths; on the contrary, such knowledge as we possess would
appear to be in favour of small depths for the habitats of those genera, which are found to produce walls and sheets of coral. The
coral rag, which seems to exhibit the remains of ancient sheets
of polypiferous matter, contains abundantly various species and
individuals of the genera *Astrea, Meandrina,* and *Caryophyllia,*
which accords with the observations of MM. Quoy and Gaimard,
that these genera are the principal architects of the coral reefs of
the South Seas. The exact relative position of numerous species
of *Achilleum, Scyphia, Manon, Tragos,* and many other polypifers,
does not appear to be so well ascertained that we can compare it
with the subdivisions of the oolitic group in England and France.
The presence of coral reefs in one part of the series more than
others would seem to point to a difference in the relative levels of

---

* The reader will recollect that this genus is stated to occur among
the cretaceous rocks of North America.

the bottom and the surface of the sea throughout parts of England, France, and Germany, which should render the existence of coral reefs possible; while at other times, circumstances were not favourable for their production. It will be observed, then, when polypifers have occurred in abundance, that the Echinital family have not been wanting, particularly the genera *Clypeus* and *Cidaris.* The crinoidal remains are principally Apiocrinites and Pentacrinites; the former occurring in the great oolite or its accompanying beds, the cornbrash, forest marble, or Bradford clay; the latter abundantly and widely distributed in the lias.

Respecting the shells, the following summary will show those that have been discovered in the same division of the oolite series *, in more than one moderately distant locality, and the places where they have been observed will be found by reference to the foregoing list.

Fig. 69.

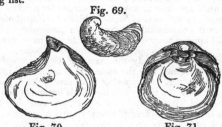

Fig. 70.   Fig. 71.

Kimmeridge Clay.—*Ostrea deltoidea* (Fig. 70.), a very characteristic shell in England; *O. Crista-galli; Gryphæa virgula* (Fig. 69.), a characteristic shell of this part of the oolitic series in France; *Pinna granulata; Trigonia clavellata; T. costata; Mya depressa; Pholadomya acuticostata; Pteroceras Ponti.*

Coral Rag.—*Ostrea gregarea; Pecten Lens; P. inæquicostatus; P. viminalis; P. vagans; Lima rudis; Plagiostoma rusticum; P. læviusculum; P. rigidum; Modiola bipartita; Gervillia aviculoides; Trigonia costata; T. clavellata; Turbo muricatus; Trochus bicarinatus; Melania Heddingtonensis; M. striata; Ammonites plicatilis; A. vertebralis.*

Oxford Clay.—*Terebratula ornithocephala; Ostrea palmetta; O Marshii; O. gregarea; Gryphæa dilatata* (Fig. 71.), a very cha-

* The student will have noticed that in the preceding general list, the same shell is stated to have been discovered in places distant from each other, but in various beds. Such shells are not here enumerated; and it may be questionable how far some of those shells stated to be found in remote situations in equivalent strata may really be so; for conclusions respecting the smaller divisions of the oolite frequently appear much forced.

racteristic shell in England and France; *Pecten fibrosus; P. Lens; Gervillia aviculoides; Trigonia clavellata; T. costata; Ammonites armatus; A. Kœnigi; A. Calloviensis; A. Duncani; A. sublævis; A. plicatilis.*

Compound Great Oolite, including Fuller's Earth, Great Oolite, Bradford Clay, Forest Marble, and Cornbrash.—*Terebratula subrotunda; T. intermedia; T. digona; T. obsoleta; T. reticulata; T. globata; T. coarctata; T. media; Ostrea Marshii; O. Cristagalli; O. costata; O. acuminata; Pecten fibrosus; P. vimineus; Plagiostoma cardiiforme; Avicula echinata; Av. costata; Lima gibbosa; Modiola imbricata; Perna quadrata; Trigonia clavellata; T. costata; Nucula variabilis; Isocardia concentrica; Patella rugosa.*

Inferior Oolite with its Sands.—*Terebratula sphæroidalis; T. ornithocephala; T. obsoleta; T. media; T. concinna; T. bullata; T. emarginata; Gryphæa Cymbium; Pecten Lens; Avicula inæquivalvis; Lima proboscidea; L. gibbosa; Plagiostoma giganteum; P. punctatum; Modiola plicata; Trigonia clavellata; T. striata; T. costata; Cardita similis; C. lunulata; Astarte excavata; Mya V scripta; Melania Heddingtonensis; M. lineata; Myoconcha crassa; Turbo ornatus; Trochus granulatus* (Sow.) *; T. fasciatus; T. sulcatus; T. ornatus; T. punctatus; T. elongatus; T. abbreviatus; T. bicarinatus; T. duplicatus; Ammonites læviusculus; A. discus; A. contractus; A. Blagdeni; A. Brocchii; A. acutus; A. Stokesii; A. Murchisonæ; A. Braikenridgii; A. elegans; A. annulatus; Nautilus lineatus; N. obesus.*

    Fig. 72.   Fig. 73.   Fig. 74.   Fig. 75.

          Fig. 76.   Fig. 77.

Lias.—*Spirifer Walcotii* (Fig. 77.), a very characteristic shell; *Terebratula ornithocephala; T. acuta; T. tetraedra; T. punctata; Gryphæa incurva* (Fig. 73.), a very characteristic shell; *G. obliquata; G. gigantea; G. Maccullochii; Plicatula spinosa; Pecten æquivalvis; P. barbatus; Plagiostoma giganteum* (Fig. 74.); *P. punctatum; P. Hermanni; Lima antiqua; Avicula inæquivalvis* (Fig. 76); *A. cygnipes; Modiola Scalprum; M. Hillana; Unio crassissimus; Pholadomya ambigua; Trochus Anglicus; T. imbricatus; Belemnites sulcatus; B. elongatus; B. apicicurvatus; B. pistilliformis; Ammonites Walcotii* (Fig. 72.), characteristic; *A. fimbriatus; A. Henleii; A. communis; A. planicostatus; A. fal-*

cifer; A. heterophyllus; A. brevispina; A. Jamesoni; A. Turneri; A. stellaris; A. Bucklandi (Fig. 75.), characteristic ; *A. obtusus; A. Stokesii* (A. Amaltheus); *A. sigmifer ; A. Conybeari ; A. concavus ; Nautilus lineatus.*

Although this list may assist the student, so far as to show the shells stated to be found in the same rock in various situations, he must be cautious in referring any particular beds, wherein he may detect any of the above remains, to the rock under the head of which such remains are here noticed ; but rather look at the general character of all the shells he may find in such beds, and thence infer their probable similarity, yet with much reserve, when the type and the rock considered equivalent to it are far distant from each other.

It has been above remarked, that the surface on which the oolitic group was deposited, was probably at very various depths beneath that of the sea ; and that even during the deposit itself, the sea varied in depth over the same point, in consequence of movements in the land. The nature of the organic remains also apparently points to the proximity of dry land in some places, while it may have been comparatively remote in others. It does not seem unphilosophical to infer that the bays, creeks, estuaries, rivers, and dry land, were tenanted by animals, each fitted to the situations where it could feed, breed, and defend itself from the attacks of its enemies. That strange reptile the Ichthyosaurus (one species of which, *I. platyodon,* was of a large size, the jaws being strong, and occasionally eight feet in length,) may, from its form, have braved the waves of the sea, dashing through them as the Porpess now does ; but the Plesiosaurus, at least the species with the long neck (*P. dolichodeirus*), would be better suited to have fished in shallow creeks and bays, defended from heavy breakers. The Crocodiles were probably, as their congeners of the present day are, lovers of rivers and estuaries, and like them destructive and voracious. Of the various reptiles of this period, the Ichthyosaurus, particularly the *I. platyodon,* seems to have been best suited to rule in the waters, its powerful and capacious jaws being an overmatch for those of the Crocodiles and Plesiosauri. Thanks to Professor Buckland, we are now acquainted with some of the food upon which these creatures lived : their fossil fæces, named *Coprolites,* having afforded evidence, not only that they devoured fish, but each other; the smaller becoming the prey of the larger, as is abundantly testified by the undigested remains.of vertebræ and other bones contained in the coprolites *. Amid such voracity, it seems wonderful that so many escaped to be imbedded in rocks, and after the lapse of ages on ages to tell the tale

---

* For an interesting account of Coprolites and their contents, see Buckland's Memoir, Geol. Trans. 2nd series, vol. iiL

of their existence as former inhabitants of our planet. And strange inhabitants they undoubtedly were; for, as Cuvier says, the Ichthyosaurus has the snout of a dolphin, the teeth of a crocodile, the head and sternum of a lizard, the extremities of cetacea (being, however, four in number), and the vertebræ of fish; while the Plesiosaurus has, with the same cetaceous extremities, the head of a lizard, and a neck resembling the body of a serpent *.

It is almost needless to remark that these two genera have disappeared from the surface of our planet; and, as the student may have collected from the various lists of organic remains, even previous to the deposit of the supercretaceous rocks, at least as far as regards Europe.

The vegetable remains have been accumulated in particular places, such, for instance, as Brora and Yorkshire. Circumstances, therefore, must have existed at such situations, during a particular part of the deposit, not common to a considerable surface, such perhaps as sheltered bays, into which the vegetables were drifted with mud and sand. The absence of remains, such as may be supposed, from analogy, to have been those of estuary animals, would seem to exclude rivers from having been the immediate cause of these carbonaceous accumulations. These deposits do not seem the result of violence; for the vegetables are well preserved, as if, like the *hortus siccus* of the botanist, for the purpose of examination. By their aid we learn that the vegetation which then clothed some parts of this portion of our planet, no longer resembles that which we now see, but one widely different. Perhaps we may, in anticipation, look forward to times when the geologist may speculate on the proximity of certain lands near places where the abundant remains of vegetables and certain animals would seem to point to such conclusions, even though, from the various movements in the land, no part of such ancient continents or islands may now appear on the surface. We of the present day, however desirous we may be to elucidate this subject, seem to possess too few data to proceed in the inquiry. One thing, however, the student should bear in mind; he must not consider that all older rocks in the vicinity of others of more recent origin, though now rising in mountains high above them, necessarily formed the dry land previous to the deposit of the newer rocks: for amid the various surface-changes that have been effected, such older rocks have frequently been upheaved after the formation of the more recent, as is shown by the mode of stratification near the junction of the two, the one being tilted up with the other. We may, perhaps, more closely approach the truth, when we find, as in Normandy, the oolitic group resting quietly

---

* Cuvier, Oss. Fossiles, t. v. This notice of the Plesiosaurus applies more particularly to *P. dolichodeirus.*

on, and surrounding, disturbed older strata; so that in that country (and the same observation applies to other situations) we may conceive the sea in which the oolitic rocks were formed, to have bathed the slaty and granitic districts of Normandy and Britanny.

Those strange flying creatures, the Pterodactyles, must have sported on dry land, probably subsisting on insects, such, among others, as that figured beneath, which was obtained by Mr. Murchison from the quarries at Solenhofen, where the remains of Pterodactyles are also discovered.

Fig. 78.

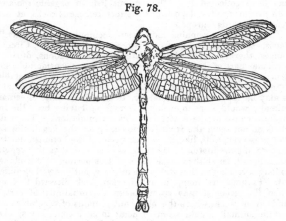

Scale = ½ of nature.

That the pterodactyles should be scarce fossils, is what we should expect, for the circumstances favourable to their preservation must have been extremely rare. Even supposing that they dashed out to sea in pursuit of their insect prey, there must have been a combination of fortunate accidents to have prevented the pterodactyles and their intended prey from being devoured by the fish and other inhabitants of the sea, among the exuviæ of which their remains are now detected.

It is curious, and seems to establish a connection between the insects and the pterodactyles, that in the spot where the remains of the latter are most abundant, the greatest quantity of fossil insects yet noticed in the oolitic group, have been detected. At Stonesfield also, where the remains of insects are stated to have been discovered, the exuviæ of pterodactyles, according to Prof. Buckland, are also observed. Not so, however, with the pterodac-

tyle of Lyme Regis, whose remains are mixed with those of ichthyosauri and other marine animals, where insects have not yet
been discovered. But when we consider the abundant exuviæ of plesiosauri, perhaps we may not err greatly, in considering dry land
not very far distant from the spot where we now find their bones
entombed.   Be the case as it may, a pterodactyle in a sea, amid
ichthyosauri and other voracious creatures, must have had but a
slight chance of escape; and geologists should be grateful that
any combination of circumstances should have so far prevailed, as
to permit the preservation of even a single individual, to show us
the strange terrestrial creatures that then existed.

    In the lias of Lyme Regis, the ichthyosauri, plesiosauri, and
many other animals, seem to have suffered a somewhat sudden
death; for in general the bones are not scattered about, and in a
detached state, as would happen if the dead animal had descended
to the bottom of the sea, to be decomposed, or devoured piece-
meal, as indeed might also happen if the creature floated for a
time on the surface, one animal devouring one part, and another
carrying off a different portion;—on the contrary, the bones
of the skeleton, though frequently compressed, as must arise
from the enormous weight to which they have so long been subjected, are tolerably connected, frequently in perfect, or nearly
perfect, order, as if prepared by the anatomist.   The skin, moreover, may sometimes be traced, and the compressed contents
of the intestines may at times be also observed;—all tending to
show that the animals were suddenly destroyed, and as suddenly
preserved.   Not only has this apparently happened to these reptiles
which, breathing air, might under favourable circumstances be
drowned simultaneously in great numbers, but also to the molluscæ, to which constant, or nearly constant, immersion in water
is absolutely necessary.   Among the multitude of ammonites discovered in the lias, I have often observed individuals, of which the
large terminating chamber of the last whorl, where the body of the
animal seems to have been placed, was hollow for half its distance
upwards towards the aperture or mouth, as if the animal, when
overwhelmed, had retreated as far as possible into this part of the
shell, so that the muddy matter was prevented from completely
filling it.   This idea is rendered more probable from the condition
of the calcareous matter filling the remaining part of the great
cavity, which is exceedingly bituminous, as would happen from the
decomposition of the animal within the remainder of the chamber.
The student should not, from what has been above remarked respecting the lias at a particular point, Lyme Regis, consider that
such observations are applicable to the same rock generally; or
even that the lias of Lyme Regis has suddenly been produced
in its whole thickness at once: on the contrary, the lias varies
materially at different points, as we should expect it to do, from

different local causes; and the lias of Lyme Regis bears evidence
of successive deposition, in part during a state of comparative
tranquillity, and partly in consequence of a series of small catastrophes, suddenly destroying the animals then existing in particular spots.   One observation is, however, necessary, and it will
be often applicable to other parts of the oolitic rocks in various
situations,—that during the formation of the lias in this part of England, there has been a certain change in the animal life of the
same place. Thus the animals and shells in the upper part of this
rock differ in the mass from those in the lower portion. Very
frequently also, particular strata afford certain organic remains,
while all others are exceedingly rare.

Notwithstanding the temptation to treat of the probable circumstances that have accompanied the deposit of a particular rock,
even within the distance of a few miles, we must abstain, as it
would lead us into detail not compatible with this little work. It
may however be remarked, that the destruction of the animals,
whose remains are known to us by the name of Belemnites, was exceedingly great at this place. When the upper part of the lias was
deposited, multitudes seem to have perished simultaneously, as is
attested by a bed composed of little else, beneath Golden Cap, a
cliff between Lyme Regis and Bridport Harbour. Not only are
millions entombed in this bed, but in the upper part of the lias
generally. The production of such a bed would seem by no means
difficult; for we have only to consider the occurrence of some circumstance destructive to molluscous creatures in the fluid containing, or otherwise carrying, the belemnites,—such as might happen
to those swarms of molluscæ which sometimes surround the navigator in warm latitudes,—and the floating animal mass, if not immediately, would eventually descend to the bottom; at least all
those that escaped the predaceous animals, which indeed might
be driven away by the circumstance, whatever it was, that destroyed the belemnitic animals. Suppose a multitude of the common cuttle-fish to be suddenly killed by the irruption of, or their
entrance into, water charged with carbonic acid; their internal
bones, as they are commonly termed, would be distributed over a
common surface after the decomposition of the animals, which
were not likely to fall a prey to other creatures; for those which
were not destroyed with the cuttle-fish, would avoid the water so
charged with carbonic acid.

The vegetables of the lias of this place occur in two different
states: the one showing that they have been scarcely injured before they were imbedded; the other seeming to point to the fracture of wood into junks, the small branches truncated, as if they
had been broken either during, or previous to, their drift. These
latter most frequently occur in argillo-calcareous nodules, often
of large size; but the nodules are not concentric concretions;

on the contrary, both these nodules, and those that frequently envelope the Ammonites and Nautili in the argillaceous beds, are
fissile, the line of the laminæ being parallel to that of the general
stratification ; so that though the nodules, particularly those containing Ammonites and Nautili, are spheroidal, their fracture is
lamellar ; and a successful blow in the line of the laminæ, through
the centre, discloses a fossil, which is sometimes a fish.

It being a very interesting inquiry to ascertain the chemical
condition of organic remains entombed in various rocks, Dr.
Turner was kind enough to analyse certain fossils from the lias of
Lyme Regis. He found that a vertebra, a rib, and a tooth of an
Ichthyosaurus examined by him, had all a highly crystalline texture, owing to the deposit of carbonate of lime, of which they chiefly
consist. The colour is nearly, and in parts quite black, in consequence of bituminous matter, which in general amounts to not
more than $\frac{1}{2}$, and he has not found it exceed $\frac{3}{4}$ per cent. The
phosphate of lime in the vertebra amounted to about 29 per cent,
while in the rib and tooth it was about 50 per cent. In fact, as
might have been expected, the phosphate of lime remains in
greater or less quantity in different specimens, probably depending
on the situation where it was preserved, and on the compactness
of the original bone.

Dr. Turner also ascertained that the scales of *Dapedium politum*,
cleared as much as possible from adhering limestone, consisted of
the same ingredients as the ichthyosaurian bones; but the phosphate of lime amounted only to 19 per cent.

Of course care was taken to select such specimens as were not
impregnated with sulphuret of iron, as sometimes happens ; and
those examined were found to be remarkably free from iron, manganese, alumina, and silex.

SECTION VII.

RED SANDSTONE GROUP.

SYN.—Red or Variegated Marls (*Marnes Irisées,* Fr.; *Keuper,* Ger.). Muschelkalk (*Calcaire Conchylien,* Al. Brong.). Red or Variegated Sandstone (*New Red Sandstone,* Eng. Auth.; *Grès Bigarré,* Fr.; *Bunter Sandstein,* Ger.). Zechstein (*Magnesian Limestone,* Eng. Auth.; *Calcaire alpin,* Fr.; *Alpenkalkstein,* Ger.). Red Conglomerate (*New Red Conglomerate, Exeter Red Conglomerate,* Eng. Auth.; *Todtliegendes, Rothe Todte Liegende,* Ger.; *Grès Rouge,* Fr.; *Pséphite Rougeâtre,* Al. Brong.).

THIS group, which is often one of very considerable thickness, succeeds, in the descending order, that previously noticed.   Perhaps very fine lines of distinction should not be drawn between the two ; for when the lower part of the one and the upper part of the other have been considerably developed, they seem in some measure to pass into each other.   This led M. Charbaut, who first observed the circumstance in the vicinity of Lons le Saulnier, to class the lias with the variegated marls, which constitute the upper portion of the group under consideration.   The rocks composing the red sandstone group occur in the following descending order:— 1. Variegated Marls; 2. Muschelkalk; 3. Red or Variegated Sandstones; 4. Zechstein; and 5. Red Conglomerate, or Todtliegendes.

Variegated Marls.—In the district of the Vosges and in the neighbouring countries, these commence beneath the sandstone named lias sandstone, into which they gradually pass ; the upper part of the variegated marls, which are green, presenting thin beds of black schistose clay, and of quartzose sandstone, nearly without cement, which latter gradually becomes the lias sandstone,—a rock which passes into the lias, and contains the same organic remains *.   M. Elie de Beaumont observes, that in many countries the variegated marls can scarcely be separated from the lias sandstone, even artificially, as is done in the Vosges; for they appear to become one deposit, as in the environs of St. Leger-sur-Dheune, and Autun, and in the *arkose* of Burgundy.   The variegated marls of the Vosges generally are, as their name implies, marked by different colours, among which the principal are wine-red and greenish or blueish gray ; they break into fragments, which have no trace of a schistose structure.   In the central portion of these marls there are beds of black schistose clay, blueish gray sandstone, and grayish or yellowish magnesian limestone.   The sandstone and clay contain vegetable impressions, and even coal.   Masses of rock-salt occur in the lower part of the marls at Vic, Dieuze,

* Elie de Beaumont, Mém. pour servir á une Désc. Géol. de la France, t. i.

and other parts of that district; and masses of gypsum are found in the upper and lower portions, but principally in the latter *. According to M. Charbaut, limestone beds, almost entirely composed of shells, are found in the upper part of this deposit.

The variegated marls, not differing considerably in their mineralogical characters, occur in various parts of the neighbouring districts of France and Germany, and according to M. Dufrénoy they crown the red sandstone rocks of the South of France. How far the variegated marls may be traced in England remains questionable; but it would appear far from improbable, that the upper part of the red sandstone deposit of this country would answer sufficiently well in its mineralogical structure to the rocks above noticed in the Vosges. There is with us no apparent passage of the lias into the red sandstone series; on the contrary, we sometimes have, as at the Old Passage near Bristol, a kind of conglomerate of pieces of limestone, bones, teeth, and other remains of saurians and fish, with their fossil fæces or *coprolites,* which would seem to mark a period when comminuted deposits ceased, and currents of water sufficient to transport pebbles were in action, accumulating bones and other substances, as at the bottom of some seas. Where seen on the southern coast of England, between Lyme Regis and Sidmouth, the upper part of the red sandstone series is so like the variegated marls of the Vosges and parts of Germany, that I have little hesitation in considering them contemporaneous deposits. In this part of England these marls contain vegetable remains, and, though rarely, scales of fish, and bones of pterodactyles (?). According to M. Rozet, the upper part of the variegated marls contains the teeth and bones of saurians, with *Pectines* and *Entrochi* †.

Organic Remains.—No great variety has yet been observed; they however seem to be, besides those enumerated above,— VEGETABLES: *Equisetum Meriani,* Al. Brong., Neuewelt, near Bâle; *E. columnare,* Al. Brong., Lorraine, Alsace, Wurtemberg, (Rozet); *Pecopterus Meriani,* Al. Brong., Neuewelt; *Tæniopteris vittata,* Al. Brong., Neuewelt (also Hör, Scania); *Pterophyllum longifolium,* Al. Brong., Neuewelt; *Pt. Meriani,* Al. Brong., Neuewelt; (according to M. Rozet, these marls also contain *Calamites arenaceus? Filices Stuttgardiensis, F. lanceolatus,* and *Pterophyllum Jägeri*). REPTILES: *Phytosaurus cylindricodon,* Jäger, Boll, Wurtemberg (Jäger); *P. cubicodon,* Jäger, Boll (Jäger); *Ichthyosaurus* and *Plesiosaurus,* species not stated, Dürrheim (Hœninghaus). The other remains are: *Posidonia Keuperina,* Voltz, Hall, Wurtemberg (Hœninghaus) ‡. *Ophiura,*

* Elie de Beaumont, Memoir above cited.
† Rozet, Cours E'lémentaire de Géognosie.
‡ This fossil is noticed as being found in slate between the muschelkalk and variegated marls.

species not determined, Vosges (Rozet); *Saxicava Blainvillii,* Ballbron (Hœninghaus).

As the lower lias sandstone passes into the variegated marls, and even seems in some measure equivalent to them, a deposit of sands having possibly taken place in one situation, while marls were produced in another, we should not, when considering the general subject, force our conclusions too far, nor carry those divisions which may be locally useful, beyond the countries where they may be advantageously employed.   Prof. Pusch, in his very interesting account of the Polish rocks, has shown that between the oolitic series of Poland and the muschelkalk there is an extensive and important deposit of sandstone, usually termed *white sandstone,* from its colour.   The deposit is divisible into two portions ; the upper being formed of the white sandstone, while the lower part is composed of alternations of fine white marly sandstone, schistose sandstone, shale, and other schistose and dark-coloured rocks, the whole inclosing beds of coal from three to twenty-five inches thick.   The white sandstone of the upper part alternates with thick beds of gray blue marls, partly red, and more rarely variegated.   Beds of limestone are also found in it; but the most valuable product is iron ore, which furnishes the largest amount of iron of any rock in Poland, twenty-seven furnaces affording annually 560,000 quintals of metal.   Fossils are rare in this deposit, with the exception of vegetable remains.   M. Pusch refers this rock to the lias sandstone, the same as it occurs in Suabia, in Scania, and in the Isle of Bornholm, in all which places it is rich both in iron and coal＊.   It seems to unite both the characters of the variegated marls and lias sandstones of the South of Europe, the two being intimately blended.

The following are the fossils mentioned as occurring in the sands known as lias sandstone : *Belemnites Aalensis,* Voltz, Aalen, Wurtemberg ; *Tellina striata,* Vic (Hœninghaus) ; *Clathropteris meniscoides,* Ad. Brong., Hör, Scania ; and St. Étienne ; Vosges, Al. Brong.; *Glossopteris Nilssoniana,* Ad. Brong., Hör, Scania, Ad. Brong.; *Pecopteris Agardhiana,* Ad. Brong., Hör; *Tæniopteris vittata,* Hör; and also in the variegated marls, Neuewelt, near Bâle, *Marantoidea arenaria,* Jäger, Stuttgard; *Lycopodites patens,* Ad. Brong., Hör; *Culmites Nilssonii,* Ad. Brong., Hör; *Pterophyllum Jägeri,* Ad. Brong., Stuttgard ; *Pt. dubium,* Ad. Brong., Hör ; *Nilssonia brevis,* Ad. Brong., Hör; and several undetermined shells.

Muschelkalk.—A limestone varying in texture, but being most frequently gray and compact.   It is occasionally dolomitic, and passes into marls above and beneath.   When very compact, with

＊ Pusch, Esquisse Géognostique du Milieu de la Pologne ; Journal de Géologie, t. ii.

numerous remains of the *Encrinites liliiformis,* Schlot. (a very characteristic fossil of at least a considerable portion of the deposit), it has much the appearance of some varieties of the carboniferous limestone of England. It is sometimes, as at Epinal (Vosges), sufficiently hard to be employed as marble. In some situations organic remains would appear to be very abundant, while in others they are somewhat rare. According to M. Alberti, salt is contained in the muschelkalk of Wurtemberg *. This rock would appear to be unknown in England and in the North of France, but on the east and south of the latter country, and in parts of Germany, it is found interposed, in its place, between the variegated marls and red or variegated sandstone. According to Prof. Pusch it occurs in Poland, and is described as being gray and yellow.

Organic Remains of the Muschelkalk.

REPTILIA.

Plesiosaurus . Boll, Wurtemberg, *Jäger.*
Ichthyosaurus . Place not stated, *Hœninghaus.*
Great Saurian, genus not determined. Lunéville, *Al. Brong.*

CRUSTACEA.

1. Palinurus Sueurii, *Desm.* Dürrheim, *Hœninghaus.*

MOLLUSCA AND CONCHIFERA.

1. Nautilus bidorsatus, *Schlot.* Weimar, *Hœn.;* Heinberg, Wurtemberg, *Al. Brong.*
1. Ammonites nodosus, *Schlot.* Weimar, *Hœn.;* Göttingen, Wurtemberg; Toulon, *Al. Brong.,* Lorraine, *Beaum.*
2. —————— bipartitus, *Gaillardot.* Lunéville, *Al. Brong.*
3. —————— Henslowi?† *Sow.* Baireuth, *Hœninghaus.*
1. Buccinum obsoletum, *Schlot.* Göttingen, *Hœn.*

---

* Alberti, Die Gebirge des Konigreihs Würtemberg, 1826.

† If this shell be really discovered in the muschelkalk of Germany, it is worthy of notice that the *Ammonites nodosus* is reported to be found in the Isle of Man, whence the *Ammonites Henslowi* was first obtained. The general character of the sutures in both shells is similar, being that represented in fig. 84. It must however be confessed that the supposed association of these two shells in the Isle of Man, which might lead us at first to consider the occurrence of muschelkalk in that island as probable, is by no means so clear as could be wished. Ammonites with a certain general resemblance may have been mistaken for each other. The organic remains, which, according to Prof. Henslow, are associated with *A. Henslowi* in the Isle of Man, are such as are commonly found in the carboniferous and grauwacke limestones (*Trilobites, Producta Scotica,* &c.)

1. Turritella terebralis, *Schlot.*   Weimar, *Hœninghaus.*
1. Dentalites torquatus, *Schlot.*   Göttingen, *Al. Brong.*
2. ———— lævis, *Schlot.*   Göttingen, *Al. Brong.*
1. Terebratula perovalis,   Jena, *Hœninghaus.*
2. ———— ——— sufflata,   Jena, *Hœninghaus.*
3. ———— ——— vulgaris, *Schlot.* Göttingen, *Hœn.;* Wurtemberg,
        Lunéville, Toulon, *Al. Brong.*
4. ———— ——— orbiculata, *Schlot.*   Dornberg, *Hœn.*
1. Trigonia vulgaris, *Schlot.*   Weimar, Erlingen, *Hœn.* Göt-
        tingen, *Al. Brong.*
2. ———— Pes-anseris, *Schlot.*   Lunéville, Mosbach, *Hœn.* Göt-
        tingen, *Al. Brong.*
1. Mytilus eduliformis, *Schlot.*   Dürrheim, *Hœn.* Göttingen,
        Lunéville, *Al. Brong.*
2. ———— socialis *, *Schlot.*   Weimar, *Hœn.;* Göttingen, Mont
        Meisner, Wurtemberg, Lunéville, *Al. Brong.*
1. Myacites musculoides, *Schlot.*   Weimar, *Hœninghaus.*
2. ———— intermedius,   Mezieres, *Hœn.*
3. ———— elongatus, *Schlot.*   Wurtemberg, *Al. Brong.*
4. ———— ventricosus, *Schlot.*   Lunéville, *Al. Brong.*
1. Pecten reticulatus, *Schlot.*   Göttingen, *Hœninghaus.*
1. Ostrea spondyloides,   . Quedlinburg, *Hœn.;* Göttingen,
        Lunéville, Toulon, *Al. Brong.*
?1. Cardium striatum,   . Wurtemberg, Göttingen, *Al. Brong.*
1. Plagiostoma lineatum,   . Mossbach, Michelstadt, Horgen,
        Hœn.; Göttingen, *Al. Brong.*
2. ———— ——— rigidum,   . Rauhthal near Jena, *Hœn.;* Göt-
        tingen, *Al. Brong.*
3. ———— ——— lævigatum,   . Mosbach, *Hœninghaus.*
4. ———— ——— punctatum †,   . Göttingen, Toulon, Gotha,
        Al. Brong.

RADIARIA.

1. Encrinites liliiformis, *Schlot.*   Göttingen, Wurtemberg, Alsace,
        &c., *var. Authors.*
2. ———— epithonius,   . Soleure, *Hœninghaus.*

VEGETABLE.

1. Neuropteris Gailliardoti, *Ad. Brong.*   Lunéville, *Al. Brong.*

To this list should be added those bodies named *Rhyncolites,*
which are apparently the beaks of the Sepia family.
Red or Variegated Sandstone.—This rock is, as its name im-

---

* M. Brongniart remarks that this shell is certainly not a *Mytilus.*
† This is the *Chamites punctatum* of Schlotheim.

plies, of different tints,—these being red, white, blue, and green ; the former, however, greatly predominating. It is principally siliceous and argillaceous, occasionally containing mica, masses of gypsum, and rock-salt. In the Vosges, the upper part of the variegated sandstone often presents, according to M. Elie de Beaumont, thin beds of marly limestone and dolomite, which gradually become more abundant ; so that, finally, they constitute the lower part of the muschelkalk*. An oolitic and calcareomagnesian rock † is found in this deposit in some parts of Germany, and conglomerates are also included in it.

A very extensive deposit, varying but little in its character, occurs in the Vosges, and has thence obtained the name of the *Grès de Vosges.* A difference of opinion seems to exist between M. Elie de Beaumont and M. Voltz respecting the exact member of the red sandstone series to which this rock should be referred ; the former considering it the equivalent of the rothe todte liegende, which occurs beneath the zechstein ; the latter, that it is the lower portion of the red or variegated sandstone which rests on the zechstein : as the zechstein is wanting in the district, there is perhaps but little essential difference in these opinions, as will be noticed in the sequel.

The *Grès de Vosges* is essentially composed of amorphous grains of quartz, commonly covered by a thin coating of red peroxide of iron ; among which are discovered others which appear fragments of felspar crystals. It is often marked by cross and diagonal laminæ so common in arenaceous rocks, the result, probably, of deposit by cross currents of water. The rock contains quartz pebbles, sometimes so abundantly as to present a conglomerate with an arenaceous cement. From the mineral character of these pebbles, M. Elie de Beaumont considers that they are derived from the destruction of the older rocks, and are merely larger portions which have better resisted trituration than the smaller grains composing the body of the sandstone.

The variegated or red sandstone of some countries affords a good building-stone, and when nearly free from colour, as at Épinal (Vosges), one of handsome appearance. In situations where it becomes schistose from mica, it is often employed, like some varieties of the old red sandstone of the English, for flag-stones, and even tiles for houses.

According to Professor Sedgwick, the red sandstone occurring above the magnesian limestone, in the North of England, represents the Bunter sandstein of Germany, the variegated marls surmounting it being the equivalent of the keuper of the same country. This sandstone is represented as of a complex character,

* Elie de Beaumont, Terrains Secondaires du Système des Vosges.

† The grains forming this oolitic rock are radiated from the centre to the circumference.

from the variable mixtures of sand, sandstone, and marl. In its range from Nottinghamshire into Yorkshire it is generally coarse, often nearly incoherent, and here and there passes into a fine conglomerate. The superincumbent marls are red and gypseous*.
Organic Remains of the Red or Variegated Sandstone.—M. Elie de Beaumont notices that at Domptail (Vosges), this rock contains abundantly the casts of shells, for the greater part, of the same genera, and even of the same species, as those discovered in the muschelkalk. And M. Voltz remarks, that the red sandstone of the Vosges presents the following shells:—*Terebratula, Trigonia, Pecten, Plagiostoma, Avicula (Mytilus) socialis, Turritella? Natica?* These different shells would appear, from the lists of MM. Al. Brong-
niart and Hœninghaus, to have received
the following names : *Turritella Schoteri*
(Sulz-les-Bains, near Strasburg) ; *Natica
Gaillardotii,* Voltz (Domptail) ; *Mytilus
eduliformis,* Schlot. (Domptail); *Trigonia vulgaris,* Schlot. (Domptail) ; *Plagiostoma lineatum* (Sulz-les-Bains) ; *P. striatum* (Sulz-les-Bains) ; *Myacites elongatus* (Sulz-les-Bains); *M. musculoides* (Sulz-
les-Bains) ;—fossils nearly the whole of
which have been enumerated above as
found in the Muschelkalk.

Fig. 79.

The following list of Vegetables, with
their localities, is from M. Adolphe Brongniart : *Calamites arenaceus,* Wasselonne
and Marmoutier (Bas-Rhin); *C. Mougeotii,* Marmoutier ; *C. remotus,* Wasselonne ; *Anomopteris Mougeotii,* Wasselonne, Sulz-les-Bains; *Nevropteris Voltzii,*
Sulz-les-Bains; *N. elegans,* Sulz-les-Bains;
Sphenopteris Myriophyllum, Sulz-les-
Bains ; *Sp. palmetta ; Filicites scolopen-
droides ; Voltzia brevifolia ; V. elegans ;
V. rigida ; V. acutifolia ; V. heterophylla ;
Convallarites erecta ; C. nutans ; Pale-
oxyris regularis ; Echinostachys oblongus ;
Æthophyllum stipulare ;*—all from Sulz-
les-Bains†.

[The above wood-cut, taken from a figure by M. Adolphe Brongniart (Ann. des Sci. Nat. t. xv. pl. 16.), represents the fructification of the *Voltzia brevifolia,* as exhibited in a specimen from Sulz-les-Bains.]

---

* Sedgwick, Geol. Trans. 2nd series, vol. iii.
† Brongniart, Prodrome d'une Histoire des Végétaux Fossiles, 1828.

Zechstein.—This name has, fortunately, been applied by Humboldt to distinguish a limestone series of a very variable character, to which different names were given, the term *zechstein* having been previously applied to only one of the varieties. The various beds were known to the German miners by the names of Asche (friable marl), Stinkstein (fetid limestone), Rauchwache, Zechstein, and Kupferschiefer (copper-slate) ; the latter and lowest deposit being worked for the copper it contains, particularly in the Mansfeld country, Thuringia, Franconia, and the Hartz. According to Daubuisson, the mean thickness of the copper-slate in these countries is about one foot. The zechstein is represented as sometimes from twenty to thirty yards thick ; the rauchwacke, when pure and compact, one yard thick, when cellular sometimes attaining fifteen to sixteen yards; the stinkstein, from one to thirty yards thick; and the asche, very variable. Notwithstanding these minor divisions, to which an extraordinary value has been attached, it does not appear that they can always be observed in the countries where they have been established; for Daubuisson observes, that the upper portions pass into each other, and even sometimes into the zechstein.

According to Professor Sedgwick, the magnesian limestone of the North of England, which is the equivalent of this deposit in Germany, is divisible into, 1. Marl slate and compact limestone, or compact and shelly limestone, and variegated marls. 2. Yellow magnesian limestone. 3. Red marl and gypsum. 4. Thin-bedded limestone. The same author considers No. 1. as equivalent to the kupferschiefer and zechstein, and Nos. 2. 3. and 4. to the rauchwacke, asche, stinkstein, &c. of Thuringia.

Organic remains of the Zechstein and Copper Slate.

REPTILIA.

Monitor of Thuringia, *Cuv.* Copper or Bituminous Slate, Mansfeld; Rothenburg on the Saale; Gluckbrunn ; Memmingen, &c. *Al. Brong.*

PISCES.

1. Palæothrissum macrocephalum, *Blain.* Cop. or Bit. Slate, Mansfeld, *Al. Brong. ;* Marl Slate, Midderidge and East Thickley, *Sedg.*
2. ———————— magnum, *Blain.* Cop. or Bit. Slate, Mansfeld, *Al. Brong. ;* Marl Slate, Midderidge and E. Thickley, *Sedg.*
3. ———————— inæquilobum, *Blain.* Bit. Slate, Autun, *Al. Brong.*
4. ———————— parvum, *Blain.* Bit. Slate, Autun, *Al. Brong.*
5. ———————— macropterum, *Bronn.* Cop. Slate, Börschweiler, Thuringia, *Hœn.*

384   *Organic Remains of the Zechstein and Copper Slate.*

6. Palæothrissum elegans,   Marl Slate, Midderidge and
        East Thickley, *Sedg.*
—————————, sp. not determined.   Marl Slate, Midderidge
        and East Thickley, *Sedg.*
Fish, genera not determined.   Marl Slate, East Thickley,
        Sedg.; Mag. Limestone, Pallion, *Winch.*

MOLLUSCA AND CONCHIFERA.

Turbo? ————. Mag. Limest., Marr and Hickleton, *Sedg.*
Pleurotomaria? ————. Mag. Limest., Humbleton, *Sedg.*
Melania? 5 species. Mag. Limest., Hawthorn Hive, *Phil.*
Ammonites, sp. not determined.   Humbleton, *Sedg.*
1. Producta aculeata*, *Al. Brong.* Büdengen; Neustadt, *Hœn;*
        Thuringia, &c., *Al. Brong.;* Durham and North-
        umberland, *Sedg.*
2. ———— rugosa, *Schlot.* Röpsen, near Gera, *Hœn.*
3. ———— speluncaria, *Al. Brong.* Röpsen, *Hœn.;* Glückbrunn,
        Al. Brong.
4. ———— antiquata, *Sow.* Midderidge, *Sedgwick.*
5. ———— calva, *Sow.* Humbleton; Midderidge, &c., *Sedg.*
6. ———— spinosa, *Sow.* Humbleton, &c., *Sedg.*
7. ———— longispina, *Sow.* Cop. Slate, Schmerbach, Thuringia,
        Hœn.
1. Spirifer trigonalis, *Sow.* Röpsen, *Hœn.*
2. ———— undulatus, *Sow.* Midderidge; Humbleton, *Sedg.*
3. ———— multiplicatus,   Humbleton, *Sedg.*
4. ———— minutus,   Humbleton, *Sedg.*
1. Terebratula intermedia,   Röpsen, *Hœn.*
2. ———— inflata, *Schlot.* Röpsen, *Hœn.;* Schmerbach, *Al.*
        Brong.
3. ———— cristata,   Röpsen, *Hœn.*
4. ———— lacunosa, *Schlot.* Cop. Slate, Schmerbach; Zech-
        stein, Röpsen, *Hœn.*
5. ———— paradoxa, *Schlot.* Schmerbach, *Al. Brong.*
6. ———— elongata, *Schlot.* Schmerbach, *Al. Brong.*
7. ———— pelargonata, *Schlot.* Schmerbach, *Al. Brong.*
8. ———— pygmæa, *Schlot.* Leimstein near Schmalkalde,
        Al. Brong.
————————, sp. not determined.   Durham, *Sedg.*
1. Axinus obscurus, *Sow.* Durham, *Sedgwick.*
1. Arca tumida, *Sow.* Humbleton, Durham, *Sedg.*
1. Cucullæa sulcata, *Sow.* Humbleton, Durham, *Sedg.*
1. Avicula gryphæoides, *Sow.* Humbleton (abundant), *Sedg.*

* M. Hœninghaus considers this shell, which is the *Gryphites aculeatus*
of Schlotheim, the same as the *Producta horrida* and *P. scabricula* of
Sowerby.

Ostrea, sp. not determined. Northumberland, *Sedg.*
Astarte? ————. Whitley, Northumberland, *Sedg.*
1. Modiola acuminata, *Sow.* Black Rocks, Durham, *Sedg.*
———, sp. not determined. Durham, *Sedg.*
1. Mytilus squamosus, *Sow.* Ferrybridge, *Sedg.*
Pecten, sp. not determined. Humbleton, &c., *Sedg.*
Plagiostoma? ————. } Humbleton, *Sedg.*
Venus? ————.

RADIARIA.

1. Cyathocrinites planus, *Miller.* Mag. Limestone, Durham and
Northumberland, *Sedg.*
1. Encrinites ramosus, *Schlot.* Glückbrunn, *Al. Brong.*
Crinoidea, genera not determined. Durham and Northumberland, *Sedg.*

ZOOPHYTA.

1. Retepora flustracea, *Phil.* Shelly Mag. Limest., Durham,
Sedg.
2. ———— virgulacea, *Phil.* Shelly Mag. Limest., Durham,
Sedg.
Polypifers, genera not determined. Durham and Northum-
berland, *Sedg.*

PLANTÆ.

1. Fucoides Brardii, *Ad. Brong.* Cop. Slate, Frankenberg,
Al. Brong.
2. ———— selaginoides, *Ad. Brong.* Cop. Slate, Mansfeld,
Al. Brong.
3. ———— lycopodioides, *Ad. Brong.* Cop. Slate, Mansfeld,
Al. Brong.
4. ———— frumentarius, *Ad. Brong.* Cop. Slate, Mansfeld,
Al. Brong.
5. ———— pectinatus, *Ad. Brong.* Cop. Slate, Mansfeld,
Al. Brong.
6. ———— digitatus, *Ad. Brong.* Cop. Slate, Mansfeld, *Al.*
Brong.
1. Cupressus Ullmanni, *Bronn.* Locality not named, *Hœn.*
Vegetables not determined. In the Marl Slate and Blue
Shelly Limestone, Durham, *Sedg.*

Todtliegendes.—This name is given to a series of red conglomerates and sandstones which occur between the zechstein or magnesian limestone and the rocks of the next group. The term is
applied to those beds of Thuringia and other adjacent countries

S

upon which the copper slate reposes, with the intervention only of portions which are white. It is for the most part a conglomerate, formed from the partial destruction of those rocks on which it rests, the fragments being sometimes angular as well as rolled, and of considerable size.

It seemed necessary to premise the above notices of the organic contents and more remarkable mineralogical structures of the various rocks of this group, known as Variegated or Red Marl, Muschelkalk, Red or Variegated Sandstone, Zechstein, Todtliegendes, in order that the student might be acquainted with the whole when fully developed. Taken as a mass, the group may be considered as a deposit of conglomerate, sandstone, and marl, in which limestones occasionally appear in certain terms of the series; sometimes one calcareous deposit being absent, as the muschelkalk is in England; sometimes the zechstein, as in the East and South of France; and sometimes both being wanting, as in Devonshire. The conglomerates, or todtliegendes, commonly occupy the lowest position, though conglomerates are occasionally noticed higher in the series; the sandstones form the central part, and the marls occur in the highest place.

When we look for the causes which have produced this mass, we may, perhaps, in some measure approach them, by observing the state of the rocks on which it rests. These are found in the greater number of instances highly inclined, contorted, or fractured;—evidences of disturbance which the inferior and older rocks have suffered previous to the deposit of the red sandstone group upon them. These appearances are not confined to particular districts, but are more or less general. From an examination of the lower beds, no doubt can exist that the fragments of rock contained in them have, for the greater part, been broken off from the older rocks of the more immediate neighbourhood. It therefore does not appear unphilosophical to conclude, that, as far at least as regards these lower conglomerate beds, we have approached to something like cause and effect, the cause being the disruption of the strata, the effect being the dispersion of fragments, consequent on this violence, over greater or less spaces by means of water, probably thrown into agitation by the disturbing forces. That these forces have, in some places at least, not been small, is attested by the large size of the fragments driven off, and the rounded condition of some of them, as may be well seen in the vicinity of Bristol, where the rolled masses of carboniferous limestone are sometimes considerable. Of the evidence of the great force employed, I know of no better and easily observed example, than that at the cliff named Petit Tor, in Babbacombe Bay, Devon, whence so large a portion of Devonshire marble is obtained. Of this the following is a section :

Fig. 80.

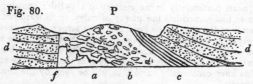

P. Petit Tor Cliff. *a.* Fractured limestone, the rents filled, when sufficiently open, with the finer matter of the conglomerate above ; when small, with carbonate of lime. *b.* A breccia composed of large blocks (some many tons in weight) of the same marble limestone as that on which it rests, mixed with others which are smaller. The cementing matter is sometimes a red sandstone, at others a reddish clay. The marble (known as Babbacombe marble) is wholly derived from these blocks, which are detached from their situations, and either partially worked on the spot or removed elsewhere. Upon this rest beds of fine conglomerate, sandstone, and red marl at *c*, which are surmounted by a considerable thickness of red conglomerate *d*, extending many miles eastward, and composed of angular pieces of limestone, numerous pieces of slate, such as is of common occurrence in the surrounding country, as also of pebbles of flinty slate, grauwacke, &c. Among these are rounded pieces of various red quartziferous porphyries. *f.* A fault or dislocation of the strata, bringing down the conglomerates on the left hand against the fractured limestones on the right. Such faults or dislocations are common in the district.

The annexed figure (81.) represents one of the fissures in the fractured limestone at Petit Tor, filled with the matter of the superincumbent conglomerate. *b, b.* Limestone. *a.* Fissure filled with the smaller matter of the red conglomerate above.

Fig. 81.

It will, I think, be scarcely doubted that the angular blocks of the conglomerate *b* (Fig. 80.) have been detached by violence from the limestone *a*, and that during the commotion they were thrown upwards in such a manner that other and smaller detrital substances were insinuated between them ; the watery mass being highly charged with sand, mud, and other substances held in mechanical suspension. It may be right, while on the subject of these Devonshire conglomerates, to adduce evidence of the unequal action of currents of water, in this vicinity, at the same period. There is perhaps no situation where better examples of this can be observed than on the line of cliffs between Babbacombe and Exmouth. The alternations of conglomerates and sandstones at the upper part of the conglomerate series are very fre-

quent, more particularly in the vicinity of Dawlish ; showing that
the water had sometimes the power of carrying forward rounded
fragments of the size of the head and even larger, while at others
it merely accomplished a transport of sand. Not only do the alternations exhibit this difference in the velocity of water, but the
structure of the beds themselves shows that the directions of the
currents have continually varied, as will be seen by the annexed
wood-cuts.

Fig. 82.   Fig. 83.

Fig. 82. in the cliff west of Dawlish.   Fig. 83. on the east of
the same place. *a.* Conglomerate. *b. b. b. b.* Sandstones deposited
by changing currents. *c.* Wavy sandstone. The velocity of
the currents must have varied considerably in the immediate
neighbourhood of these sections; for amidst sandstones and moderately sized conglomerates on the west side of Little Haldon
Hill, there are blocks of quartziferous porphyry, generally rounded, of a ton or more in weight. Being scattered on the side of
the hill, they might be mistaken for superficial erratic blocks,
did we not find them in their proper situations on the sea cliffs,
imbedded in the mass of rock. The transport of these must have
required water moving with considerable velocity, so great, possibly, as to grind down by attrition against each other, the rock
fragments of inferior hardness, while the pieces of quartziferous
porphyry being exceedingly hard and of very difficult fracture
have better resisted attrition.

The presence of these porphyries in the red conglomerate of
South Devon, where they are abundant, is somewhat remarkable,
for they have not yet been satisfactorily traced to their parent
rocks.   The absence of such rocks on the exposed surface is certainly no proof that they may not be near ; for when we consider
the area covered by the red sandstone series in that district, there
is ample space for the abundant occurrence of such rocks beneath
the sandstone; and there are also many unexplored situations,
where they may yet be detected among the rocks now uncovered by the sandstone series. The student must be careful
not too hastily to generalize on such facts as have been above
noticed in Devonshire, for the appearances may be more or less
local. When however we extend our observations, we find that
conglomerates are very characteristic of deposits of the same age
in other parts of Britain, France, and Germany, and they most
frequently, though not always, rest on disturbed strata. As we

can scarcely conceive such a general and simultaneous movement in the inferior strata, immediately preceding the first deposits of the red sandstone series, that every point on which it reposes was convulsed and threw off fragments of rocks at the same moment, we should rather look to certain foci of disturbance for the dispersion of fragments or the sudden elevation of lines of strata, sometimes, perhaps, producing lines of mountains, in accordance with the views of M. Elie de Beaumont. The accumulation of the larger fragments, and the relative amount of conglomerate, would, under this hypothesis, be greatest nearest to the disturbing cause ; and amid such turmoil we might anticipate the occurrence of igneous rocks thrown up at the same period. If we return for the moment to that part of Devonshire with which we commenced these remarks, we shall observe facts which seem to afford support to this view ; for where the conglomerates are abundant, there is no want of trappean rocks in the vicinity, such as various greenstones and porphyries, which have cut and broken through the slates, limestones, and other older rocks, in various directions. But notwithstanding the abundance of these trap rocks, not a fragment of them has yet been discovered among the conglomerates, though the latter are extensively laid open to examination by coast and other sections. This fact seems to attest that the trap rocks in question did not exist in such a state, when fragments of slate, limestone, &c., were broken off, that they could be fractured and transported with the rest; though it does not show that the trap rocks may not have been protruded at the time of the convulsion, thus aiding the confusion, and in a great measure causing it. Another fact observable in this district, namely, the occurrence of trappean rocks (at Heavitree near Exeter, and other places,) so blended with the conglomerates that lines of separation cannot be drawn between them, also seems to point to a similar conclusion; for if igneous rocks were ejected—a conclusion which the facts appear to justify—at the time of the production of the conglomerate, there would seem no reason why, under favourable circumstances, the two should not be in some measure blended with each other. Another circumstance also lends probability to this view, and that is the occurrence of pebbles cemented in certain inferior beds, (well observed on the coast and in-land between Babbacombe Bay and Teignmouth, at the Corbons, Torbay, in the vicinity of Exeter, and other situations,) by a kind of semi-trappean paste, containing crystals of that variety of felspar named Murchisonite by Mr. Levi. Such a cement might possibly have resulted from the upburst of igneous rocks, accompanied by various gases beneath a mass of water, when some of the erupted matter may have so combined as to form a cement, in which crystals of Murchisonite became developed : without some such hypothesis this cement seems of very difficult explanation.

We must now turn from this scene of disturbance, which may
be one of the extreme cases, to that state of things where no violent
disrupting cause is to be surmised, but where, on the contrary,
the causes which produced the arenaceous rocks which constitute
the upper portion of the next, and inferior group, have not been
interrupted by any sudden violence, one series of rocks passing
into the other, so that the exact lines of demarcation are imaginary. Such a state of things is perfectly consistent with local and
violent disturbances; for the consequences of a violent disruption
of the inferior rocks would extend no further than to distances
proportioned to the agitating cause; and the effects would gradually
become less, until finally the deposits at remote places would not
be interrupted, though the disturbing causes may have produced
such a general state of things in the fluid mass, and in the relative positions of land and water, that future deposits would have
an altered character;—one more common over a large area.

This supposed passage of certain lower parts of the red sandstone group into the upper part of the coal measures, seems also
supported by facts; for such is stated to be the case in certain parts
of the continent of Europe (Thuringia, &c.), so that some geologists, and among them Humboldt, Daubuisson, and others, consider the two rocks as one.

Between such extremes there would be every variety of deposit, produced either by difference in the intensity of the disturbing forces, or by local circumstances. Thus, sands and little
or no conglomerate might be found resting unconformably upon
older rocks, even in the vicinity of greatly disturbed situations, as
may be ocasionally observed in the district first noticed.

After the causes, whatever they were, which produced the conglomerates and sandstones known by the name of *Todtliegendes*, had in some measure been modified, a considerable deposit of carbonate of lime, often charged with carbonate of magnesia, took place over certain parts of Europe. This is the *Zechstein*, which, though somewhat extensively developed in certain
parts of Germany and England, seems little known in France.
The causes which produced this limestone have therefore not been
so general as those which have furnished the limestones formerly
noticed under the head of the Oolitic Group, which are distributed
over a far larger area. A deposit of bituminous or marly slate
appears to have been contemporaneous at distant places, in parts of
Germany and in the North of England, containing the remains of a
marked genus of fishes, *Palæothrissum*. There is nothing in itself
remarkable that the same fish should be discovered in rocks formed
within the same geological epoch, at such distances as Mansfeld
and Durham; for if these districts were now beneath a common
sea, no naturalist would be surprised that cod-fish, turbots, and
many other fish, should be caught at the two places, being aware

that cod-fish are found on the shores of North America and Europe, and that salmon ascend the rivers of both continents. The geologist, therefore, should expect to find the remains of similar fish entombed in contemporaneous deposits within certain reasonable limits of latitude and longitude.

As yet, these fish seem only to have been observed in the copper slate, or its equivalent marly slate, and they have apparently perished by some common cause; what that cause was, is by no means clear; but certainly waters which held the component parts of the copper slate of Thuringia either in chemical solution or mechanical suspension, would be far from favourable to their existence; and if the fish should by any chance be enveloped by, or enter into, such a medium, they are little likely to escape from it alive. When we consider the numerous marine animals always ready to prey upon fish either dead or alive, and the small chance that any part of them will remain undevoured, their occurrence in a fossil state would seem to show that the fossil individuals have been so circumstanced that the creatures which preyed on them were either destroyed with them, avoided those situations which had been fatal to the fish, or were otherwise unable to get at them.

By reference to the lists of organic remains, it will be observed that marine vegetables occur with the fish in the copper slate. Now certainly these could no more exist in a medium impregnated with copper, than the fish; and therefore one would suppose they existed prior to the presence of such medium: but as we cannot be certain that these grew near the spot where now entombed, (for marine plants may, like the Gulf Weed in the Atlantic, be drifted considerable distances,) they do not afford direct proof that the copper slate was of sudden formation. The remains of the Monitor seem to indicate a certain proximity of land, while the mass of fossils in the copper slate of Thuringia point to a peculiarly marine origin.

The remainder of the zechstein deposit is of a very mixed character; part being such as we may consider mechanical, while much seems a deposit from a solution of carbonate of lime, carbonate of magnesia, and sulphate of lime. The very frequent occurrence of the two latter in rocks that have apparently originated from some common causes, is very remarkable, and has not yet received any satisfactory explanation.

In Somersetshire and the neighbouring districts, as will be found detailed in the valuable memoir of Prof. Buckland and Mr. Conybeare, the lower part of the red sandstone group is very frequently a conglomerate, composed of the broken fragments of inferior and older rocks, united by a cement containing much magnesia, whence the term Magnesian or Dolomitic conglomerate. This rock sometimes graduates into a limestone of a more homogeneous character, apparently containing also much magnesia. This conglomerate seems the result of violent action on the carboniferous rocks of the district, detaching various portions of them; in fact, producing effects similar to those noticed under the head of Todt-liegendes. How far it may be the exact equivalent of the latter deposit, that is, how far the epoch of disturbance may be precisely contemporaneous, may admit of doubt; for the disturbance which caused the deposit of the todtliegendes in Germany may have pre-ceded that in Somersetshire, so that the latter may have been brought more within the influence of a spread of calcareous and magnesian matter. Still, however, the production of the magne-sian conglomerates of Somersetshire and the todtliegendes of Thuringia would not appear to be widely separated from each other as to time; they both constitute the lower part of the red sandstone group in their respective situations, and both contain fragments of rocks commonly derived from their more immediate vicinities.

The organic character of the zechstein approaches, as far as re-searches have yet gone, that of the next, or carboniferous group; *Productæ*, which abound in the carboniferous limestone, being not only discovered in the zechstein, (and the student will observe that these shells are now introduced for the first time to his atten-tion,) but also *Spirifers*, shells which also abound in the carboni-ferous limestone.

This resemblance in organic character will at all times render the determination of the two rocks difficult, when their geological position cannot be ascertained with certainty, as it may be in Ger-many and England; and this difficulty may in some cases be con-sidered as insurmountable, should the deposit of the two groups have been continuous, without a violent break, the limestones of the carboniferous group being dispersed through the coal measures (the upper part of the next group) in such a manner that they should approach the upper terms of the series on the one hand, while the zechstein should descend towards the lowest parts of the red sandstone group on the other. We might under these circumstances have a series of arenaceous and limestone rocks re-presenting the carboniferous group and the lower part of the red sandstone group, with one common, or nearly common, organic character.

The zechstein is surmounted by a mass of rocks for the most part arenaceous, though occasionally argillaceous, gypseous and saliferous. The predominant colour is red, though it is not unfre-quently variegated, whence the names *Bunter sandstein, Grès bigarré.* Where the zechstein is wanting, this sandstone graduates into the inferior conglomerates; and when the muschelkalk is ab-sent, as is commonly the case in England, into the red or variegated marls : therefore where both calcareous deposits are wanting, as in Devonshire, the whole group is composed of conglomerates in the lower part, sandstones in the central, and marls in the upper part; an arrangement which suggests the possibility of the whole, in that district, being the result of some violent commotion, which, as the disturbing causes ceased, deposited the various matters held by water in mechanical suspension, somewhat in the order of their specific gravities; always considering the deposit in the mass: for not only are there alternations of conglomerates and sandstones, sandstones and marls, where these pass into each other on the large scale, but frequent mixtures of them also occur on the small scale, a circumstance easily accounted for by the various directions and velocities of the currents produced.

Viewed in the mass, circumstances appear to have been unfavourable in those parts of Europe which have been best examined, if not to the existence of animal and vegetable life, at least to their envelopment and preservation; for, with the exception of Alsace and Lorraine, few or no organic remains have been detected in it. The vegetables have been enumerated in the foregoing lists, from the descriptions of M. Adolphe Brongniart, as also the remains of shells noticed by M. Voltz and others. It will be observed that these shells are not analogous to those found in the zechstein, but to those discovered in the muschelkalk, a rock well developed in the same district; and it is further important to observe, that the remains discovered by M. Elie de Beaumont in the sandstone under consideration in the district of the Vosges, were not far beneath the muschelkalk. I obtained numerous fragments of vegetables from the sandstone near Epinal, Vosges, and the quarrymen informed me that they very commonly discovered them.

We next arrive, in the ascending order, to the Muschelkalk, a limestone the general characters and known extent of which have been above noticed. We here have evidence, that probably at the same epoch, a deposit of calcareous matter, mixed sometimes with carbonate of magnesia, took place, if not continuously, at least at various places, from Poland to the South of France inclusive, and that the marine animal life distributed over this surface was nearly of the same kind. But it is a remarkable circumstance, that this life was not of the same kind as that which existed at the time when the zechstein was formed; the organic character of the two rocks is distinct, and therefore those who found their divisions of strata solely on this character, do well in drawing a line between the zechstein and muschelkalk. If, however, we look at these rocks on the large scale, and see the mineralogical passages which exist between the muschelkalk and the rocks above and beneath it, and observe that the latter is far from being a constant rock in the series, and that, when it is absent, those beds between which it is interposed graduate into one another, there seems, theoretically, a difficulty in separating them, and, practically, a very great inconvenience in doing so. In whatever manner this may be considered, the fact appears certain, that circumstances had arisen, changing the character of marine life over certain portions of Europe; that certain animals abounding previously, and apparently for a great length of time, (for, as will be seen in the sequel, they are enveloped in various thick and older deposits,) have disappeared never to reappear, at least as far as we can judge from our knowledge of organic remains.

Among the organic remains of the muschelkalk, two of the most characteristic of which are considered to be the *Ammonites nodosus* (Fig. 84.) and *Encrinites liliiformis* (*E. moniliformis* Miller), are reptiles of various forms. That extraordinary genus the Plesiosaurus, and perhaps also his common fossil companion the Ichthyosaurus, then existed near what now constitutes the eastern part of France, and the

Fig. 84.

adjoining portion of Germany. How far these singular Saurians now first appeared in any numbers on this part of the globe, it would be premature to say; for it must always be recollected that the preservation of such remains would seem in some measure to depend on local circumstances, possibly also, in some cases, on the proximity of land, and the chance that if drifted about at sea when dead, they escaped the other predaceous animals of the deep, all, great and small, ready to devour them.

The Red or Variegated Marls, which surmount the muschelkalk, possess a common mineralogical character over very considerable surfaces, such as would lead us to suppose some cause or causes exerting an influence of a similar kind over a large area. At least some of the deposit would appear chemical, more particularly the masses of gypsum and rock salt which exist in certain situations. How far the mass of marls may be partly chemical or wholly mechanical, may in the present state of science admit of a doubt; but the sandstones with which they are in some countries connected, and which even seem to replace them in others, (as has been above noticed,) are of mechanical origin, inclosing beds of coal, the result probably of accumulated vegetable matter. Some of the vegetable remains are still sufficiently preserved to be determined, as in the red or variegated marls of the Vosges.

If we now abstract our attention from these divisions, and regard the group as a mass, it would seem to constitute the base of a great system of rocks, which when not deranged by local accidents has filled numerous hollows and inequalities of land over considerable parts of Europe. Such a hollow is well seen in our own island, where the central counties are occupied by the red sandstone series, apparently filling up a previously existing depression in that situation; but it is here without that great capping of the oolitic group, which for the most part rests so conformably upon it; so that taken as a whole, and abstraction being made of minor derangements, they would both seem to fill up great depressions in Europe; sometimes, as is the case in Normandy, the oolitic rocks overlapping and coming in contact with strata older than the red sandstone group, upon which latter they nevertheless rest so conformably that the one seems a tranquil deposit on the other. We must of course consider that numerous local disturbances would produce a marked difference in the deposits, even amounting to a perfectly unconformable position; yet the conformable nature of the two groups taken in the mass is somewhat striking. During their deposit, great and remarkable changes were effected in animal and perhaps vegetable life; and it seems somewhat necessary to admit that considerable differences in the relative levels of sea and land were produced at various times, causing changes in the character of the inhabitants of the sea, from variations of pressure and other circumstances, while no small difference might be effected from the filling up and rise of the bottom.

It would appear, more particularly from the descriptions of Humboldt, that very extensive tracts of red sandstones and conglomerates exist in Mexico and South America: how far these may be of contemporaneous production with the red sandstone series of Europe, the state of science does not permit us very satisfactorily to determine. The porphyries and slates of New Spain are surmounted by red conglomerates and sandstones, forming the plains of Celaya, Salamanca, and Burras, and supporting a limestone which mineralogically resembles those of the Jura. The conglomerates contain fragments of pre-existing rocks, cemented by an argillo-ferruginous, and yellowish brown or brick red, paste. In Venezuela, the vast plains are in a great measure covered by red sandstones and conglomerates, with limestones and gypsum; the former being deposited in a concave manner between the coast mountains of the Caracas and the mountains of Parima, resting on slates, termed transition, on the north, while on the south they repose upon granite. This arenaceous deposit is covered, at Tisnao, by compact whitish gray limestone. An immense extent of red sandstone is described as "not only covering, nearly without interruption, the southern plains of New Grenada, between Mompox, Mahates, and the mountains of Tolu, and Maria, but also the basin of the Rio de la Magdalena, between Teneriffe and Melgar, and that of the Rio Cauca, between Carthago and Cali." The conglomerates of this country are composed of angular fragments of lydian stone, clay slate, gneiss and quartz, cemented by argillaceous and ferruginous matter. These conglomerates alternate with schistose and quartzose sandstones *. According to Humboldt, the Cordilleras of Quito presented him with the greatest extent of red sandstone which he had observed, covering the whole plateau of Tarqui and Cuenca for twenty-five leagues. The sandstone is generally very argillaceous, with small grains of slightly rounded quartz; but it is sometimes schistose, and alternates with a conglomerate containing fragments of porphyry from three to nine inches in diameter. The same author considers that the red sandstone of Cuença also occurs in High Peru, and remarks on the resemblance of these rocks of New Grenada, Peru and Quito, to the red sandstone or *todtliegendes* of Germany †.

A series of red sandstones, intermixed with conglomerates, occurs extensively in Jamaica, particularly in the Port Royal and St. Andrew's mountains, stretching thence north-west towards the north side of the island. The sandstone is generally siliceous and compact, intermixed with marly red sandstone and marl, and, though rarely, with gypsum (Hope Valley). The conglomerate is formed of pebbles (from an inch to four inches in diameter) of granite, large-grained greenstone, syenite, quartz, hornstone, &c. Beds of a gray colour are intermixed with these rocks, and subordinate to them are strata of compact gray limestone and of shale, and schistose sandstone intermixed with coal. The higher portion of the mass is formed of a conglomerate in a great measure composed of pieces of trap rocks, principally porphyry, the cementing matter being most frequently reddish brown and argillaceous, varying in induration, sometimes so obscure that the pebbles seem joined by a trappean cement. Mixed more particularly with this superior portion there is a great variety of trappean rocks, such as syenite, greenstone, porphyries, &c. appearing as if an upburst of igneous matter had accompanied the production of the conglomerates. The red conglomerates and sandstones pass beneath into a rock, which at first differs from them only in colour, and finally presents the mineralogical character of grauwacke. The aggregate thickness of the whole is considerable, amounting to several thousand feet. These rocks appear to me the equivalent of those named red sandstone in the neighbouring continent of America ‡.

The mere mineralogical resemblance of this deposit, in America and Jamaica, with the sandstones and conglomerates of the red sandstone group of Europe, is in itself of no great value, and

---

* Humboldt, Gisement des Roches dans les deux Hémisphères.
† Ibid.
‡ For a more detailed account of these and other Jamaica rocks, with sections, consult my Remarks on the Geology of Jamaica, Geol. Trans., 2nd series, vol. ii.

therefore we can only at present conclude that considerable forces have been exerted in both parts of the world (whether contemporaneous or not remains to be determined), which have dispersed fragments of pre-existing rocks, scattering them, most probably by the medium of water violently agitated, in various directions, the transporting powers being unequal, so that sandstones and marls alternate with conglomerates. These sandstones and conglomerates would appear, from the descriptions of geologists and intelligent travellers, to extend from Mexico further into the heart of North America, so that if different deposits have not been confounded under one head, as might easily happen in England if it were an uncultivated country and rapidly examined, (the old red sandstone of English geologists being confounded with their new red sandstone,) these sandstones and conglomerates of America would appear not the result of a limited disturbance, but of one common to a considerable surface.

Section VIII.

CARBONIFEROUS GROUP.

Syn.—Coal measures, Engl. Auth. (*Terrain Houiller,* Fr. Auth. *Stein-kohlen-gebirge,* Germ. Auth.). Carboniferous limestone, *Conyb.* (*Mountain limestone,* Engl. Auth. *Calcaire carbonifère, Calcaire an-thraxifère, Calcaire de Transition,* Fr. Auth. *Bergkalk,* Hœn. *Ueber-gangskalk,* and *Neuere Uebergangskalk,* Germ. Auth.). Old red sand-stone, Engl. Auth. (*Grès rouge intermediaire,* Fr. Auth. *Jüngeres Grauwackengebirge,* Germ. Auth.).

Coal Measures.

These are composed of various beds of sandstone, shale, and coal, irregularly interstratified, and in some countries intermixed with conglomerates; the whole showing a mechanical origin. The coal measures abound in vegetable remains, and the coal itself is now, by very general consent, referred to a vegetable origin, being considered the accumulation of an immense mass of plants. It is by no means uncommon to describe the coal deposits as basins; but it may be doubted how far this term is generally correct, for admitting that many accumulations of these beds have been deposited within depressions of the surface, it would by no means seem to follow, reasoning at least from the mode in which vegetable accumulations are now formed, that all coal measures have been thus produced. Suppose plants to be carried down rivers, such as now happens at the mouth or delta of the Mississippi, we should ill characterize the deposit, by the term basin-shaped, which would seem to imply a hollow or depression, bounded by a circumference of nearly equal elevation.

Coal varies considerably in the quantity of bitumen it contains, and is more or less valuable for economical purposes according to the admixture of this substance. The quantity of coal raised in the British isles is very considerable, and it may be said that to this substance and the iron-ore found in the same deposit, England owes a great part of her commercial prosperity; for to the abundance and cheapness of both these substances in various districts, we are indebted for a large proportion of our manufactures, the same series of beds not only furnishing fuel for working the steam-engines, but also iron for their construction.

In the present condition of the coal measures, opportunities of observing design seem to be afforded us, even when these rocks are so disposed that at first sight such design does not appear very obvious.

The accumulation of vegetable matter at a remote epoch in the history of the world, for the consumption of creatures which should afterwards exist on its surface, must strike the least inquiring; but when the upturned, twisted, and shattered strata, so common in the districts composed of the coal measures, are before us, design is not so apparent, more particularly when the miner complains of the dislocations (faults) which interrupt his progress *. We might therefore regard this apparent confusion as a bar to the ingenuity and industry of man in extracting the combustible so valuable to him.   When, however, we look more closely into this subject, we find that the shattered and contorted condition of the rocks, though it may embarrass mining operations for a time, is in reality highly advantageous.   The fractures, termed faults, frequently so cross each other, that the surface, if it could be examined without its covering of vegetation and detritus, would present much the same appearance on the great scale, as the frozen surface of a lake broken to pieces and reunited by subsequent frost. Masses of fractured strata are thus often bounded by faults which prevent the passage of subterraneous waters from one mass into the other; and the miners, in collieries situated in one particular mass, have only to contend with the waters in it; whereas if the strata were always horizontal, unbroken, and continuous, the abundance of water that would flow into the workings would render them so difficult and expensive, that the extraction of the coal must be abandoned †.

It will be obvious that in a district where the coal measures are greatly contorted, the relative position of strata alone would prevent the abundant percolation of water from one situation to another.

* For sections of faults in coal measures, see Sections and Views illustrative of Geological Phenomena, pl. 5. 6. 7.; Geol. Trans. 2nd series, vol. i. pl. 32. La Richesse Minerale, by M. Heron de Villefosse, &c.

† It should be observed, that though the two sides of a fault often come into close contact, there is very frequently an interposed clayey substance impervious to water, and it rarely happens that water on the one side passes to water on the other, so as to form a continuous and abundant percolation in one direction.   On the contrary, the water is commonly thrown out along the line of the fissure, particularly on mountain sides, in the shape of springs,—often good guides to the geologist, not only in tracing faults among the coal measures, but in other rocks. The appearance of springs along lines of fault is what we should expect; for not only do they act as main drains to the strata which they traverse, but as Artesian wells, producing the same effects as these artificial perforations.   For supposing faults to be abundant, in countries where Artesian wells are now so valuable, such countries would possess an abundant supply of water, upon the same principle that these wells now act. Knowing, therefore, that faults are abundant on the surface of our planet, we may infer that no inconsiderable portion of water is conveyed by them to that surface.

Organic Remains of the Coal Measures.

PLANTÆ.

[The following list of plants discovered fossil in the coal measures, is principally compiled from the labours of M. Adolphe Brongniart. To have abridged it would have deprived the student not only of a valuable catalogue of localities, but also of a general idea of the situations where plants of a similar general character probably existed. The names of the plants, when not otherwise noticed, are those assigned by M. Ad. Brongniart, and the localities such as are given by the same author from his own observations and the works of Sternberg, Schlotheim, Artis, and other authorities.]

Equisetaceæ.

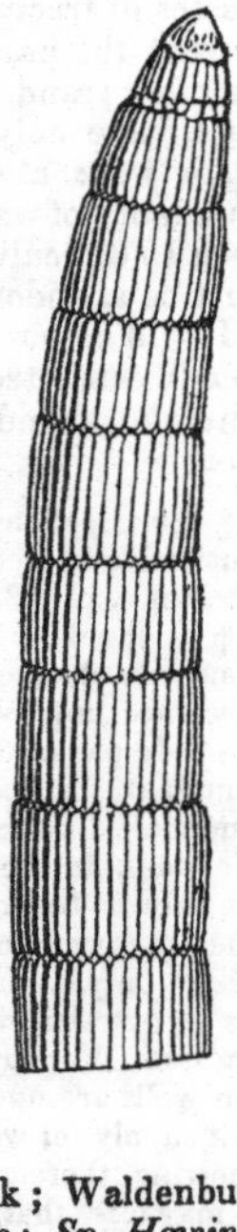

EQUISETUM *infundibuliforme* (Bronn.), Saarbruck; *E. dubium*, Wigan, Lancashire.

Fig. 85.

CALAMITES *Suckowii*, Newcastle; Saarbruck; Liége; Wilkesbarre, Pennsylvania; Richmond, Virginia; *C. decoratus*, Yorkshire; Saarbruck; *C. undulatus*, Yorkshire; Radnitz, Bohemia; *C. ramosus* (Artis), Yorkshire; Mannebach; Wettin, Germany; *C. cruciatus* (Sternb.), Litry; Saarbruck; *C. Cistii*, Montrelais; Saarbruck; Wilkesbarre, Pennsylvania ; *C. dubius* (Artis), Yorkshire; Zanesville, Ohio; *C. cannæformis* (Fig. 85.), Langeac, Haute-Loire; Alais; Yorkshire; Mannebach; Wettin; Radnitz, Germany; *C. Pachyderma*, St. Etienne; Ireland ; *C. nodosus* (Schlot.), Newcastle; Le Lardin, Dordogne; *C. approximatus* (Sternb.), Alais; Liége; St. Etienne; Kilkenny; *C. Steinhaueri*, Yorkshire.

Filices.

SPHENOPTERIS *furcata*, Newcastle; Charleroi ; Silesia; Saarbruck; *Sp. elegans*, Waldenburg in Silesia ; *Sp. stricta* (Sternb.), Northumberland ; Glasgow; *Sp. artemisæfolia* (Sternb.), Newcastle; *Sp. delicatula* (Sternb.), Saarbruck ; Radnitz ; *Sp. dissecta*, Montrelais ; St. Hippolyte, Vosges; *Sp. linearis* (Sternb.), Swina, Bohemia; Engl.; *Sp. Brardii*, Le Lardin ; *Sp. trifoliolata*, Anzin near Valenciennes; Mons; Silesia; Yorkshire ; *Sp. Schlotheimii* (Sternb.), Doutweiler near Saarbruck ; Waldenburg and Breitenbach, Silesia; *Sp. fragilis*, Breitenbach; *Sp. Hœning hausii*, Newcastle; Werden; *Sp. Dubuissonis*, Montrelais; *Sp. distans* (Sternb.), Ilmenau, Silesia; *Sp. gracilis*, Newcastle; *Sp. latifolia*, Newcastle; Saarbruck; *Sp. Virletii*, St.-Georges-Châtellaison; *Sp. Gravenhorstii*, Silesia; Anglesea; *Sp. Loshii*, Newcastle; *Sp. tenuifolia*, St.-Georges-Châtellaison; *Sp. rigida*, Waldenberg; *Sp. acuta*, Werden; *Sp. trichomanoides*, Anzin; *Sp. tenella*, Yorkshire; *Sp. alata*, Geislautern.

CYCLOPTERIS *orbicularis*, St. Etienne; Liége; *C. trichomanoides*, St. Etienne; *C. obliqua*, Yorkshire.

NEVROPTERIS *acuminata*, Klein-Schmalkalden (Schlot.); *N.Villiersii*, Alais, Gard; *N. rotundifolia*, Plessis, Calvados; Yorkshire; *N. Loshii*, Newcastle; Anzin; Liége; Wilkesbarre; *N. tenuifolia* (Sternb.), Saarbruck; Miereschau; Bohemia; Waldenburg, Silesia; Montrelais; *N.heterophylla*, Saarbruck; Valenciennes; Newcastle; *N. flexuosa* (Sternb.), Environs of Bath; Saarbruck (Sternb.); *N. gigantea* (Sternb.), Saarbruck; *N. oblongata*, Paulton, Somerset; *N. cordata*, Alais; St. Etienne; *N. Scheuchzeri* (Hoffmann), England; Osnabruck; Wilkesbarre; *N. angustifolia*, Env. of Bath; Wilkesbarre; *N. acutifolia*, Env. of Bath; Wilkesbarre; *N. crenulata*, Saarbruck; *N.macrophylla*, Dunkerton, Somerset; *N. auriculata*, St. Etienne.

PECOPTERIS *blechnoides*, Werden near Dusseldorf; St. Priest, Loire; *P. Candolliana*, Alais, Gard; *P. cyathea*, St. Etienne; *P. arborescens*, St. Etienne; Aubin, Aveyron; Anzin; Mannebach; *P. platyrachis*, St. Etienne; *P. polymorpha*, St. Etienne; Alais; Litry; Wilkesbarre; *P. Oreopteridis* (Sternb.), Le Lardin; Mannebach; Wettin (Schlot.); *P. Bucklandi*, Env. of Bath; *P. aquilina* (Sternb.), Mannebach and Wettin (Schlot.); *P. Schlotheimii*, Mannebach (Schlot.); Geislautern; *P. pteroides*, Mannebach; Aubin; *P. Davreuxii*, Liége; Valenciennes; *P. Mantelli*, Newcastle; Liége; *P. conchitica*, Newcastle; Saarbruck; Silesia; Namur; *P. Serlii*, Env. of Bath; St. Etienne; Geislautern; Wilkesbarre; *P. Grandini*, Geislautern; *P. crenulata*, Geislautern; *P. marginata*, Alais; *P. gigantea,* Abascherhütte; Trèves; Saarbruck; Liége; Wilkesbarre; *P. nervosa*, Wales; Waldenburg; Roiduc; Liége; *P. obliqua*, Valenciennes; *P. Brardii*, Le Lardin; *P. Defrancii*, Saarbruck; *P. ovata*, St. Etienne; *P. Plukenetii*, Alais; St. Etienne; *P. arguta* (Sternb.), St. Etienne; Saarbruck (Schlot.); Rhode Island, United States; *P. cristata*, Saarbruck; *P. aspera*, Montrelais; *P. Miltoni* (Artis); Yorkshire; Saarbruck; *P. abbreviata*, Valenciennes; *P microphylla*, Saarbruck; *P. æqualis*, Fresnes and Vieux-Condé near Valenciennes; Silesia; *P. acuta*, Saarbruck; Ronchamp, Haute-Saone; *P. unita*, Geislautern; St. Etienne; *P. debilis,* Ronchamp; *P. dentata*, Valenciennes; Doutweiler; *P. angustissima* (Sternb.), Swina, Bohemia; Saarbruck; *P. gracilis*, Geislautern; Valenciennes; *P. pinnæformis*, Fresnes and Vieux-Condé;

Saarbruck; *P. triangularis,* Fresnes and Vieux-Conde; *P. pec-tinata,* Geislautern; *P. plumosa,* Saarbruck; Valenciennes; York-shire (Artis)*.

LONCHOPTERIS *Dournaisii,* Valenciennes.

ODONTOPTERIS *Brardii,* Le Lardin and Terrasson, Dordogne; St. Etienne; *O. crenatula,* Terrasson; *O. minor,* Le Lardin; St. Etienne; *O. obtusa,* Terrasson; *O. Schlotheimii,* Mannebach; Wettin.

SCHIZOPTERIS *anomala,* Saarbruck.

SIGILLARIA *punctata,* Bohemia; *S. appendiculata,* Bohemia; Yorkshire; *S. peltigera,* Alais; *S. lævis,* Liége; *S. canaliculata,* Saarbruck; *S. Cortei,* Essen; *S. elongata,* Charleroi; Liége; *S reniformis,* Mons; Essen; *S. Hippocrepis,* Mons; *S. Davreuxii,* Liége; *S. Candollii,* Alais; *S. oculata,* Bohemia; *S. orbicularis,* St. Etienne; Saarbruck; *S. tessellata,* Env. of Bath; Alais; Esch-weiler; Wilkesbarre; *S. Boblayi,* Anzin; *S. Knorrii,* Saarbruck; *S. elliptica,* St. Etienne; *S. transversalis,* Eschweiler near Aix-la-Chapelle; *S. subrotunda,* Doutweiler near Saarbruck; *S. cuspi-data,* St. Etienne; *S. notata,* Saarbruck; Silesia; Liége; *S. Dour-naisii,* Charleroi; Valenciennes; *S. trigona,* Radnitz, Bohemia (Sternb.); *S. mammillaris,* Charleroi; *S. alveolaris,* Saarbruck; *S. hexagona,* Eschweiler; Bochum; *S. elegans,* Bochum; *S. Brardii,* Terrasson; *S. lævigata,* Montrelais; *S. Serlii,* Paulton, Somerset.

Marsilliaceæ.

SPHENOPHYLLUM *Schlotheimii,* Waldenburg, Silesia; *Sph. emar-ginatum,* Env. of Bath; Wilkesbarre; *Sph. truncatum,* Somerset-shire; *Sph. dentatum,* Newcastle; Anzin; Geislautern; *Sph. qua-drifidum,* Terrasson; *Sph. dissectum,* Montrelais.

Lycopodiaceæ.

LYCOPODITES *piniformis,* Saxe-Gotha; St. Etienne; *L. Graven-horstii,* Silesia; *L. Hœninghausii,* Eisleben; *L. imbricatus,* St. Georges-Châtellaison; *L. phlegmarioides,* Newcastle; Silesia; *L. tenuifolius,* St. Georges-Châtellaison, *L.? filiciformis,* Wettin; *L.? affinis,* Wettin.

SELAGINITES *patens,* Edinburgh; *S. erectus,* Mont Jean near Angers.

* To this list should be added the other species enumerated by Count Sternberg, which however may, as M. Ad. Brongniart remarks, be the same with some of those above enumerated. *Pecopteris orbiculata,* Swina, Bohemia; *P. discreta,* Swina; *P. orbiculata,* Swina; *P. cordata,* Swina; *P. varians,* Swina; *P. obtusata,* Radnitz, Bohemia; *P. undulata,* Radnitz; *P. repanda,* Radnitz; *P. antiqua,* Radnitz; *P. crenata,* Minitz, Bohemia; *P. elegans,* Schatzlar, Bohemia; *P. incisa,* Waldenburg, Si-lesia; Schatzlar; *P. dubia,* Bohemia.

Lepidodendron *selaginoides,* Bohemia; Silesia; *Lep. elegans,* Swina, Bohemia; *Lep. Bucklandi,* Colebrookdale; *Lep. Ophiurus,* Newcastle; Charleroi; *Lep. rugosum,* Charleroi; Valenciennes; *Lep. Underwoodii,* Anglesea; *Lep. taxifolium,* Ilmenau; *Lep. insigne,* St. Ingbert, Bavaria; *Lep. Sternbergii,* Swina; *Lep. longifolium,* Swina; *Lep. ornatissimum* (Sternb.), Edinburgh; Yorkshire; Silesia; *Lep. tetragonum* (Sternb.), Newcastle; *Lep. venosum,* Waldenburg; *Lep. transversum,* Glasgow; *Lep. Volkmannianum* (Sternb.), Silesia; *Lep. Rhodianum* (Sternb.), Yorkshire; Valenciennes; Silesia; *Lep. cordatum,* Durham; *Lep. obovatum* (Sternb.), Radnitz, Bohemia; Silesia; Fresnes and Vieux-Condé; *Lep. dubium,* Newcastle; *Lep. læve,* county of La Marck; *Lep. pulchellum,* Alais; Liége; *Lep. cœlatum,* Yorkshire; *Lep. varians,* Saarbruck; Wilkesbarre; *Lep. carinatum,* Montrelais; St. Georges-Châtellaison; *Lep. crenatum* (Sternb.), Bohemia; Eschweiler; Essen; Zanesville; *Lep. aculeatum* (Sternb.), Essen; Bohemia; Silesia; Wilkesbarre; *Lep. distans,* St. Etienne; *Lep. laricinum* (Sternb.), Bohemia; Silesia; *Lep. rimosum* (Sternb.), Bohemia; *Lep. undulatum* (Sternb.), Bohemia; *Lep. confluens* (Sternb.), Silesia; Eschweiler; *Lep. imbricatum* (Sternb.), Eschweiler, Wettin; *Lep. majus,* Geislautern; *Lep. lanceolatum,* Montrelais; *Lep. Boblayi,* Valenciennes; *Lep. trinerve,* Montrelais; *Lep. lineare,* Alais; *Lep. ornatum,* Shropshire; *Lep. undulatum,* England; *Lep. emarginatum,* Yorkshire.

Cardiocarpon *majus,* St. Etienne; Langeac; *C. Pomieri,* Langeac; *C. cordiforme,* Langeac; *C. ovatum,* Langeac; *C. acutum,* Langeac.

Stigmaria *reticulata,* England; *S. Weltheimiana,* Magdeburg; *S. intermedia,* St. Georges-Châtellaison; Montrelais; Wilkesbarre; *S. ficoides,* St. Georges-Châtellaison; Montrelais; St. Etienne; Liége; Charleroi; Valenciennes; Muhlheim near Dusseldorf; Dudley; Silesia; Bavaria; *S. tuberculosa,* Montrelais; Wilkesbarre; *S. rigida,* Anzin near Valenciennes; *S. minima,* Anglesea; Charleroi.

Palmæ.

Flabellaria? *borassifolia,* Swina.
Nœggerathia *foliosa,* Bohemia.

Cannæ.

Cannophyllites *Virletii,* St. Georges-Châtellaison.

Monocotyledons of uncertain families.—*Sternbergia angulosa,* Yorkshire; *St. approximata,* Langeac, St. Etienne; *St. distans,* Edinburgh: *Poacites æqualis,* Terrasson; *P. striata,* Terrasson: *Trigonocarpum Parkinsoni,* England and Scotland; *Tr. Nœggerathii,* Langeac, coal measures near the Rhine; *Tr. ovatum,* Langeac; *Tr. cylindricum,* Langeac: *Musocarpum prismaticum,*

Langeac; *M. difforme,* Langeac ; *M. contractum,* Oldham, Lancashire.

Vegetables of which the class is uncertain.—*Annularia minuta,* Terrasson; *A. brevifolia,* Alais; Geislautern ; *A. fertilis,* Env. of Bath ; St. Etienne ; Wilkesbarre ; *A. floribunda,* Saarbruck (Sternb.) ; *A. longifolia,* Env. of Bath; Geislautern; Silesia; Alais; Wilkesbarre ; (*var.*) Charleroi; Terrasson ; *A. spinulosa,* Saxony (Sternb.) ; *A. radiata,* Saarbruck : *Asterophyllites equisetiformis,* Manneback; Saxony; Rhode Island ; *As. rigida,* Alais ; Valenciennes; Charleroi; Bohemia; *As. hippuroides,* Alais ; *As. longifolia,* Eschweiler (Sternb.) ; *As. tenuifolia,* Newcastle; Silesia; *As. delicatula,* Charleroi; Anzin ; *As. Brardii,* Terrasson ; *As. diffusa,* Radnitz: *Volkmannia polystachya,* Waldenburg in Silesia ; *V. distachya,* Swina; *V. erosa,* Terrasson *.

Conchifera.

1. Pentamerus Knightii, *Sow.* Between the Coal Measures and Inf. Rocks, Bochum, *Hœn.*
1. Lingula striata, Werden, *Hœn.*
1. Vulsella elongata, *Blain.* Werden, *Hœn.*
2. —— brevis, *Blain.* Werden, *Hœn.*
1. Pecten papyraceus, *Sow.* Werden, *Hœn.; Bradford, *Hailstone.*
2. —— dissimilis, *Flem.* Locality not stated.
1. Mytilus crassus, *Flem.* Scotland, *Flem.; Werden? *Hœn.*
1. Unio acutus, *Sow.* Tanne near Bochum, *Hœn.*
2. —— Urii, *Flem.* Rutherglen, Scotl., *Flem.*
1. Nucula attenuata, *Flem.* Rutherglen, *Flem.*
2. —— gibbosa, *Flem.* Rutherglen, *Flem.*
1. Saxicava Blainvillii, *Hœn.* Nieder-Stauffenbach, *Hœn.*

---

* Whether or not certain of the coal strata of N. America be precisely identical with those of Europe, or may, like some in Ireland described by Mr. Weaver, be of the grauwacke series, it appears that many of the same plants are found, according to the foregoing list, in both Europe and America. These are : Calamites 3 species, Nevropteris 3, Pecopteris 4, Sigillaria 1, Sphenophyllum 1, Lepidodendron 3, Stigmaria 2, Annularia 2, Asterophyllites 1.

Vegetables discovered as yet only in America.—*Nevropteris Cistii,* Wilkesbarre; *N. Grangeri,* Zanesville; *N. macrophylla,* Wilkesbarre : *Sigillaria Cistii,* Wilkesbarre ; *S. rugosa,* Wilkesbarre; *S. Sillimanni,* Wilkesbarre; *S. obliqua,* Wilkesbarre; *S. dubia,* Wilkesbarre : *Lycopodites Sillimanni,* Hadley, Connecticut: *Lepidodendron mammillare,* Wilkesbarre; *Lep. Cistii,* Wilkesbarre : *Poacites lanceolata,* Zanesville. *Pecopteris punctulata,* discovered at Wilkesbarre, is, according to M. Ad. Brongniart, found at the Montagne des Rousses, in Oisans.

1. Hyatella carbonaria, Nieder-Stauffenbach near Cassel, *Hœn.*
1. Mya? tellinaria, Lüttich, *Hœn.*
2. ——? ventricosa, Lüttich, *Hœn.*
3. ——? minuta, Camerberg near Ilmenau, *Hœn.*

MOLLUSCA.

1. Euomphalus pentangularis, *Sow.* Werden, *Hœn.*
1. Turritella Urii, *Flem.* Rutherglen, Scotl., *Flem.*
2. ———— elongata, *Flem.* Rutherglen, *Flem.*
1. Bellerophon decussatus, *Flem.* Linlithgowshire, *Flem.*
2. ———— striatus, *Flem.* Linlithgowshire, *Flem.*
1. Orthoceratites Steinhaueri, *Sow.* Coal Measure Limestone, Choquier near Lüttich, *Hœn.;* Yorkshire, *Sow.*
2. ———— cylindraceus, *Flem.* Linlithgowshire, *Flem.*
3. ———— attenuatus, *Flem.* Linlithgowshire, *Flem.*
4. ———— sulcatus, *Flem.* Linlithgowshire, *Flem.*
5. ———— undatus, *Flem.* Linlithgowshire, *Flem.*
Nautilus? ——. Bituminous Shale, Werden, *Hœn.*
1. Ammonites Listeri *, *Sow.* Bit. Shale, Werden, *Hœn.;* Yorkshire, *Steinhauer.*
2. ———— primordialis, *Sow.* Bit. Shale, Werden, *Hœn.*
3. ———— sacer, Lüttich, *Hœn.*

PISCES.

Ichthyodorulites, *Buckl. & De la B.* Shale, Felling Colliery, Durham, *Taylor;* Rutherglen, *Ure;* Sunderland, *Sow.*
Fish palates, Coal. Tong, near Leeds, *George* †.

Carboniferous Limestone.

This rock in the South of England, Wales, the North of France, and Belgium, seems to possess a somewhat similar general character, being a compact limestone, frequently traversed by veins of calcareous spar, at times appearing to be in a great measure composed of organic remains, while at others not a trace of these remains can be observed. The colours are mostly gray, varying in intensity of shade; other tints are however observed in it, and in certain situations it affords good marble. It is occasionally of an oolitic structure, as near Bristol; and sometimes contains parts of encrinital columns in such abundance, that the rock is in a great measure made up of them, whence the name *Encrinal limestone.* It has also been known by the name of *Metalliferous limestone,* in

* Considered the same with *Ammonites subcrenatus,* Schlöt. by M. Hœninghaus.
† Zoological Journal, vol. ii. pl. 2. fig. 2.

consequence of the quantity of lead ore obtained from it, more
particularly in the central and northern parts of our island.

Organic Remains of the Carboniferous Limestone.

ZOOPHYTA.

1. Millepora madreporiformis, *Wahl.* Gottland, *A.*
2. ———— cervicornis, *Linn.* Gottland, *A.*
3. ———— repens, *Wahl.* Gottland, *A.*
4. ————? foliacea, *Wahl.* Gottland, *A.*
5. ————? Retepora, *Wahl.* Gottland, *A.*
1. Cellepora Urii, *Flem.* Rutherglen, *Flem.*
1 Retepora elongata, *Flem.* Rutherglen, *Flem.*
1. Caryophyllia stellaris, *Linn.* Gottland.
2. ———— articulata, *Wahl.* Gottland.
3. ———— truncata, *Linn.* Gottland, *A.*
4. ———— calycularis, *Linn.* Gottland, *A.*
5. ———— duplicata,   Derbyshire, *Mart.*
6. ———— affinis,   Derbyshire, *Mart.*
7. ———— juncea, *Flem.* Rutherglen, *Flem.*
1. Fungites patellaris, *Lam.* Gottland, *A.*
2. ———— deformis, *Schlot.* Gottland, *A.*
1. Turbinolia turbinata, *Linn.* Gottland, *A.*
2. ———— echinata, *Hisinger.* Gottland, *A.*
3. ———— pyramidalis, *Hisinger.* Gottland, *A.*
4. ———— mitrata, *Schlot.* Gottland, *A.*
5. ———— furcata, *Hisinger.* Gottland, *A.*
1. Cyathophyllum excentricum, *Goldf.* Ratingen, near Dussel-
            dorf, *Goldf.*
  Meandrina, sp. not determined. Visby, Gottland, *A.*
1. Astrea favosa, *Linn.* Gottland, *A.*
2. ———— Ananas, *Linn.* Gottland, *A.*
3. ———— interstincta, *Wahl.* Gottland, *A.*
4. ———— undulata,   Bristol, *Park.*
1. Catenipora escharoides, *Lam.* Gottland, *A.*
2. ———— axillaris, *Lam.* Gottland, *A.*
3. ———— Strues, *Wahl.* Gottland, *A.*
4. ———— serpula, *Wahl.* Gottland, *A.*
5. ———— fascicularis, *Wahl.* Gottland, *A.*
1. Tupipora tubularia, *Lam.* Theux near Liège, *Al. Brong.*
1. Syringopora cæspitosa, *Goldf.* Paffrath near Cologne, *Goldf.*
1. Calamopora polymorpha, *Goldf.* Namur ; Paffrath, *Goldf.*
1. Favosites Gothlandica, *Lam.* Gottland, *A. ;* Dublin, *Al.*
            Brong.
2. ———— Alcyonium, *Defr.* Gottland, *A.*
3. ———— septosus, *Flem.* Scotland, *Flem.*
4. ———— depressus, *Flem.* Scotland, *Flem.*

1. Lithostrotion striatum, . Wales, *Park.*
2. ————— floriforme, *Mart.* Bristol, *Woodward.*
3. ————— marginatum, *Flem.* Scotland, *Flem.*
1. Amplexus coralloides, *Sow.* King's County; Limerick, *Weav.*
Polypifers, genera undetermined. Very numerous in the
British Isles.

RADIARIA.

1. Pentremites Derbiensis, *Sow.* Derbyshire, *Watson.*
2. ————— ellipticus, *Sow.* Preston, Lancashire, *Kenyon.*
3. ————— ovalis, *Goldf.* Ratingen, Dusseldorf, *Goldf.*
1. Poteriocrinites crassus, *Miller.* Somerset; Yorkshire, *Miller.*
2. ————— tenuis, *Miller.* Mendip Hills; Bristol, *Miller.*
1. Platycrinites lævis, *Miller.* Dublin ; Bristol, *Miller.*
2. ————— rugosus, *Miller.* Mendip Hills; Caldy Island,
Miller.
3. ————— tuberculatus, *Miller.* Mendip Hills, *Miller.*
4. ————— granulatus, *Miller.* Mendip Hills, *Miller.*
5. ————— striatus, *Miller.* Bristol, *Miller.*
6. ————— pentangularis, *Miller.* Mendip Hills; Bristol,
Miller.
1. Actinocrinites triacontadactylus, *Miller.* Yorkshire; Bristol;
Mendip Hills, *Miller.*
2. ————— polydactylus, Mendip Hills ; Caldy Island,
Miller.
1. Rhodocrinites verus, *Miller.* Bristol; Mendip Hills, *Miller.*
1. Cyathocrinites planus, *Miller.* Clevedon, Bristol, *Miller.*
2. ————— quinquangularis, *Miller.* Bristol, *Miller.*

ANNULATA.

1. Serpula Lithuus, *Schlot.* Klinteberg, Gottland, *A.*
2. ————— compressa, *Sow.* Lothian.

MOLLUSCA.

1. Pentamerus Aylesfordii, *Sow.* Colebrooke Dale, *Farey.*
2. ————— Knightii, *Sow.* Downton, Croft Arbery, *Park.*
? 3. ————— lævis, *Sow.* Shropshire, *Aikin.*
1. Spirifer alatus (or in the zechstein), Pöseneck, *Hœn.*
2. ————— ambiguus, *Sow.* Ratingen, *Hœn.;* Derbyshire,
Watson.
3. ————— bisulcatus, *Sow.* Visé, *Hœn.;* Dublin, *Sow.*
4. ————— glaber, *Sow.* Ratingen, *Hœn.;* Derbyshire, *Martin;*
Ireland, *Sow.*
5. ————— oblatus, *Sow.* Visé, *Hœn.;* Derbyshire; Flintshire,
Farey.
6. ————— obtusus, *Sow.* Ratingen, *Hœn.;* Yorkshire, *Ducket.*
7. ————— pinguis, *Sow.* Dublin, *Sow.*

8. Spirifer plicatus, *Hœn.* Ratingen, *Hœn.*
9. ——— rotundatus, *Hœn.* Visé, *Hœn.* ; Dublin, *Sow.*
10. ——— trigonalis, *Sow.* Ratingen, Visé, *Hœn.;* Derbyshire, *Martin;* Rutherglen, *Flem.*
11. ——— triangularis, *Sow.* Derbyshire, *Martin.*
12. ——— striatus, *Sow.* Derbyshire, *Martin ;* Namur, *Al. Brong.*
13. ——— lineatus, *Sow.*
14. ——— attenuatus, *Sow.* Dublin, *Sow.*
15. ——— distans, *Sow.* Dublin, *Sow.*
16. ——— resupinatus, *Sow.* Derbyshire, *Martin ;* Rutherglen, *Flem.*
17. ——— Martini, *Sow.* Derbyshire, *Sow.*
18. ——— Urii, *Flem.* Rutherglen, *Ure.*
19. ——— exaratus, *Flem.* West Lothian, *Flem.*
20. ——— cuspidatus, *Sow.* Bristol, *Beeke ;* Derbyshire, *Martin.*
21. ——— minimus, *Sow.* Derbyshire, *Watson.*
22. ——— octoplicatus, *Sow.* Derbyshire, *Martin.*
1. Terebratula acuminata, *Sow.* Ratingen, *Hœn. ;* Yorkshire ; Derbyshire, *Sow. ;* Rutherglen, *Flem. ;* (var.) Clitheroe, Lancashire, *Stokes ;* (var.) Ireland, *Sow.*
2. ——— crumena, *Sow.* Visé, *Hœn. ;* Derbyshire, *Martin.*
3. ——— hastata, *Sow.* Visé, *Hœn.;* Dublin, Limerick, *Wright ;* Bristol, *Sow.*
4. ——— indentata, *Sow.* Ratingen, Visé, *Hœn.*
5. ——— lævigata, *Schlot.* Visé, Norway, *Hœn.*
6. ——— monticulata, *Schlot.* Visé, *Hœn.*
7. ——— obliqua, *Sow.* Ratingen, *Hœn.*
8. ——— resupinata, *Sow.* Ratingen, *Hœn.;* Derbyshire, *Martin.*
9. ——— striatula, Ratingen, *Hœn.*
10. ——— vestita (var.), *Schlot.* Visé, *Hœn.*
11. ——— cuneata, *Dalman.* Isle of Gottland, *A.*
12. ——— diodonta, *Dalman.* Isle of Gottland, *A.*
13. ——— bidentata, *Hisinger.* Djupviken, Gottland, *A.*
14. ——— marginalis, *Dalman.* Klinteberg, Gottland, *A.*
15. ——— didyma, *Dalman.* Isle of Gottland, *A.*
16. ——— affinis, *Sow.* Derbyshire, *Martin.*
17. ——— ? lineata, *Sow.* Derbyshire, *Martin.*
18. ——— ? imbricata, *Sow.* Derbyshire ; Yorkshire, *Sow.*
19. ——— Sacculus, *Sow.* Derbyshire, *Martin ;* Rutherglen, *Flem.*
20. ——— lateralis, *Sow.* Dublin, *Moore.*
21. ——— Wilsoni, *Sow.* Mordeford, E. S. E. of Hereford.
22. ——— Mantiæ, *Sow.* Ireland.
23. ——— cordiformis, *Sow.* Ireland, *Sow.*

'24. Terebratula platyloba, *Sow.* Clitheroe, *Stokes.*
25. ———— Pugnus, *Sow.* Derbyshire, *Martin;* Ireland, *Sow.*
26. ———— Fimbria, *Sow.* Gloucestershire, *Taylor.*
27. ———— reniformis, *Sow.* Dublin, *Sow.*
28. ———— lateralis, *Sow.* Dublin, *Weav.*
1. Crania prisca, *Hœn.* Ratingen, *Hœn.*
1. Producta antiquata, *Sow.* Visé; Ratingen, *Hœn.;* Derby-
        shire, *Sow.;* Cloghran, Dublin, *Humphreys.*
2. ———— comoides, *Sow.* Visé; Ratingen, *Hœn.;* Llangeveni,
        Anglesea, *Farey.*
3. ———— concinna, *Sow.* Visé; Ratingen, *Hœn.;* Derbyshire;
        Yorkshire, *Sow.*
4. ———— fimbriata, *Sow.* Visé; Ratingen, *Hœn.;* Derby-
        shire, *Stokes.*
.5. ———— formicata, *Sow.* Ratingen, *Hœn.*
6. ———— hemisphærica, *Sow.* Ratingen, *Hœn.;* Caermarthen-
        shire, *Taylor.*
7. ———— humerosa, *Sow.* Ratingen, *Hœn.*
8. ———— latissima, *Sow.* Ratingen; Visé, *Hœn.;* Tydmawr,
        Anglesea, *Farey.*
9. ———— lobata, *Sow.* Ratingen; Visé, *Hœn.;* Northumber-
        land; Derbyshire, *Sow.;* Arran, *Leach.*
10. ———— Martini, *Sow.* Ratingen; Visé, *Hœn.;* Derbyshire,
        Martin; Yorkshire, *Danby.*
11. ———— personata, *Sow.* Ratingen, *Hœn.;* Derbyshire; Ken-
        dal, *Sow.*
12. ———— plicatilis, *Sow.* Ratingen; Visé, *Hœn.;* Derbyshire,
        Stokes.
13. ———— punctata, *Sow.* Visé; Ratingen, *Hœn.;* Derbyshire,
        Martin; Rutherglen, *Flem.*
14. ———— rugosa, . Visé, *Hœn.*
15. ———— sarcinulata, . Visé, *Hœn.*
16. ———— splinulosa, *Sow.* Ratingen; Visé, *Hœn.;* Linlith-
        gowshire, *Flem.*
17. ———— sulcata, *Sow.* Visé, *Hœn.;* Derbyshire, *Salt.*
18. ———— transversa, . Visé, *Hœn.*
19. ———— Flemingii, *Sow.* Rutherglen, *Ure;* Linlithgow, *Flem.*
20. ———— longispina *, *Sow.* Linlithgow, *Flem.*
21. ———— crassa, *Flem.* Derbyshire, *Martin.*
22. ———— aculeata, *Sow.* Derbyshire, *Martin.*
23. ———— scabricula, *Sow.* Derbyshire, *Martin;* Vizet, *Al.
        Brong.*
24. ———— spinosa, *Sow.* Scotland, *Flem.*
25. ———— Scotica, *Sow.* Scotland, *Flem.;* Liége, *Al. Brong.*
26. ———— gigantea, *Sow.* Derbyshire, *Martin;* Yorks., *Sow.*

* Considered the same with *P. Flemingii* by Dr. Fleming.
T

27. Producta costata, *Sow.* Glasgow, *Murch.* *
 1. Vulsella lingulata, *Hœn.* Visé, *Hœn.*
 1. Ostrea prisca, *Hœn.* Visé, *Hœn.*
 1. Hinnites Blainvillii, *Hœn.* Ratingen, *Hœn.*
 1. Pecten priscus, *Schlot.* Ratingen, *Hœn.*
 2. —— granosus, *Sow.* Queen's County, Ireland, *Sow.*
 3. ——— plicatus, *Sow.* Queen's County, *Sow.*
 1. Mytilus minimus, *Hœn.* Paffrath, *Hœn.*
 1. Modiola Goldfussii, *Hœn.* Ratingen, *Hœn.*
 1. Nucula Palmæ, *Sow.* Derbyshire, *Martin.*
 1. Arca cancellata, *Sow.* Derbyshire, *Martin.*
 1. Chama? antiqua, *Hœn.* Ratingen, *Hœn.*
 1. Hippopodium abbreviatum, *Goldf.* Paffrath, *Hœn.*
 1. Cypricardia? annulata, *Hœn.* Visé, *Hœn.*
 1. Cardium elongatum, *Sow.* Ratingen, *Hœn.;* Derbyshire,
           Martin.
 2. —— hibernicum, *Sow.* Ratingen, *Hœn.;* Queen's County,
           Sow.; Limerick, *Weav.*
 3. —— alæforme, *Sow.* Queen's County, *Sow.*
 1. Tellina lineata, *Hœn.* Ratingen, *Hœn.*
 1. Sanguinolaria gibbosa, *Sow.* Queen's County, *Sow.*

CONCHIFERA.

 1. Patella Primigenus, *Schlot.* Ratingen, *Hœn.*

* The Swedish naturalists having established divisions in the Terebratulite family (including *Producta* and *Spirifer*) different from those
generally employed, and used in this list, it was thought advisable to
present the catalogue of these shells discovered in the upper of two Swedish limestones (one often considered equivalent to the carboniferous
limestone of the British Isles), such as it is given in the Tableau des Petrifications de la Suède, Stockholm, 1829. The following are all from
the Isle of Gottland.
  1. LEPTÆNA (Producta) rugosa, *Hisinger ;* 2. L. depressa, *Sow.;*
       3. L. ruglypha, *Dalman ;* 4. L. transversalis, *Wahl.*
  1. ORTHIS Pecten ; 2. O. striatella, *Dalman ;* 3. O. basalis, *Dalman ;*
       4. O. elegantula, *Dalman.*
  1. CYRTIA exporrecta, *Wahl.;* 2. C. trapezoidalis, *Hisinger.*
  1. DELTHYRIS (Spirifer) elevata, *Dalman ;* 2. D. cyrtæna, *Dalman ;*
       3. D. crispa, *Dalman ;* 4. D. sulcata, *Hisinger ;* 5. D. ptycodes,
       Dalman ; 6. D. cardiospermiformis, *Hisinger ;* 7. Pusio? *Hi-
       singer.*
  1. GYPIDIA Conchidium.
  1. ATRYPA reticularis, *Wahl. ;* 2. A. alata (var. β), *Hisinger ;* 3. A.
       aspera, *Schlot. ;* 4. A. galeata, *Dalman ;* 5. A. Prunum, *Dalman ;*
       6. A. tumida, *Dalman ;* 7. A. tumidula? *Hisinger.*
  1. TEREBRATULA lacunosa; 2. T. Plicatella; 3, T. cuneata, *Dalman ;*
       4. T. diodonta, *Dalman ;* 5. T. bidentata, *Hisinger ;* 6. T. mar-
       ginalis, *Dalman ;* 7. T. didyma, *Dalman.*

1. Planorbis æqualis, *Sow.* Kendal, *Sow.*
1. Natica elongata, *Hœn.* Ratingen, *Hœn.*
2. ――― Gaillardotii, . Ratingen, *Hœn.*
3. ――― globosa, . Visé, *Hœn.*
4. ――― patula, . Ratingen, *Hœn.*
1. Melania bilineata, *Goldf.* Paffrath, *Hœn.*
2. ――― constricta, *Sow.* Derbyshire, *Martin.*
1. Ampullaria helicoides, *Sow.* Queen's County, *Sow.*
2. ――― nobilis, *Sow.* Queen's County, *Sow.*
1. Melanopsis coronata, *Hœn.* Paffrath, *Hœn.*
1. Nerita sinuosa, *Sow.* Paffrath, *Hœn.*
2. ――― striata *, *Flem.* Corry, Arran, *Flem.*
3. ――― spirata, *Sow.* Bristol, *Beeke;* Derbyshire, *Sow.*
1. Pyramidella antiqua, *Hœn.* Ratingen, *Hœn.*
1. Delphinula canalifera, . Paffrath, *Hœn.*
2. ――――― alata, *Wahl.* Gottland, *A.*
3. ――――― catenulata, *Wahl.* Gottland, *A.*
4. ――――― Cornu Arietis, *Wahl.* Gottland, *A.*
5. ――――― æquilatera, *Wahl.* Gottland, *A.*
6. ――――― funata, *Sow.* Gottland, *A.*
7. ――――― subsulcata, *Hisinger.* Gottland, *A.*
8. ――――― tuberculata, *Flem.* West Lothian, *Flem.*
? 1. Cirrus acutus, *Sow.* Derbyshire, *Martin.*
2. ――― rotundatus, *Sow.* Yorkshire, *Sow.*
1. Euomphalus nodosus, *Sow.* Ratingen, *Hœn.;* Derbyshire, *Martin.*
2. ――――― angulosus, *Sow.* Ratingen, *Hœn.* Benthnall Edge, *Flem.*
3. ――――― catillus, *Sow.* Ratingen, *Hœn.;* Derbyshire, *Martin.*
4. ――――― delphinularis †, *Hœn.* Ratingen, *Hœn.*
5. ――――― pentangulatus, *Sow.* Ratingen, *Hœn.;* Dublin, *Weav.;* Namur, *Al. Brong.*
6. ――――― coronatus, . Visé; Ratingen, *Hœn.*
7. ――――― rotundatus, . Ratingen; Visé, Paffrath, *Hœn.*
8. ――――― rugosus, *Sow.* Colebrooke Dale, *Sow.*
9. ――――― discus, *Sow.* Colebrooke Dale, *Sow.*
10. ――――― centrifugus, *Wahl.* Isle of Gottland, *A.*
11. ――――― angulatus, *Wahl.* Isle of Gottland, *A.*
12. ――――― substriatus, *Hisinger.* Gottland, *A.*
13. ――――― costatus, *Hisinger* (Ammonites?). Gottland, *A.*
――――――, species not determined. Argenteau, Belgium; Vizet, *Al. Brong.*

* According to Dr. Fleming this shell closely approaches the recent *Nerita polita.*
† *Cirrus delphinularis*, Goldfuss; *Helicites delphinularis*, Schlotheim.

1. Pleurotomaria delphinulata,   . Ratingen, *Hœn.*
1. Trochus catenulatus,   . Ratingen, *Hœn.*
1. Turbo carinatus *, *Hœn.* Visé; Ratingen, *Hœn.;* Yorkshire, *Sow.*
2. ——— helicinæformis,   Ratingen, *Hœn.*
3. ——— Tiara, *Sow.* Preston, Lancashire, *Gilbertson.*
4. ——— muricatus, *Sow.* Visé, *Hœn.*
5. ——— striatus †, *Hœn.* Visé, *Hœn.;* Derbyshire, *Martin.*
1. Helix? cirriformis, *Sow.* Derbyshire, *Martin.*
1. Turritella angulata, *Sow.* Ratingen, *Hœn.*
2. ——— cingulatus, *Hisinger.* Isle of Gottland, *A.*
3. ——— constricta ‡, *Flem.* Derbyshire, *Martin.*
1. Buccinum arculatum, *Schlot.* Paffrath, *Hœn.*
2. ——— subcostatum, *Schlot.* Paffrath, *Hœn.*
3. ——— cribrarium, *Hœn.* Ratingen, *Hœn.*
4. ——— lævissimum,   Ratingen, *Hœn.*
5. ——— acutum, *Sow.* Queen's County, Ireland, *Sow.*
1. Bellerophon hiulcus, *Sow.* Visé; Ratingen; Paffrath, *Hœn.;* Derbyshire, *Martin.*
2. ——— apertus, *Sow.* Ratingen, *Hœn.;* Kendal; Bristol; Yorks., *Sow.*
3. ——— tenuifascia, *Sow.* Visé; Ratingen, *Hœn.;* Kendal; Derbyshire, and Yorkshire, *Sow.*
4. ——— costatus, *Sow.* Visé, *Hœn.;* Derbyshire, *Sow.*
5. ——— depressus, *Montfort.* Ratingen, *Hœn.*
6. ——— Cornu-Arietis, *Sow.* Kendal, *Sow.;* Linlithgowshire, *Flem.*
7. ——— Urii, *Flem.* Rutherglen, *Ure.*
8. ——— vasulites, *Montf.* Namur, *Holl.*
1. Conularia quadrisulcata, *Miller.* Bristol, *Miller;* Rutherglen, *Flem.*
2. ——— teres, *Sow.* Scotland, *Sow.*
1. Orthoceratites undulatus, *Sow.* Scalebar, Yorkshire, *Ducket,* (*Schlot.*); Vizet, near Liége, *Al. Brong.*
2. ——— Breynii, *Sow.* Ashford, Derbyshire, *Sow.*
3. ——— annulatus, *Sow.* Colebrooke Dale, Shropshire, *Sow.* Gottland, *A.* King's County, *Weav.*
4. ——— paradoxicus, *Sow.* Ireland, *Ogilby.*
5. ——— fusiformis, *Sow.* Queen's County, Ireland, *Sow.;* Lancashire, *Gilbertson.*
6. ——— cinctus, *Sow.* Preston, Lancashire, *Moore.*
7. ——— imbricatus, *Wahl.* Gottland, *A.*
8. ——— angulatus, *Wahl.* Gottland, *A.*

* *Helix carinatus* of Sowerby.
† Probably *Helix striatus* of Sowerby.
‡ *Melanea constricta* of Sowerby.

9. Orthoceratites undulatus, *Hisinger* (not *Sow.*).　Gottland, *A.*
10. ──────── crassiventris, *Wahl.*　Gottland, *A.*
11. ──────── lineatus, *Hisinger.*　Gottland, *A.*
12. ──────── Gesneri,　.　Derbyshire, *Martin.*
13. ──────── lævis, *Flem.*　Linlithgowshire, *Flem.*
14. ──────── pyramidalis, *Flem.*　Linlithgowshire, *Flem.*
15. ──────── convexus, *Flem.*　Linlithgowshire, *Flem.*
16. ──────── annularis, *Flem.*　Linlithgowshire, *Flem.*
17. ──────── rugosus, *Flem.*　Linlithgowshire, *Flem.*
18. ──────── angularis, *Flem.*　Linlithgowshire, *Flem.*
1. Nautilus globatus, *Sow.*　Ratingen, *Hœn.*
2. ──────── discus, *Sow.*　Kendal, *Sow.*
3. ──────── ingens,　.　Derbyshire, *Martin.*
4. ──────── marginatus, *Flem.*　Bathgate, Scotland, *Flem.*
5. ──────── quadratus, *Flem.*　West Lothian, *Flem.*
6. ──────── biangulatus, *Sow.*　Bristol, *Beeke.*
7. ──────── sulcatus, *Sow.*　Derbyshire, *Sow.*
8. ──────── Woodwardii, *Sow.*　Derbyshire, *Martin.*
9. ──────── excavatus, *Flem.*　Limerick, *Wright.*
1. Ammonites sphæricus, *Sow.*　Visé, *Hœn.*; Derbyshire, *Sow.*
2. ──────── Dalmanni, *Hisinger.*　Gottland, *A.*
3. ──────── striatus, *Sow.*　Derbyshire, *Sow.*

CRUSTACEA.

1. Calymene Blumenbachii, *Al. Brong.*　Gottland, *A.*
2. ──────── variolaris, *Al. Brong.*　Ratingen, *Hœn.*
3. ──────── punctata, *Wahl.*　Island of Gottland.
4. ──────── concinna, *Dalman.*　Gottland, *A.*
1. Asaphus cordatus, *Al. Brong.*　Gottland, *A.*
1. Paradoxicus spinulosus, *Al. Brong.*　Ratingen, *Hœn.*
Trilobites, genera not determined.　Bristol, *De la B.*　Llangeveni, Anglesea, *Farey;* Linlithgowshire, *Flem.*

PISCES.

Ichthyodorulites, *Buckl.* & *De la B.*　Bristol, *De la B.*
Tritores, or fish palates, Bristol, *De la B.*　Northumberland, *Phil.*

Old Red Sandstone.

This rock is of very variable thickness, sometimes consisting of a few conglomerate beds, while at others it swells out to the depth of several thousand feet.　As might be expected, this variation in thickness is accompanied by differences in mineralogical structure, conglomerates being abundant in some situations, while in others they are exceedingly rare.　The sandstone possesses different degrees of induration, and is not unfrequently schistose and micaceous, affording flag-stones and coarse materials for roofing.　The prevalent colour is red, generally dull, which, as commonly occurs in the red marls and sandstones of all ages, is occasionally intermixed with different tints of greenish blue (Pembrokeshire, &c.). The conglomerates of course vary in their contents, but pieces of quartz are very common, so much so in the southern parts of England and Wales, that the greater proportion of such beds are wholly composed of them. The sandstones also are principally siliceous, so that if the mass be considered as wholly of mechanical origin, it must have resulted from a considerable destruction of pre-existing siliceous rocks.

Few organic remains have been discovered in this rock, and those that have been observed would appear to be the same as in grauwacke beneath, or the carboniferous limestone above. According to Dr. Fleming, *Orthoceratites cordiformis, O. giganteus, Nautilus bilobatus,* and *N. pentagonus,* are found in a limestone associated with the old red sandstone of Dumfriesshire *.

The student being now acquainted with the general zoological and botanical characters, and with the more marked mineralogical composition of the three rocks comprised within this group, we will proceed to a more general notice of the same rocks taken in the mass.

It has been above remarked, that the todtliegendes is considered, in certain parts of Europe, to pass into the coal-measures, so that the two rocks constitute the upper and lower portion of the same mass. Some geologists have gone further, and considered the coal-measures as subordinate to the todtliegendes, the carboniferous portion bearing the same relation to the general mass, which certain lignites, such for instance as those noticed by M. Elie de Beaumont in Dauphiné and Provence, bear to the transported matter in which they are found included.

This apparent subordinate character of the coal-measures to the todtliegendes in Thuringia, has induced M. Hoffmann to divide the rothe-todte-liegende of that country into three parts : *a.* red sandstone, with schist, schistose sandstone, and conglomerate (500 feet thick), referred to the old red sandstone ; *b.* carboniferous rocks with limestone (250 feet thick), considered equivalent to the carboniferous limestone, coal-measures, and millstone grit of the English series ; and *c.* red sandstone, with schist, conglomerate, and porphyry-breccia (2590 feet thick), referred to the red conglomerate of Exeter †. By consulting Mr. Weaver's observations respecting the views of the German geologists previous to those of M. Hoffmann ‡, the student will observe that he was led to the

---

* Fleming, History of British Animals.

† Hoffmann, Uebersicht der orographischen und geognostischen Verhältnisse vom Nordwestlichen Deutschland, 1830, p. 504.

‡ Weaver, Annals of Philosophy, 1821.

same conclusions as to the equivalent character of the old red sandstone and the lower part of the rothe-todte-liegende.

Perhaps by following up the views of Mr. Weaver, and considering that the coal is not necessarily constant to a particular part of the series, and that the mutual relations of different portions of the whole mass vary materially, we may approach towards a solution of this apparent difficulty. In the first place, we obtain no help from organic remains; for it will have been observed that the general zoological character of the marine exuviæ is the same in the zechstein (above the todtliegendes), in the carboniferous limestone, and, as will be seen in the sequel, in the grauwacke series. The general character of the vegetable remains was probably also similar, as we know it was in the descending order. Assuming, therefore, and it does not appear unphilosophical to do so, that organic remains will not aid us in the investigation, we can only appeal to mineralogical structure and relative geological position. Our first inquiry therefore should be, are these constantly the same, allowances being only made for smaller variations? We can only reply to this question by a statement of facts.

In the southern part of our island the three divisions of old red sandstone, carboniferous limestone, and coal-measures, are well marked, and there is clearly no passage of the latter into the red sandstone (commonly termed new red sandstone) above it; on the contrary, the coal-measures and the inferior rocks have been upset prior to the deposition of the magnesian conglomerates and limestones with their associated red sandstones and conglomerates; and it seems more than probable that the lower portions of the latter series of rocks resulted from the disturbance produced by the fracture, contortion, and elevation of the coal-measures and older rocks. With respect also to the carboniferous group itself, the masses of the old red sandstone, carboniferous limestone, and coal-measures are well separated from each other, though there may be small alternations at their contact, as the student can observe at Clifton Gorge near Bristol, and other places in that district*.

As we advance northwards into the central part of England, we find that the lower part of the coal-measures and the upper part of the carboniferous limestone, which in the south only alternated at their contact in a comparatively moderate degree, have now assumed a new character as they approach each other, pre-

---

* For the necessary details respecting the coal-measures and the carboniferous series generally, consult the labours of Mr. Conybeare in his Outlines of the Geology of England and Wales; and for that of the southern part of our island in particular, the Observations on the South-western Coal District of England, by Dr. Buckland and Mr. Conybeare, Geol. Trans., 2nd series, vol. i.

senting a mass of shales, sandstones (most frequently coarse), and limestones, with occasional seams of coal, the whole being of very considerable thickness, and known as Millstone Grit.

Prof. Sedgwick has shown, that still further north in England the great lines of distinction between the carboniferous limestone and the coal-measures are broken up, and that the one is lost in the other. As the student can have no better or more condensed view of the subject than in Prof. Sedgwick's own words, I shall offer no apology for inserting them here.

"On the re-appearance of the carboniferous limestone at the base of the Yorkshire chain, we still find the same general analogies of structure : enormous masses of limestone form the lowest part, and the rich coal-fields the highest part of the whole series; and we also find the millstone grit occupying an intermediate position. The millstone grit, however, becomes a very complex deposit, with several subordinate beds of coal; and is separated from the great inferior calcareous group (known in the North of England by the name of *scar limestone*), not merely by the great shale and shale-limestone, as in Derbyshire, but by a still more complex deposit, in some places not less than 1000 feet thick, in which five groups of limestone strata, extraordinary for their perfect continuity and unvarying thickness, alternate with great masses of sandstone and shale, containing innumerable impressions of coal plants, and three or four thin seams of good coal extensively worked for domestic use.

"In the range of the carboniferous chain from Stainmoor, through the ridge of Cross Fell to the confines of Northumberland, we have a repetition of the same general phænomena. On its eastern flanks, and superior to all its component groups, is the rich coal-field of Durham. Under the coal-field we have, in regular descending order, the millstone-grit, the alternations of limestones and coal-measures nearly identical with those of the Yorkshire chain, and at the base of all is the great *scar limestone*. The *scar limestone* begins however to be subdivided by thick masses of sandstone and carbonaceous shale, of which we had hardly a trace in Yorkshire, and gradually passes into a complex deposit, not distinguishable from the next superior division of the series. Along with this gradual change is a greater development of the inferior coal-beds alternating with the limestone, some of which, on the north-eastern skirts of Cumberland, are three or four feet in thickness, and are now worked for domestic use, with all the accompaniments of rail-roads and steam-engines.

"The alternating beds of sandstone and shale expand more and more as we advance towards the north, at the expense of all the calcareous groups, which gradually thin off and cease to produce any impress on the features of the country. And thus it is that the lowest portion of the whole carboniferous system, from

Bewcastle Forest along the skirts of Cheviot Hills to the valley of
the Tweed, has hardly a single feature in common with the inferior part of the Yorkshire chain; but, on the contrary, has all
the most ordinary external characters of a coal-formation. Corresponding to this change is also a gradual thickening of carbonaceous matter in some of the lower groups. Many coal-works have
been opened upon this line, and near the right bank of the Tweed
(almost on a parallel with the great *scar limestone*) is a coal-field
with five or six good seams, some of which are worked, not merely
for the use of the neighbouring districts, but also for the supply
of the capital [*]."

We thus observe that a very material change has been effected
in the carboniferous rocks, the limestone beds having become
mixed up with, and even disappearing among, the arenaceous and
shaly coal-measures. Two rocks of the series are therefore as it
were amalgamated, and no line of distinction can be drawn between them Not only have the separate characters of coal-measures and carboniferous limestone disappeared, but the remaining
rock, the old red sandstone, no longer presents the common arenaceous aspect which it possesses in the south of England and in Wales.
It is here a conglomerate, which, instead of offering the appearance
of a passage into the grauwacke strata beneath, actually rests on the
upturned edges of those strata, and is frequently absent, so that,
as has been shown by Prof. Sedgwick and Mr. J. Phillips, the
carboniferous limestones repose directly on the previously disturbed and upset grauwacke rocks [†].

If we now proceed to Scotland, to that mass of conglomerate
and arenaceous deposits intermixed with limestone and coal described by Dr. Fleming, Prof. Jameson, Dr. MacCulloch, Mr.
Bald, Dr. Boué, Prof. Sedgwick, Mr. Murchison, and other geologists, there would appear to be some difficulty in establishing
distinctions such as can readily be made in the southern parts of
our island; and this difficulty is increased by the presence of rocks,
referable, at least in part, to the (new) red sandstone group. In
the northern English districts noticed by Prof. Sedgwick, the red
sandstone rocks have clearly been deposited on the carboniferous
limestone and coal-measures after the two latter rocks had suffered great disturbance and violent dislocations; but it may be questionable, at least in parts of Scotland, how far fine lines of dis-

[*] Sedgwick, Address to the Geological Society, 1831. Phil. Mag.
and Annals, vol. ix. p. 286, 287.

[†] See Phillips's Sections in the Geol. Trans., 2nd series, vol. iii.; and
Sedgwick, Proceedings of the Geol. Soc.1831. When the sections and descriptions of the latter author shall have been made public, geologists
will be in possession of a highly valuable and illustrative series of documents on this subject.

T 5

tinction can be drawn between the upper part of the coal-measures and the lower portion of the red sandstone group. Organic remains will be of little assistance, for reasons before stated, neither is the mineralogical character of much avail, for it will have been seen that this also changes, and there is nothing, that we are aware of, which should prevent the zechstein, if produced under general similar circumstances, from assuming the character of the carboniferous limestone, such more particularly as the latter appears when divided and included in the coal-measures. The colour of the rocks is, if possible, of still less importance, for the coal-measures are not unfrequently red; so that should the whole get mixed up together, more particularly without discordant stratification, it would appear highly theoretical to distinguish the various portions by particular names, each portion being considered the decided equivalent of divisions that may be established elsewhere. It is by no means intended to infer that during any considerable deposit, such as the one under consideration, there should not be equivalents in age: such there must always have been: but the contemporaneous effects produced by different causes may have varied most materially; so that distinctions which mark particular events in one situation are not always useful when applied generally; for, perhaps without being aware that we are doing so, we theoretically consider circumstances to have been generally the same at a given time, whereas we should, in the first place, consider them as more local, resulting from the operation of more limited causes. I am fully aware that this view may be carried too far, and that the minor divisions of rocks should be established as much as possible; but we should also avoid extremes, and not pass that point where the distinctions may admit of very great doubt, for by doing so we seem in a great measure to preclude ourselves from tracing the causes, that have produced the great changes in the mineralogical and zoological characters of rocks on the surface of the earth generally.

Dr. Boué considers the conglomerates, sandstones, limestones and coal of the great arenaceous deposit of Scotland, as subordinate portions of one great whole, which he believes equivalent to the red sandstone (*grès rouge*); an opinion coinciding with the views of M. Hoffmann respecting the coal-deposit of Thuringia. To what extent this opinion may be correct would not yet appear to be well decided; and it no doubt may startle English geologists to compare, in any manner, the old red sandstone with a system of rocks containing the todtliegendes: but supposing a series of conglomerate, sandstone, and other rocks of a certain common character, to have been produced, so that during the deposit the inferior should not in any manner have been disturbed, but the strata to have been laid regularly on one another, and that we obtain little or no aid from organic remains,—it would seem difficult to regard the mass in any other light than as resulting from the operation of nearly similar and uninterrupted causes. I am far from stating that this actually is the case in Scotland, being merely desirous to show that an union of the todtliegendes and zechstein with the carboniferous group may not be impossible elsewhere.

In certain districts, such as Pembrokeshire, the old red sandstone has the appearance of passing into the grauwacke series beneath; in such situations, therefore, we may regard this rock as resulting from the continuance of causes similar to those which have produced the grauwacke; for the zoological character of the two deposits is the same, as is also their mineralogical structure; the difference between them is in the colour,—a circumstance of no importance, for red rocks are often present in the body of the grauwacke group itself.

In the North of England, the old red sandstone, as has been above noticed, rests on upturned grauwacke: causes therefore have acted violently in one situation at a given period, which have scarcely, if at all, produced marked effects in another at no very considerable distance; leading us to infer that at still greater distances the differences observable in deposits of this period may be more remarkable.

The carboniferous group occupies the surface of a large portion of Ireland, the limestones being exceedingly abundant. Mr. Weaver describes sandstones and conglomerates as frequently, though not constantly, interposed between the older deposits and the carboniferous limestone, and refers them to the old red sandstone. The Gaultees mountains are mentioned as wholly composed of them. They occur along the flanks of the clay slate districts, and isolated caps of the sandstone often rest on these older deposits. The red sandstone emerges from beneath the interior of the great limestone plain at Moat, Ballymahon, and Slievegoldry Hill. The same author notices that the strata of this sandstone deposit are most inclined as they approach, and are in contact with, the older rocks; but that as they accumulate and recede from the latter, they become more and more horizontal

The carboniferous limestone may be considered as the prevalent rock in Ireland; for, as Mr. Weaver observes, all its counties, with the exception of Derry, Antrim and Wicklow, are more or less composed of it. This limestone is described as coming in contact with, and sweeping round, various mountain chains, "filling up every interval and hollow between them." It supports the coal-measures, properly so called; and thus the analogy between the carboniferous series of central and southern England, and that of the corresponding portions of Ireland, is complete, the arenaceous and conglomerate deposits of the old red sandstone in the latter country being surmounted by a sheet of limestone, varying in thickness, sometimes attaining a depth of 700 or 800 feet, but

generally averaging 200 or 300 feet; and being in its turn covered by coal-measures *.

The carboniferous rocks of the North of France and Belgium
have a direction from east-north-east to west-south-west from the
vicinity of Aix-la-Chapelle to and beyond Valenciennes, and rise
from beneath cretaceous or newer rocks. The carboniferous
limestone and coal-measures of the Boulognais can be considered
only as a continuous portion of the same deposit.

These carboniferous rocks occur much in the same manner as
in the southern parts of England. Red arenaceous deposits
equivalent to the old red sandstone occupy the lowest part, carboniferous limestone the central, and coal-measures the higher
portion of the series. According to M. de Villeneuve the coal-
measures and limestone alternate at their contact with each
other between Liége and Chaude Fontaine. The limestones are
metalliferous, blueish and compact, and contain subordinate
conglomerates of blue limestone. The alternating sandstones are
sometimes reddish, and at others greenish brown ; they are sometimes compact, at others fissile with mica, and the lines of cleavage are in some beds not the same with those of stratification.
The upper part of the limestone and sandstone contains aluminous
shale, worked for profitable purposes (Huy and other places)†.

According to the same author the coal-measures, which are
composed of the usual mixture of sandstones, shales, and coal
beds, present at the Montagne de St. Gilles no less than sixty-
one beds of the latter, varying from six feet to a few inches in
thickness. The strata of the district are greatly disturbed, as is
well seen at Mons, and are traversed by faults, as may be observed at St. Gilles. Coal is worked far down in the lower beds,
and even amid the limestones at Mons, which circumstance, however, M. de Villeneuve attributes to the contortions of the strata.

The carboniferous rocks of the Netherlands would appear to
be continued into Germany, to the deposits between Essen, Werden, Bochum, Hattingen, Wetter and Dortmund, which repose
on the north-west corner of the great exposure of grauwacke
rocks in that part of Europe. To the north of these deposits, on
the northern side of the great gulf of cretaceous and supercretaceous rocks which enters easterly into Germany, and on which
stands Münster, there is, according to M. Hoffmann, an outcrop
of carboniferous strata at Ibbenbühren, between Osnabrück and
Rheine ‡. Coal-measures occur at Seefeld, in Saxony. At Wet-

* Weaver, On the Geological Relations of the East of Ireland, Geol.
Trans., vol. v.—Also consult Griffith's Account of the Connaught and
Leinster Coal Districts ; and Sections and Views illustrative of Geolo-
gical Phænomena, pl. 29.

† De Villeneuve, Ann. des Sci. Nat. t. xvi. 1829.

‡ For the localities of the coal in the north-west of Germany consult

ten, north of Halle, is another deposit; and at Saarbrück, and the neighbouring country, the coal-measures are abundant, and rest, when trappean rocks are not interposed, upon part of the grauwacke mass previously mentioned *.

M. Pusch describes the coal-measures in Poland as extending from Hultschin to Krzeszowice, the more ancient beds passing into the grauwacke on which they rest; but the same author remarks, that in the rocky valleys of Czerna Szklary, and near Debnik, not far from Krzeszonice, a black marble, employed in the arts, supports the coal-measures. M. Pusch considers this marble as equivalent to the carboniferous limestone of the English geologists, and observes that the calcareous conglomerates which accompany the coal sandstones and shales in the gorges of Miekina and Filipowice are referable to the same marble beds. The same author states that the coal-measures contain the plants so commonly observed elsewhere in similar deposits, and that he has identified thirty-six species with those noticed in the works of MM. Sternberg and Ad. Brongniart †. According to M. Sternberg, part of the coal strata of Bohemia follows the line of the so called transition rocks from Merklin in the circle of Klattau, to Mühlhausen on the Moldau, having a length of fifteen leagues, and a breadth of from four to five leagues. He also remarks that the coal formation of Silesia, which stretches the length of seventeen leagues, reaches Schatzlar in the Riesengebirge on the one side, and Schwadowitz in the lordship of Nachod, in Bohemia, on the other; red sandstone and red porphyry accompanying both these deposits. In the western part of Bohemia, coal-measures occur in the circles of Klattau, Beraun, Pilsen, and Rakowitz ‡.

M. Hoffmann's map of that country; and for descriptions, Uebersicht der orographischen und geognostischen Verhältnisse vom Nordwestlichen Deutschland, by the same author.

* The student will find instructive plans and sections of the coal mines at Werden, Essen, Eschweiler, Valenciennes, Mons, Fuchsgrube (Silesia), and Saarbruck, in the Atlas to la Richesse Minerale, by M. Heron de Villefosse, pl. 24, 25, 26, 27, and 28. He should also consult the geological map and sections of the countries bordering the Rhine, by MM. Oeynhausen, la Roche and Von Decken, for the coal-measures of Saarbruck and the adjacent country. Parts of these sections are inserted in Sections and Views illustrative of Geological Phænomena, pl. 18. figs. 1. and 2.

† Pusch, Journal de Geologie, t. ii. The limestones of the Isle of Gottland, many fossils from which are enumerated above among those of the carboniferous limestone, are referred to this series in accordance with the views of M. Hisinger.

‡ Sternberg, Versuch einer geognostisch-botanischen Darstellung der Flora der Volwelt.

The coal deposits of central France repose on granite, gneiss, mica slate, &c., without the intervention of any limestones, sandstones, or slates, which can be distinctly referred to the carboniferous limestone, old red sandstone, or grauwacke : such are the coal-fields of St. Etienne, Rive de Gier, Brassac, Fuis, &c., At St. Georges-Châtellaison the coal-measures also rest on gneiss and mica slate *.

The carboniferons deposits of the United States are, according to Prof. Eaton, of different ages ; one being contained in the argillaceous slates (argillite) of Worcester (Mass.) and Newport; another being considered equivalent to the coal-measures of Europe ; and a third being of a more recent epoch, though older than certain lignites. The deposit referred to the same epoch as the carboniferous series of Europe, occurs at Carbondale, Lehigh, Lackawaxen, Wilkesbarre, and other places †.

Mr. Cist describes the coal of Wilkesbarre as alternating with various sandstones and shales, the latter containing a great abundance of fossil plants ‡, many of which it will have been seen by one of the foregoing lists are identical with some discovered in the coal-measures of Europe, and all are of the same general character with those obtained from the carboniferous and grauwacke series. The sandstone beds vary from five to one hundred feet in depth, and the coal is sometimes from thirty to forty feet thick, its general thickness being from twelve to fifteen feet. Prof. Silliman states that the beds at Mauch Chunk, (Pennsylvania) consist of conglomerates, sandstones, and argillaceous slate. The pebbles in the conglomerate are described as pieces of quartz rounded by attrition, and the cementing matter of the conglomerates and sandstones as siliceous §. According to Prof. Eaton, the limestone which supports the strata containing the Pennsylvanian coal extends along the foot of the Catskill Mountains, and is continued from the southern part of Pennsylvania to Sackett's Harbour on Lake Ontario ‖.

Mr. Hitchcock informs us that coal is associated with trappean rocks, fetid, siliceous, and bituminous limestones, red and gray sandstones, and conglomerates, in Connecticut. It is described as bituminous, whereas the Wilkesbarre coal is frequently termed anthracite by the American geologists ¶. It occurs at Durham,

* Mr. Grammer remarks that the coal deposit of Virginia rests on granite. American Journal of Science, vol. i.
† Eaton, Amer. Journ. of Science, vol. xix.
‡ Cist, American Journ. of Science, vol. iv. where a map of the deposit will be seen.
§ Silliman, American Journ. of Science, vol. xix.
‖ Eaton, ib. vol. xix.
¶ This distinction would not, in itself, appear to be of any great im-

Chatham, Berlin, Enfield, and other places in Connecticut, and is described as passing into the so called old red sandstone of the country, which is composed of a series of sandstones and conglomerates generally of a dark red colour. An excellent section of this coal deposit, described in great detail by Mr. Hitchcock, is exposed where the Connecticut river cuts through it between Gill and Montague.

Fossil fish are obtained from associated bituminous shale at Westfield (Connecticut) and Sunderland (Massachusetts), and one species is considered referable to the genus *Palæothrissum* of Blainville *, a genus which the student has seen noticed under the head of Zechstein. The presence of this genus would however by no means necessarily mark the deposit as alone referable to the todtliegendes or zechstein, even admitting, for the sake of the argument, that the latter were recognizable as minor divisions in America; for the genus *Palæothrissum* is quite as likely to have formed a part of the animals existing during the deposit of the coal-measures and carboniferous limestone, as the *Productæ* during the formation of the zechstein.

If we, for the moment, abstract the limestone beds, there would appear little doubt that the carboniferous series is of mechanical formation, a deposit from water varying in its transporting powers. Thus, at one time the velocities were sufficient to force forward gravels, while at others they only accomplished the transport of silt or mud. If proportional sections be made of coal deposits, it will be observed that the coal beds occur at very unequal intervals, showing that the causes which produced them have acted irregularly. From the careful examination of the Forest of Dean by Mr. Mushet, we have a detailed list of the various beds of the coal-measures, carboniferous limestone and old red sandstone, the whole constituting a collective thickness of about 8700 feet; the coal-measures being 3060 feet in depth, and the limestone 705. The mass reposes on the grauwacke (transition) limestone of Long Hope and Huntley †. The sandstones of the old red sandstone in Gloucestershire, Somersetshire, and the neighbouring parts of England afford us no great evidence of a quick deposit, more particularly as the conglomerates are not common ; the latter are however sufficient to show that the velocities of the transporting waters were not constant, but liable to variation. A great change in the depositing and transporting powers was subse-

portance ; for the continuous coal deposit of South Wales is anthracitic in Pembrokeshire, and bituminous in its eastern prolongation through Monmouthshire.

* Hitchcock, American Journal of Science, vol. vi.
† Mushet, Geol. Trans. 2nd Series, vol. i. p. 288.

quently produced, and instead of the siliceous and arenaceous sediment, carbonate of lime, often enveloping a variety of marine animal remains, was produced; and this not for a short time, but apparently during a long period; for the carboniferous limestone of this district bears marks of slow formation, many beds being composed of a mass of fossils, the remains of myriads of animals, which have apparently lived and died where we now find them entombed. It must however be admitted that the origin of many beds, which do not present a trace of animal exuviæ, remains obscure, and we have no direct evidence that they may not have been produced more suddenly by deposits from water, either holding carbonate of lime in chemical solution or in mechanical suspension. After a thickness of seven or eight hundred feet of calcareous rock had been formed, another great change in the matter deposited was effected; not however so suddenly but that the arenaceous sediment whichafter wards became so abundant, and the calcareous matter, were alternately produced for a comparatively limited period. An immense mass of sandstone, shales, and coal was then accumulated in beds one above another, which, though irregular with regard to the relative periods of deposit, are frequently persistent over considerable areas.

By general consent the coal is considered as resulting from the distribution of a body of vegetable remains over areas of greater or less extent, upon a previously deposited surface of sand, argillaceous silt, or mud, but principally the latter, now compressed into shale. After the distribution of the vegetables, other sands, silt, or mud, were accumulated upon them; and this kind of operation was continued irregularly for a considerable time, during which there was an abundant growth of similar vegetables at no very distant place, to be suddenly, at least in part, destroyed and distributed over considerable areas on the more common detritus.

Great length of time would be requisite for this accumulation, because the phænomena observed would lead us to consider the transporting power, though variable, to have been generally moderate : moreover a very considerable growth of vegetables requiring time, would be necessary at distinct intervals; for coal beds only now six or ten feet thick, must, before pressure was exerted upon them, have occupied a much greater depth. It is a remarkable circumstance connected with the coal measures of the South of England, that marine remains have not been detected in them, which, though it does not prove the deposit of coal to have been effected in fresh water, does appear to show that there was something which prevented the presence of marine animals,—a circumstance the more remarkable, as we have seen that such animals swarmed during the formation of the carboniferous limestone.

These remarks are not only applicable to the small district above nóticed, but to a large extent of country, one stretching from Belgium, through the North of France, and the southern parts of England and Wales, into Ireland; and for the most part concealed beneath masses of newer rocks. As we advance northward, however, the marked distinctions, as before noticed, disappear; so that the causes, whatever they were, which produced such a separation of arenaceous and calcareous rocks on the south, became modified, and the limestones were more intimately blended, in alternating beds, with the sandstones and shales, affording a greater mixture of marine with terrestrial remains. It has long been known that the Yorkshire coal measures presented a bed containing the remains of *Ammonites* and *Pectines*, and that the fossils of the carboniferous limestone and coal measures were detected in the millstone grit; or, in other words, that there was an alternation of terrestrial with marine remains, showing that the causes which effected the deposit of calcareous matter and envelopment of marine remains sometimes predominated, while at others a transport of mud and sand entombed an abundance of vegetables. The occurrence of marine remains amid the coal measures, it will be observed by reference to the foregoing lists, is not confined to Great Britain, but is also remarked in different parts of Germany; so that the same modification of circumstances which has produced a mixture, or rather alternation, of marine and terrestrial remains in Great Britain, has extended into the continent of Europe.

There is another class of appearances connected with these rocks which demands our attention. From a considerable mixture of porphyry in certain situations with the coal measures, it has sometimes been considered that this rock was an essential and component part of the group under consideration. From all analogy it may be concluded that porphyries are of igneous origin; and for the same reason it is inferred that the coal measures and their accompanying beds were produced by aqueous deposition. We therefore should be led *a priori* to consider, that two substances of such different origin did not necessarily constitute parts of a common whole, but that their admixture was accidental. And we may consider this at once proved, by the abundant occurrence of coal measures without porphyry, such as is so commonly the case in England.

In the sections which M. Hoffmann has presented us of the coal measures of Wettin, and other places in north-western Germany, it is easy to conceive, although porphyry occurs both above and beneath the coal strata, that the latter are not necessarily of contemporaneous formation; on the contrary, the fractured and contorted state of the beds shows that great violence has been exercised upon them, precisely such as would be expected if igneous rocks had burst in amidst them, when among other accidents we should expect to find large masses of coal measures caught up and included in the porphyry, as we find masses of chalk caught up and enveloped by basalt on the north of Ireland. As we shall return to the subject of the igneous rocks found among the carboniferous group in another place, the above notice has been introduced merely to show that the supposed connection of porphyry and coal strata has not been overlooked.

From the similarity of general circumstances attendant on the coal strata, we have reason to conclude, although the series may contain more limestone at one place than at another, that in Poland, Western Germany, Northern France, Belgium, and the British Isles, there were some common causes in operation at the same epoch, producing the envelopment of a great abundance of terrestrial vegetables, of a nature that could not, from the want of the necessary heat, now flourish in the same latitudes.

Proceeding to the central part of France, we find several smaller coal deposits, which, more particularly from their organic character, are referred to the carboniferous epoch of which we are treating. How far they may have been once more extensive and continuous, and how far they may have suffered from movements in the land, dislocations, and denudation, we are not certain ; but we are certain that they were directly deposited on granite, mica slate, gneiss, and other rocks of that character. The causes therefore which produced the calcareous beds, sometimes very abundantly, in the countries above noticed, have not extended to them. The observed phænomena are however sufficient to show that a vegetation similar to that of the more northern carboniferous rocks is there entombed, though we are not quite assured to what precise period their formation can be referred ; for, as will be seen in the sequel, similar vegetables are detected in the grauwacke series, and it is also possible that they might be discovered in the todtliegendes under the zechstein. The precise period of any particular deposit of similar vegetables may thus be sought through a considerable lapse of time, and it becomes hazardous to fix, without very good evidence, on any relative portion of that time. The conglomerates usually referred to the old red sandstone in Northern England, sometimes intervening between the upturned grauwacke rocks and the carboniferous limestone beds resting upon them, and noticed by Prof. Sedgwick and other geologists, may have been followed by a coal deposit where circumstances were favourable ; and the result would have been a formation similar to the deposits of central France, except that the subjacent rocks would be, perhaps, more ancient in the latter case. It may however have also happened, that during the grauwacke deposit, to be noticed in the next section, circumstances favoured a production of rocks similar to those of St. Etienne and other places ; as also that during a subsequent period, one equivalent to the lower part of the red sandstone group, a similar deposit might be effected; for as rocks may be violently disturbed in one place and not in another, so may they also be quietly formed in one situation, while a few hundred miles distant, disruptions of strata and the trituration of organic remains may have taken place, so that no trace of organic life may be left.

Let us now consider the mode in which the remains of terrestrial vegetables, so abundantly preserved in the coal strata, occur. They are for the most part laid flat, and the leaves and stems parallel to the line of stratification; but there are other cases where they repose at various angles in the beds; and finally they are found vertical, with their roots downwards. The student will recollect that this is precisely the manner in which the vegetables of the submarine forests are found; and if several submarine forests, such as those which are discovered around the shores of Great Britain, occurred above each other, with the intervention of sandy and clay beds, we should have a series of deposits not very unlike the coal strata, so far as regards the position of the vegetable remains. If we are to consider parts of the coal measures as in any way resulting from a series of similar deposits, we are certainly called upon to admit a very remarkable series of changes in the relative surface levels of land and water: but there are also very great difficulties attending the supposition that the vegetables have been swept by strong currents of water into the positions where we now find them; for not only have similar effects been produced over considerable areas, but the vegetables have suffered very little injury, their delicate leaves being most beautifully preserved. Now though we know that vegetables are abundantly borne down by river floods into the sea, they by no means remain uninjured; and if they be of a soft nature, such as the bulk of the coal plants are considered to have been, the damage done them by transport is considerable, as I have had occasion to remark on the coast of Jamaica, where arborescent ferns and other tropical productions are sometimes, though very rarely, carried by floods from the neighbouring mountains into the sea. In the few cases which passed under my observation, the fern-trees were so damaged in the river-courses as to be with difficulty recognisable *.

* The height at which arborescent ferns are found, would seem much to depend on local causes. Thus, on the southern side of Jamaica they do not flourish much under an elevation of 2000 feet above the sea; while on the northern side of the same island I have seen them at not more than 400 or 500 feet above the same level. The cause would seem to be the greater moisture of the northern side. It would therefore appear that a considerably moist climate would be necessary for the abundant production of this class of plants in the low situations, such as it has been imagined the lands were which produced the mass of the coal plants.

We have now so many cases in France, Germany, and Great Britain, of the occurrence of some coal plants in a vertical position, with their roots downwards, that such cases can scarcely be considered as accidental, but as similar to the vertical stems in the submarine forests, and therefore, in some measure, as characteristic of the deposit in particular situations.

Mr. Witham has brought forward some good examples of vertical stems in the carboniferous rocks of Durham and Newcastle. Two stumps or stems of *Sigillariæ* (Ferns) are described as standing erect, with their roots imbedded in bituminous shale, in the Derwent Mines, near Blanchford, Durham : the space round them was cleared out to obtain the lead ore; and one plant is stated to have been about five feet high, and two feet in diameter A more curious case was observed by the same author in the Newcastle district, where, in sandstone beneath the High Main coal, numbers of fossil vegetables, chiefly *Sigillariæ*, are discovered erect, their roots imbedded in a small seam of coal under the sandstone, while they are all truncated on the line of the High Main coal bed, to the formation of which their higher ends have in all probability partly contributed *.

Such cases as these, and that long since noticed by M. Al. Brongniart at St. Etienne †, where numerous stems are also included upright in coal sandstone, without however being truncated by a coal bed, are sufficient to show the very great analogy which exists between them, certain submarine forests, and the *dirt bed* at Portland, inasmuch as they all apparently point to a quiet submergence‡.

We may have some difficulty in considering the deposition of sand to have been effected so quietly amid the stumps of trees as not to have washed away the substances in which they were imbedded; but we have only to recollect, that among the submarine forests round our shores, if once any of them were at such a depth beneath the surface of the sea as to be sufficiently beyond the influence of the waves, they would become quietly covered by sand; for the

---

* Witham, Observations on Fossil Vegetables, 1831, p. 7, where there is an illustrative section.

† Annales des Mines, 1821.

‡ It cannot be denied that under particular circumstances, stems of trees preserving a vertical position may be forced onwards by river inundations. Thus, *snags*, or trees with their roots downwards and only forced by the current from a vertical position, are common, so as to be very dangerous, in the Mississippi; and trees were forced down the valley, during the debacle of the Vallée de Bagnes, and left standing with their roots downwards, at Martigny. These facts admit of easy explanation; for if trees be suddenly detached from the soil, and their roots loaded with stones and other heavy matter, they would naturally float with the branches upwards.

velocity of water sufficient to transport this sand, would scarcely disturb the trees. The principal difficulty attending this explanation arises from the repeated oscillations of the ground that it would seem to require, and the possible decay of the trees before they could be covered. We can scarcely, ever as an hypothesis, consider all coal beds as having been thus produced, for a large proportion seem to have been otherwise formed; but the difficulty of obtaining an abundance of vertical stems over a large area, unless by quiet submersion, is considerable; and an explanation merely requiring a wash or drift of vegetables accompanied by sand and silt, such as might take place at the delta of a great river, appears insufficient for the phænomena observed, more particularly when there are repeated alternations of marine remains with vegetable exuviæ, the former, as far as we can judge from analogy (the *Encrinites* and *Corals* for example), not being such as would be found in estuaries. The occurrence of limestone strata, continuous over considerable areas, and alternating with the shales and sandstones, would seem to require comparative tranquillity for their production, more particularly as the marine shells entombed in them have evidently not been subjected to violence, but appear to have been imbedded at no great distances from the places where they lived and died.

The vegetable remains are often of considerable size. M. Brongniart observes that in the coal strata of Dortmund, Essen, and Bochum, stems are found in the planes of the strata, more than fifty or sixty feet long, and that they may be traced in some of the galleries for more than forty feet without observing their natural extremities *. Vegetables of large size have also been detected in Great Britain. Mr. Witham mentions one in Craigleith quarry as being forty-seven feet in length from the highest part discovered to the root. The bark is described as converted into coal †.

Respecting the general character of the vegetation of this period, such as we find it entombed in the carboniferous rocks of the northern hemisphere, M. Ad. Brongniart observes, that it is remarkable; 1. for the considerable proportion of the vascular cryptogamic plants, such as the *Equisetaceæ, Filices, Marsileaceæ,* and *Lycopodiaceæ ;* 2. for the great development of the vegetables of this class, so that they have attained a magnitude far beyond those of the same class now existing; thus proving that circumstances were particularly favourable to their production during the period under consideration.

As, in the opinion of botanists, islands in warm countries are favourable to the growth of Ferns and other plants of the same

* Brongniart, Tableau des Terrains qui composent l'Ecorce du Globe.
† Witham, Edinburgh Journal of Natural and Geographical Science : April, 1831.

natural class, not only from the presence of the necessary heat, but from the moisture so congenial to them, it has been considered by MM. Sternberg, Boué, and Ad. Brongniart, that the vegetation of of this period, such as we find it in the carboniferous deposits of Europe and North America, was the growth of islands scattered in archipelagos. If, in accordance with this view, we suppose the islands to have been low, such as is the case with numerous coral islands in the Pacific Ocean, we may imagine that by oscillations of the soil, produced by movements in the earth, the surface covered by a dense vegetation might alternately be submerged and raised above the level of the sea.

When we come narrowly to look into the structure of the coal measures, the vast accumulations of shale and sandstones, sometimes amounting to the depth of 460 feet (Forest of Dean), do not precisely accord with the mere oscillations of islands above and beneath the level of the sea; for these accumulations of detritus require considerable drift, and must have resulted from the destruction of pre-existing rocks, mostly siliceous, and therefore, if solid, requiring much time for their degradation, with the assistance of other forces than the mere battering of the surf on clusters of low islands, perhaps defended, like those in the Pacific, by coral reefs.

The presence of larger masses of land, with mountains, rivers, and other physical features necessary for the production of a larger amount of detritus, would seem requisite, independent of volcanic eruptions, and other exertions of internal force, for the accumulations we observe. The oscillations of low islands is, therefore, merely brought forward as the possible explanation of some of the observed phænomena, and the student must be careful to consider it only in that light. While on this subject, however, it may be as well to notice a possible explanation of some of the minor alternations of limestones with marine remains, with shales and coal containing terrestrial remains such as are found in the millstone grit, because such hints, without attributing any particular value to them, very frequently lead to further inquiry. Suppose a tract of low land covered with a dense vegetation, such as is found in the tropics, to be, by a movement in the earth, an earthquake for instance, submerged a few feet beneath the sea, marine animals would establish themselves on the submerged surface, which would become in the condition of the submarine forests previously noticed; and the consequence would probably be, that not only millions of testaceous creatures would leave their exuviæ, but that the corals would also swarm, and might eventually produce coral islands, upon which vegetation might again establish itself, to be again submerged. That coral islands are sometimes raised above the sea, is what we should expect; and evidence of it has been adduced by Captain Beechey, who describes Henderson s Island (in the Pacific) as apparently upheaved by one effort of nature to the height of eighty feet. It is composed of dead coral, bounded by perpendicular cliffs, which are nearly encompassed by a reef of living coral, so that the cliffs are beyond the reach of the spray *. Now depression to this amount might as easily have taken place, in which case the vegetation of the island would have been submerged eighty feet, and the amount of destruction it would suffer would depend on the greater or less suddenness of the movement. Such movements cannot be considered great when regarded, as they always should be, with reference to the mass of the world, for we have proof that far greater have occurred, and the differences which have been produced in the relative levels of land and water are, when viewed on the great scale, of very trifling importance.

According to M. Ad. Brongniart, if we look at the arborescent ferns and the mass of the other plants, we must consider the vegetation of the carboniferous group to have been produced in climates at least as warm as those of the tropics; and as we now find plants of the same class increase in size as we advance towards warm latitudes, and as the coal-measure plants exceed the general size of their existing congeners, he concludes, with much apparent probability, that the climates in which the coal plants existed were even warmer than those of our equinoxial regions.

This view leads us to another consideration. There certainly was a similar vegetation about the same period (for whether the American coal measures may be, like those of Ireland, somewhat older, does not alter the question) over parts of Europe and North America; we may therefore infer a similar climate over a large portion of the northern hemisphere, such as we have not at present, for it was at least tropical, and very probably ultra-tropical. The question naturally arises, Is there any evidence to show that the same temperature existed at the same period in the southern hemisphere? for if there is, there must have been some common cause to produce such an equality of climate, at present unknown to us. Unfortunately, the actual state of our knowledge will not permit an answer to this question; but by it we learn the importance of ascertaining the botanical character of the various rocks in the southern hemisphere, more particularly those of the earliest formation, such as may be considered the equivalents of the carboniferous and grauwacke groups of the North.

With respect to the testaceous remains, the limestones contain a great abundance, not only of species, but of individuals of the

* Beechey, Voyage to the Pacific Ocean and Behring's Straits, p. 194. Descriptions of other coral islands, with sections of their general structure, will be found, pp. 160 and 186 of the same work.

genera *Spirifer* and *Producta.* Of the form of these shells, and
of the *Cardium hibernicum* and *C. alæforme,* (the latter by no
means a rare fossil in the limestones of the next group,) the following figures will afford examples.

Fig. 86.   Fig. 87.   Fig. 88.

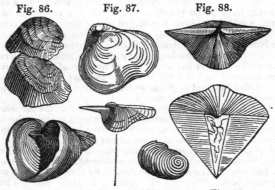

Fig. 91.   Fig. 92.   Fig. 90.   Fig. 89.

Fig. 86. *Producta Martini ;* Fig. 87. *Spirifer glaber ;* Fig. 88.
Spirifer attenuatus; Fig. 89. *Spirifer cuspidatus;* Fig. 90. one
of the two spiral appendages contained in *Spirifer trigonalis* * ;
Fig. 91. *Cardium hibernicum ;* and Fig. 92. *Cardium alæforme.*

Of the vertebrated creatures which may have existed at this
period our knowledge is very limited ; but it may be observed that
the Tritores or palates of fish, still retain phosphate of lime; for
Dr. Turner ascertained that a palate from the carboniferous lime-
limestone of Bristol, contained 24·4 per cent. of phosphate of lime,
the remainder being carbonate of lime and bituminous matter, the
latter abundant. A palate from the chalk, examined for the purpose of comparison, was found to contain 18·8 per cent. of phosphate of lime, the remainder being carbonate of lime, with traces
of bituminous matter.

* Their position in the shell will be seen by reference to Sowerby's
Mineral Conchology, pl. 265, fig. 1.

SECTION IX.

GRAUWACKE GROUP.

SYN.—Grauwacke (*Traumate*, Daubuisson). Grauwacke Slate (*Grauwacke schistoide*, Fr.; *Schiste Traumatique*, Daubuisson; *Grauwackenschiefer*, Germ.). Grauwacke Limestone (*Transition Limestone*, Engl. Authors; *Calcaire de Transition, Calcaire Intermédiaire*, Fr. Authors; *Uebergangskalkstein*, Germ. Authors.)

IT has been observed that the old red sandstone of some countries graduates into grauwacke, whence it may be inferred that the causes, whatever they may have been, which produced the latter deposit, were not violently interrupted in such situations, but that they were gradually modified,—if indeed it be necessary to consider the old red sandstone in any other light, taken generally, than the upper portion of the grauwacke series. That it is so, is the opinion of most contivental geologists; and where the one graduates into the other, such an opinion seems well founded. Variations in the classification of the old red sandstone would appear solely to arise from its mode of occurrence in the particular countries where geologists have been accustomed to observe it. When accidents have happened to the grauwacke, throwing the strata on their edges, and a red sandstone or conglomerate deposit intervenes between the carboniferous limestone and the upturned beds, classifications made in the countries where such phænomena prevail, would naturally be framed so as to separate the (old) red sandstone from the grauwacke: but when such accidents have not happened, and the carboniferous limestone, the intervening red sandstone, and the grauwacke are so circumstanced that the two former rest conformably on the latter, and they all graduate into one another, it is altogether as natural that the old red sandstone should be pronounced the upper part of the grauwacke series. Nor should we be surprised that the carboniferous limestone should also be included in the group; for the general organic character of the whole is similar, and does not differ more, if so much, as the upper part of the oolitic group from the lower portion of the same deposit, or as the chalk from the green sand.

Viewed on the large scale, the grauwacke series consists of a large stratified mass of arenaceous and slaty rocks intermingled with patches of limestone, which are often continuous for considerable distances. The arenaceous and slate beds, considered generally, bear evident marks of mechanical origin, but that of the included limestones may be more questionable. The arena-

U

ceous rocks occur both in thick and schistose beds; the latter state being frequently owing to the presence of mica disposed in the lines of the laminæ. Their mineralogical character varies materially; and while they sometimes, though rarely, pass into a conglomerate, they very frequently graduate into slates, which become of so fine a texture as to lose the arenaceous character altogether. Roofing-slate is not rare among the grauwacke rocks; and if we consider it of mechanical origin, like the mass of the strata among which it is included, we must suppose it to have originated from the deposition of a highly comminuted detritus.

If the size of transported substances be considered as the necessary evidence of rapid currents of water, the grauwacke rocks, taken as a mass, have been slowly deposited; for though evidences of cross currents are sufficiently abundant in the various directions of the laminæ, and in the mode in which arenaceous and slaty beds are associated with each other, the substances are generally fine-grained, rarely passing into conglomerates. A rapid current would however not appear to require large pebbles as the necessary evidences of its existence at any given period, although when large pebbles are present we infer that small currents of water could not have transported them; for the size of the substances transported by a current moving with considerable velocity will greatly depend on the surface over which it passes and the nature of the substances carried onwards. Thus, if sandstones having no great induration be rapidly transported over a hard surface, which the moving mass cannot tear up, but merely erode, the sandstones will, by trituration, be converted into sand, and be deposited in the first favourable situation, with the addition of the small detritus derived from the erosion of the hard rock. The same may to a certain extent happen when more compact fragments are washed about upon a hard surface of rock for a considerable time, so that they may be ground down into sand and mud. Perhaps from the absence of organic remains in a large proportion of the arenaceous part of this deposit, and their abundance in some of the included limestones, we might infer that there had been something in the transport and deposit of the sands unfavourable to their preservation, such as trituration in waters moving with rapidity. There is, however, a general appearance in the mass of the grauwacke which would lead us rather to consider a great portion of it of slow deposition.

It is by no means an uncommon circumstance for the laminæ of the slates of this group to be so arranged as to form various angles with other lines, which may be considered as those of the beds, or of stratification. Of this structure the annexed section of grauwacke slates at Bovey Sand Bay on the east side of Plymouth Sound, affords us an instructive example.

Fig. 93.

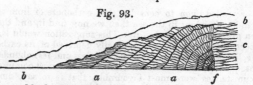

a a, curved beds of slate, the laminæ of which meet the apparent lines of stratification at various angles, being even perpendicular to them. The beds are cut off by the fault (*f*) from the slates *c,* the laminæ of which are more confusedly disposed, having however a general horizontal arrangement. The whole is covered by a detritus (*b b*) composed of fragments of the same kind of slate as that on which it reposes, and of the various grauwacke rocks of the hill behind.

The origin of the limestones is of far more difficult explanation than the sandstones and slates in which they are included. We cannot well seek it in the destruction of pre-existing calcareous rocks; for as far as our knowledge extends, such rocks are of comparative rarity among the older strata. In fact, the quantity of calcareous matter present in the grauwacke group greatly exceeds that discovered in the older rocks; and the same remark applies to many of the newer deposits when considered with reference to the grauwacke series. If we take the mass of deposits up to the chalk inclusive, we shall find that, instead of a decrease of carbonate of lime, such as we should expect if that contained in each deposit originated solely from the destruction of pre-existing limestones, the calcareous matter is more abundant in the upper than in the lower parts of the mass; and we may hence conclude that this explanation is insufficient.

If, as has been done with other limestones, we attribute the origin of the grauwacke limestones in a great measure to the exuviæ of testaceous animals and polypifers, we must grant the animals carbonate of lime with which to construct their shells and solid habitations. This they may have obtained either in their food or from the medium in which they existed. The marine vegetables are not likely to have supplied them with a greater abundance of carbonate of lime at that time than at present. Those that were carnivorous might acquire much carbonate of lime by devouring other animals more or less possessed of this substance : but the difficulty is by no means lessened by this explanation; for the creatures devoured must have procured the lime somewhere. It would appear that we should look to the medium in which testaceous animals and polypifers existed, for the greater proportion, if not all, of the carbonate of lime with which they constructed their shells and habitations. Now if we consider the mass of limestone rocks to have originated from the exuviæ of marine animals, we are called upon to consider that carbonate of lime was once far more abundant in the sea than we now find it, and that it has been gradually deprived of it. This supposition would lead us to expect, that as the sea was gradually deprived of its carbonate of lime, limestone deposits would become less and less abundant; and consequently, that calcareous rocks would be most common, when circumstances were most favourable, that is to say, during the formation of the older rocks. This, however, is precisely the reverse of what has happened. Hence we may infer that the origin of the mass of limestone deposits must be sought otherwise than in the attrition or solution of older and stratified rocks, or from the exuviæ of marine animals deriving their solid parts from a sea, which has gradually been deprived of nearly all its carbonate of lime. Both these causes may have eventually produced important modifications on the surface of the earth; but the great proportion of lime necessary for the formation of the calcareous masses covering a considerable part of it, would appear to have been otherwise obtained.

It has been usual to consider the lime of calcareous deposits as derived from limestone rocks, through which waters charged with carbonic acid percolated, the carbonic acid dissolving a certain portion of the lime, which is thus held in solution by the water until it reaches the surface, where it is thrown down in the shape of limestone. This explanation may suffice for the small deposits we observe in calcareous countries, but is insufficient for the productions of limestones generally; for it assumes that the solution of a small quantity of lime obtained from older rocks is, as previously noticed, capable of producing an immense deposit of the same substance. We know that carbonic acid is now discharged into the atmosphere from the earth by means of volcanos, fissures, and springs, and we have no reason to doubt that this has been the case during a long succession of ages; indeed we have every reason to believe that such discharge of carbonic acid formed a part of the great economy of nature, for without this aid we should have much difficulty in explaining the abundance of carbon and carbonic acid now locked up in coal deposits and limestones, all of which have clearly been produced successively on the earth's surface. Lime has been derived somewhere; and we have reason to believe from the interior of the earth, otherwise there is a difficulty in explaining the observed phænomena. The reason why extensive tracts of carbonate of lime have been produced at one time more than at another is not quite so apparent; but it may be observed, as a mere conjecture, that as this substance is not very unfrequent in volcanic regions, great disruptions of strata may have produced circumstances favourable for its deposition, and that without disturbances, carbonate of lime may have been thrown upwards in water, through fissures, more abundantly at

one time than at another, from causes unknown to us. Be this as
it may, the limestones in the grauwacke series most frequently
run in lines parallel to the general direction of the beds; and
although the calcareous matter may not be altogether continuous, there has evidently been some cause in operation at the
same time, within a given district, more favourable to the production of limestone than at another. It is also worthy of attention
that when the limestones occur, then also do the organic remains
generally become more abundant, appearing as if the calcareous
rocks and the organic remains were connected with each other.
That the animals, by secreting carbonate of lime from the medium
in which they lived, sometimes contributed considerably to the
mass, we are certain, as their remains now constitute a large
portion of it; but that they were the means through which all
the carbonate of lime was derived from the waters, may very
justly be doubted, more particularly as in certain districts not a
trace of animal exuviæ can be detected in such limestones. If
carbonate of lime were present in some situations and not in
others, animals, such as the *Crinoidea, Testacea,* and *Polypifers,*
would naturally flourish more in the former than in the latter, as
they could there more readily obtain the lime necessary for them,
and we should consequently expect to find their remains more
common there than elsewhere. In limestones devoid of organic
remains we appear to have evidence of carbonate of lime being
abundant in such situations unconnected with animal life, and we
may consider it derived from the interior and dispersed through the
waters over a given space, where it has been gradually deposited.
When, however, the remains of shells and corals are present, and
nearly constitute the mass of the rock, other causes may have produced the effects required, precisely as coral reefs and accumulations of shells now occur in one place and not in another, either
in consequence of shelter, proximity to the surface of the sea, or
other favourable circumstances.

Be the general origin of the grauwacke limestones what it may,
the causes which produced them were destined to cease during
the deposit of the grauwacke itself, and a series of sandstones and
slates, similar for the most part to those beneath, were accumulated upon them. In some districts, such as the North of Devon,
there has been a return of causes favourable to the deposit of
limestone, and two bands parallel to each other have been produced.

In other districts more limestones have been formed, while in
some they are nearly absent; a state of things we should expect from variations produced by local circumstances on similar
general causes in operation over a considerable area.

The grauwacke sometimes assumes a red colour in the midst of
beds of the usual gray and brown tints (South of Devon, Pembroke-

shire, Normandy, &c.), and is then undistinguishable from the old red sandstone of English geologists *.

Beds and even accumulations of strata are sometimes mingled with the common grauwacke and grauwacke slate, which at least show a variation in the mode of deposit. Thus, flinty slate sometimes associated in this series (Devonshire, &c.), is exceedingly compact, and, as its name implies, is principally composed of silex, the rock having much the appearance of a deposit from water in which silica was chemically dissolved †.

We have sometimes also beds which, in mineralogical composition, greatly resemble certain igneous rocks, known by the name of greenstones, corneans, &c. ; and though we may feel some hesitation in admitting such compounds as forming an original portion of the grauwacke series, not injected amid divisions of the strata by violence after deposition, such beds are nevertheless sometimes so continuous, without an apparent connection with a mass of trappean or igneous rocks, that, to say the least of it, their origin is very problematical. From the facility with which beds of this nature may often be traced to masses of similar rocks, as for example in Devonshire and Pembrokeshire, we may be more inclined to refer such included beds generally to the mere filling up of a fissure by igneous matter, which may, on the surface exposed to us, appear interstratified with the grauwacke. But as in the next group we shall observe stratified and similar compounds, which would appear to have been originally arranged in beds, we are not always certain that the beds in question have not also been contemporaneously produced with those among which they are included.

From the advance of geology, many districts which were formerly considered as composed of grauwacke, are now referred to less ancient deposits, and consequently the surface occupied by grauwacke is much less extensive than was formerly supposed. Thus large portions of the Alps and of Italy have been deprived of their supposed antiquity, which had been founded on the mineralogical structure of the deposits.

The grauwacke group occurs in Norway, Sweden, and Russia. It forms a portion of southern Scotland, whence it ranges, with

---

* This circumstance renders the determination of those limestones of Southern Devonshire which are much broken by faults, greatly disturbed and contorted, or much concealed by superincumbent (new) red sandstones, exceedingly difficult. This difficulty is particularly felt in the vicinity of Tor Quay, where, however, the limestones of the southern side of Tor Bay would certainly appear to be included in the grauwacke series, as is shown by coast sections, and their prolongation to the Dart.

† The student will recollect that under the head of Deposits from Springs, siliceous beds were noticed as having been produced by deposition from thermal waters in Iceland and the Azores.

breaks formed by newer deposits or the sea, down western England, into Normandy and Britanny. It appears abundantly in Ireland. A large mass of it is exposed in the district constituting the Ardennes, the Eifel, the Westerwald, and the Taunus. Another mass constitutes a large portion of the Hartz mountains, while smaller patches emerge in other parts of Germany, on the north of Magdeburg, and other places. In all these situations there is, notwithstanding small variations, a general and prevailing mineralogical character, which points to a common mode of formation over a considerable area. From all the accounts also that have been presented to us by Dr. Bigsby and the American geologists, we have every reason to consider that a deposit closely agreeing in relative antiquity, and in its general mineralogical and zoological characters, exists extensively in North America: so that there is evidence to show that some general causes were in operation over a large portion of the northern hemisphere, and that the result was the production of a thick and extensive deposit enveloping animals of similar organic structure over a considerable surface*.

Organic Remains of the Grauwacke Group.

PLANTÆ.

Algæ.

1. Fucoides antiquus, *Ad. Brong.* Christiania, Sweden, *Ad. Brong.*
2. ———— circinatus, *Ad. Brong.* Kinnekulle, Sweden, *Ad. Brong.*
   ————, sp. not determined. S. of Ireland, *Weav.*

Equisetaceæ.

1. Calamites radiatus, *Ad. Brong.* Bitschweiler, Haut Rhin, *Ad. Brong.*
2. ———— Voltzii, *Ad. Brong.* Zundsweiher, Baden, *Ad. Brong.*
   ————, sp. not determined. S. of Ireland, *Weav.;* Val St. Amarin, Haut Rhin, *Hœn.*

Filices.

1. Sphenopteris dissecta, *Ad. Brong.* Berghaupten, Baden, *Ad. Brong.*

* It was considered useless to present a long detail of the exact areas occupied by the grauwacke rocks, as the student will comprehend more by a single glance at good geological maps of any given country,—such as Greenough's map of England, Hoffmann's North-Western Germany, Oeynhausen, La Roche, and Von Decken's Countries near the Rhine, and De Beaumont and Dufrénoy's France,—than by long and tedious descriptions.

1 Cyclopteris flabellata, *Ad. Brong.* Berghaupten, Baden, *Ad Brong.*
1. Pecopteris aspera, *Ad. Brong.* Berghaupten, *Ad. Brong.*
1. Sigillaria tessellata, *Ad. Brong.* Berghaupten, *Ad. Brong.*
2. ——— Voltzii, *Ad. Brong.* Zundsweiher, *Ad. Brong.*

Lycopodiaceæ.

Lepidodendron, several species not determined. Berghaupten and Bitschweiler, *Ad. Brong.*
1. Stigmaria ficoides, *Ad. Brong.* Bitschweiler, *Ad. Brong.*

Class doubtful.

1. Asterophyllites pygmea, *Ad. Brong.* Berghaupten, *Ad. Brong.*

ZOOPHYTA.

1. Manon cribrosum, *Goldf.* Rebinghausen, Eifel, *Goldf.*
2. ——— favosum, *Goldf.* Eifel, *Goldf.*
1. Scyphia conoidea, *Goldf.* Nieder-Ehe, Eifel, *Goldf.*
2. ——— costata, *Goldf.* Eifel, *Goldf.*
3. ——— turbinata, *Goldf.* Eifel, *Goldf.*
4. ——— clathrata, *Goldf.* Eifel, *Goldf.*
1. Tragos acetabulum, *Goldf.* Keldenich, Eifel, *Goldf.*
2. ——— capitatum, *Goldf.* Bensberg, Prussian Prov., *Goldf*
1. Gorgonia antiqua, *Goldf.* Eifel; Ural, *Goldf.*
1. Stromatopora concentrica, *Goldf.* Eifel, *Goldf.*
   Madrepora, sp. not determined. Gloucestershire; Herefordshire; S. of Ireland, *Weav.*
1. Cellepora antiqua, *Goldf.* Heisterstein, Eifel, *Goldf.*
   ———, sp. not determined. Gloucestershire; Herefordshire, *Weav.*
1. Retepora antiqua, *Goldf.* Heisterstein, Eifel, *Goldf.*
2. ——— prisca, *Goldf.* Eifel, *Goldf.*
   ———, sp. not determined. Gloucestershire; Herefordshire; S. of Ireland, *Weav.*
   Flustra, sp. not determined. Gloucestershire; Herefordshire; S. of Ireland, *Weav.*
1. Ceriopora verrucosa, *Goldf.* Bensberg, Pruss. Prov., *Goldf.*
1. Agaracia lobata, *Goldf.* Eifel, *Goldf.*
1. Lithodendron cæspitosum, *Goldf.* Bensberg, *Goldf.*
   Caryophyllia, sp. not determined. Gloucestershire; Herefordshire, *Weav.*
1. Anthophyllum bicostatum, Heisterstein, Eifel, *Goldf.*
   Turbinolia, sp. not determined. Gloucestershire; Herefordshire; S. of Ireland, *Weav.*
1. Cyathophyllum Dianthus, *Goldf.* Eifel, *Goldf.*

2. Cyathophyllum radicans, *Goldf.* Eifel, *Goldf.*
3. ———————— marginatum, *Goldf.* Bensberg, *Goldf.*
4. ———————— explanatum, *Goldf.* Bensberg, *Goldf.*
5. ———————— turbinatum, *Goldf.* Eifel, *Goldf.*
6. ———————— hypocrateriforme, *Goldf.* Eifel, *Goldf.*
7. ———————— Ceratites, *Goldf.* Bensberg; Eifel, *Goldf.*
8. ———————— flexuosum, *Goldf.* Eifel, *Goldf.*
9. ———————— vermiculare, *Goldf.* Eifel, *Goldf.*
10. ———————— vesiculosum, *Goldf.* Eifel, *Goldf.*
11. ———————— secundum, *Goldf.* Eifel, *Goldf.*
12. ———————— lamellosum, *Goldf.* Eifel, *Goldf.*
13. ———————— placentiforme, *Goldf.* Eifel, *Goldf.*
14. ———————— quadrigeminum, *Goldf.* Eifel; Bensberg, *Goldf.*
15. ———————— cæspitosum, *Goldf.* Bensberg; Eifel, *Goldf.*
16. ———————— hexagonum, *Goldf.* Bensberg; Eifel, *Goldf.*
17. ———————— helianthoides, *Goldf.* Eifel; Lake Huron, *Goldf.*

1. Strombodes pentagonus, *Goldf.* Drummond Island, Lake Huron, *Goldf.*

1. Astrea porosa, *Goldf.* Eifel; Bensberg, *Goldf.*
——, sp. not determined. Gloucestershire; Herefordshire; S. of Ireland, *Weav.*

1. Catenipora escharoides, *Lam.* Eifel; Norway; Drummond Isl., *Goldf.;* Ratoska, Gov. of Moscow, *Fischer.*
2. ———————— labyrinthica, *Goldf.* Groningen; Drummond Island, *Goldf.*
3. ———————— tubulosa, *Lam.* Christiania, *Al. Brong.*
————————, sp. not determined. Gloucestershire; Herefordshire, *Weav.*

1. Syringopora verticillata, *Goldf.* Drummond Isl., *Goldf.*
Tubipora, sp. not determined. Gloucestershire; Herefordshire, *Weav.*

1. Calamopora alveolaris, *Goldf.* Eifel, *Goldf.*
2. ———————— favosa, *Goldf.* Drummond Isl., *Goldf.*
3. ———————— Gothlandica, *Goldf.* Eifel, *Goldf.*
4. ———————— basaltica, *Goldf.* Eifel; Gothland; Lake Erie, *Goldf.*
5. ———————— infundibulifera, *Goldf.* Eifel; Bensberg, *Goldf.*
6. ———————— polymorpha, *Goldf.* Eifel; Bensberg, *Goldf.*
7. ———————— spongites, *Goldf.* Eifel; Bensberg; Sweden, *Goldf.*

1. Aulopora serpens, *Goldf.* Eifel, *Goldf.;* Christiania, *Al. Brong.*
2. ———————— tubiformis, *Goldf.* Eifel, *Goldf.*
3. ———————— spicata, *Goldf.* Eifel; Bensberg, *Goldf.*
4. ———————— conglomerata, *Goldf.* Bensberg, *Goldf.*

442   *Organic Remains of the Grauwacke Group.*

1. Favosites Gothlandica, *Lam.*   Sloeben-Aker; Christiania; Eifel; Catskill; Batavia, New York, *Al. Brong.*
2. ———— Bromelli, *Ménard de la Groye.* Nehou, *Al. Brong.*
3. ———— truncata, *Rafinesque.* Kentucky, *Al. Brong.*
4. ———— Kentuckensis, *Raf.* Kentucky, *Al. Brong.*
5. ———— Boletus, *Ménard de la Groye.* Christiania, *Al. Brong.*
1. Mastrema pentagona, *Raf.* Garrard, Kentucky, *Al. Brong.*
1. Amplexus coralloides, *Miller.* South of Ireland, *Weav.;* Montechaton, near Coutances; Catskill, New York, *Al. Brong.*
———————, sp. not determined. Plymouth, *Hennah.*

RADIARIA.

1. Actinocrinites moniliformis, *Miller.* S. of Ireland, *Weav.*
2. ———————— triacontadactylus, *Miller.* South of Ireland, *Weav.*
———————, sp. not determined. Gloucestershire; Herefordshire, *Weav.*
1. Cyathocrinites tuberculatus, *Miller.* South of Ireland, *Weav.;* Dudley? *Miller.*
2. ———————— rugosus, *Miller.* Shropshire; Herefordshire; Isl. of Oeland; Dalecarlia, *Miller.*
———————, sp. not determined. Gloucestershire; Herefordshire, *Weav.*
1. Platycrinites lævis, *Miller.* Cork, *Weav.*
2. ———————— pentangularis, *Miller.* Dudley; Dinevar Park, Wales, *Miller.*
1. Rhodocrinites verus, *Miller.* Dudley, *Miller.*
1. Sphæronites* Pomum, *Wahl.* Isl. of Oeland; Kinnekulle, in Vestrogothia; Dalecarlia, *A.;* Tzarko-Sselo, near St. Petersburgh, *Al. Brong.*
2. ———————— Aurantium, *Wahl.* Mösseburg, Vestrogothia, *A.*
3. ———————— granatum, *Wahl.* Furndal, Dalecarlia; Boedahamn, Isl. of Oeland, *A.*
4. ———————— Wahlenbergii, *Esmark.* Gulf of Christiania, *Al. Brong.*

CONCHIFERA.

1. Thecidea? antiqua, *Hœn.* Gerolstein, *Hœn.*
1. Spirifer speciosus, *Bronn.* Eifel, *Holl.*
2. ———— cuspidatus, *Sow.* Eifel, *Holl.;* S. of Ireland, *Weav.;* Bensberg; Blankenheim, *Hœn.;* Plymouth, *Hennah.*

* Sphæronites, *Hisinger ;* Echinosphærites, *Wahlenberg.*

3. Spirifer glaber, *Sow.*   S. of Ireland, *Weav.;* Plymouth? *Hennah.*
4. ———— obtusus, *Sow.*   S. of Ireland, *Weav.*
5. ———— striatus, *Sow.*   S. of Ireland, *Weav.*
6. ———— pinguis, *Sow.*   S. of Ireland, *Weav.*
7. ———— intermedius *, *Schlot.*   Gloucestershire; Hereford-shire, *Weav.;* Eifel; Alleghany Mountains, *Al. Brong.*
8. ———— alatus, *Sow.*   Env. of Coblentz, *Al. Brong.*
9. ———— sarcinulatus †, *Schlot.* Coblentz; Malmö; Mösseberg, Sweden; Catskill, New York, *Al. Brong.*
10. ———— rotundatus, *Sow.*   Cork, *Wright;* Newton Bushel? Devon, *De la B.*
11. ———— lineatus, *Sow.*   Dudley, *Stokes.*
12. ———— ambiguus, *Sow.*   Blankenheim, *Hœn.*
13. ———— attenuatus, *Sow.*   Bensberg, *Hœn.*
14. ———— minimus, *Sow.*   Blankenheim, *Hœn.*
15. ———— Sowerbii,   Eifel, *Hœn.*
16. ———— decurrens, *Sow.*   Newton Bushel, Devon, *De la B.*
17. ———— distans, *Sow.*   Plymouth, *Hennah.*
18. ———— octoplicatus, *Sow.*   Plymouth, *Hennah.*
1. Terebratula crumena, *Sow.*   S. of Ireland, *Weav.*
2. ———— cordiformis, *Sow.*   S. of Ireland, *Weav.*
3. ———— Pugnus, *Sow.*   S. of Ireland, *Weav.;* Plymouth, *Hennah.*
4. ———— rostrata, *Schlot.*   S. of Ireland, *Weav.*
5. ———— prisca ‡, *Schlot.*   S. of Ireland, *Weav.;* Bensberg, *Schlot.;* Eifel; Urft, *Hœn.;* Plymouth, *Hennah.*
6. ———— affinis, *Sow.*   Dudley, *Ryan;* Eifel, *Hœn.*
7. ———— lævigata, *Schlot.*   S. of Ireland, *Weav.*
8. ———— elongata, *Schlot.*   S. of Ireland, *Weav.*
9. ———— plicatella, *Linn.*   Borenhult and Husbyfjoell, Os-trogothia, *A.*
10. ———— lacunosa §, *Schlot.*   S. of Ireland, *Weav.;* Ply-mouth? *Hennah.*
11. ———— osteolata, *Schlot.*   Eifel, *Schlot.*
12. ———— aperturata, *Schlot.*   Bensberg, *Schlot.*
13. ———— lenticularis, *Wahl.*   Westrogothia; Andrarum, Scania, *Al. Brong.*
14. ———— acuminata, *Sow.*   Cork, *Wright.*

---

* Terebratula, *Schlotheim.*   † Terebratula, *Schlotheim.*
‡ Considered by Mr. Sowerby the same as *T. affinis* of the Mineral Conchology.
§ Considered by Mr. Sowerby the same with the *T. Pugnus* of Min. Con.

15. Terebratula lateralis, *Sow.* Cork, *Wright;* Blankenheim, *Hœn.*
16. ———— reniformis, *Sow.* Cork, *Sow.*
17. ———— alata, *Lam.* Eifel, *Hœn.*
18. ———— aspera, *Schlot.* Eifel; Bensberg; Christiania, *Hœn.*
19. ———— comprimata, *Schlot.* Eifel, *Hœn.*
20. ———— curvata, *Schlot.* Gerolstein, *Hœn.*
21. ———— excisa, *Schlot.* Eifel, *Hœn.*
22. ———— explanata, *Schlot.* Blankenheim, *Hœn.*
23. ———— imbricata, *Sow.* Eifel, *Hœn.;* Plymouth, *Hennah.*
24. ———— intermedia, *Lam.* Eifel and America, *Hœn.*
25. ———— Mantiæ, *Sow.* Blankenheim, *Hœn.*
26. ———— monticulata, *Schlot.* Blankenheim, *Hœn.*
27. ———— resupinata, *Sow.* Blankenheim, *Hœn.*
28. ———— striatula, . Blankenheim; Christiania; Trenton Falls, *Hœn.*
29. ———— speciosa, *Schlot.* Eifel, *Hœn.*
30. ———— Sacculus, *Sow.* Blankenheim, *Hœn.*
31. ———— Wilsoni, *Sow.* Porsgrund, Norway, *Hœn.*
32. ———— hysterolita*, *Hœn.* Hickeswagen, *Hœn.;* Coblentz; Oberlahnstein, near Mayence, *Schlot.*
33. ———— paradoxa †, *Hœn.* Lahnstein; Crefeld; Catskill Mountains, America, *Hœn.;* Kaisersternal, &c., *Schlot.*
34. ———— porrecta, *Sow.* Newton Bushel, Devon, *De la B.*
35. ———— platyloba (jun.), *Sow.* Plymouth, *Hennah.*
1. Strygocephalus Burtini, *Defr.* Bensberg, *Hœn.*
2. ———— elongatus, *Goldf.* Bensberg, *Hœn.*
1. Calceola sandalina, *Lam.* Eifel, *Bronn;* Gerolstein, Blankenheim, *Hœn.*
2. ———— heteroclita, *Defr.* Blankenheim, *Hœn.*
1. Strophomena Goldfussii, *Hœn.* Blankenheim, *Hœn.*
2. ———— rugosa, *Raff.* Catskill Mountains; Trenton, America; Dudley; Eifel; Crefeld, *Hœn.*
3. ———— euglypha, *Hœn.* Eifel, *Hœn.*
4. ———— pileopsis, *Raf.* Kentucky, *Al. Brong.*
5. ———— umbraculum ‡, *Schlot.* Eifel; Christiania, *Hœn.*
6. ———— marsupita §, *Defr.* Catskill Mountains; Lorkport; Eifel, *Hœn.*
1. Producta Scotica, *Sow.* S. of Ireland, *Weav.;* Eifel, *Hœn.;* Isle of Man, *Henslow.*
2. ———— Martini, *Sow.* S. of Ireland, *Weav.*
3. ———— concinna, *Sow.* S. of Ireland, *Weav.*

* *Hysterolites vulvarius*, Schlot.
† *Hysterolites hystericus*, Schlot.
‡ M. Brongniart considers this may be the same with *Str. pileopsis*
§ *Leptæna depressa*, Dalman.

4. Producta lobata, *Sow.* S. of Ireland, *Weav.*
5. ———— longispina, *Sow.* Blankenheim, *Hœn.*
6. ———— punctata, *Sow.* Blackrock, Cork, *Wright.*
7. ———— fimbriata, *Sow.* S. of Ireland, *Weav.*
8. ———— depressa, *Sow.* S. of Ireland, *Weav.*; Dudley, *Sow.*;
        Plymouth, *Hennah.*
9. ———— hemisphærica, *Sow.* Eifel ; Catskill Mountains ;
        Albany, Lexington, *Hœn.*
10. ———— rostrata, *Sow.* Bensberg, *Hœn.*
11. ———— sarcinulata, *Goldf.* Eifel ; Catskill Mountains,
        Hœn.
12. ———— sulcata, *Sow.* Catskill Mountains, *Hœn.**
    Gryphæa, sp. not determined. Keswick, near Kirby Lons-
        dale, *Phil.*
    Pecten, sp. not determined. Keswick, *Phil.*; Plymouth,
        Hennah ; S. of Ireland, *Weav.*
    Plagiostoma, sp. not determined. Keswick, *Phil.*
1. Megalodon cucullatus, *Sow.* Newton Bushel, Devon, *De
        la B.*
    Trigonia, sp. not determined. Keswick, *Phil.*
1. Cardium costellatum, *Munst.* Elbersreuth ; Prague, *Hœn.*
2. ———— hybridum, *Munst.* Elbersreuth, *Hœn.*
3. ———— lineare, *Munst.* Elbersreuth, *Hœn.*
4. ———— priscum, *Munst.* Elbersreuth ; Prague, *Hœn.*
5. ———— striatum, *Munst.* Elbersreuth, *Hœn.*

* The following are the fossils of the Terebratulite family, according to
the divisions of the Swedish naturalists, discovered in the grauwacke
rocks of Sweden.

1. Leptæna rugosa, *Hisinger.* Borenshult, Ostrogothia ; Westrogothia.
2. ———— deflexa, *Dalman.* Ostrogothia.
3. ———— transversalis, *Wahl.* Osmundsberg, Dalecarlia.
1. Orthis pecten, . Borenshult, Ostrogothia ; Westrogothia.
2. ———— zonata, *Dalman.* Borenshult.
3. ———— callactes, *Dalman.* Husbyfjoel ; (var.) Ulanda, Westrogothia.
4. ———— calligramma, *Dalman.* Skarpäsen, Ostrogothia.
5. ———— testudinaria, *Dalman.* Borenshult (also at Blankenheim,
        Hœn.).
6. ———— demissa, *Dalman.* Boeda, I. of Oeland.
?7. ———— novemradiata, *Wahl.* I. of Oeland, Darlecarlia.
8. ———— elegantula, *Dalman.* Blankenheim, *Hœn.*
1. Delthyris subsulcata, *Dalman.* Boeda, I. of Oeland.
?2. ———— Psittacina, *Wahl.* Osmundsberg, Dalecarlia.
?3. ———— jugata, *Wahl.* Osmundsberg, Dalecarlia.
1. Atrypa reticularis, *Wahl.* Westrogothia.
2. ———— canaliculata, *Dalman.* Borenshult, Ostrogothia.
3. ———— Nucella, *Dalman.* Husbyfjoel, Ostrogothia.
4. ———— cassidea, *Dalman.* Borenshult, Ostrogothia.
?5. ———— crassicostis, *Dalman.* Westrogothia.

6. Cardium alæforme, *Sow.* Scarlet, Isle of Man, *Henslow;* Plymouth, *Hennah;* Newton Bushel, Devon, *De la B.*
1. Cardita costellata, *Munst.* Elbersreuth, *Hœn.*
2. ———— gracilis, *Munst.* Elbersreuth, *Hœn.*
3. ———— plicata *Munst.* Elbersreuth, *Hœn.*
4. ———— tripartita, *Munst.* Elbersreuth, *Hœn.*
1. Isocardia Humboldtii, *Hœn.* Wissenbach, near Dillenburg, *Hœn.*
2. ———— oblonga, *Sow.* Cork, *Flem.*
1. Cypricardia? ————. Bensburg; Eifel, *Hœn.*
1. Posidonia Becheri, *Bronn.* Geistl. Berg, near Herborn, *Hœn.*

Mollusca.

Patella, sp. not determined. Keswick, near Kirby Lonsdale, *Phil.*
1. ————? conica, *Wahl.* Kinnekulle, Westrogothia, *A.*
2. ————? pennicostis, *Wahl.* Ulanda, Westrogothia, *A.*
3. ————? concentrica, *Wahl.* Mösseberg, &c., Westrogothia, *A.*
1. Peleopsis vetusta, *Sow.* South of Ireland, *Weav.* Plymouth, *Hennah.*
1. Melanopsis coronata, *Hœn.* Bensberg, *Hœn.*
1. Melania constricta, *Sow.* South of Ireland, *Weav.*
2. ———— bilineata, *Goldf.* Bensberg, *Hœn.*
1. Natica canrena, *Lam.* Hudson City; Catskill Mountains, United States, *Hœn.*
————, sp. not determined. Plymouth, *Hennah.* Newton Bushel? *De la B.*
1. Nerita spirata? *Sow.* Plymouth, *Hennah.*
————, sp. not determined. Herefordshire; Gloucestershire; South of Ireland, *Weav.*
1. Solarium fasciatum, . Bensberg, *Hœn.*
1. Delphinula æquilatera, *Wahl.* Westrogothia, *A.*
1. Cirrus acutus, *Sow.* S. of Ireland, *Weav.* Plymouth, *Hennah.*
1. Pleurotomaria cirriformis, *Sow.* Plymouth, *Hennah.*
1. Euomphalus catillus, *Sow.* S. of Ireland, *Weav.* Blankenheim; Lake Erie, *Hœn.*
2. ———— centrifugus, *Wahl.* Wikarby, Dalecarlia, *A.*
3. ———— dubius, *Goldf.* Dillenburg, *Hœn.*
4. ———— funatus, *Sow.* Dudley, *Johnstone.*
————, sp. not determined. Newton Bushel, Devon, *De la Beche.*
1. Trochus ellipticus, *Hisinger.* Furudal, Dalecarlia, *A.*
1. Turbo bicarinatus, *Wahl.* Wikarby, Dalecarlia; Borenhult, Ostrogothia, *A.*

2. Turbo Tiara, *Sow.*   Plymouth, *Hennah.*
3. —— antiquus, *Goldf.*   Bensberg, *Hœn.*
1. Turritella abbreviata, *Sow.*   Newton Bushel, Devon, *De la B.*
   ——, sp. not determined.   Beckfoot, near Kirby Lons-
      dale, *Phil.*
   Pleurotoma, sp. not determined.   Newton Bushel, *De la B.*
1. Murex? Harpula, *Sow.*   Newton Bushel, *De la B.;*   Ply-
      mouth, *Hennah.*
1. Buccinum spinosum, *Sow.*   Plymouth, *Hennah;*   Newton
      Bushel, *De la B.*
2. —————— acutum, *Sow.*   Plymouth, *Hennah.*
3. —————— breve, *Sow.*   Newton Bushel, Devon, *De la B.*
4. —————— imbricatum, *Sow.*   Newton Bushel, *De la B.;*   Ply-
      mouth, *Hennah.*
1. Bellerophon tenuifascia, *Sow.*   S. of Ireland, *Weav.;*   New-
      ton Bushel, Devon, *De la B.*
2. —————— ovatus, *Sow.*   S. of Ireland, *Weav.*
3. —————— hiulcus*, *Sow.*   Blankenburg, *Hœn.*
4. —————— Hüpschii, *Defr.*   Chimay; Blankenburg, *Hœn.*
5. —————— nodulosus, *Goldf.*   Bensberg, *Hœn.*
6. —————— Cornu Arietis, *Sow.*   Catskill Mountains, *Hœn.*
7. —————— apertus, *Sow.*   Plattsburg, New York, *Hœn.*
8. —————— costatus, *Sow.*   Plymouth, *Hennah.*
   ——, sp. not determined.   Plymouth, *Hennah.*
1. Conularia quadrisulcata, *Miller.*   Gloucestershire, *Weav.;*   Bo-
      renshult, Ostrogothia, *A.;* Montmorency Falls,
      Quebec, *Hœn.*
2. —————— pyramidata,   . May, near Caen, *Deslongchamps.*
3. —————— teres, *Sow.*   Lockport, N. America, *Hœn.*
   ——, sp. not determined.   May, Calvados, *Deslong.*
1. Orthoceratites striatus, *Sow.*   S. of Ireland, *Weav.;*   Malmoe,
      Christiania, *Al. Brong.;*   Trenton Falls,
      New York, *Hœn.*
2. —————— undulatus, *Sow.*   S. of Ireland, *Weav.* Tzarko-
      Sselo, near St. Petersburg, *Al. Brong.*
3. —————— paradoxicus, *Sow.*   S. of Ireland, *Weav.*
4. —————— circularis, *Sow.*   Gloucestershire; Hereford-
      shire, *Weav.;* Plymouth, *Hennah.*
5. —————— annulatus, *Sow.* Gloucestershire, *Weav.;* Ge-
      rolstein, Eifel, *Schlot.*
6. —————— flexuosus, *Schlot.*   Oeland; Gerolstein, Eifel,
      Holl.   Black River, New York, *Hœn.*
7. —————— communis, *Wahl.*   Common in Sweden, *A.*
8. —————— duplex, *Wahl.*   Kinnekulle, Sweden, *A.*
      Black River, New York, *Hœn.*

         * *B. striatus,* Goldfuss.

9. Orthoceratites trochlearis, *Dalman.* Solleroe, Dalecarlia, *A.*
10. ——————— turbinatus, *Dalman.* Dalecarlia; Isle of
        Oeland, *A.*
11. ——————— centralis, *Dalman.* Solleroe, Dalecarlia, *A.*
12. ——————— gracilis, *Schlot.* Hellenburg, Nassau, *Al.*
        Brong.; Wissenbach, *Hœn.*
13. ——————— crassiventer, *Wahl.* N.W. side of Lake Hu-
        ron, *Hœn.*
14. ——————— duplex, *Wahl.* Black River, New York, *Hœn.*
15. ——————— falcata,   Trenton Falls, *Hœn.*
16. ——————— tenuis, *Wahl.* Geistlichen Berg, near Herborn,
        Hœn.
17. ——————— rectus, *Bosc.* Kuchel, near Prague, *Hœn.*
18. ——————— regularis, *Schlot.* Oeland, *Hœn.*
19. ——————— gigantea, *Sow.* Gerolstein, *Hœn.*
20. ——————— excepticus, *Goldf.* Bensberg; Gledbach, near
        Mülheim, *Hœn.*
    ———————, sp. not determined. Gloucestershire; Here-
        fordshire, *Weav.;* Plymouth, *Hennah;* Env.
        of St. Petersburg, *Strangways.*
 1. Cyrtoceratites ammonius, *Goldf.* Montmorency Falls, Lower
        Canada, *Hœn.*
 2. ——————— compressus, *Goldf.* Eifel, *Hœn.*
 3. ——————— depressus, *Goldf.* Gerolstein, *Hœn.*
 4. ——————— ornatus, *Goldf.* Bensberg, *Hœn.*
 1. Lituites perfectus, *Wahl.* Mösseberg, Sweden, *Al. Brong.;*
        Revel, *Hœn.*
 2. ——————— imperfectus, *Wahl.* Jungby, Sweden, *Al. Brong.*
 1. Nautilus globatus*, *Sow.* S. of Ireland, *Weav.*
 2. ——————— multicarinatus, *Sow.* S. of Ireland, *Weav.*
 3. ——————— complanatus, *Sowerby.* Scarlet, Isle of Man,
        Henslow.
 4. ——————— cariniferus, *Sow.* Black Rock, Cork, *Sow.*
 5. ——————— divisus, *Munst.* Geistlichen-Berg, near Herborn,
        Hœn.
 6. ——————— Wrightii, *Flem.* Cork, *Wright.*
†7. ——————— funatus, *Flem.* Cork, *Sow.*
†8. ——————— compressus, *Flem.* Cork, *Sow.*
†9. ——————— ovatus, *Flem.* S. of Ireland, *Weav.*
 1. Ammonites Henslowi, *Sow.* I. of Man, *Henslow.*
 2. ——————— subnautilinus, *Schlot.* Wissenbach, near Dillen-
        burg, *Hœn.*
 3. ——————— Becheri,   Dillenburg, *Hœn.*

* Dr. Fleming considers that this shell may probably be his *Nautilus
Wrightii.*
† *Ellipsolites* of Sowerby.

Ammonites, sp. not determined. Gloucestershire; Herefordshire; S. of Ireland, *Weav.*

CRUSTACEA.

1. Calymene Blumenbachii, *Al. Brong.* Dudley; Lebanon,
Ohio; Newport, Utica, United States, *Al.
Brong.;* Gloucestershire; Herefordshire, *Weav.;*
Skartofta, Scania; Ostrogothia, *A.;* Blankenheim, *Hœn.*
2. ———— macrophthalma, *Al. Brong.* United States; Crom-
ford, near Dusseldorf, *Al. Brong.;* Dudley,
Weav.; Shropshire, *Stokes;* Dillenburg, *Hœn.*
3. ———— variolaris, *Al. Brong.* Dudley, *Stokes;* Gloucestershire; Herefordshire, *Weav.*
4. ———— Tristani, *Al. Brong.* Breuville, Cotentin; Falaise;
La Hunandière; Bain, near Rennes, *Al.
Brong.;* Angers; Genesee, *Hœn.*
5. ———— bellatula, *Dalman.* Husbyfjoel, Ostrogothia, *A.*
6. ———— ornata, *Dalman.* Husbyfjoel, Ostrogothia, *A.*
7. ———— verrucosa, *Dalman.* Varving, near the mountain
of Billingen, Westrogothia, *A.*
8. ———— polytoma, *Dalman.* Ljung, Ostrogothia, *A.*
9. ———— actinura, *Dalman.* Berg, Ostrogothia, *A.*
10. ———— Schrops, *Dalman.* Furudal, Dalecarlia; Ostrogothia, *A.*
11. ———— Schlotheimi, *Bronn.* Blankenheim, *Hœn.*
12. ———— latiferus, *Bronn.* Blankenheim, *Hœn.*
1. Asaphus cornigerus, *Al. Brong.* Env. of St. Petersburg, *Al.
Brong.;* Revel, *Schlot.;* Blankenheim, *Hœn.*
2. ———— cordigerus, *Al. Brong.* Dudley, *Stokes.*
3. ———— Hausmanni, *Al. Brong.* Nehou (La Manche);
Prague, *Al. Brong.;* Canada; Catskill Mountains; Karlstein; Kugel, *Hœn.*
4. ———— de Buchii, *Al. Brong.* Dinevawr Park, Wales;
Cyer, Norway, *Al. Brong.;* Eifel, *Hœn.*
5. ———— Brongniartii, *Deslongchamps.* May; Nehou, Normandy; Eifel, *Al. Brong.*
6. ———— extenuatus, *Wahl.* Husbyfjoel, Heda, Ostrogothia,
A.
7. ———— granulatus, *Wahl.* Varving, Olleberg, Westrogothia; Furudal, Dalecarlia, *A.*
8. ———— angustifrons, *Dalman.* Husbyfjoel, Ostrogothia, *A.*
9. ———— Heros, *Dalman.* Kinnekulle, Westrogothia; Vi-
karby, Dalecarlia, *A.*
10. ———— expansus, *Wahl.* Common in Sweden, *A.*
11. ———— platynotus, *Dalman.* Westrogothia, *A.*

12. Asaphus frontalis, *Dalman.* Ljung, Ostrogothia, *A.*
13. ———— læviceps, *Dalman.* Husbyfjöl, Ostrogothia, *A.*
14. ———— palpebrosus, *Dalman.* Husbyfjöl, Ostrogothia, *A.*
15. ———— crassacanda, *Wahl.* Husbyfjöl; Christiania; Bain; Tzarko-Sselo, *Al. Brong.*
16. ———— Sulzeri,   Ginez, Bohemia, *Hœn.*
1. Ogygia Guettardii, *Al. Brong.* Angers, *Al. Brong.*
2. ———— Desmaresti, *Al. Brong.* Angers, *Al. Brong.*
3. ———— Wahlenbergii, *Al. Brong.* Angers, *Al. Brong.*
4. ———— Sillimani, *Al. Brong.* Banks of the Mohawk, near Schenectady, *Al. Brong.*
1. Paradoxides Tessini, *Al. Brong.* Olstorp, Westrogothia, *Al. Brong.;* Ginez, Bohemia, *Hœn.*
2. ———— spinulosus*, *Al. Brong.* Andrarum, Scania, *Al. Brong.;* Westrogothia, *A.*
3. ———— gibbosus†, *Al. Brong.* Kinnekulle, *Al. Brong.*
4. ———— scaraboides‡, *Al. Brong.* Falköping, *Al. Brong.;* Ostrogothia; Westrogothia, *A.*
5. ———— Hoffii, *Goldf.* Braatz, near Ginez, Bohemia, *Al. Brong.*
1. Nileus Armadillo, *Dalman.* Husbyfjöl and Skarpäsen, Ostrogothia; Tomarp, Scania; Furudal, Dalecarlia, *A.*
2. ———— Glomerinus, *Dalman.* Husbyfjöl, Ostrogothia, *A.*
1. Illænus Centaurus, *Dalman.* Isle of Oeland, *A.*
2. ———— centrotus, *Dalman.* Husbyfjöl, Ostrogothia, *A.*
3. ———— latecauda, *Wahl.* Osmundsberg, Dalecarlia, *A.*
1. Ampyx nasutus, *Dalman.* Skarpäsen and Husbyfjöl, Ostrogothia; Varving, Westrogothia, *A.*
1. Olenus Bucephalus, *Wahl.* Olstorp, Westrogothia, *A.*
1. Agnostus pisiformis, *Al. Brong.* Kinnekulle, Mösseberg.; Westrogothia, *Al. Brong.*
1. Isotelus Gigas, *Dekay* §. Trenton Falls.
2. ———— planus, *Dekay.* Trenton Falls.
Trilobites, species not determined. Env. of St. Petersburg, *Strangways;* Isle of Man, *Henslow;* Brixham, Devon, *De la B.*

Pisces.

Ichthyodorulites, *Buckl.* and *De la B.* Dudley, *Clayfield;* Herefordshire, *Phil.*
Fish bones and a tooth, Whitefield quarry and Skeay's Grove, Tortworth, Gloucestershire, *Weav.*
Casts referable to the vertebræ of fish, S. of Ireland, *Weav.* ‖

* *Olenus spinulosus,* Wahlenberg.
† *Olenus gibbosus,* Wahlenberg.
‡ *Olenus scaraboides,* Wahlenberg.
§ *Asaphus platycephalus,* Stokes.
‖ It should be observed that there are several hitherto undescribed

The student will have perceived from an inspection of the foregoing list, that the organic remains of the grauwacke series are by no means deficient in variety of form. Of Zoophytes we find the genera *Manon, Scyphia, Tragos, Gorgonia, Stromatopora, Madrepora, Cellepora, Retepora, Flustra, Ceriopora, Lithodendron* (including *Caryophyllia*), *Anthophyllum, Turbinolia, Cyathophyllum, Strombodes, Astrea, Catenipora, Syringopora, Calamopora, Aulopora, Favosites, Mastrema,* and *Amplexus.* Of Radiaria, the genera *Actinocrinites, Cyathocrinites, Platycrinites, Rhodocrinites,* and *Sphæronites.* Of Conchifera, the genera *Thecidea? Spirifer (Dalthyris,* Dalman), *Terebratula, Strygocephalus, Calceola, Strophomena, Producta, Gryphæa, Pecten, Plagiostoma, Megalodon, Trigonia, Cardium, Cardita, Isocardia, Cypricardia?* and *Posidonia.* Of Mollusca, the genera *Patella, Pileopsis, Melanopsis, Melania, Natica, Nerita, Solarium, Delphinula, Cirrus, Euomphalus, Trochus, Turbo, Turritella, Pleurotoma, Murex? Buccinum, Bellerophon, Conularia, Orthoceratites, Cyrtoceratites, Lituolites, Nautilus,* and *Ammonites.* Of Crustacea, the various Trilobites divided into the genera *Calymene, Asaphus,*

fossils of the grauwacke limestone of Plymouth, in the collection of the Rev. R. Hennah; among these Mr. Sowerby has noticed the following:

Conchifera.

1. Spirifer reticulatus, *Sow. MS.,* also from Ireland, *Sow.*
2. ———— pentagonus, *Sow. MS.*
1. Terebratula Hennahiana, *Sow.* Oblong, rather square, convex, and smooth: a wide furrow runs along the middle of the larger valve, the beak of which is much produced.
2. ———— gigantea, *Sow.* Oval, the front rather straight; valves equally convex, a little flattened towards the front; beak of the large valve moderately produced, not incurved. Five and a half inches long, four inches wide.
3. ———— rotundata, *Sow.* Globose, smooth; beaks large, touching.
4. ———— like *T. affinis,* but has finer striæ, a produced beak to the larger valve, and a greater length; it is also rather more flat.
5. ———— Lachryma, *Sow. MS.*
1. Producta anomala, *Sow. MS.,* also from Ireland and Preston, *Sow.*

Mollusca.

1. Turbo cirriformis, *Sow.* Spire short, of three very convex whorls, smooth; length and breadth equal.
1. Natica? ————. Nearly globose; spire pointed; whorls few; the last large; smooth.
1. Terebra Hennahiana, *Sow.* Turrited, sides nearly straight; whorls flat, crossed by slightly curved deep striæ. Also from Preston, *Sow.*

Ogygia, Paradoxides, Nileus, Illænus, Amphyx, Olenus, Agnostus, and *Isotelus.* Of fish, the remains of bones, teeth, and the defensive fin-bones named *Ichthyodorulites.*

In this catalogue we find a mixture of existing and extinct genera, which is remarkable when we consider the great antiquity of the rocks containing them. It may be doubtful whether all the genera have been correctly determined, for possibly some of them may have been rather hastily referred to those now existing; but even admitting this, we have evidence that there was not that poverty of organic structure which was once supposed.

From the various forms of the fossils imbedded in the grauwacke, we may infer that the animals, of which they constituted the solid parts, occupied situations as different as those of the present day ; some preferring deep waters, while others were fitted for shallow seas, and not a few swam freely in the open ocean ; certain creatures frequenting one kind of bottom, while others sought another of a different description. The most abundant shells belong to the genera *Orthoceratites, Producta, Spirifer,* and *Terebratula.* The former often attain a large size, even reaching a yard or more in length ; so that if they really once constituted a part of swimming mollusca, analogous to the Nautilus of the present day, some of such creatures must have far exceeded the size of the animals of that kind now known to us. The three latter most abundant genera constitute a natural group which the Swedish naturalists have arranged under the heads of *Leptæna (Producta), Orthis, Cyrtia, Delthyris (Spirifer), Gypidia, Atrypa, Rhynchora,* and *Terebratula ;* the characters being considered such as to justify the formation of the genera. Supposing this arrangement to be well founded, it would appear from the lists of those who have proposed it, that *Terebratulæ* are rare in the older rocks, such as those under consideration, while they are abundant in the newer strata.

Productæ are, as has been seen, common in this and the carboniferous groups, and existed during the deposit of the zechstein. Spirifers, which also abounded during the deposit of the grauwacke and carboniferous series, have been observed as high up as the lias, where three species of the genus Spirifer have been detected, one (*Spirifer Walcotii*) being a very common and characteristic shell. The Terebratulæ, which, even admitting the Swedish divisions, are found in the preceding series, if not in the higher part of this, extend upwards to the present day, many species being now known. Taking, therefore, this natural group as it existed at this early period, in which we should probably include the carboniferous limestone, and tracing it upwards through the various rocks, we find that the Productæ first disappeared, and then the Spirifers, while the Terebratulæ have been preserved through all the changes which have taken place on the surface of our planet.

The family of the Trilobites was one, the individuals of which must have swarmed in particular places during the deposit of the grauwacke. In some parts of Wales the *Asaphus Debuchii* (Fig. 94.) is so abundant that the laminæ of the slates are charged with them, so that millions have probably lived and died not far distant from those places where we now discover their remains. This species has not been confined to Wales, though it is there very abundant, but has also been discovered in Norway and Germany. The Trilobite long known in museums as the Dudley Trilobite, because found

Fig. 94.

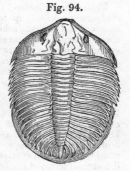

so commonly at that place, is the *Calymene Blumenbachii* of M. Al. Brongniart (Fig. 95.). This species existed over a considerable area, having not only been discovered in England, Germany, and Sweden, but also in North America. Although many parts of these creatures are found distributed in such a manner that we may conclude they were separated by decomposition after the death of the animal, the perfect preservation of others, and their frequent contracted attitudes, such as we should expect creatures of this structure to assume when disturbed, would lead us to conjecture that they had been often suddenly destroyed, and as suddenly enveloped in that matter which subsequently became hard rock; thus preventing the separation of the harder parts by decomposition. The forms of the Trilobite family

Fig. 95.

vary more considerably than might be supposed from the *Asaphus* and *Calymene* represented above, as will be seen by the annexed figure of *Agnostus pisiformis,* Fig. 96. being the natural size of the animal, and Fig. 97. a magnified representation of it. The Trilobite family seem now to have entirely disappeared from among existing animals, and we may perhaps venture to infer, from our present information respecting organic remains, that it became extinct before the Productæ; and we are nearly certain it ceased to exist long before the Spirifers, for neither in the muschelkalk nor in the lias has the smallest trace of them ever been detected.

Fig. 96.

Fig. 97.

Unlike the Trilobites, the Crinoidea common in this early period are continued up to the present day, though the genera observed in the grauwacke series and in the carboniferous group seem to have disappeared previous to the deposit of the oolitic series, when other genera were called into existence, one of which, *Pentacrinites*, is discovered in the present seas. It was even at one time considered that a particular species, *Pentacrinites Caput Medusæ*, was common to the actual ocean and the lias, but this is now doubted.

The discovery of the defensive fin bones, named Ichthyodorulites, in the grauwacke series is worthy of attention, as it shows that the class of animals to which they belong was among the earliest inhabitants of the globe, and that it continued to exist over what now constitutes Europe, up to the cretaceous rocks inclusive, though differing in species, as far at least as we can judge from the various forms of the bones. The Ichthyodorulites are usually accompanied by palates : these latter have not been detected in the grauwacke; but this need not surprise us, as the specimens of the defensive fin-bones, as yet noticed in this rock, only amount to two.

Fig. 98.

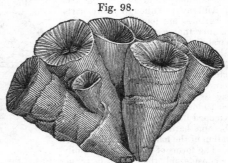

Cyathophyllum turbinatum, Goldf.

Among the corals will be found several genera now existing; and it deserves notice, that throughout the series of fossiliferous rocks, wherever there is an accumulation of polypifers, such as would justify the supposition of coral banks or reefs, the genera Astrea and Caryophyllia are present,—genera which, according to the more recent observations of naturalists, in addition to Meandrina and one or two others, are the principal architects of coral reefs at the present day.

Our knowledge of the kind of vegetation existing at, and entombed during, the epoch of the grauwacke group, is insufficient to warrant any general conclusions respecting it, further than it was apparently much the same as that, the remains of which are abundantly preserved in the carboniferous series. It was long known that anthracite was found in the grauwacke of North Devon, but it was not until the researches of Mr. Weaver in the South of Ireland that the value of this fact was justly appreciated, for not until then was it considered that carboniferous deposits of considerable development might constitute a part of the grauwacke group. This author observes, that all the coal in the province of Munster, excepting that in the county of Clare, is of this age. "At Knockasartnet, near Killarney, and on the north of Tralee, thin anthracitic beds, inclined at various angles from 70° to verticality, are included in grauwacke and slate. In the county of Cork this old coal is more extensively developed, particularly near Kanturk, extending from the north of the Blackwater to the Allow." The anthracite is described as employed in burning the limestone of the adjoining country; and the amount raised at Dronagh collieries is estimated at 25,000 tons per annum. "The coal and accompanying pyritiferous strata are abundantly charged with the remains or impressions of plants, belonging chiefly to *Equiseta* and *Calamites*, with some indications of *Fucoides*." Mr. Weaver has noticed this coal in various places, among which he enumerates beds in the county of Limerick, on the left bank of the Shannon, north of Abbeyfeale and at Longhill *.

We have here evidence that the accumulation of vegetables sufficient to produce beds of coal commenced at a very early period in Europe, and as it would also appear in America; for according to Professor Eaton, anthracite is observed in equivalent deposits at Worcester (Mass.) and Newport †. This is important, as it proves the existence of dry land, with vegetables upon it, contemporaneously, or nearly so, with the first appearance of animal life.

Although when we regard the mass of the grauwacke rocks we are struck with the minute proportion that organic remains bear to the whole, we must still perceive that the atmosphere was capable of supporting vegetation, and the seas of sustaining Zoophytes, Crinoidea, Conchifera, Mollusca, Crustacea, and Fish. What other creatures existed we are unable, from the absence of their remains, to judge: it may however be by no means unphilosophical to conclude that vegetation did not exist alone on dry land, but that, consistently with the general harmony of nature, it afforded food to terrestrial creatures suited to the circumstances under which they were placed.

* Weaver, Proceedings of the Geological Society, June 4, 1830.
† Eaton, American Journal of Science, vol. xix.

Section X.

LOWEST FOSSILIFEROUS GROUP.

This group should be considered as little more than one of convenience, in which rocks containing a few organic remains are sometimes mixed with strata of the same character as those enumerated under the head of Non-fossiliferous Rocks; so that we seem to have arrived, in the descending order, at a state of the world when there was a combination of those causes which have produced fossiliferous and non-fossiliferous strata. That there should be a transition or passage, even effected by the alternate operation of particular causes, from that condition of the world's surface when chemical action prevailed to that when mechanical action became more abundant, is what we should expect, since it is in accordance with our knowledge of rock deposits generally; for we observe, however sudden certain changes may have been produced in particular situations, that viewed on the large scale, a general change of circumstances attending rock formations has been more or less gradual.

These alternations or mixtures of substances, which are partly mechanical and partly the result of chemical action, have already been observed in the grauwacke series; and the only difference would appear to be, that they are more frequent, as might be expected, on approaching the great mass of crystalline rocks. In point of fact, there would appear little reason for considering this group as more than the lower part of the grauwacke series: the organic remains are, as far as we know, similar, and the mineralogical character of that portion which may have had a mechanical origin the same, excepting perhaps that argillaceous slates are more abundant, and arenaceous rocks rare, a distinction merely pointing to a more gentle transport by water; if indeed these slates, often in considerable mass, really were produced by deposit from water in motion, carrying forward a mass of detritus held in mechanical suspension. As, however, some geologists appear to consider that these rocks may be separated from the grauwacke mass, there can be no objection to retain this group for the present, as it enables us to observe the passage of one great class of stratified rocks into the other.

The Tintagel (Cornwall) and Snowdonian slates, both containing organic remains, have been sometimes considered as of an older date than the common grauwacke. The former are argillaceous slates passing into that variety known as roofing-slate. The latter are also argillaceous slates, but associated with them are some ambiguous rocks, the composition of which is not quite so apparent, Messrs. Phillips and Woods have employed the name of Steachist for it, as a provisional term: steatite, however, it can scarcely be; for according to the analysis made of these rocks by Mr. R. Phillips, they were found to consist chiefly of alumine and silex, with a small proportion of lime, and only a minute trace of magnesia, a substance usually constituting a considerable portion of steatitic compounds *.

The organic remains obtained both at Tintagel and on the summit of Snowdon are far from being well preserved. Those of the latter are most determinable, and are found chiefly to consist of shells, and among these shells some appear referable to the genus *Producta* †.

MM. Brongniart and Omalius d'Halloy were the first to point out the alternations of the granitic and schistose rocks of the Cotentin and Britanny, and also the probability that the deposits thus associated with the granitic compounds were fossiliferous ‡. The grauwacke of the Cotentin is certainly associated, more particularly in its lowest parts, with rocks, the mechanical origin of which is far from evident. Even the decidedly crystalline compounds, such as some varieties of a syenitic rock, are so mixed up with them that it would be very hazardous to affirm where the series, in which confusedly crystalline compounds prevail, may commence, or where the mechanical and fossiliferous deposits may terminate. The highly indurated sandstones also, which are clearly included among the fossiliferous rocks of Normandy, so pass into a quartzose rock, that, as M. Brongniart has observed, they often present the appearance of having been produced by confused crystallization.

The study of this part of our subject must always be attended with great difficulty; for independently of the mixture of chlorite, talcose, and other slates with the lowest fossiliferous deposits, caused by circumstances above noticed, we have to contend with the presence of igneous rocks injected among these deposits in the line of their stratification, thus producing the most deceptive appearances. No small difficulties are also caused by the alteration of rocks, arising from the protrusion of granite and other igneous products among them, such products often causing, when the masses are large, very remarkable changes, the grauwacke argillaceous slates assuming the appearance of a variety of older rocks.

To describe the precise limits of the fossiliferous deposits, and

---

* Phillips and Woods, Annals of Philosophy, 1822.

† Figures of the Organic Remains obtained from Snowdon, by Messrs. Phillips and Woods, will be found in the Annals of Philosophy, vol. iv. (1822) pl. 17. new series. Shells from Tintagel and Snowdonia are also figured in the Geol. Trans. vol. iv. pl. 25.

‡ Journal des Mines, tom. xxxv. 1814.

draw fine lines of distinction between them and the non-fossilife-rous rocks, is obviously impossible. We can only infer that the remains of organic life were generally entombed in deposits of mechanical origin, as well at this early period as subsequent to it. Respecting the abundance and different structures of the animals first called into existence, we shall never perhaps have any defi-nite ideas ; for the preservation of any portion of their more solid parts must always have depended on a great variety of circum-stances, not likely to have been most favourable during a state of things, in which a change was effected from the formation of such rocks as gneiss, mica slate, and others of the same descrip-tion, to the deposit of those evidently of mechanical origin.

Whatever the kind of animal life may have been which first appeared on the surface of our planet, we may be certain that it was consistent with the wisdom and design which has always pre-vailed throughout nature, and that each creature was peculiarly adapted to that situation destined to be occupied by it. Bearing therefore in mind this general adaptation of animals to the cir-cumstances under which they are placed, we may be led so far to speculate at this early condition of life, as to inquire, what kind of creatures, judging from the general character of those known to us, might flourish at a period when there might have been a compara-tive difficulty in procuring carbonate of lime for their solid parts. It will be obvious that fleshy and gelatinous creatures, such as *Me-dusæ* and other animals of the like kind, might have abounded, as far as regards a comparative scarcity of this substance. Hence it would be possible to have the seas swarming with these and simi-lar animals, while testaceous creatures and others with solid parts were rare.

These remarks are merely intended to show, that the scarcity of organic remains observed in the lowest fossiliferous deposits by no means proves a scarcity of animal life at the same period, though from it we may infer that testaceous and other animals with solid parts were not abundant. Mere fleshy creatures may have existed in myriads without a trace of them having been transmitted to us. In proof of this, if any were requisite, we may inquire what portion of those myriads of fleshy animals, which now swarm in some seas, could be transmitted, as organic remains, to future ages *

* Dr. Turner has suggested to me, that under this supposition of an abundance of Medusæ, or of analogous creatures among the early inha-bitants of our globe, we may perhaps account for the bituminous nature of some of the earlier limestones, more particularly of the carboniferous series, in which not a trace of solid organic remains can be observed ; for the decomposition of a mass of such creatures would produce much bitumi-nous matter, which may have entered largely into the composition of lime-stones then forming.

It may be remarked, while on this subject, that though an extensive distribution of carbonate of lime is essential to a great variety of animals, it is surprising how little may supply the wants of some, even those with vertebræ, such as sharks and cartilaginous fish generally. To consider that there may have been some connection between the animals with solid parts and a facility of procuring carbonate of lime on the surface of the globe, appears perfectly consistent with the design manifested in the creation, because it assumes such design at all periods, and constant harmony between the forms of creatures and their mode of existence. If we imagine a mass of animals to be suddenly called into life, each properly provided with its solid parts, the carbonate of lime contained in these bodies would no doubt be sufficient for a constant quantity of the same animal life during a succession of ages; for, by devouring each other, this necessary substance would be transmitted from one creature to another. We are however certain that this has not been the case; for the solid parts of animals which have been successively imbedded in various rocks, constitute a very large proportion of certain of those rocks, and if withdrawn from the fossiliferous deposits generally, would very considerably diminish their thickness. Therefore if the exuviæ of animals had not been entombed, and if the supply of carbonate of lime had not been greater than that which could have been derived from the mere destruction of one animal by another, for the purpose of food, the surface of our planet would not have been what it now is; and consequently, the fitness of things for the end proposed being constant in creation, the general condition of animal and vegetable life would not have been such as we now find it.

SECTION XI.

INFERIOR STRATIFIED OR NON-FOSSILIFEROUS ROCKS.

SYN.—Clay slate (*Schiste Argilleux*, Fr.; *Phyllade*, Daubuisson; *Thonschiefer*, Germ.). Aluminous slate (*Ampelite Alumineux*, Brong. *Schiste Alumineux*, Fr.; *Alaunschiefer*, Germ.). Whetstone slate (*Schiste coticulé*, Brong.; *Wetzschiefer*, Germ.). Flinty slate (*Schiste siliceux*, Fr.; *Jaspe Schistoide*, Brong.; *Kieselschiefer*, Germ.). Chlorite slate (*Schiste Chloriteux*, Fr.; *Chloritschiefer*, Germ.). Talcose slate (*Schiste Talqueux*, Fr.; *Talkschiefer*, Germ.). Steachist. Hornblende slate (*Amphibolite Schistoide*, Fr.; *Hornblendschiefer*, Germ.). Hornblende rock (*Amphibolite*, Daubuisson). Quartz rock (*Quartzite*, Brong.; *Quarzfels*, Germ.). Serpentine (*Ophiolite*, Brong.; *Serpentin*, Germ.). Diallage rock (*Euphotide*, Haüy; *Schillerfels*, Germ.). Whitestone (*Eurite*, Daubuisson; *Weisstein*, Germ.). Mica slate (*Schiste Micacé; Micaschiste*, Fr.; *Glimmerschiefer*, Germ.). Gneiss (*Gneiss*, Fr.; *Gneuss*, Germ.). Protogine.

WE have now arrived at that early condition of our planet, when, as far as our knowledge extends, neither animal nor vegetable life existed on its surface. The student, instead of wandering in imagination amid forests and over lands and seas, surrounded by strange vegetables and still stranger animals, should now direct his attention to those laws which govern inorganic matter. This may not at first sight be so attractive as the contemplation of the varied forms of organic life and the possible conditions under which it may have existed, but it will nevertheless be found equally, if not more delightful, as the inquirer obtains more certain results, from the investigation being conducted through the medium of the exact sciences.

It must, on the outset, be confessed that little has yet been accomplished respecting the causes which may have produced gneiss, mica slate, and other rocks of the same character. Names of the various compound and confusedly crystalline rocks we have in abundance, and if the investigation required no other aid we might sit down satisfied; but unfortunately the abundance of these names has confused the subject, and the student has more frequently contented himself with arranging and disarranging particular mineral compounds in a cabinet, than in investigating their general relations to each other, and the occurrence of the whole in the mass.

It will readily be admitted, that the difficulty of the subject is very considerable, requiring considerable insight into the exact sciences; but the subject being difficult would seem a good reason why the more advanced cultivators of those sciences should attack it, offering as it does such an ample field for the exertion of their abilities.

The inferior stratified rocks are of various compositions, sometimes so passing into each other, that it is almost impossible to affix definite names to the different mixtures. The strata rarely present a simple mineral substance constituting a large tract of country, without the admixture of other substances, unless we consider clay slate as such. Before however we proceed further, the student should become acquainted with the following rocks, which more particularly appear to deserve distinguishing names.

Argillaceous or Clay Slate.

This rock, as its name implies, is schistose, and contains a considerable portion of argillaceous matter. It varies materially as to induration, fissility, and composition; and is commonly undistinguishable, except in its geological relations, from the argillaceous slates of the grauwacke series. Its origin therefore becomes very ambiguous, and is not the less so from often containing cubical and other regularly formed crystals of iron pyrites, affording evidence that the rock was once in that condition to permit the free arrangement of sulphuret of iron into crystals,—a fact observable in the argillaceous deposits of all ages, some decidedly of mechanical origin; therefore we have no direct evidence to show that the argillaceous slates of this epoch may not also have been mechanically produced; for the fineness of grain will by no means assist us, the texture of the roofing slates obtained from the grauwacke series being altogether as fine, as that of the argillaceous slates associated with the mica slate or gneiss. Like, also, the argillaceous slates of the same series, the lines of cleavage are frequently not the same with those which appear to be lines of stratification, but meet them at various angles. Argillaceous schist passes gradually into chlorite slate, talcose slate, and other rocks, by gradually acquiring particular minerals, which finally replace the matter of the argillaceous slate.

Chlorite Slate.

This is by no means an unfrequent associate of the preceding, into which it passes on the one hand, while it graduates into mica slate, &c., on the other. It is of course essentially composed of chlorite, which occurs alone or mixed with quartz, felspar, hornblende and mica, in various proportions.

Talcose Slate.

This is also a rock into which argillaceous slate graduates, at first acquiring a few plates of talc, and afterwards becoming replaced by that mineral, generally associated with quartz, or

quartz and felspar.   There is not unfrequently a transition from
this rock into mica slate.

Quartz Rock.

According to Dr. Macculloch, to whom we are indebted for our
more exact knowledge of this substance, quartz rock so varies in
its texture, that sometimes it appears of a chemical, and at others
of a mechanical origin,—a circumstance from whence it derives
much interest, being associated as it is with a large proportion of
the rocks under consideration.   The same author observes that it
rarely occurs compact and crystalline throughout, like quartz, as
it usually appears in veins; its general aspect, when pure, being
obscurely granular, " which by degrees becomes somewhat lax
and arenaceous; the grains varying in their size and the intimacy
of their union.   In some of these examples it appears to be a
granular crystalline mass ; in others it possesses a mixed mechanical and chemical texture; while in a third the rounded aspect
of the grains, and the small number of the points of contact, appear to indicate an origin chiefly mechanical, and resulting from
the agglutination of sand *." It should be remarked respecting
this definition of quartz rock, that Dr. Macculloch has included the
grauwacke among his primary class; therefore if quartz rock be associated with grauwacke, as it is in Normandy and other countries,
an arenaceous texture would be quite in accordance with the
mechanical origin of the mass of grauwacke generally ; while if
quartz rock should be commonly more crystalline when mixed
with gneiss or mica slate, this also would harmonize with the texture of the associated rocks.   This however is mere theory; but
as the subject involves the question of the occurrence of mechanical rocks throughout the mass of confusedly crystalline compounds, it should not be lightly passed over ; consequently we
should neither hastily admit nor reject the possibility of such a
mixture or alternation.   This rock is well known in Scotland and
its isles ; and according to MM. Humboldt and Eschwege, it
is of an extent and thickness in the Cordilleras of the Andes and
in Brazil, far exceeding what we are acquainted with in Europe.   Some of these Brazilian rocks are auriferous, and M.
Eschwege attributes the auriferous and platiniferous deposits of
that country to their decomposition or destruction.

Hornblende Rock and Slate.

Under this head are included, following the suggestions of Dr.
Macculloch, all those compounds, clearly contemporaneous with
the rocks among which they occur, of which hornblende constitutes an essential and prevailing ingredient.   Much of this rock
has been known by the names of primitive greenstone, and green-

* Macculloch, Geological Classification of Rocks.

stone slate, being composed of hornblende and felspar. The hornblende sometimes so predominates as to exclude other minerals. As the names imply, these rocks occur both compact and fissile ; in the latter case the felspar is frequently green.

Saccharine Limestone.

This rock occurs variously associated among the inferior stratified rocks, but is by no means confined to them; for, as has already been noticed, it is discovered among the fossiliferous deposits, as for instance, amid the Belemnitic rocks of the Western Alps. It is of various colours, but principally white, affording the well known statuary marbles of Greece and Italy. It is sometimes large-grained, as, for example, that included in mica slate on the lake of Como, which afforded the mass of materials for the construction of the celebrated Duomo at Milan. From a mixture of talc or mica, it sometimes becomes schistose. It is more than probable that some of the crystalline dolomites are associated with these marbles and others of the rocks under consideration. The limestones not only vary in their crystalline character, but pass into compact substances, and become mixed with various minerals, such as hornblende, augite, quartz, &c. A remarkable compound, consisting of nearly compact limestone with small crystals of felspar, and thus forming a kind of porphyry with a calcareous base, occurs at the Col de Bonhomme, near Mont Blanc, constituting the *calciphyre felspathique* of M. Brongniart.

Eurite.

A rock principally, and in many cases entirely, composed of the substance named compact felspar. It does not appear to constitute any extensive tracts in nature, but to be generally subordinate to gneiss or mica slate.

Mica Slate.

This rock is essentially composed of mica and quartz, and forms extensive tracts of country, as well as thin beds included among other rocks. Mica slate sometimes contains garnets so abundantly that they may almost be regarded a regular component part of the rock. It graduates on the one hand into gneiss, and on the other into talcose slate, chlorite slate, and other compounds.

Gneiss.

This rock is either schistose or divided into beds which vary in thickness. It is composed of quartz, felspar, mica, and hornblende, with the occasional mixture of other minerals. Sometimes one of these minerals is absent, sometimes another: from this loss of either the quartz, felspar, mica or hornblende, and from the occasional absence of even two of them, as well as the admixture of other substances, there results a very variable general compound. When it occurs confusedly crystallized in regular beds, the mica not being distributed in plates parallel to the strata, as is the case in the fissile and schistose gneiss, it is really, as far as mineralogical characters are concerned, nothing but that much disputed substance, stratified granite. And this is rendered even more apparent, when, as happens in the Alps, Scotland, and other situations, large crystals of felspar are disseminated through it as in the granite of Dartmoor, &c. When blocks have been detached from this gneiss, as has happened with many of the erratic blocks of the Alps, they cannot be distinguished from those of true granite. Gneiss, with its variations, constitutes very considerable tracts of country.

Protogine may conveniently be arranged with gneiss, the only difference between its decidedly stratified varieties and the gneiss being the substitution of talc and steatite for the mica. Protogine is the well known granitic rock of Mont Blanc, which certainly has the appearance of graduating into a more massive compound; but in this it does not differ from gneiss, which also seems to pass into granite in a similar manner.

Although the above are the most remarkable of the inferior stratified rocks, they are far from being the whole of them. The varieties and transitions of one to the other appear endless, and, occurring in no determinate order, set classifications utterly at defiance. It was at one time considered that gneiss was the inferior rock, and was succeeded by mica slate; but this is found to be by no means the case, the two being intimately blended with each other as well as with other compounds. It must however be confessed that the mass of the gneiss frequently appears to occupy an inferior position.

All this apparent confusion, and this passage of one rock into another, though it may embarrass arrangements, may be precisely the circumstances which may lead to some knowledge of the causes that have produced the lowest stratified rocks. These irregular passages, and the possibility of discovering any given rock at the top as well as at the bottom of the series, show that the causes, whatever they may have been, which produced this variety in the substances, were secondary, and that there was some general cause upon which the formation of the whole depended.

If we also consider what minerals have entered most largely into the composition of the whole mass, we find that quartz, felspar, mica, and hornblende, are those with which it most abounds, and which impress their characters upon its various portions; chlorite, talc, and carbonate of lime, are certainly not wanting; but if we, as it were, withdraw ourselves from the earth and look down upon such parts of its surface as are geologically known, we find that these latter mineral substances constitute a very small portion of the whole. The inferior stratified rocks which form the largest part of the exposed surface of our planet are gneiss and mica slate, and when viewed on the great scale, the others are more or less subordinate to them.

Supposing this view on approximation to the truth, we arrive at another and important conclusion; namely, that the minerals which compose the mass of these stratified rocks are precisely those which constitute the mass of the unstratified rocks, rocks which, from the phænomena attending them, are referred to an igneous origin. We may here inquire what are the circumstances which have determined the arrangement of these minerals into stratified masses in the one instance, and into unstratified masses in the other? This question is by no means of easy solution in the present state of our knowledge: but while we wait for information, it may be observed that the conditions, under which the two classes of rocks were produced, must, to a certain extent, have been very distinct. Yet we find, still viewing the subject in the mass, that the same elementary substances have produced the same minerals in both, the only difference between them being their general difference of arrangement relatively to each other, so that they should constitute a stratified compound in the one case, and not in the other. Looking into the structure of gneiss, mica slate, chlorite slate, talc slate, &c. we find, if we except the thick-bedded gneiss or stratified granite, that it is the arrangement of the mica, chlorite, or talc in certain general planes which has produced the fissile and schistose structure. This, however, has not been the only cause of stratification, (if it may be so termed, the lines of fissility not being necessarily those of stratification,) for we find in the thick-bedded gneiss, the hornblende rock, the quartz rock, the eurite, and the saccharine limestone, that other causes must have produced thick beds of confusedly crystallized substances.

There is, nevertheless, so much apparent mineralogical resemblance between these two classes of rocks, that we can scarcely refrain from conjecturing the remote origin of the one and of the other to be in some manner connected, modifying circumstances having impressed certain characters on each. It must be confessed this is a mere hypothesis, and the student must be careful only to consider it in that light; but it may be asked, what essential difference there is between thick-bedded gneiss, particularly that with imbedded crystals of felspar, and granite, between some hornblende rocks and greenstone,—except that the one occurs quietly interstratified in beds, while the other is unstratified, even sometimes cutting through stratified and similar compounds? We may here also notice serpentine and diallage rock, of which there is often good evidence (as will be seen in the next section) for con-

X 5

sidering igneous and injected rocks, cutting strata in the manner
of granite and greenstone. I have never myself observed these
rocks stratified, but Dr. Macculloch appears to be certain that
they are so in the Scottish Isles. *A priori*, we should imagine that
there was as much probability in finding stratified rocks, whose mineralogical composition should render them serpentine, and its common associate diallage rock, as that we should find stratified rocks
mineralogically the same with granite and greenstone : therefore
we should be disposed to admit them into the catalogue of inferior
stratified rocks, even if we had not the direct opinions of Dr.
Macculloch and some other geologists on the subject. As the
question is one of some interest, it should be stated that the localities where the stratification may be observed, and which are
pointed out by this author, are ; for diallage rock, Unst, Balta,
and Fetlar ; and for serpentine, also the Shetland Islands. The
stratification is described as often obscure; but the diallage rock is
stated to be associated with gneiss, mica slate, chlorite slate, and
argillaceous slate, alternating with them ; and when occurring distinct, presenting the same dip and direction as the neighbouring
rocks. According to Dr. Macculloch, there can be no doubt that
serpentine is stratified in Unst; as also appears to be the case
in Fetlar, though the strata are not there so regular.

It must not be inferred, from the small space here dedicated to
the inferior stratified rocks, that they are of little importance ; for
they are found to occupy a large portion of the earth's surface,
wherever, from denudations and disruptions of strata, or from the
original absence of superincumbent rocks, they are exposed to
our observation. As whenever they are observed, whether in
Asia, North America, or Europe, they appear with constant general characters, we may assume that common causes have produced them over the surface of the globe, and that these common
causes are principally chemical, inasmuch as the prevalent mineralogical character of the mass is confusedly crystalline.

We may therefore infer, from finding these rocks with constant general characters, whenever circumstances permit us to
observe them emerging from beneath the mass of strata in which
organic remains are entombed, that general chemical laws have
been in operation contemporaneously over the surface of our
planet, and previously to the existence of animal and vegetable
life upon it, producing rocks of great collective thickness. Hence
the student may always consider, that, whatever may be the nature of the deposits on which he stands, such strata exist beneath
them, unless in cases where masses of igneous rocks have, by pro
trusion, forced them asunder, and left no stratified substances intermediate between the surface and the interior of the globe.

It would be tedious to enumerate the various situations where
these rocks may be found; it will suffice to state that there is

scarcely any very large extent of country, where from some accident or other they are not exposed on the surface. They abound in Norway, Sweden, and Northern Russia; they are common in the North of Scotland, whence they stretch over into Ireland. In the Alps and some other mountains they occupy the central lines of elevation, as if brought to light by the movements which have thrown up the different chains. They abound in the Brazils, and occur extensively in the United States. Our navigators have shown that they are sufficiently common in the various remote parts of North America visited by them. They are found extensively in the great range of the Himalaya. Ceylon is in a great measure composed of them; and they would not appear to be scarce in various other parts of Asia. In Africa also we know that they are not wanting, though but so small a part of that continent has been yet explored with scientific views.

SECTION XII.

UNSTRATIFIED ROCKS.

THE rocks constituting this natural group are widely distributed over the surface of the world, are found mixed with almost all the stratified rocks, and bear every mark of having been ejected from beneath. They commonly occur either as protruded masses, as overlapping masses, resulting from the spread of matter after ejection, or as veinstones filling fissures, apparently consequent on some violence to which the strata have been subjected.

The aspect of the unstratified rocks is exceedingly various as far as respects their texture, and the absence or presence of the few minerals which essentially enter into their composition. These variations would however in general appear the result of the circumstances to which they have been exposed; and not unfrequently the same mass, if of tolerable extent, will present a great variety of compounds, to which separate names might be (and indeed have been) assigned, if, instead of directing attention to the mass, the small changes in mineralogical structure are alone observed.

In the earlier days of geology, granite was considered the fundamental rock on which all others were accumulated; but this opinion, like many others, has now given way before facts; for, as will be seen in the sequel, we have examples of granite resting upon stratified and fossiliferous rocks of no very great comparative antiquity. It must however be confessed, that granite appears sometimes to alternate in considerable thickness with the inferior stratified rocks, and that the separation of it from gneiss, particularly thick-bedded gneiss, is very ambiguous. Before, however, we proceed further with the consideration of the unstratified rocks, it will be necessary to premise a sketch of their mineralogical characters, omitting those of the rocks usually termed volcanic, which have been already noticed.

Granite

Is a confusedly crystalline compound of quartz, felspar, mica, and hornblende. It is not essential that all these four minerals should be present; on the contrary, rocks have been termed granite when only felspar and mica, felspar and quartz, felspar and hornblende, and quartz and hornblende, have been the constituent minerals. Such an employment of the term granite must be used with much caution, as for instance in the case of the compound of felspar and hornblende, which in fact is mineralogical greenstone, and should not be named granite unless it constitutes a very subordinate portion of a mass to which the term may be more properly applied, and results from the accidental absence of one or two of the above-named minerals for a limited space. The most prevalent compound is one with quartz, felspar, and mica; when hornblende replaces the mica, it is sometimes termed syenite. Other minerals, such as chlorite, talc, steatite, &c. are sometimes arranged with those above enumerated in various ways and proportions; but such compounds can only be considered as accidental varieties. When the quartz and felspar occur alone, and the crystallization is such that the former appears disseminated in the latter, it is termed *graphic granite*, from the supposed resemblance it bears to antique characters. Granite is occasionally porphyritic, as is the case in Cornwall and Devonshire, large crystals of felspar being disseminated through the mass, showing that however confused the general crystallization may have been, circumstances were such as to permit the production of distinct crystals of felspar.

Diallage Rock (Euphotide, Haüy; *Schillerfels,* Germ.). *Serpentine (Ophiolite,* Al. Brong.; *Serpentin,* Germ.)

These are so intimately connected, that to separate them seems impossible, passing, as they sometimes do, in all directions into each other. Diallage rock when pure is composed of diallage and felspar. Serpentine when pure is generally considered as a simple mineral substance, and forms large masses in that state, but seldom prevails to any extent without acquiring diallage. These rocks are sometimes blended with compounds of the greenstone class, and apparently pass so insensibly into them that they can only be considered as parts of a common mass, though the serpentine and diallage rock generally prevail in such cases.

Greenstone (Grünstein, Germ.; *Diabase,* Al. Brong.), *and the other rocks usually termed Trappean.*

These also so pass one into the other, that frequently in a mass of inconsiderable extent a great variety may readily be obtained. They vary in texture from an apparently simple rock to a confusedly crystalline compound, in which crystals of felspar are disseminated. It has long since been observed by Dr. Macculloch, that "the predominant substance in the members of this family is a simple rock, of which indurated clay or wacké may be placed at one extreme, and compact felspar at the other; the intermediate member being claystone and clinkstone. In some cases it forms the whole mass; in others it is mixed with other minerals, in various proportions and in various manners; thus producing great diversities of aspect, without any material variations in the fundamental character*." As may be readily imagined, no exact de-

* Macculloch, Geological Classification of Rocks, 1821. p. 480.

finition can be given of that which is constantly changing in
nature. Claystone, as its name implies, resembles clay under
different degrees of induration; and not unfrequently, when in
mass, acquires a columnar structure. Clinkstone appears an intermediate step to compact felspar, which, according to Dr. Macculloch, contains both potash and soda, while common felspar
contains potash only. Where substances are so continually varying, it is clearly not very possible exactly to define what compact
felspar should geologically be. I have elsewhere * applied the
term *cornean* to designate some of the more simple forms of that
kind of rock known as hornstone, which would appear in some
cases to be nothing else than compact felspar; in others, however,
it partakes of the characters of other minerals. Thus, in Pembrokeshire, where there is a remarkable variety of trappean rocks,
the corneans may be divided into felspathic, quartzose, and hornblendic, as those minerals appear to prevail in the mass; the
quartzose variety, which is the most rare, even appearing like
some kinds of quartz rock, with the exception that it is unstratified. These more simple forms of trap-rock very frequently become porphyritic by the admixture of either quartz or felspar
crystals, and sometimes of both in the same mass, as in the red
quartziferous porphyries, rocks which not unfrequently pass into
granite. Porphyries are generally known by the name of the base
or paste which includes the disseminated crystals; thus, we have
claystone porphyry (*Thonstein porphyr*, Germ.; *Argillophyre*,
Brongniart); felspathic porphyry (*True porphyry* of Brongniart;
Porphyre Euritique, Fr.; *Hornstein porphyr*, *felspath porphyr*,
Germ.); and clinkstone porphyry (*Klingstein porphyr*).

It very frequently appears as if the elements of quartz, felspar,
and hornblende composed the mass, and various circumstances
determined their union in such a manner as to produce a large
proportion of the various compounds known as trap-rocks, sometimes the hornblende being in mass, at others the felspar, while
the quartz rarely predominates. In other situations confusedly
crystalline compounds have been the result; quartz, felspar, and
hornblende united form syenite, or felspar and hornblende without
the quartz constitute greenstone. The granular structure of these
compounds varies materially, and finally becomes somewhat imaginary; at least this texture is rather inferred than seen. The compounds occasionally contain disseminated crystals of felspar, and
thus become what are commonly known as greenstone porphyries
(*Diabase porphyroide*, Fr.; *Grunstein porphyr*, Germ.). A paste
of green hornblendic cornean containing crystals of felspar constitutes the *ophite* of Brongniart, the antique green porphyry.

* Geology of Southern Pembrokeshire; Geol. Trans. 2nd Series,
vol. ii.

Some of the rocks of this family are not unfrequently vesicular, in the manner of modern lavas, the vesicles however being generally filled up by some mineral substances which have since been infiltrated into them. Such substances are not unfrequently agates, and those employed in the arts are principally thus derived. From these cavities being frequently of an almond shape, or rather from the appearance of their solid contents resembling almonds in form, the term *Amygdaloid* has been applied to rocks of this class. It will be readily understood that the base or paste of the amygdaloids is not constantly the same, but varies materially. A trappean rock is sometimes both amygdaloidal and porphyritic at the same time (Devonshire, Scotland, &c.). The amygdaloidal cavities afford the mineralogist a great abundance of siliceous, calcareous, zeolitic, and other minerals.

Other minerals than those above enumerated occur in the trappean rocks, but cannot be considered as forming an essential part of them, with the exception of augite and hypersthene, which with the mixture of either common, compact, or glassy felspar, constitute the *augite* and *hypersthene rocks* of Dr. Macculloch. It would be endless to attempt a notice of the various aspects under which these rocks present themselves; it should however be remarked that the term basalt is applied to substances which are not precisely the same, being sometimes given to a fine compound of augite and compact felspar, at others to a minute mixture of hornblende and compact felspar, sometimes to dark indurated claystones, and finally to a compound of felspar, augite, and titaniferous iron. The last mixture seems that now most commonly termed basalt.

Such are the rocks commonly considered unstratified. It will have been seen that they so pass into one another that distinctions are not easily established between them. Mineralogical granite passes through various stages, and graduates into the compounds named greenstone, and others of the trappean class*. Instead also

---

* Dr. Hibbert notices the passage of granite into one of those compounds named basalt, (in this case formed of an intimate mixture of hornblende with a small proportion of felspar,) as taking place in the Shetland Islands. As this author's account is illustrative of such changes in general, it may advantageously find a place here. The basalt extends from the Island of Mickle Voe northwards to Roeness Voe, a distance of twelve miles. On the west of this is a considerable mass of granite. The transition is thus described : " Not far from the junction we may find, dispersed through the basalt, very minute particles of quartz. This is the first indication of an approaching change in the nature of the rock. In again tracing it still nearer the granite, we find the particles of quartz dispersed through the basalt becoming still more distinct, more numerous, and larger, an increase of magnitude even extending to every other description of particles. The rock may now be observed to consist

of being solely mixed with rocks of the oldest date, it is found among the Montagnes de l'Oisans (Western Alps), cutting through and superincumbent upon deposits, referable, according to M. Elie de Beaumont, to the oolitic series *. This phænomenon is not confined to the Western Alps, but has also been observed by M. Hugi in the Swiss Alps, numerous sections of which by this author will be viewed with much interest †. Among the places where the superposition of granite is remarked, may be mentioned the Tosenhorn, the Tristenhorn, the Bötzberg in the Ursernthal, and the Jungfrau, in which latter mountain limestones and slates, referred to the lias, appear, as it were, entangled in the granitic rock. The analogy between the facts noticed by M. Elie de Beaumont in one part of the Alps, and by M. Hugi in another, is striking; for not only does the granite occur above these rocks in both situations, but also beneath them, as is shown in the sections of both these authors.

The following figure, representing the Bötzberg, and taken from M. Hugi's work, will afford the student an idea of this superposition.

a, limestones and slates, referred to the lias; *b*, mica slate;
c, gneiss; *d*, granite. The superposition in this section is evident; and it may be further inquired how far the rocks, marked
as mica slate and gneiss, may not
be the slate of the lias altered
by the presence of the granite
above it.

Fig. 99.

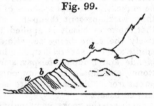

Not only have granite rocks been thus referred to an epoch pos-

---

of separate ingredients of quartz, hornblende, felspar, and greenstone ; the latter substance (greenstone) being a homogeneous commixture of hornblende and felspar. Again, as we approach still nearer the granite, the disseminated portions of greenstone disappear, their place being supplied by an additional quantity of felspar and quartz. The rock now consists of three ingredients, felspar, quartz, and hornblende. The last change which takes place results from the still increasing accumulation of quartz and felspar, and from the proportionate diminution of horn-blende. The hornblende eventually disappears, and we have a well characterized granite, consisting of two ingredients of felspar and quartz." Hibbert, Brewster's Edin. Journal of Science, vol. i. p. 107. The same author also notices a passage of felspar porphyry into granite near Hills-wick Ness.

* Elie de Beaumont, sur les Montagnes de l'Oisans; Mém. de la Soc. d'Hist. Nat. de Paris, t. v.; as also Sections and Views illustrative of Geo-logical Phænomena, pl. 15.

† Hugi, Naturhistorische Alpenreise; Soleure, 1830.

terior to the oolitic group, but it has been already seen that such are also found above the chalk of Weinbohla, whence it may be inferred that granite was produced at the supercretaceous epoch. Assuming therefore that the evidence is good, we should expect to find granitic rocks traversing or superincumbent upon rocks of all ages, from the inferior stratified to the cretaceous inclusive. The superposition of granitic rocks to fossiliferous limestone has long since been remarked by Von Buch in Norway, and by Dr. Macculloch in the Isle of Sky. Similar rocks have also been noticed as incumbent on those of the age of either the oolitic or cretaceous series at Predazzo. Respecting the latter, Mr. Herschel observes, that where the dolomite plunges beneath the granitic rock at Canzocoli, at an angle of 50° to 60°, both rocks appear altered, and that there are laminæ of serpentine between the two *.

The contact of the granite with the oolitic rocks of Brora is attributed by Prof. Sedgwick and Mr. Murchison to the elevation of the granite in mass, which is supposed to have turned up the edges of the oolitic deposit †. The same authors have also remarked a very curious occurrence of granite and limestone on the north coast of Caithness, near Sandside, where the granite appears thrust up among the limestones, and a breccia has been produced containing fragments of limestone and granite. The cement of this breccia is described as generally granitic, though it is calcareous in some places, and approaches a sandstone in others. One great block of limestone is noticed as apparently entangled in the granite. The limestone beds on the eastern side are stated not to be much disturbed, while those on the western side are in the utmost confusion, and, which are important circumstances, crystalline and cellular ‡.

Thus far we have only seen granite rising through and covering other rocks in considerable masses, but we have also evidence in granite veins, that the matter of the rock was in such a state of igneous fusion, as to penetrate into thin clefts opened in stratified and older rocks by some violence, such as probably resulted from the upburst of the igneous matter accompanied by elastic vapours. If we imagine fractures to be suddenly produced in contact with a mass of rock in fusion, such as we may presume granite to have been, the natural result would be the injection of the substance in fusion into all the crevices, in consequence of the great pressure exerted on one side; the intruding substance breaking off and including in it all loose fragments, and those projecting portions which opposed the fury of the injection. This is precisely the condition of granite veins, which though much doubted during the reign of the Wernerian theory, are now known to be abundant in nature.

* Herschel, Edinburgh Journal of Science, vol. iii.
† Sedgwick and Murchison, Geol. Trans. vol. ii. pl. 34.
‡ Ibid. vol. iii. p. 132.

Glen Tilt, which is reported to have produced such delight in Hutton when viewed by him for the first time, presents excellent examples of the intrusion of granite veins into other and stratified rocks. The great features consist of a mass of granite on the northern side of the glen, and of schist and limestone on the southern; from the former, veins issue in all directions, disturbing and intermingling with the latter in such a complicated manner, as to render a description useless without the aid of maps and sections, for which, and for a detail of the various singular phænomena observable in Glen Tilt, the student must be referred to the memoirs of Lord Webb Seymour, Prof. Playfair *, and Dr. Macculloch †.

Granite veins traversing the stratified rocks are now known in various parts of the world. Some fine examples are to be observed in the district of the Land's End; among them one at Cape Cornwall shows that there has been a shift or fault in the slate rocks, for a quartz vein has been cut through, and elevated more on one side than the other, thus proving that force has been employed ‡. At Mousehole the veins can be seen to proceed from the main body of the granite §. In the Alps they also proceed from masses of granite, which appear to have much influenced the present position of strata in parts of those mountains, as has been shown by M. Necker de Saussure ‖. They traverse gneiss in the Vállée de Vallorsine, as also at the head of the lake of Como.

They are not confined to Europe, but are found cutting and including portions of slate rocks at the Cape of Good Hope, as has been shown by Captain Basil Hall and Dr. Clarke Abel ¶. In America also they have been observed by Mr. Hitchcock traversing mica slate, hornblende slate, limestone (described as of a peculiar character), gneiss, and granite in Connecticut; the veins frequently branching out in various directions **. Granite veins therefore cannot be considered as rare; on the contrary, they would appear sufficiently common when circumstances permit good sections of the junctions of the granitic mass, and of the rocks among which they appear intruded. We should expect these veins to be

* Trans. of Royal Soc. of Edinburgh, vol. vii.

† Geol. Transactions, 1st series, vol. iii.

‡ Oeynhausen and Von Decken, Phil. Mag. and Annals of Philosophy, 1829; also Sections and Views illustrative of Geological Phænomena, pl. 17. fig. 4.

§ Sections and Views illustrative of Geological Phænomena, fig. 5.

‖ Necker de Saussure, sur le Vallée de Vallorsine; Mém de la Soc. de Physique et d'Hist. Nat. de Genève.

¶ Basil Hall, Transactions of the Royal Soc. of Edinburgh, vol. vii.; and Clarke Abel's Voyage to China.

** Hitchcock On the Geology of Connecticut, American Journal of Science, vol. vi.

of various dates, and accordingly we find that masses of granite are themselves traversed by veins, also of granite.

The exact composition of the granite in these veins must naturally vary, depending much on local circumstances; for if we suppose a substance in igneous fusion to be injected into fissures of rocks, such injected matter will be subjected to different conditions. Where the fused substance cooled more suddenly, as was likely to be the case in the distant and smaller fissures, the result would be less crystalline; while in the wider clefts, and near the great heated mass, the crystallization would be more perfect, and bear the greatest resemblance to the parent mass. Consequently, in a system of granite veins, we should expect a great diversity in the aspect of the granitic matter, which generally appears to be the case.

The trappean rocks, though there is much difficulty in separating them from the granitic, may for convenience be considered separately from them. They also form considerable masses, and constitute dykes and veins. When considered in the mass, they may be regarded as containing much less mica than the granitic rocks, while hornblende has become much more abundant; they also, when viewed on the large scale, appear more abundantly among the comparatively modern deposits than the granites, though it cannot be denied that they run into the latter in a remarkable manner. If this opinion of the greater prevalence of the granitic rocks over the trappean at the earliest periods be correct, it would seem to point to a certain condition of things at such periods, which subsequently became so modified that the igneous eruptions became altered. What that condition of things may have been, we do not as yet appear to have any very definite ideas, and we obtain little help on the subject from the phænomena of modern volcanos, granite never having been known to flow from them. We however learn from this circumstance that igneous eruptions into the atmosphere are not favourable to the production of granites, and we may consequently infer, that the conditions under which granite was produced were not similar to those which we now observe on the surface of the earth, at least so far as relates to those phænomena which occur in the atmosphere. What igneous matter ejected beneath a great pressure of sea may form we are unable to determine, but that it would be greatly modified by such pressure cannot be doubted. Still, however, we do not exactly see why a difference of pressure should cause mica to be generally less abundant, and the quantity of hornblende to be so greatly increased; and therefore we may infer that there was something in the then condition of our planet's surface, which permitted the production of that great abundance of granite so commonly associated with the earliest stratified rocks with which we are acquainted, and which frequently differ from it only in being stratified, or having the component minerals arranged in laminæ.

Admitting this prevalence of granitic compounds at the earliest periods, their production at more recent epochs shows that the conditions necessary for their formation continued up to such epochs, though they may have been infinitely more rare, having in a great measure given place to those under which the more common trappean rocks were produced.

Trappean rocks, under their various modifications, are so common in nature, that to attempt a notice of localities would be entirely useless. They occur mingled with the stratified rocks in every possible way,—injected among the beds for considerable distances, so that sections are exhibited in which the igneous rock appears quietly interstratified with the aqueous deposits; constituting caps of hills, thus appearing like a stratified and quiet deposit on other beds, the continuous parts that once connected these caps into a sheet of matter ejected from beneath having been removed by denudation; or as dykes or veins filling fissures, previously produced, in some instances showing that the igneous matter entered the fissure with such force as to tear away portions of the sides, while at others it seems to have risen more slowly, gradually filling the rent.

There would appear no more convenient place for observing all these modes of occurrence, or indeed the various mineral aspects of the rocks themselves, than the coast and islands of Scotland, which have been described by Dr. Macculloch * and other geologists, and where the student possesses the great advantage of innumerable coast sections, those invaluable aids in all geological investigations.

Apparent interstratifications of igneous rocks with beds which have had a different origin may be observed in many places, but may be well studied in High Teesdale, where the igneous matter has been injected among strata of limestone, sandstone, and shale, forming part of the carboniferous limestone series, in such a manner that a great apparent bed, commonly known as the Great Whin Sill, was considered as constituting a regularly stratified portion of a common whole, before the investigations of Prof. Sedgwick showed that it had evidently been injected among the aqueous deposits, and was connected with a mass of igneous matter which had disturbed and altered a continuation of the same rocks†. In Derbyshire, trappean rocks, generally known by the provincial term *toadstones,* from the aspect of the prevailing amygdaloid, are apparently interstratified with the carboniferous limestone. These we may, from all analogy, consider as injected among the lime-

* Macculloch's Western Islands.

† Sedgwick, Trans. of the Cambridge Phil. Soc. vol. ii. p. 139; and Sections and Views illustrative of Geological Phænomena, pl. 13. In some situations the limestone and slate have been turned up by the trap, and the former has become granular, and the latter indurated.

stones, the strata of which would easily be separated by the application of the proper force, in the manner already noticed under the head of Volcanic Rocks (p. 125). The student will find ample details of this association of trap-rocks and limestones in Mr. Conybeare's account of the rocks of Derbyshire *.

According to Mr. Aikin, a good example of the apparent interstratification of greenstone with the coal-measures is observable at Birch Hill colliery, Staffordshire. The bed seems to be connected with a mass of trap on one side, whence it has been injected among the coal strata, altering the coal where it covers it, by depriving it of its bitumen†.

The connection of trap rocks with coal-measures has often been insisted on, and certainly in some countries they are much associated; but when the facts are narrowly examined, it generally appears that the igneous rocks have been introduced among the sandstones, shales, and coal, subsequently to the deposit, and even consolidation, of the latter. It does not however necessarily follow that igneous eruptions and coal deposits may not have been contemporaneous; on the contrary, we may inquire if the violent movements of land, which probably accompanied such igneous eruptions, did not assist in the destruction of the vegetation, by depressing it beneath the waters; and even if accompanying and violent agitations of the atmosphere, similar to that which happened during the great eruption of Sumbawa, might not also contribute something towards the transport of the different parts of plants in particular cases ‡. Difficulties very frequently arise from the want of good natural sections; for it is clear that a mass of injected trap may be so spread over, or injected among, the coal strata, in a district wanting such sections, and only explored by means of the miner's galleries, that ambiguous appearances abound, more particularly when the whole mass has been traversed by faults, as is often the case.

Among the various trap dykes noticed by Mr. Winch as traversing the coal-measures in the vicinity of Newcastle, there is one described by Mr. Hill as occurring at Walker colliery, which has converted the coal contiguous to it into coke. This dyke is stated not to alter the level of the coal-measures, though it cuts through them; but in the plan which accompanies the memoir

---

* Outlines of the Geology of England and Wales, Book III. chap. v.

† Aikin, Geol. Trans. vol. iii. In the section which illustrates this memoir, a fault is seen to have traversed the beds after the injection of the trap, for it is dislocated with the rest;—a fact also observable in Prof. Sedgwick's sections of High Teesdale, where dislocations are represented to have affected all the rocks equally.

‡ During tropical hurricanes among islands, such as those in the West Indies, plants, more particularly their lighter parts, arē carried abundantly out to sea.

there is a fault marked on the south side of the dyke, parallel to it, and on the eastern part, producing a dislocation to the amount of nine feet, so that the fracture does not appear to have been quite simple *.

Trap dykes are to be found in all parts of the world, the composition of the rock varying materially, even in the dyke itself, as we might expect from differences in the cooling and pressure, so that the central parts are not unfrequently more crystalline than the sides †.

There is good evidence of the great mechanical force which has been exerted on the stratified rocks, as has been already pointed out in the North of Ireland, where large disrupted masses have been caught up in the igneous matter; similar phænomena have also been noticed by Dr. Macculloch and Mr. Murchison in the Western Islands of Scotland, the disrupted and included rocks being in the latter cases of an older date than those fractured in Northern Ireland ‡.

Fig. 100.

The annexed figure will illustrate a considerable frac- ture and alteration in the lime-stones at the Black Head, Bahbacombe Bay, Devon, effected by the eruption of greenstone, which, though it overlies the limestones in this section, oc-

curs beneath them not far distant. *a*, argillo-calcareous slate, traversed by veins of calcareous spar, and occasionally indurated. *b b*, limestones which have become semi-crystalline; they have also, judging from lines of colour, been once, more fissile than at present. *c*, a slate, with a thin bed of reddish limestone (*e*). This slate is apparently much altered. *d d d*, greenstone and its varieties, constituting the mass of the hill, and traversed by calcareous

---

* Geol. Trans., 1st series, vol. iv.

† One of the longest dykes with which we are acquainted is that described by Prof. Sedgwick in his memoir on those of Yorkshire and Durham. It is very probably continued from "High Teesdale to the confines of the eastern coast, a distance of more than sixty miles." During this traverse it cuts the coal-measures, red sandstone and lias.—Cambridge Phil. Trans., vol. ii. p. 31; and Sections and Views illustrative of Geological Phænomena, pl. 14.

‡ Macculloch, On the Western Islands of Scotland. Several of the sections contained in this work are copied into Sections and Views illustrative of Geological Phænomena, for the purpose of exhibiting the various modes in which trappean are associated with stratified rocks in the Hebrides. Mr. Murchison has represented fragments of the oolite series of the same islands as caught up in the trap of the southern side of Mull. Geol. Trans., 2d series, vol. ii. pl. 35.

veins near the limestones. *f f*, lines of fracture which divide the limestones and slates into three masses. The slates and limestones have evidently suffered, not only from the mechanical action of the erupted greenstone, but also chemically from the proximity of the mass in a state of igneous fusion. Notwithstanding the general pressure preventing the disengagement of the carbonic acid contained in the limestones, some carbonate of lime has been given off, and filled crevices and cracks in the trappean mass above. The alterations of limestone in contact with trappean rocks is sufficiently common, producing a greater or less amount of crystallization, in accordance with the well known experiments of Sir James Hall *, who has proved that carbonate of lime when subjected to great heat beneath sufficient pressure, does not part with its carbonic acid, but that it is fused and is rendered crystalline,—a fact previously doubted.

The alteration of limestones and trap at their contact would not always appear to be confined to a crystalline arrangement in the parts of the former, for Dr. Macculloch has observed trap to become changed into serpentine at the point of contact with limestone, at Clunie in Perthshire. A trap vein traverses limestone, and the rock is noticed as a kind of greenstone, the greater part of which is moderately coarse, passing into lamellar towards the sides. This (lamellar) structure " gradually becomes more distinct towards the edges of the vein, where it frequently splits off by the process of decomposition into laminæ, resembling, on a cursory view, black slate. These laminæ are often intersected by other cross fissures, dividing the whole into cuboidal masses, which sometimes decompose still further into spheroidal forms, during the approximation to the calcareous boundary; and whether the laminæ are present or not, the texture becomes gradually finer and softer, the rock still retaining its black colour, or sometimes assuming a greenish cast. At length the observer finds the vein converted under his hand into serpentine, without being at first aware that any change has been brought about †." Thus the transition from greenstone to serpentine can gradually be traced, but only where the vein traverses the limestone; for where the continuation of the same vein cuts through schist and conglomerate, no such change is observable. The trap is much entangled, in the small scale, with the limestone, and the minuter ramifications from the vein are entirely composed of serpentine. The limestone does not pass into the serpentine; on the contrary, the line of separation is described as well defined. Veins of green asbestos and steatite occur in the serpentine. Dr. Macculloch also states that the trap veins traversing the calcareous sandstone at Strathaird abound in

* Sir James Hall, Trans. of the Royal Soc. of Edinburgh, vol. vi.
† Macculloch, Brewster's Edin. Journ. of Science, vol. i. p. 1.

steatite, found in the outer parts of the vein and approaching the calcareous rock. The same author observes, that the trap vein which traverses the white marble of Strath passes into serpentine at its outer edges, as at Clunie. " At the line of contact, a zone of transparent serpentine of a fine oil-green colour, is found intermixed with the limestone *."

The above is sufficient to show that trap under certain conditions may pass into serpentine. We have now to consider dykes and masses of serpentine and diallage rock which occur under circumstances analogous to those of the trap rocks. Mr. Lyell has described a serpentine dyke which cuts through a sandstone (equivalent either to grauwacke or the old red sandstone), near West Balloch Farm, in Forfarshire. The phænomena can be well observed where the dyke traverses the Carity. The serpentine dyke is ninety feet thick, nearly vertical, and ranges east and west. It is stated to be flanked in part by a hard compact rock, about three yards thick, standing vertically, and forming a parting wall between the sandstone and the serpentine. " This rock consists of equal parts of green serpentine and an indurated brick-coloured rock, harder than serpentine, and sometimes passing into jasper." The serpentine is also described as bounded, on the left bank of the Carity, by " a vertical mass of sandstone conglomerate, evidently much altered, about five yards thick. Some parts of this rock approach jasper in hardness and appearance." But the most interesting fact connected with this altered conglomerate is, that the quartz pebbles contained in it have been fractured and reunited,—a circumstance also noticed by Mr. Lyell in a conglomerate flanking a greenstone dyke on the Isla, also in Forfarshire. This fracture of the quartz pebbles is precisely what we should expect from a sudden application of heat, and would speak strongly in favour of the once igneous fusion of the serpentine in the dyke, if any evidence were wanting. That common association of serpentine with greenstone is here also observable, the dyke being bordered on the right side of the Carity by a fine-grained rock of that kind. The dyke can be traced at intervals for at least fourteen miles, stretching in a straight line from Cortachie to Banff †.

The serpentine and diallage rocks of Liguria are particularly instructive, as they occur under a variety of forms, and appear connected with the disturbed condition of the strata in that country. These rocks pass into each other in all directions, and into those of a trappean character (Levanto). Between Braco and Matanara the student may observe them with perfect ease, on the high road from Genoa to Florence, cutting though the limestone and slate as a dyke, insinuated between the strata,

---

* Macculloch, Brewster's Edin. Journ. of Science, vol. i.
† Lyell, Brewster's Edin. Journ. of Science, vol. iii.

so as to seem interstratified, and constituting an enormous mass, apparently thrust upwards. The whole district is full of interesting facts of this nature.

If it has been correctly determined that the limestones of La Spezia are of the age of the oolitic groups of England, France, and Germany, the serpentine and diallage rocks of southern Liguria have been erupted since that period; for the La Spezia limestones and their associated deposits have been upheaved, contorted, and cut through by them. Possibly also the date of their intrusion may even be later, and of the supercretaceous period; for the supercretaceous lignite deposits of Caniparola, near Sarzana, are thrown into a vertical position, and I did not detect any serpentine or diallage rock pebbles among their associated conglomerates; this latter date, however, must be considered uncertain, for the serpentine rocks are not observed to be actually intruded among the supercretaceous deposits.

At Capo Mesco, between Levanto and Monte Rosso, gray schist and a compact calcareo-siliceous sandstone (one of the Italian *mæ-eignos*) are broken into faults by a mass of serpentine and diallage rock, which branches from a larger mass at Levanto. The valley of Rochetta, near Borghetto, has attracted much attention since it was noticed by M. Brongniart *. It shows the intrusion of the serpentine and diallage rocks (which here also pass into each other in various ways) among stratified rocks, similar to those which are observed at Capo Mesco. At the entrance into the valley, the sandstone is seen dipping at a considerable angle and resting upon gray limestone and schist, which are supported by serpentine. The serpentine then passes over contorted gray limestone and schist, and occupies a considerable portion of the valley, mixed with diallage rock, until the latter predominating to the exclusion of the serpentine, the mass rests on beds of red and green jasper, having the same dip as the sandstones at the entrance of the valley. These jasper beds repose on contorted gray limestone and schist, on the left bank of the river and opposite Rochetta. The jasper beds have sometimes been considered as a subordinate part of the serpentine: that it may be an altered rock is very possible; but I do not imagine it can be regarded as a portion of the unstratified mass of the diallage rock and serpentine, more particularly as similar jaspers occur among the limestones in the gulf of La Spezia, not far from Lerici, interstratified with them, and distant from either serpentine or diallage rock.

The mass of serpentine and diallage rock which constitutes the Monte Ferrato, north of Prato, Tuscany, rests also upon stratified jasper on the west, and this again upon a schistose rock based on limestone: this also appears an accidental circum-

* Annales des Mines, 1821.

Y

stance, for jasper is interstratified with brown shale at Paciana on the opposite side of the mountain, where it is not in contact with the serpentine. The diallage rock and serpentine here also pass into each other in all directions, and one variety of the former is worked for millstones. The whole seems a mass ejected from beneath, which has overflowed the stratified rocks, appearing to cut through them on the northward, beyond the north-west knoll, where there is a good section of the serpentinous mass resting on the jaspers, slates, and limestones.

At the Lizard, Cornwall, there is a well known mass of serpentine, which seems intimately connected with greenstones: unfortunately, however, from its position, we do not obtain any clear idea of its relative date *.

The volcanic rocks, at least such as have been considered the products of what are commonly termed modern and extinct volcanos, have already been noticed; therefore a statement of their general characters need not be repeated.

If we regard these various igneous products as a mass of matter which has successively, and during the lapse of all that time comprehended between the earliest formation of the stratified rocks and the present day, been ejected from the interior of the earth, we shall be struck with certain differences of these rocks on the great scale, which has led to their practical arrangement under the heads of granitic, trappean, serpentinous, and volcanic products, as above noticed. The two former and the last occur most abundantly, whilst the third is comparatively more scarce, though sufficiently common in nature As yet we are unacquainted with the conditions necessary for the production of these different compounds, and it would be a highly interesting inquiry, and one well worthy the attention of the chemist, to ascertain, as nearly as may be, the essential average differences which may exist as to the ultimate elementary substances constituting the rocks of this nature, thus approaching towards a knowledge of the possible circumstances, which may have determined such substances to arrange themselves in one manner rather than in another. Possibly the quantity and proportion of the elementary substances might not vary so much as we might, from the general mineral character alone, be led to expect; but at first sight we may imagine that silica predominated more in the granitic rocks than in the others, while magnesia abounded in those parts of the earth which vomited forth the serpentinous deposits. It is however obviously premature to speculate upon that which can only be learned through the medium of careful and exact investigation, and the subject is only introduced

* For descriptions of this district, consult the memoirs of Prof. Sedgwick, Cambridge Phil. Trans., vol. i. ; of Mr. Magendie, Trans. Geol. Soc. of Cornwall, vol. i. ; and of Mr. Rogers, same work, vol. ii.

for the purpose of promoting inquiry, and the possibility of attract-
ing the attention of those chemists who may be induced to enter
on the hitherto little explored, though vast field of chemical geology.

It has been generally considered that the mineralogical character of igneous rocks has changed during the deposit of the stratified
rocks, through which they have more or les sforced their way; that
is, we do not find granite and serpentine flowing from modern
volcanos, nor trachite nor leucitic lavas intimately associated with
the oldest strata in such a manner, that their relative differences
of age could not be very considerable. Admitting that true mineralogical granite may even be reckoned among the products of the
supercretaceous period, the mass of granite is associated with the
oldest rocks, even omitting all consideration of the gneiss, composed of the same minerals, and probably of exactly the same
amount of elementary substances. The same with those igneous
compounds into which augite largely enters, which abound in the
more recent products, while certainly they are scarce, if they be
not altogether absent, among the older rocks of an igneous origin;
and we have no stratified rocks of similar mineralogical composition constituting extensive districts, as is the case with gneiss. We
are compelled therefore to admit, that the conditions, under which
the two kinds of igneous rocks have been formed, have not been the
same. What those conditions may have been is a separate question,
and one, as above noticed, requiring investigation; but it will be at
once obvious, that the ejection of a mass, in a state of igneous fusion,
into the atmosphere would be likely to have its constituent parts
arranged in a different manner from those in a similar mass forced
out beneath great pressure, such as we may consider to exist beneath deep seas. Independently, however, of this consideration,
there appears to have been something in the condition of the world
at the earliest times, causing certain compounds to be formed in
great abundance, which does not now continue in such force as to
permit the production of similar compounds.

We cannot conclude this sketch of the unstratified rocks without
adverting to the concretionary and columnar structure which they
frequently assume. The most familiar examples of the columnar
structure are those of the basalt in the Giant's Causeway and at
Staffa, in the latter place constituting the sides of the justly celebrated Fingal's Cave*. The concretionary or globular structure
is often visible in the decomposition of trappean and volcanic rocks,
and is remarkable in a solid rock named the orbicular granite of
Corsica (*diorite orbiculaire*, Al. Brong.), in which balls or sphe-

* See Macculloch's Western Islands of Scotland; and Sections and
Views illustrative of Geological Phænomena, pl. 11 and 19.

roids of concentric and alternate coats of hornblende and compact felspar are disseminated in the mass of the rock.

We are indebted to Mr. Gregory Watt for our first great advance towards a knowledge of the circumstances which have produced this structure. This author fused seven hundred weight of an amorphous basalt named Rowley Rag, described as fine-grained and of a confused crystalline texture; the fire was maintained for more than six hours, and the fused mass was suffered to cool very gradually, so that eight days elapsed before it was removed from the furnace. The fused mass was then three feet and a half long, two feet and a half wide, about four inches thick at one end, and above eighteen inches at the other. This irregularity of form, resulting from the shape of the furnace, was highly advantageous, showing the arrangement of the bodies passing from a vitreous to a stony state. A portion taken out while the basalt was in fusion became perfect glass. The most important result observed was the formation of spheroids, sometimes extending to a diameter of two inches. They were radiated with distinct fibres, the latter also forming concentric coats, when circumstances were only favourable to such an arrangement; but this structure gradually disappeared when the temperature was sufficiently continued, the centres of most of the spheroids becoming compact before they attained the diameter of half an inch. This structure gradually pervades the whole body of the spheroid. " A continuation of the temperature favourable to arrangement speedily induces another change. The texture of the mass becomes more granular, its colour rather more gray, and the brilliant points larger and more numerous; nor is it long before these brilliant molecules arrange themselves in regular forms, and finally the whole mass becomes pervaded by thin crystalline laminæ which intersect it in every direction, and form projecting crystals in the cavities."

Mr. Gregory Watt applied the facts here noticed in explanation of the globular structure of many decomposing basaltic rocks, in which, after a certain stage of disintegration, the included balls resist decomposition with great obstinacy. He moreover extended his remarks to the columnar structure, and observed, that when in his experiments "two spheroids came into contact, no penetration ensued, but the two bodies became mutually compressed and separated by a plane, well defined and invested with a rusty colour;" and when several met, they formed prisms. His inferences from this arrangement were as follows:—

"In a stratum composed of an indefinite number in superficial extent, but only one in height, of impenetrable spheroids, with nearly equidistant centres, if their peripheries should come in contact on the same plane, it seems obvious that their mutual action would form them into hexagons; and if these were resisted below and there was no opposing cause above them, it seems equally clear that they would extend their dimensions upwards and thus form hexagonal prisms, whose length might be indefinitely greater than their diameters. The further the extremities of the radii were removed from the centre, the nearer would be their approach to parallelism; and the structure would be finally propagated by nearly parallel fibres, still keeping within the limits of the hexagonal prism with which their incipient formation commenced; and the prisms might thus shoot to an indefinite length into the undisturbed central mass of the fluid, till their structure was deranged by the superior influence of a counteracting cause *."

Basaltic columns are often curved, and sometimes there is a somewhat confused arrangement of them, so that the disturbing causes have been considerable. They are also frequently articulated, which Mr. Watt considered might be produced by the same cause which determined the concentric fractures of the fibres of the spheroids. Supposing the general theory of the formation of columns correct, the irregularities of the prisms would obviously depend upon the unequal distances of the centres of the spheroids, and the consequent unequal pressure. Mr. Watt accounts for the horizontal arrangement of basalt in some perpendicular dykes, such for example as those at the Giant's Causeway, from the refrigerating or absorbing cause operating on each side of the vein, so that columns should strike out from it, but would not coincide so as to form continuous prisms across the vein, as there would be confusion where the two sets of columns met, if indeed circumstances were sufficiently favourable to produce a meeting.

The columnar arrangement is not confined to basalt, but is more or less observed in all the trappean rocks, the magnitude of the columns being often very considerable. Granite also assumes a prismatic form, as has already been remarked by Mr. Carne respecting that of the western part of Cornwall †, where it is well seen near the Land's End; but instead of assuming an hexagonal arrangement, such as we might presume to be the figure, if the theory respecting basalt was altogether applicable to it, it is quadrangular, and so divided into joints that the resulting solids are parallelopipeds and even cubes. If the student should pass round the Land's End, Cornwall, in a boat, he will be particularly struck with the general arrangement of granite into columns, and the picturesque effect is considerably heightened by the varied disin-

* Gregory Watt, Observations on Basalt, and on the Transition from the vitreous to the stony Texture which occurs in the gradual Refrigeration of melted Basalt; Phil. Trans. 1804.

† Carne on the Granite on the western part of Cornwall; Geol. Trans. of Cornwall, vol. iii. p. 208.

tegration of the blocks from the united action of the sea and atmosphere *.

* Stratified rocks are sometimes found prismatic, and have been rendered so from other circumstances than those above mentioned. Dr. Macculloch notices the prismatic arrangement in a hearthstone taken down from a blast furnace at the Old Park Iron-Works, near Schiffnall, which had been sixteen or eighteen years in constant work. The prisms sometimes extended through the whole thickness of the stone, about ten inches, while in other instances they only penetrated a certain depth. The prismatic structure was considered to have been produced by a long continued action of great heat upon the slab of sandstone. The same author applies this discovery in explanation of the prismatic sandstone found beneath basaltic rock at the hill of Scuirmore, in Rum, as also of the columnar rocks at Dunbar, where he states the prismatic configuration of the sandstone to be assumed in a very gradual manner.—Quarterly Journ. of Science, 1829.

Section XIII.

On the Mineralogical Differences in Contemporaneous Rocks, either Original, or resulting from Alteration after Deposition.

Among the variety of stratified rocks which have been noticed above, it will have been observed that there was much difference as well in the mineralogical composition as in the zoological character of the deposits. Some rocks are evidently formed from the destruction of others; some are chemical; while others appear as if they had suffered alteration subsequently to deposition. The rocks which have been produced by deposit from water, in which mud, sand, gravel, and great blocks were for a time mechanically suspended, have already been sufficiently discussed; and that they should not precisely resemble each other over considerable areas is only what would have been expected, as we cannot imagine any detritus so uniform as to be the same over considerable spaces, for it assumes a perfect uniformity in the transporting power, a constant, equal, and uniform supply of detritus, and a surface, over which the transported substances were carried, so constituted as to offer an equal resistance throughout.

When a stratified rock is crystalline, it has evidently been chemically and not mechanically produced; but it remains to inquire whether such structure be original or consequent on circumstances which have permitted an alteration of the rock. This investigation is one of considerable difficulty, as we cannot always obtain the data necessary for decision, since it is obvious that the same substance may often be obtained in different ways: thus crystalline carbonate of lime may either be produced directly by deposition from an aqueous solution of that substance, or common limestone may be fused by heat under pressure, and the results be similar. The same with many other substances. It therefore becomes a very difficult though always interesting question to discover, whether such stratified and crystalline substances have been produced in the one way or the other.

There are certain generalities on which we may base our investigations. If crystalline and stratified substances occur as sheets of matter, included among beds evidently of mechanical origin, and igneous rocks are not intimately connected with the whole, and there has been no violent disturbance of strata, permitting the possible influence of gaseous compounds on the beds, we may fairly infer that the crystalline rock was formed by aqueous chemical deposition, and that its occurrence among decidedly mechanical compounds, merely shows a difference in the condition of the medium from which they have both resulted; there being solution in the one case, and mere mechanical suspension in the other.

When we find uncrystalline strata assuming a crystalline structure in the immediate vicinity of igneous rocks, so that they constitute different parts of a common whole, the question assumes another character, and we have to inquire if this difference arises from an alteration of a part of the whole, subsequently to deposition, or whether it is the result of certain causes which have operated only on parts of the same whole during deposition.

It has been seen that dolomite, a crystalline compound of carbonate of lime and carbonate of magnesia, occurs in the oolitic series of Poland and Germany; therefore we should not be surprised that it occurred in the same series in the Alps, and apparently among the same rocks in Dalmatia and Greece. From this presence of a particular crystalline compound in a given rock, and over a considerable area, we should be led to consider, that circumstances existed during the formation of the rock which produced this compound over the area, and consequently, that it was original, and did not result from the subsequent application of heat, or any other chemical agent.

Supposing compounds of this nature to have resulted from an aqueous solution of the carbonates of lime and magnesia, we should not be surprised at the absence of organic remains; for it would scarcely be a mixture in which animals would flourish, if even they could exist. Organic remains are however not absent from dolomite, though they are rare in it, for I have seen them in the dolomite of Nice, and they have been noticed elsewhere. The occurrence of organic remains in dolomite does not, it must be confessed, well accord with the supposition that it has been a limestone on which chemical agents have subsequently so acted that it became crystalline and charged with magnesia; for we cannot well understand in the new arrangement of particles how the form of the organic remains could be preserved, more particularly when they are of the same substance with the rock, or solely carbonate of lime. The fossiliferous dolomites would therefore appear to be excluded from the altered rocks, and reduced to those of original and chemical deposition. There are however masses of dolomite not so easily reconcileable with the supposition of aqueous deposition, which occur in patches among limestones, and not far removed from igneous rocks, in such a manner that Von Buch considers them to have resulted from the action of chemical agents on the limestones, subsequently to the deposition and consolidation of the latter, and at the time when igneous rocks were intruded among the stratified mass. To convert a series of beds into a crystalline mass to a certain distance from a rock in a

state of igneous fusion, provided there was sufficient pressure to
prevent the escape of the carbonic acid, would, we know, not be
difficult: but it is difficult to obtain the magnesia necessary to
produce the dolomite, unless it was insinuated into the altered mass
at the time when the various particles were arranging themselves
conformably to the laws of crystallization; in fact, when all the
elementary substances were so circumstanced that they could
freely unite according to their proper affinities. Von Buch considers this was effected by the escape of magnesia from the augite
porphyries or *melaphyres* (augite containing, according to Klap-
roth, 8·75 per cent of that substance) at the period when such
porphyries were protruded through limestones, as in the Tyrol and
other places. He is of opinion that the gas evolved at the time
these igneous rocks were upheaved, entered among the fissures of
the limestone, and converted a considerable proportion of it into
dolomite. This celebrated author adduces the mountain of San
Salvador, on the lake of Lugano, as confirming the truth of his
theory. A red conglomerate, of a similar character to that which
occurs on the lake of Como, separates the mica slate on which Lugano is situated from the limestones and dolomite. "These beds
dip rapidly at 70° to the south, and form a promontory on which
the chapel of San Martino is built. This rock appears in place for
about ten minutes walk, the dip of the beds diminishing to 60°.
It is then covered by a compact smoke-gray limestone, in beds
about a foot thick. These dip as the beds on which they rest, and
have the same inclination on the side of the mountain; but in their
prolongation towards the lake the dip continually diminishes, until,
at its level, it is scarcely 20°. The beds as they rise describe a
curve that somewhat resembles a parabola. The further we advance on the road, the more we find these beds traversed by small
veins, the sides of which are covered by rhombs of dolomite. Similar crystals are also observable in small cavities of the rocks.
As we advance, the rock appears divided into fissures, and the stratification ceases to be distinct. Lastly, where the face of the mountain becomes nearly perpendicular, it is found to be entirely formed
of dolomite. There is no marked separation between the limestone and the latter rock. By the increase of the veins and geodes
the calcareous rock entirely disappears, and pure dolomite occurs
in its place. * * * As we advance along the high road, the purer
we find the dolomite, and at the same time the more white and
granular. * * * From hence to beyond Melide the mountains
are composed of dark augite porphyry mixed with epidote, the
same as it appears at Campione, Bissone, and Rovio*."

* Von Buch, Ann. des Sci. Nat. 1827, where there are sections and a
map of the district of the lakes Orta, Maggiore, and Lugano; as also in
Sections and Views illustrative of Geological Phænomena, pl. 8. fig. 2.
and pl. 30.

This is undoubtedly a remarkable case, as the mass of augitic rock is on the side of the dolomite, and as crystals of dolomite are found in the cracks of the limestones, because the latter circumstance shows that such crystals were not contemporaneous with the deposition of the limestone, but were formed subsequently, after cracks had been produced in it, while the former circumstance is precisely in accordance with the theory. According to Von Buch's sections, however, a small quantity of red porphyry and mica slate intervenes between the mass of the augite porphyry and the dolomite: it certainly does not follow that they should constantly intervene because they may do so in one situation and not in another, therefore this is no great objection, indeed according to the map, they do not always separate the dolomite from the igneous rock. Other masses of dolomite occur round a large patch of granite extending westward from the south-western branch of the Lago di Lugano, on which are situated Casco al Monte and Porto. One of these masses at Monte Schieri is connected with tufaceous augite rock; while others are only in contact with the granite, as far at least as regards the surface; but this proves little, for the augite porphyry may be beneath them, as it is seen to cut through the granite at Brincio.

From its vicinity to these places the dolomite of the lakes of Como and Lecco acquires considerable interest, although augite rocks have not yet been detected among them. They certainly occur intermingled with the limestones, which are compact and gray, while at other times they also appear the prolongation of limestone beds which have gradually lost their compact texture, and at the same time have acquired magnesia and become crystalline. In countries like these, where so much confusion prevails, and where we may expect to find extensive faults, a perfect continuance of a given series of beds is most difficult to trace; but with ample allowance for these difficulties, there seems every probability that the continuations of some of the limestone beds become dolomite *.

The north part of the lake of Como is composed of gneiss and mica slate, which correspond on either shore, and dip southwards; the lake of Lecco and the southern part of that of Como are formed of limestones and dolomite. Between the two masses of rock are conglomerates and sandstones which have the same dip and direction as the gneiss and mica slate. On the south of these latter rocks the sides of the lake cease to correspond. Thus, on the eastern shore, after passing a small portion of dolomite, we find limestones, among which are the black marbles of Varenna,

* See Geological Map and Sections of the shores of the lake of Como, in Sections and Views illustrative of Geological Phænomena, pl. 31. and 32.

and these continue as far as opposite to Bellaggio Point, while on the other and western side dolomite prevails during the whole corresponding distance, with the exception of a few limestone beds on the south of Menaggio. Here then is no correspondence; on the contrary we have limestones on the one side and dolomite on the other, the latter containing a mass of gypsum at Nobiallo. If we proceed down the lake of Como, from Bellaggio to Como, we find nothing but limestones, producing the fine scenery for which this lake is so celebrated, after having passed the promontory of Dosso d'Albido and the opposite shores of the Croci Galle: but if we go down the lake of Lecco to the town of the same name, also from Bellaggio, there is scarcely anything to be seen but dolomite, if we except a mass of gypsum included in it at Limonta, and a long strip of limestone between Lierna and Mandello. Here also we have no correspondence, though the general direction of the beds in both lakes would lead us to suspect, that those in the one were continued from those in the other. And this view is strengthened if we ascend the Monte San Primo, a mountain already noticed as strewed over with thousands of erratic blocks, for the highest crest is composed of limestone ranging W.N.W. and E.S.E., with a dip to the S.S.W. If we follow the direction of these beds to the lake of Lecco on the east, we find dolomite, so that the change in this place appears somewhat sudden.

Notwithstanding this apparent conversion of limestone into dolomite in the direction of the beds, which might lead us to suppose that some cause had produced a change in the rock after its consolidation, it must be confessed that there is also an interstratification of the dolomite with the limestone (Fig. 27. p. 168.), and moreover, the dolomite rests upon limestone in the lake of Lecco; facts which are at variance with the supposition that all the dolomites of this part of Italy have been altered rocks. Some of them at least appear to have been original deposits; and this supposition, as far as it affects the rocks which are apparently limestones in one place and dolomites in another, involves a curious question ; for if we admit a contemporaneous deposition of the two to have taken place, it follows, that carbonate of lime was thrown down in one place, while a mixture of the carbonates of lime and magnesia was deposited close to it in another, and that the two depositions were so far influenced by circumstances, that the one was compact while the other was either wholly or semicrystalline. These observations do not apply to the alternations, for there we have to suppose a change of circumstances over the same place, and this somewhat gradual, for the calcareous beds seem gradually to acquire magnesia, as may be seen on the western shores of the lake of Lecco.

Some good examples of a mixture of dolomite and limestone

may be observed near Nice, where again the continuation of limestone beds apparently becomes dolomitic, which here, as elsewhere, generally loses its stratification when most pure, while the
less pure semicrystalline compounds are distinctly stratified.
This is however not a general rule, for I have seen some nearly
pure beds stratified; and supposing such beds to be original deposits, their division into strata is no more remarkable, than that the
saccharine marble of Carrara should be stratified.

In the vicinity of Nice, gypsum also accompanies the dolomite,
and the connection of the two is so intimate that the gypsum of
Sospello contains rhombs of dolomite, a circumstance also observed
in the gypsum accompanying the dolomites in the Tyrol. This
frequent association of gypsum with dolomite has not yet been
satisfactorily accounted for.

Gypsum is not a rare accompaniment of rocks, even those of a
decidedly mechanical origin, yet it must be considered a chemical deposit; its occurrence therefore in such situations shows that
other causes were in force than a mere drift of detritus; and when
gypsum is to a certain extent characteristic of a deposit over a
considerable area, it proves that the operation of such causes has
not been local, but that during such period, and over such area,
the circumstances permitting those deposits prevailed extensively.
Gypsum has been considered characteristic of the upper part of
the red sandstone series, generally known as red or variegated
marls. Without however asserting that it is a necessary part of
the rock, or that it is constantly present, it is remarkable how
very frequently it is discovered in this deposit, in England, France,
and Germany, proving that circumstances were then favourable
to its production, if it proves nothing more.

When we recollect that the intrusion of igneous rocks has been
sufficient to convert chalk into granular limestone in the north
of Ireland, we need not be surprised that other rocks have been
altered by the intrusion of similar substances. The slates for instance in many parts of the country surrounding the granite of
Dartmoor, Devon, have suffered from its intrusion, some being simply micaceous, others more indurated and with somewhat of the
characters of mica slate and gneiss, while others appear converted
into a hard zoned rock strongly impregnated with felspar. These
changes are no more than we should expect from the intrusion of
a mass in a state of igneous fusion; for when intruded in such a
mass as that of Dartmoor, the affinities must have become exceedingly loose in the substances in contact with it, and the various
changes observed would readily be produced. Cases of induration and alteration of rocks in contact with igneous products are
so common that it would be useless to enumerate them; but the
student must carefully distinguish between igneous rocks which
have evidently been intruded among the others, and those which

are the older rocks, upon which the others have been deposited; for it may happen that the older rocks may have been disintegrated previous to deposition of the superincumbent substance; in which case, if the latter be arenaceous, there may be an apparent passage of the arenaceous into the igneous rock, producing the false appearance of an altered substance.

The change in the mineralogical character of certain calcareous rocks, at different points of the area which they may happen to cover, has been previously adverted to, and it has been shown that in the oolitic group there was a probability, from the zoological character of the deposit, of a portion having been produced at the bottom of a deep sea, while other parts were formed in shallower waters. The physical circumstances under which the different parts of the deposit would be placed, could scarcely do otherwise than influence the product; but what that precise influence may have been we are as yet ignorant: it may, however, be anticipated that the differences of pressure, and the liability to be disturbed by currents in one situation, while the latter might be scarcely felt, if not entirely absent, in another, would alone cause a great variation in the mineralogical texture.

It might also happen, that in a deep part of an ocean successive depositions were effected during periods when frequent changes were produced in other and remote situations, so that though contemporaneous there should be no mineralogical agreement between them; and if in the course of events, the continuous and quiet deposits were upheaved, as might even happen by a very moderate thermometrical expansion of a portion of our globe, and a continent be the result, the difficulty of identifying clear divisions in the one place with the mass in the other would be insurmountable. It is more than probable that this supposition has been realized on the surface of our planet, and that eventually geologists will show less determination in identifying deposits, more particularly those of moderate comparative antiquity, over very considerable distances. It is much more desirable, for instance, that India should be described with reference to itself, so that when its geology shall have been sufficiently advanced, Europe may be fairly compared with it, than that there should be a determination to find nothing but European equivalents in that quarter of the world.

On the Elevation of Mountains.

Although the direction of the various chains of mountains has long engaged the attention of geologists and geographers, and although the direction of upturned strata has also long been noticed by the former, and found generally to coincide with that of the mountain chains which they constitute, it has only been recently, and since the labours of M. Elie de Beaumont, that the subject

has acquired a new interest, and will henceforth form an important branch of geological investigation, whether the theory of this
distinguished geologist shall eventually be found tenable to the extent supposed, or require very material modifications.

Von Buch appears some time since to have discovered that the
mountain-systems of Germany were not contemporaneous, but
were of distinct dates; and geologists had long been in the habit
of noticing the unconformable position of strata, an older and inferior rock having been upheaved, while a newer rock rested
upon the edges of the upturned strata. Here, however, the subject seemed to rest, until M. Elie de Beaumont, from a series of
very exact observations in certain parts of France and the Alps,
remarked, that the dislocations of the strata were not only referable
to distinct epochs, but that there was a parallelism between dislocations and upheaved mountains of the same date; and he
further considered that these events produced breaks in the rocks
then in the course of deposition, so that those subsequently formed
rested unconformably on the disturbed strata of the older rocks.

To determine the general unconformable position of two rocks
sometimes requires very great care, though at first sight it may
appear extremely easy to observe whether one rock rests on the
upturned edges of another, or not; and so it undoubtedly is in many
cases; but when they meet at small angles *, or the one rests on
the contortions of the other, the inquiry becomes more difficult,
and it requires numerous observations to be certain of the general
fact. The difficulty in the case of contortions will be seen in the
annexed cut.

If a section be obtained   Fig. 101.
on the left or right, it will
very easily be seen that the
beds *a a* rest upon the contorted strata *b b*; but if one
were only observed at *c*, the

unconformable position of the two rocks might be doubtful. The

---

* A general unconformability does not always prove a movement in the
inferior rocks prior to the deposition of the superior; for supposing a
given series to be so produced that the newer rocks may be formed within
successively diminishing areas, and another deposit to cover the whole,
it is evident that the higher mass will so far rest unconformably on the
inferior rocks that it will cover them all in succession. Now this is what
has happened with the chalk and oolite groups in England, the former
overlapping the various members of the latter as they successively fine
off. See Sections and Views illustrative of Geological Phænomena, for
an overlap of the chalk on the coasts of Dorset and Devon. The angles
at which the cretaceous and other rocks meet is there so small, that their
unconformability could scarcely be determined at any particular point,
though in the mass it is evident.

student may perhaps consider, that from a little research in the same place he would soon find evidence of the disturbed condition of the strata beneath; and if the contortions were always on the small scale, and natural sections common, such would be the case: but when the latter are scarce or the undulations and contortions on the large scale, the great bends being measured by miles instead of fathoms, the subject is not so easy. It may be stated as an example, that the mass of the calcareous Alps is considered to rest unconformably on the mass of those composed of protogine, gneiss, &c.; but the situations where the contrary opinion may be formed are very numerous, the sections there exposing perfect conformability. It also requires great care in tracing strata up to a mountain range, for the purpose of ascertaining its relative antiquity, to distinguish between those beds which have been decidedly upturned subsequently to deposition, and those which may have originally taken a small angle during their formation on the flanks of a chain, previously elevated to a certain extent.

Perhaps the annexed diagram may assist the student in comprehending the manner in which the relative age of mountain ranges is determined from the position of strata.

Fig. 102.

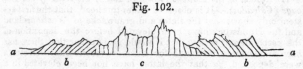

If the rocks *a a* are found resting quietly on the upturned strata *b b*, it is inferred that *b b* have been disturbed previous to the deposition of *a a*; and consequently, if *a a* be a known rock occupying a certain place in the series, we obtain a relative date for the elevation of *b b*, so far as relates to *a a*; but if it should chance that there are no commonly known deposits absent between them, we obtain the exact relative date of the elevation of so much of the mountains as do not exhibit any other unconformability of strata, and of the whole range, if no such unconformability can be detected. When however this does occur, as in the above diagram, we learn, that more than one elevation of strata has taken place in the same chain, for *b b* resting in discordant stratification on *c*, proves that *c* was tilted up prior to the deposition of *b b*; so that in this case there would be evidence of two disturbances in the same chain, and the relative dates of both would be obtained on the same principles. It will be obvious if two lines of elevated strata cross each other, there would be much confusion where they traverse; and should great violence have been employed, so that the strata be even thrown over, it will require much caution to determine the relative age of the fractures at those points.

From the obliging communications of M. Elie de Beaumont, it appears that he now considers he can distinguish twelve systems of mountains in Europe, each regarded as characterized by the relative direction and elevation of its strata, and each elevation as corresponding with a solution observed in the continuity of the sedimentary deposits, also of Europe *. These various systems are as follows, commencing with that of the greatest relative antiquity:

I. *System of Westmoreland and the Hundsruck.*—The direction of slate rocks in Westmoreland is, according to Prof. Sedgwick, N.E. by E. and S.W. by W.; and that of the slates and grauwacke of the Eifel, the Hundsruck, and of the Nassau Mountains, about N.E. and S.W. The slates, grauwacke, and grauwacke limestones of the northern and central part of the Vosges have the same direction. The carboniferous rocks rest on the upturned rocks of the North of England; coal measures repose upon the edges of those of the Vosges; and the carboniferous rocks of Belgium and Saarbruck were probably deposited at the foot of the Eifel, Hundsruck, &c.

II. *System of the Ballons (Vosges), and of the Hills of the Bocage (Calvados).*—It is observed under this head, that the first system only shows that the slates and grauwacke of Westmoreland and the Hundsruck have been elevated before the deposition of the carboniferous series; but it would also appear that there has been an elevation of strata before the more recent transition strata were deposited, so that the latter have not been elevated in a N.E. and S.W. direction, but would on the contrary seem to have been formed on upturned beds of the former. Such are the calcareous, marly, and arenaceous deposits, with *Orthoceratites, Trilobites,* &c., in Podolia, of the environs of St. Petersburg, and of Sweden, where they are but slightly removed from their original horizontal position. Such are also the transition beds of Dudley and Gloucestershire; and possibly also the transition beds of the South of Ireland may be included in the same list, which M. Elie de Beaumont considers may also contain certain slate and grau-

* In point of fact, M. Elie de Beaumont was kind enough to send me, at my request, a condensed account of his views, corrected up to the present time, relating to the elevation of mountains, for insertion in this work; but unfortunately it reached me so late, that I had only time to replace an abstract of his former and printed labours on this subject by that above given; the original memoir, though greatly condensed, would also have been too long for this little work, already swollen to a larger size than was at first intended; it was therefore considered most advisable to present the above brief notice to the attention of the reader, referring him to the memoir itself, which will be found in the Phil. Mag. and Annals of Philosophy, for October, 1831.

wacke beds with anthracite, forming the south-east angle of the Vosges. When these beds are not horizontal, they are dislocated in directions the most marked of which is comprised between an E. and W. line, and one E. 15° S. and W. 15° N.

III. *System of the North of England.*—This is the north and south range of the carboniferous series, noticed by Prof. Sedgwick. It is considered to have been produced immediately previous to the deposit of the red conglomerate (todteliegendes). M. Elie de Beaumont remarks; " The elevation of the chain in the North of England is not probably an isolated phænomenon ; but if we glance at the geological map of England by Mr. Greenough, or that accompanying the memoir of Prof. Buckland and Mr. Conybeare on the Environs of Bristol, we are naturally led to remark, that the problematical rocks which penetrate and dislocate the coal deposits of Shrewsbury and Colebrooke Dale, and those which form the Malvern Hills, appear to be connected with a series of fractures which run nearly N. and S., and are prolonged across the more recent transition and the carboniferous rocks to the environs of Bristol." It is also considered probable that the form of the west coast of the department of La Manche, which runs nearly N. and S., may be due to a fracture of this age.

IV. *System of the Netherlands and South Wales.*—This is the great E. and W. range of the carboniferous rocks from the environs of Aix-la-Chapelle to St. Bride's Bay, Pembrokeshire, which whenever visible from beneath other deposits, exhibits this general direction for about 400 miles. It is considered that the beds of the (new) red sandstone series which repose on this dislocation are not so ancient as those noticed in the previous group.

V. *System of the Rhine.*—The Vosges and the Swarzwald terminate opposite one another in two long cliffs parallel to each other and to the course of the Rhine. These are apparently due to great faults, having a direction S. 15° W. and N. 15° E. These fractures preceded the deposit of the rocks in the basin of Alsace, among which are the red or variegated sandstone (grès bigarré), the muschelkalk, and the variegated marls.

VI. *System of the S.W. coast of Britanny and of La Vendée, of Morvan, of the Bohmerwaldgebirge, and of the Thuringerwald.*—The general direction of this system is from N.W. to S.E., and while the red or variegated marls (marnes irisées), the muschelkalk, and all older strata have been thrown out of their original positions, the oolitic series, comprehending the lias and its lower sandstone, have remained undisturbed in these various situations.

VII. *System of the Pilas, the Côte d'Or, and the Erzgebirge.*—In this system, which also contains a portion of the Cevennes, the strata are disturbed up to the oolitic rocks inclusive, while the cretaceous series (green sand and chalk) remain apparently in the position in which they were deposited. The direction of this system is considered to be N.E. and S.W.

VIII. *System of Mount Viso.*—" The French Alps and the south-west extremity of the Jura, from the environs of Antibes to those of Pont d'Ain, and Lons le Saulnier, present a series of crests and dislocations with a direction towards the N.N.W., in which the older rocks of the Wealden formation, the green sand, and chalk, are found upheaved, as well as those of the oolitic series. The pyramid of primæval rocks composing Mont Viso is traversed by enormous faults, which, from their direction, evidently belong to this system of fractures. The eastern crests of the Devolny, on the north of Gap, are formed of the oldest beds of the green sand and chalk thrown up in the direction in question, and raised more than 4700 English feet above the sea. At the feet of these enormous escarpments are, horizontally deposited and at more than 2000 feet lower down, those upper beds of the cretaceous system which are distinguished from the rest by the presence of *Nummulites, Cerithia, Ampullariæ,* and other shells, the genera of which were long considered as exclusively found in the tertiary (supercretaceous) rocks. Thus it was between the two portions of that which is commonly termed the series of the Wealden formation, green sand, and chalk, that the beds of the Mont Viso system have been thrown up."

IX. *Pyreneo-Apennine System.*—" This includes the whole chain of the Pyrenees, the northern and some other ridges of the Apennines, the calcareous chains on the north-east of the Adriatic, those of the Morea, nearly the whole Carpathian chain, and a great series of inequalities continued from that chain through the north-east escarpment of the Hartz Mountains to northern Germany." The general direction of this system is about W.N.W. and E.S.E.

X. *System of the Islands of Corsica and Sardinia.*—This elevation is considered to have taken place during the supercretaceous period, and it is remarked that the north and south direction of the system in Corsica and Sardinia is observed " in many small valleys and ridges of mountains in the Apennines, and in Istria, and in the disposition of many volcanic masses and metalliferous sites of Hungary."

" It is worthy of remark that the directions of the system of the Pilas and the Côte d'Or, of that of the Pyrenees, and that of the islands of Corsica and Sardinia, are respectively nearly parallel to those of the system of Westmoreland and the Hundsruck; of the system of the Ballons and of the Bocage, and of the system of the North of England. The corresponding directions differ but a small number of degrees, and the corresponding systems of the two series have succeeded each other in the same order; leading to the supposition that there has been *a kind of periodical recurrence* of the same, or nearly the same, directions of elevation."

XI. *System of the Western Alps.*—The mean direction of this system is about N.N.E. and S.S.W., and the elevation is considered to have taken place after the deposit of those recent super-cretaceous beds named Shelly Molasse (*molasse coquillière*), beds contemporaneous with the *fahluns* of Touraine.

The direction of strata is of a complicated character where this system and that to be next mentioned cross each other, as they do around the Mont Blanc, Mont Rose, and Finsteraarhorn, at about an angle of 45° or 50°.

XII. *System of the principal Chain of the Alps (from the Valais into Austria), comprising also the Chains of the Ventoux, the Lebaron and the Sainte Baume (Provence).*—The direction of this system is about E. ¼ N.E. and W. ¼ S.W., and the strata are considered to have been elevated previous to the dispersion of the erratic blocks and those gravels which have been termed *diluvium*, but which in the vicinity of the Alps are found to have been deposited upon other gravels, often of considerable thickness.

M. Elie de Beaumont concludes with the following observations: " The independence of successive sedimentary formations is the most important result obtained from the study of the superficial beds of our globe ; and one of the principal objects of my researches has been to show, that this great fact is the consequence and even a proof of the independence of mountain systems having different directions.

" The fact of a general uniformity in the direction of all the beds upheaved at the same epoch, and consequently in the crests formed by these beds, is perhaps as important in the study of mountains, as the independence of successive formations is in the study of superimposed beds. The sudden change of direction in passing from one group to another has permitted European mountains to be divided into a certain number of distinct systems, which penetrate and sometimes cross each other without becoming confounded. I have recognized from various examples, of which the number now amounts to twelve, that there is a coincidence between the sudden changes established by the lines of demarcation observed in certain consecutive stages of the sedimentary rocks, and the elevation of the beds of the same number of mountain systems.

" Pursuing the subject as far as my means of observation and induction will permit, it has appeared to me, that the different systems, at least those which are at the same time the most striking and recent, are composed of a certain number of small chains, ranged parallel to the demi-circumference of the surface of the globe, and occupying a zone of much greater length than breadth , and of which the length embraces a considerable fraction of one of the great circles of the terrestrial sphere. It may be observed respecting the hypothesis of each of these mountain-systems being the product of a single epoch of dislocation, that it is easier geometrically to conceive the manner in which the solid crust of the globe may be elevated into ridges along a considerable portion of one of its great circles, than that a similar effect may have been produced in a more restricted space.

" However well established it may be by facts, the assemblage of which constitutes positive geology, that the surface of the globe has presented a long series of tranquil periods, each separated from that which followed it by a sudden and violent convulsion, in which a portion of the earth's crust was dislocated, that, in a word, this surface was ridged at intervals in different directions; the mind would not rest satisfied if it did not perceive, among those causes now in action, an element, fitted from time to time to produce disturbances different from the ordinary march of the phænomena which we now witness.

" The idea of *volcanic action* naturally presents itself when we search, in the existing state of things, for a term of comparison with these great phænomena. They nevertheless do not appear susceptible of being referred to volcanic action, unless we define it, with M. Humboldt, as being *the influence exercised by the interior of a planet on its exterior covering during its different stages of refrigeration.*

" Volcanos are frequently arranged in lines following fractures parallel to mountain chains, and which originate in the elevation of such chains ; but it does not appear to me that we can thence regard the elevation of the chains themselves as due to the action of *volcanic foci*, taking the words in their ordinary and restricted sense. We can easily conceive how a *volcanic focus* may produce accidents circularly and in the form of rays from a central point, but we cannot conceive how even many united *foci* could produce those ridges which follow a common direction through several degrees.

" Volcanic action, such as it is commonly understood, could not therefore be itself the first cause of these great phænomena, but volcanic action appears to be related (and this is a subject which has long occupied M. Cordier, though he has considered it under another point of view) with the high temperature now existing in the interior of the globe; and the analogy which at first sight would lead us to seek the cause of the revolutions on the globe's surface in volcanic action, properly so called, should, if I do not deceive myself, lead us finally to seek this same cause in the high internal temperature of the earth.

" Now the secular refrigeration, that is to say, the slow diffusion of the primitive heat to which the planets owe their spheroidal form, and the generally regular disposition of these beds from the centre to the circumference, in the order of specific gravity,—the secular refrigeration, on the march of which M. Fourier has thrown so much light, does offer an element to which these extraordinary effects may be referred. This element is the relation which a refrigeration so advanced as that of the planetary bodies establishes between the capacity of their solid crusts and the volume of their internal masses. In a given time, the temperature of the interior of the planets is lowered by a much greater quantity than that on their surfaces, of which the refrigeration is now nearly insensible. We are, undoubtedly, ignorant of the physical properties of the matter composing the interior of these bodies; but analogy leads us to consider, that the inequality of cooling above noticed would place their crusts under the necessity of continually diminishing their capacity, notwithstanding the nearly rigorous constancy of their temperature, in order that they should not cease to embrace their internal masses exactly, the temperature of which diminishes sensibly. They must therefore depart in a slight and progressive manner from the spheroidal figure proper to them, and corresponding to a maximum of capacity; and the gradually increasing tendency to revert to that figure, whether it acts alone, or whether it combines with other internal causes of change which the planets may contain, may, with great probability, completely account for the ridges and protuberances which have been formed at intervals, on the external crust of the earth, and probably also of all the other planets *."

From this sketch of M. Elie de Beaumont's theory, in which his views respecting the connection of distant mountains with those of Europe have been omitted, because the necessary detail could scarely be abridged, it will be evident that it will require much time and very exact observation in various parts of the world before we can fairly ascertain what are exceptions and what the general rules. M. Elie de Beaumont has already remarked on the parallelism, or near parallelism, of European mountain chains of different dates; therefore this character alone is insufficient to determine the relative age of an elevated range of strata; a conclusion that may be still further strengthened by observing certain lines of disturbed strata in the British Isles, which, when we regard the general surface of the world, are not far distant from each other.

The disturbed strata in the Isle of Wight range east and west, as do those also in the Weymouth district, in the Mendip Hills, in a large part of Devonshire, and in South Wales. The date of the elevation of the Isle of Wight beds was certainly posterior to the deposition of the London clay, and there would appear little reason to doubt that the disturbed and fractured condition of the

* Elie de Beaumont, MSS.

Weymouth district was effected at the same time. But when we continue our researches into Devonshire, we find that the east and west arrangement of a large proportion of the grauwacke in that country was produced anterior to the deposit of the (new) red sandstone series, since the latter rests upon the upturned edges of the former *. If we now proceed northwards to the carboniferous rocks of the Mendips and South Wales, we find they also have suffered an elevation in an east and west direction, prior to the formation of the (new) red sandstone; so that in the southern parts of England the strata have been twice elevated in a given direction at different dates; and if we continue our researches, we find, from the observations of Mr. Weaver, that the grauwacke in the South of Ireland was elevated previous to the deposition of the old red sandstone, also in an east and west direction; thus affording three elevations of strata (not far distant from each other) in the same direction, but at different dates †.

In offering these observations it is by no means intended to combat the general principle of the contemporaneous elevation of strata at various and distant places, resulting from the gradual refrigeration of the globe; beds of various kinds having been subsequently and quietly deposited over large areas, on such disturbed strata; but simply to remark, that parallelism may not always be a necessary condition of such elevated strata; for perhaps by laying too much stress upon this point, we not only are in danger, in the present state of our knowledge, of permitting theory to take the lead of facts, but of shutting ourselves out from a consideration of other possible lines which contemporaneously elevated strata may follow. And should it eventually be discovered that contemporaneously disturbed beds are by no means parallel though still in straight lines, it does not appear that the main principle of M. Elie de Beaumont's theory would be affected by it. What lines may eventually be found to prevail, will, as previously remarked, require much time and great patience to discover; but let the event be as it may, geologists will not the less have reason to feel grateful to M. Elie de Beaumont for having rescued the subject from the state in which he found it; it being impossible but that the investigations, to which this theory will necessarily give rise, must end in the most important additions to geological knowledge.

It has already been noticed by Prof. Sedgwick, that the change in the zoological character of deposits has not always coincided

* Both the one and the other have been subsequently fractured, and many of the faults have somewhat of east and west direction.

† It has been considered that the north and south line of the carboniferous rocks in Northern England was elevated at a different epoch from the east and west line of South Wales and Somersetshire, but it must be confessed that this point is far from being proved.

with disruptions of strata; and the student will have collected
from the foregoing pages, as indeed is also remarked by Prof.
Sedgwick, that there was, in Europe, no important change in the
general zoological character of deposits up to the zechstein inclusive; the first great alteration, as far as we can at present see our
way, being observed in the remains entombed in the variegated
sandstone (grès bigarré), and muschelkalk. It has already been
observed, but may be conveniently repeated here, that the effect
produced on animal and vegetable life by an upburst of a line of
rocks sufficient to produce a range of mountains, might destroy
all terrestrial animals and even a large proportion, if not the whole,
of the vegetation within the influence of the disturbing cause, not
only by producing a deluge over the land, which might wash off
the animals and carry away a great proportion of the vegetation,
but by elevating such vegetation into colder regions of the atmosphere where it could no longer exist. In this case we suppose
land so situated as to produce plants and to support terrestrial creatures, but it will be evident that if we admit a contemporaneous
disruption of strata at different points, it would take place under
various conditions. In one place it may be effected in the atmosphere, in another in shallow seas, and in a third beneath a great
depth and pressure of the ocean; consequently the resulting phænomena would be as various as the conditions under which the
disruption and elevation of strata were produced: but it will be
obvious that the destruction of marine life would be very difficult,
and we can scarcely, from known facts, consider that there has
been a disruption of the rocks composing the earth's surface so
general as to annihilate all marine creatures at a given time, even
with every allowance for the operations of powerful and destructive currents; though we can conceive that near the centres of
every great disturbance there might be a very great, and as far as
related to certain areas complete destruction of marine creatures.

On the Occurrence of Metals in Rocks.

To enter fully into this subject would require a volume; the
following notice is therefore solely intended to call the attention
of the student to a few circumstances which may be generally interesting.

Metals occur in rocks either disseminated; in bunches; in a
net-work of strings or small veins; in beds; or in veins filling fissures, which traverse beds or masses of rock. When metals are
disseminated through a rock, as tin often is in granite, and iron
pyrites in many trap rocks and clay slates, there can be little
doubt that they constituted original portions of the rock, and that
they were chemically separated from the mass during consolidation. When metals occur in bunches, as the copper in Ecton,

Staffordshire, or the lead in the Sierra Nevada, in Spain, there is a difficulty in considering them otherwise than contemporaneous with the rocks in which they are included. The occurrence also of metals in strings or small veins crossing each other in all directions, so that in a section they appear like net-work, reminds us strongly of the small strings or veins of carbonate of lime in many limestones, as has already been observed by Mr. Weaver respecting those of copper in Ross Island, Lake of Killarney; so that if not precisely contemporaneous with the original formation of the including rock, they were, like the calcareous veins in the limestone, secreted from the rock into small cracks possibly produced during consolidation. The occurrence of metals in beds has been much disputed or commented on, but it must be admitted that iron ore frequently occurs in beds, and we must regard the copper slate of Thuringia and other adjacent countries as to a certain extent a metallic bed, though it does not strictly come under the head of a bed of solid ore. The appearance of metals in beds is often deceptive, being nothing more than a continuation of a vein laterally between strata; thus in the rich copper mine of Allihies, in the South of Ireland, " the ore occurs in a large quartz vein, which generally intersects the slaty rocks of the country from north to south, but in some cases runs parallel to the stratification *." Mr. Taylor informs me that the lead at Nent Head in Alston Moor, Cumberland, shoots out laterally among the strata, and the same fact is observable in different mines in Yorkshire and Flintshire.

The most common occurrence of metals is however in veins, or, as they are termed in Cornwall, *lodes*. These are in part filled up, but in various proportions, with metallic substances, and have the general appearance of fissures. They dip at various angles, not unfrequently approaching a vertical position. It was at one time much disputed whether these fissures had been filled from above or beneath; but from facts that have been noticed within a few years, more particularly by Mr. Taylor and Mr. Carne, there is much difficulty in considering that either hypothesis is generally correct. It now appears that the mineral character of a metalliferous vein greatly depends upon the rock which it traverses, that is, when a vein traverses two rocks, as for instance granite and slate, the contents of the vein are not generally the same in the two rocks, but will be different in the one and the other.

Mr. Carne has observed respecting the metalliferous veins of Cornwall, that it is a rare circumstance when a vein which has been productive in one rock continues rich long after it has entered into another. The same author has also remarked that a similar change will be observed even in the same rock, should such rock

---

* Weaver, Proceedings of the Geol. Soc. June 4, 1830.

become harder or softer, more slaty or more compact. He admits that such changes are sometimes small, but states that the general fact is sufficiently apparent, and often very striking *.

Such facts are not confined to Cornwall, but have been observed elsewhere; thus the lead veins traversing the carboniferous limestone of Derbyshire, which is in some places much associated with trap rocks, are found to be so altered in their passage through the trap, which, from the mode of association, presents the appearance of interstratification, that it was once considered the trap cut off the lead veins; this is however now well known not to be the case.

This fact of the alteration of metallic veins in their passage from one kind of rock to another, or in the same rock, should that become changed, would lead us to consider, with Mr. Fox, that their formation has been in a great measure due to the silent though powerful influence of electricity. This inquiry may yet be considered in its infancy; but the experiments of Mr. Fox on the electro-magnetic properties of the metalliferous veins of Cornwall will be read with great interest †.

That many of these veins are fissures produced by dislocations similar to those which are commonly found in various countries, and are supposed to abound more in the coal measures only because opportunities of detecting them are there more frequent, seems highly probable; indeed if veins are of different ages, and by cutting one another shift each other, as has been shown to be frequently the case in Cornwall, we can scarcely doubt it. The following is, according to Mr. Carne, the relative ages of the veins in Cornwall: 1. oldest tin lodes; 2. the more recent tin lodes; 3. the oldest east and west copper lodes; 4. the contra copper lodes; 5. cross courses; 6. the more recent copper lodes; 7. the cross flukans (clay veins); and 8. the slides (faults with clay in the fissures) ‡.

Now if this relative antiquity of veins be generally correct as

---

* Carne, Trans. Geol. Soc. of Cornwall, vol. iii. p. 81.

† Fox, Philosophical Transactions, 1830, p. 399. This author considers that the relative power of conducting galvanic electricity is in the following order in some of the metalliferous minerals. *Conductors:* Copper nickel, purple copper, yellow sulphuret of copper, vitreous copper, sulphuret of iron, arsenical pyrites, sulphuret of lead, arsenical cobalt, crystallized black oxide of manganese, Tennantite, Fahlerz. *Very imperfect conductors:* Sulphuret of molybdenum, sulphuret of tin, or rather bell-metal ore. *Non-conductors:* Sulphuret of silver, sulphuret of mercury, sulphuret of antimony, sulphuret of bismuth, cupriferous bismuth, realgar, sulphuret of manganese, sulphuret of zinc, and mineral combinations of metals with oxygen, and with acids.

‡ On the relative Age of the Veins in Cornwall; Carne, Geol. Trans. of Cornwall, vol. ii.

Z

far as respects Cornwall, it becomes a curious question, why, if similar causes have produced them, similar results should not be the consequence. If we admit the possibility of secreting the contents of veins from the rocks by electrical means, we cannot so readily understand why different metals should fill the veins in the same rocks, though the direction of the veins might have considerable influence on the conditions and mineralogical combinations of the *same metal.* While again if we consider them ejected from beneath, we are at a loss to understand why the metallic veins should be so much altered in their passage through different rocks. We are certainly not prepared to say what effect may have been produced on the vein, and on the including rocks, from the continued passage of electricity through the vein during an immense lapse of time, or from the arrangement of rocks on the large scale, producing, when properly connected, the effects of a grand galvanic battery ; but as the information at present stands, the history of metalliferous veins is anything but clear. It is quite certain from the dissemination of metals in rocks, that they may constitute an original portion of them ; the small strings also which cross each other, and are unconnected with great veins, have all the appearance of chemical separations from the including rock ; therefore a given rock may contain the necessary elements for secreting substances into a fissure, in the same manner that carbonate of lime frequently fills fissures in limestones, and quartzose veins are common in rocks where silex is abundant.

If the theory of internal heat be well founded, it will be evident that the two ends of a metallic vein will be differently heated, and therefore we should have a thermo-electrical apparatus on the large scale, producing effects which, though slow, might be very considerable. How far such really exist in nature remains questionable ; but it may be observed that the experiments of Mr. Fox show the possibility of their occurrence ; and should his further researches in this highly interesting subject merely so divide it, that some of its present apparent complexity may disappear, a great advance will be made in this now obscure branch of geological inquiry.

APPENDIX.

A. *On some of the Terms employed in Geology.*

East.   Fig. 103.   West.

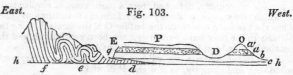

Stratum.—Although, perhaps, this term should only be applied to a bed of rock, the upper and under surfaces of which should be parallel planes, it is also employed to designate beds the upper and under surfaces of which are irregular. Hence rocks are termed stratified, even when the planes of the beds are not precisely parallel to each other.

Seam.—This term is employed to designate a thin stratum.

Dip.—Strata are said to dip when they form an angle with the horizon; the point towards which they plunge being considered the dip. The amount of the dip is estimated by the size of the angle. In Fig. 103. the strata *f* dip at a considerable angle to the west, because they form a considerable angle with the horizontal line *h h*; the strata *d* do the same towards the east. The strata *a b c* are nearly horizontal, having only a slight dip to the west.

As it is evident that mere vertical sections, viewed only in one direction, may afford a false idea of the real dip, and as the planes of the strata may be irregular, the student must be careful to ascertain the general and real dip of such planes.

Anticlinal line—is that line from which strata dip on either side : the ridge of a house-top will convey an idea of this line, the slope of the roof representing the dip of the strata. This line is often extremely useful in tracing disturbances of strata over a country.

Contorted Strata.—Strata are said to be contorted when they are twisted and bent, as at *e*, Fig. 103. These contortions are sometimes on the large scale, as for example in the Alps, where whole mountains are thus twisted.

Conformable Strata.—Strata are termed conformable when their general planes are parallel to each other. Thus *a* rests conformably on *b*.

Z 2

Unconformable Strata.—Strata are said to be unconformable when they rest as *a b* and *c* do on the edges of the beds *d*, Fig. 103.

Outcrop.—Beds are stated to crop-out when they make their appearance on the surface from beneath others. Thus the strata *d* crop out at *g*, as do also the beds at *a b* at the same place.

Outlier.—Strata are said to form outliers, when they constitute a portion of country, detached from a main mass of similar beds, of which they have evidently once formed a continuous part. Thus the beds *a' a* constitute the outlier O of the strata forming the plateau P, for they have evidently once been continuous, and the continuity has been interrupted by the valley D.

Escarpment.—Strata are said to terminate in an escarpment, when they end abruptly, as *a' a* and *b'* do at E.

Fault—is such a dislocation of strata that not only is their continuity destroyed, but the mass of beds on the one side of the fracture or on the other, and sometimes both, are heaved out of their original position. Thus the beds in Fig. 104. have been dislocated or broken into a fault at *f*, the parts of the stratum *a* being no longer on the same plane.

Fig. 104.

Dyke.—This is a wall of rock intermediate between two sides of a dislocation, interrupting the continuity of the beds on either side. Sometimes the latter exhibit marks of having been shoved up by the intrusion of the rock in the dyke, as in Fig. 104, where the dyke *d* has turned up the edges of the beds which it traverses. At other times, there has been a simple fracture and separation, permitting the presence of the matter in the dyke.

Rock.—This term is used by geologists not only for the hard substances usually thus termed, but also for sands, clays, &c. It is also employed to express a general collection of such substances ; thus the expression "the rocks of a country ;" or a particular series of mineral substances, such as "the carboniferous rocks," "the cretaceous rocks," &c.

Formation.—A certain series of rocks supposed to have been produced under similar general circumstances, and at about the same epoch.

B. *Osseous Breccia of Australia.*

From a communication by Major Mitchel to the Geological Society of London (see Proceedings of that Society), it appears that the osseous breccia of Australia bears a great resemblance in its general mode of occurrence to that observed in Europe. The principal ossiferous cavity is near a large cave in Wellington Valley, about 170 miles west of Newcastle, through which valley flows the river Bell, one of the principal sources of the Macquarrie. This cavity is described as a wide and irregular kind of well or fissure, accessible only by ladders and ropes, and the breccia is a mixture of limestone fragments of various sizes, and bones enveloped in an earthy red calcareous stone. Such of the bones found in it as were sent to Europe, and inspected by Mr. Clift, were referred by that anatomist to the *Kangaroo, Wombat, Dasyurus, Koala,* and *Phalangista,* all animals at present existing in Australia. With these were found two others; one of which, considered to be that of an elephant, was obtained in a singular manner by Mr. Rankin, who first visited this fissure; for, supposing it to be a projecting portion of the rock, he fastened the rope by which he descended to it, and was only undeceived by the support breaking, and showing itself to be a large bone. According to Mr. Pentland, the bones from the Australian breccia, forwarded to Paris, and examined by Baron Cuvier and himself, belong to " eight species of animals, referable to the following genera : *Dasyurus* or *Thylacinus ; Hypsiprymnus* or *Kangaroo Rat,* one species; *Phascolomys,* one species; *Kangaroo,* two if not three species; *Halmaturus,* two species; and *Elephant,* one species. Of these eight species, four appear to belong to animals unknown to zoologists of the present day : viz. two species of *Halmaturus ;* one species of *Hypsiprymnus ;* and the *Elephant.*" It is further stated that another collection from Wellington Valley "contains the remains of a species of *Kangaroo* exceeding by one-third the largest known species of that genus."

Major Mitchel notices other and similar breccias on the Macquarrie, eight miles N.E. from the Wellington cavity ; as also at Buree, fifty miles to the S.E., and at Molony, thirty-six miles to the E.; the latter, according to this author, containing bones apparently larger than those of the animals now existing in the country *.

C. *List of the Fossil Shells contained in the Supercretaceous Rocks of Bordeaux and Dax, and enumerated by M. De Basterot*†.

NAUTILUS Aturi, *Bast.,* Dax, Houdan; not considered identical with the N. Pompilius existing in the eastern seas.

LENTICULITES complanata, *Defr.,* Dax, Léognan, Antwerp, Pontoise, Montpellier, and Italy ; common at Saucats.

NUMMULITES lævigata, *Lam.,* common in many supercretaceous deposits ; N. complanata, *Lam.,* Dax, Soissons?

---

* See Jameson's Edin. Phil. Journal, 1831; and Phil. Mag. and Annals, June 1831.

† Description Géologique du Bassin Tertiaire du Sud-Ouest de la France : Mém. de la Soc. d'Hist Nat. de Paris, t. ii.

Lycophoris lenticularis, *Defr.*, common near Bordeaux, Claudiopolis in Transylvania.

Vaginella depressa, *Bosc*, Léognan, Saucats.

Bulla lignaria, *Linn.* (var.), analogous to the existing species, Dax, Léognan, Piedmont, England, Paris; B. cylindrica, *Brug.*, Grignon, Piedmont, Vienna, Dax, Bordeaux; B. Utriculus, *Broc.*, analogous to the existing species, Piacenza, Bordeaux, Dax; B. Labrella, *Fér.*, Dax; B. clathrata, *Defr.*, Dax; B. truncatula, *Brug.*, analogous to the existing species, environs of Paris, Sienna, Riluogo, Dax.

Bullina Lajonkaireana, *Bast.*, abundant at Saucats, Léognan, and Mérignac.

Helix nemoralis, analogous to the existing species, fresh-water limestone at Saucats; H. variabilis, *Drap.*, analogous to the existing species, fresh-water limestone, Saucats.

Bulimus? terebellatus, *Lam.*, analogous to the existing species, Grignon, Placentine, Dax.

Planorbis corneus, *Drap.*, analogous to the existing species, Saucats.

Limnea palustris, *Drap.*, analogous to the existing species, in many supercretaceous deposits, Saucats.

Auricula ringens, *Lam.*, analogous to the existing species, Paris, Nice, Italy, Touraine, Bordeaux, Dax; A. hordeola, *Lam.*, Grignon, Léognan.

Tornatella sulcata (Auricula sulcata, *Lam.*), Grignon, Dax, Bordeaux; T. inflata, *Fér.*, Champagne, Dax; T. semistriata, *Defr.*, Léognan; T. punctulata, *Fér.*, Léognan, Saucats, Dax; T. papyracea, *Bast.*, Dax; T. Dargelasi, *Bast.*, Léognan, Saucats.

Pyramidella Mitrula, *Fér.*, Léognan, Mérignac; P. terebellata (Auricula terebellata, *Lam.*), Grignon, Volterra, Bordeaux, Dax.

Turbo Parkinsoni, *Bast.*, Dax; T. Fittoni, *Bast.*, Dax; T. Lachesis, *Bast.*, common at Bordeaux and Dax.

Delphinula marginata, *Lam.*, Grignon, Dax; D. Scobina (Turbo Scobina, *Al. Brong.*), Castelgomberto, Dax, and near Valognes; D. sulcata, *Lam.*, Grignon, var. at Léognan; D. trigonostoma, *Bast.*, Dax.

Turritella terebralis, *Lam.*, common at Dax, Léognan, and Saucats; T. Archimedis, *Al. Brong.* (var. Burdigalensis, *Bast.*), Ronca, var. α Bassano, var. β Anjou, var. γ Bordeaux; T. asperula, *Al. Brong.*, Ronca, Dax; T. Turris, *Bast.*, analogous to the existing species, Dax; T. quadriplicata, *Bast.*, above the fresh-water limestone at Saucats; T. cathedralis, *Al. Brong.*, Turin, Léognan, Saucats; T. Proto, *Bast.*, Saucats; T. Desmarestina, *Bast.*, Dax.

Scalaria communis (var.), analogous to the existing species, Placentine, Volterra, Bramerton, Dax; S. acuta, *Sow.*, Barton, Dax; S. multilamella, *Bast.*, Parnes, Léognan.

CYCLOSTOMA Lemani, *Bast.*, fresh-water limestone, Saucats, Dax, and Tongres, near Maëstricht.

PALUDINA pusilla (Bulimus pusillus, *Al. Brong.*), analogous to the existing species, Paris, Bordeaux.

MONODONTA elegans, *Faujas de St. F.*, Léognan, rare at Bordeaux; M. Modulus, *Lam.*, analogous to the existing species, Dax; M. Araonis, *Bast.*, analogous to the existing species? Mérignac, Touraine, Dax.

TROCHUS Benetti, *Sow.*, Stubbington, Turin, Léognan, Saucats; T. patulus, *Broc.*, Placentine, Bologna, Vienna, Bordeaux, Dax; T. Boscianus, *Al. Brong.*, Castelgomberto, Dax; T. Labarum, *Bast.*, Dax; T. turgidulus? *Broc.*, Italy, Mérignac; T. Bucklandi, *Bast.*, above the fresh-water limestone, Saucats; T. Audebardi, *Bast.*, Léognan.

ROTELLA Defrancii, *Bast.*, Léognan.

SOLARIUM carocollatum, *Lam.*, Léognan, Dax.

AMPULLARIA compressa, *Bast.*, Dax; A. crassatina, *Lam.*, Pontchartrain, var. at Dax.

MELANIA costellata, *Lam.*, Grignon, Ronca, Sangonini, Dax; M. subulata, Volterra, Léognan, Dax; M. hordeacea, *Lam.*, Houdan, Pierrelaye, Beauchamp, Isle of Wight, var. at Saucats; M. clathrata, *Bast.*, Dax; M. nitida, *Lam.*, Grignon, Placentine, Parnes, Dax; M. distorta (Turbo politus, *Montagu*), analogous to the existing species, Thorigne, Bordeaux.

MELANOPSIS Dufourii, *Fér.*, Dax.

RISSOA Cochlearella (Melania Coohlearella, *Lam.*), Grignon, Mérignac, var. at Dax; R. Cimex (Turbo Cimex, *Broc.*), analogous to the existing species, Bologna, Isle of Ischia, Mérignac, var. at Dax; R. varicosa, *Bast*, Mérignac; R.? Grateloupi, *Bast.*, Mérignac.

PHASIANELLA turbinoides, *Lam.*, Grignon, Mérignac, Dax; P. Prevostina, *Bast.*, Léognan, Saucats.

NATICA Canrena, *Broc.*, analogous to the existing species, Italy, England, Léognan, Saucats, Dax; N. glaucina, *Broc.*, analogous to the existing species, Italy, Léognan, Dax.

NERITA Plutonis, *Bast.*, Mérignac.

NERITINA fluviatilis, *Lam.*, analogous to the existing species, Tuscany, Mérignac, Dax (often preserves its colours).

CONUS deperditus, *Lam.*, analogous to the existing species at Owhyhee, Grignon, Ronca, Turin, Bordeaux, Dax; C. alsiosus, *Al. Brong.*, Ronca, Dax, Bordeaux; C. Mercati, Vienna, San Miniato, Saucats.

CYPRÆA Coccinella, *Lam.*, Grignon, Suffolk, Angers, Nantes, Dax; Cy. annulus, *Broc.*, analogous to the existing species, Piedmont, Ronca, Bordeaux; Cy. annularia, *Al. Brong.*, Turin, Bordeaux; Cy. leporina, *Lam.*, Dax; Cy. lyncoides, *Al. Brong.*, Turin, Bordeaux; Cy. Duclosiana, *Bast.*, Dax.

OLIVA plicaria, *Lam.*, Mérignac, Léognan, Dax, Saucats; O.

CLAVULA, *Lam.*, Mérignac, Dax; O. Dufresnii, *Bast.*, Mérignac, Dax, Saucats.

ANCILLARIA canalifera, *Lam.* (A. turrellata, *Sow.*), Grignon, Barton, Dax, Bordeaux; A. inflata (Anolax inflata, *Al. Brong.*), Turin, Vienna, Léognan, Mérignac, Dax, Saucats.

VOLUTA Lamberti, *Sow.*, analogous to the existing species, Suffolk, Anjou, Léognan; V. rarispina, *Lam.*, Dax, Bordeaux; V. affinis, *Broc.*, Ronca, Turin, Léognan.

MARGINELLA cypræola, Placentine, Touraine, Dax.

MITRA Dufresnii, *Bast.*, rare at Léognan; M. scrobiculata, Placentine, Piedmont, Sienna, var. at Bordeaux; M. incognita, *Bast.*, Mérignac, Dax.

CANCELLARIA acutangula (C. acutangularis, *Lam.*), Léognan, Saucats; C. trochlearis, *Lam.*, Léognan, Saucats; C. doliolaris, *Bast.*, rare, Léognan; C. Geslinii, *Bast.*, Léognan, Saucats; C. buccinula, *Lam.*, Vienna, Crépy in Valois, Bordeaux; C. contorta, *Bast.*, Italy, Saucats; C. cancellata, *Lam.*, anal. exist. species; Piedmont, Placentine, Sienna, Bordeaux.

BUCCINUM Veneris, *Faujas de St. F.*, Léognan, Saucats; B. baccatum, *Bast.*, Saucats, Léognan, Mérignae, var. α Dax, var. β Saucats, var. γ Vienna; B. politum, *Bast.*, Piedmont, Saucats.

EBURNA spirata (Buccinum spirata, *Brug.*), anal. exist. species, Rennes, Dax, Saucats.

NASSA reticulata (Buccinum reticulatum, *Broc.*), anal. exist. species, San Miniato, Castel-Arquato, Sienna, var. α Dax, var. β Saucats and Léognan; N. asperula, *Broc.*, Placentine, Sienna, var. α Dax, var. β Léognan and Saucats; N. angulata, Volterra, Saucats; N. columbelloides, *Bast.*, Vienna, Angers, Touraine, Dax, Léognan, Saucats (approaches a living species); N. Desnoyersi, *Bast.*, Dax, Saucats; N. cancellaroides, *Bast.*, Dax; N. Andrei, *Bast.*, Bordeaux.

PURPURA costata (Nerita costata, *Broc.*), Placentine, Dax, Bordeaux; P. Lassaignei, *Bast.*, Léognan.

CASSIS Saburon (Cassidea Saburon, *Brug.*), analogous to the existing species, Calabria, Placentine, Vienna, Léognan, Saucats, Dax; C. Rondeleti, *Bast.*, Léognan, Dax.

CASSIDARIA Cythara, Italy, Bordeaux.

TEREBRA plicaria, analogous to the existing species, Saucats; T. plicatula, *Lam.*, Grignon, Saucats, Léognan, Dax; T. cinerea (T. aciculina, *Lam.*), analogous to the existing species, Piedmont, Léognan, Saucats; T. striata, analogous to the existing species, Saucats; T. duplicata, *Lam.*, analogous to the existing species? Sienna, Piedmont, Rome; T. pertusa (var.), analogous to the existing species, Saucats; T. murina, Dax.

CERITHIUM margaritaceum, *Al. Brong.*, Sienna, Mayence; C. corrugatum, *Al. Brong.*, Ronca, Saucats; C. inconstans, *Bast.*, Saucats; C. ampullosum, *Al. Brong.*, Castelgomberto, Vienna,

Mérignac, Dax; C. plicatum, *Lam.*, Montpellier, Pontchartrain,
Mayence, Castelgomberto, Saucats; C. cinctum, *Lam.*, Montpellier, Pontchartrain, Beynes, Houdan, Saucats, C. Charpentieri,
Bast., Dax; C. papaveraceum, *Bast.*, Touraine, Mérignan; C.
lemniscatum, *Al. Brong.*, Ronca, Dax; C. Salmo, *Bast.*, Léognan, Mérignac; C. pictum, *Bast.*, Vienna, Mérignac, Soucats;
C. lamellosum, *Lam.*, Courtagnon, Grignon, var. Dax; C. angu-
losum, Grignon, Saucats; C. Diaboli, *Al. Brong.*, the Diablerets,
Switzerland, Dax; C. resectum, *Defr.*, Hauteville, Dax, Mérignac; C. calculosum, *Bast.*, Dax, Léognan; C. pupæforme, *Bast.*,
rare, Mérignac; C. granulosum (Murex granulosus, *Broc.*), analogous to the existing species, Volterra, Mérignac; C. scaber,
analogous to the existing species, Italy, Mérignac, Léognan.

Murex Pomum, *Linn.*, analogous to the existing species, Placentine, Saucats, Mérignac; M. sublavatus, *Bast.*, rare, Mérignac, Léognan, Saucats; M. Lingua-Bovis, *Bast.*, Saucats, Léognan; M. suberinaceus, *Bast.*, Bordeaux.

Typhis tubifer (Murex tubifer, *Lam.*), analogous to the existing
species, Grignon, Barton, Highgate, Léognan.

Triton doliare (Murex doliaris, *Broc.*), Placentine, Pisa, Sienna, Léognan.

Ranella marginata, *Al. Brong.*, Piedmont, Pisa, Placentine,
Volterra, Turin, Léognan, Mérignac; R. leucostoma, *Lam.*,
analogous to the existing species, Placentine, Bordeaux.

Fusus lavatus, *Sow.*, Barton, Paris, Léognan, Saucats, Dax;
F. buccinoides, *Bast.* (Buccinum subulatum, *Broc.*), Placentine,
Saucats, Mérignac (a Mediterranean shell approaches this spe-
cies); F. rugosus, *Lam.*, Grignon, Valognes, Dax (different from
the Fusus rugosus of Sowerby); F. clavatus, Placentine, var.
Bordeaux.

Pleurotoma tuberculosa, *Bast.*, Vienna, Saucats, Léognan;
P. Pannus, *Bast.*, Léognan, Saucats, Dax; P. denticulata, *Bast.*,
Touraine, Saucats, Léognan, Mérignac, Dax; P. ramosa, *Bast.*,
Thorigné, Angers, Léognan, Saucats; P. Borsoni, *Bast.*, Sau-
cats, Léognan, Mérignac; P. plicata, *Lam.*, Grignon, Dax; P.
undata, *Lam.*, Grignon, Epernay, Dax; P. multinoda, *Lam.*,
Bordeaux; P. Turrella, *Lam.*, Grignon, var. Dax; P. crenulata,
Lam., Grignon, var. Léognan; P. cataphracta, Placentine,
Sienna, Bologna, Bordeaux; P. purpurea, *Bast.*, analogous to
the existing species, Léognan; P. terebra, *Bast.*, Léognan,
Saucats, Dax; P. costellata, *Lam.*, Grignon, Léognan, Dax; P.
cheilotoma, *Bast.*, Bordeaux.

Fasciolaria Burdigalensis, *Defr.*, Léognan, Saucats, Mérignac; F. uniplicata (Fusus uniplicatus, *Lam.*), Grignon, Epernay,
Dax.

Pyrula condita, *Al. Brong.*, Turin, Léognan, Saucats; P.
Clava, *Bast.*, Dax, Bordeaux; P. Lainei, *Bast.*, Saucats, Léognan,

Mérignac, Dax ; P. Melongena, analogous to the existing species,
Courtagnon, Saucats, Dax, Mérignac ; P. rusticula, *Bast.*, Dax,
Bordeaux.

TURBINELLA Lynchi, *Bast.*, rare, Léognan.

STROMBUS decussatus, Dax ; S. Bonelli, *Al. Brong.*, Turin, Dax.

ROSTELLARIA Pes-Pelicani, analogous to the existing species,
common in many supercretaceous deposits, Léognan, Dax ; R.
curvirostris, *Lam.*, analogous to the existing species, Dax.

SIGARETUS canaliculatus, *Sow.*, Hordwell, Paris, Bordeaux, Dax.

CAPULUS (Pileopsis) sulcosus (Nerita sulcosa, *Broc.*), Placentine, Mérignac.

CREPIDULA unguiformis (C. Italica, *Defr.*), analogous to the
existing species, Placentine, Sienna, Vienna, Saucats ; C. cochlearia, *Bast.*, analogous to the existing species, Mérignac.

FISSURELLA costaria, *Desh.*, Grignon ? Dax.

CALYPTRÆA deformis, *Lam.*, Dax, Bordeaux ; C. depressa,
Lam., abundant at Bordeaux ; C. muricata (Patella muricata,
Broc.), analogous to the existing species, Piedmont, Placentine,
Castel-Arquato, Léognan, Saucats ; C. ornata, *Bast.*, Dax.

HIPPONYX granulatus, *Bast.*, Dax.

OSTREA flabellula, *Lam.*, Grignon, Hordwell, Barton, Brussels,
Saucats, Léognan ; O. undata, *Lam.*, Dax, Bordeaux ; O. Cymbula, *Lam.*, Grignon, Barton, Saucats.

PECTEN scabrellus, *Lam.* (Ostrea dubia, *Broc.*), Val Andone,
Piedmont, Saucats ; P. Burdigalensis, *Lam.*, Saucats, approaches
P. Pleuronectes, P. obliteratus, and P. Laurenti ; P. multiradia-
tus, *Lam.*, Italy, Saucats.

SPONDYLUS, fragments.

PERNA Ephippium, *Lam.*, analogous to the existing species,
Bordeaux.

AVICULA phalænacea, *Lam.*, Léognan. PINNA, fragments.

ARCA biangula, *Lam.*, Grignon, Léognan ; A. scapulina, *Lam.*,
Grignon, Mérignac ; A. clathrata, *Defr.*, analogous to the exist-
ing pecies, Angers, Thorigné, Nice, Mérignac ; A. Diluvii, *Lam.*
(A. pectinata, *Broc.*), Houdan, Touraine, Placentine, Sienna,
Turin, Vienna, Bordeaux ; A. Breislaki, *Bast.*, Dax.

PECTUNCULUS Cor, *Lam.*, Saucats, Mérignac, Léognan ; P.
pulvinatus, *Lam.*, Paris, Touraine, var. *α* Dax and Bordeaux,
var. *β* Léognan ; M. de Basterot considers this shell the same
with that found at Walton.

NUCULA emarginata, *Lam.*, Léognan, Saucats ; N. margaritacea, *Lam.* (Arca Nucleus, *Broc.*), analogous to the existing species, Grignon, Placentine Barton, Highgate, Léognan, Dax.

MYTILUS antiquorum, *Sow.*, Suffolk, var. Saucats, Mérignac ;
M. Brardii, *Al. Brong.*, Mayence, Dax, Mérignac ; M. edulis,
Linn., analogous to the existing species, Piedmont, Placentine,
Sienna, Volterra, Saucats.

MODIOLA cordata, *Lam.*, Paris, Domfront, Saucats.
CARDITA hippopea, *Bast.*, Saucats.
VENERICARDIA Pinnula, *Bast.*, beds above the fresh-water limestone, Saucats, Dax; V. Jouanneti, *Bast.*, Italy, Vienna, Bordeaux; V. intermedia (Cardita intermedia, *Lam.*), Placentine, Sienna, Dax.
ERYCINA elliptica, *Lam.*, Ecouen, Senlis, Saucats.
CHAMA gryphoides, *Broc.*, analogous to the existing species. Piedmont, Placentine, Sienna, Dax, Léognan, Saucats, Merignac.
CARDIUM edule, *Linn.*, analogous to the existing species, Placentine, Piedmont, Sienna, Bramerton, Ipswich, Dax; C. Burgalinum, *Lam.*, Dax, Bordeaux; C. serrigerum, *Lam.*, Grignon, Bordeaux; C. echinatum, *Brug.*, analogous to the existing species, Placentine, Touraine, Bordeaux, var. Vienna; C. Pallassianum, *Bast.*, Dax; C. multicostatum, *Broc.*, Placentine, var. Léognan; C. discrepans, *Bast.*, bed above the fresh-water limestone, Saucats, Dax.
DONAX anatinum, *Lam.*, analogous to the existing species; var. Dax, Bordeaux; D. elongata, *Lam.*, analogous to the existing species, Mérignac; D. triangularis, *Bast.*, approaches an existing species, Saucats; D. irregularis, *Bast.*, Dax; D.? difficilis *Bast.*, Dax.
CYRENA Brongniartii, *Bast.*, Ronca, Mérignac, Saucats; C. Sowerbii, *Bast.*, Paris, Saucats.
TELLINA zonaria, *Lam.*, Dax, Saucats, Léognan, Mérignac (preserves its colours); T. elegans, *Desh.*, Hauteville, Grignon, above the fresh-water beds, Saucats; T. bipartita, *Bast.*, Saucats; T. biangularis, *Desh.*, Paris, var. Dax.
LUCINA Columbella, *Lam.*, Touraine, Léognan, Saucats, Dax, Mérignac; L. divaricata, *Lam.*, analogous to the existing species, Grignon, Léognan, Mérignac, common at Hordwell and Saucats; L. scopulorum, *Al. Brong.*, Ronca, Turin, Mérignac, Saucats; L. dentata, *Bast.*, Dax, Saucats; L. digitalis, *Lam.*, analogous to the existing species, rare, Saucats; L. hiatelloides, *Bast.*, rare, Léognan; L. gibbosula, *Lam.*, analogous to the existing species; Grignon, Dax; L. rerulata, *Lam.*, analogous to the existing species, Grignon, Bordeaux; L. neglecta, *Bast.*, Dax, Bordeaux.
VENUS Dysera, *Linn.*, analogous to the existing species, Piedmont, Placentine, Dax, Saucats; V. casinoides, *Lam.*, Vienna, Léognan, Saucats; V. vetula, *Bast.*, Saucats, Léognan; V. radiata, *Broc.*, analogous to the existing species, Italy, Saucats, Léognan, Dax.
CYTHERÆA erycinoides, *Lam.*, analogous to the existing species, Paris, Turin, Rome, Saucats, Léognan, Dax; C. Deshayesiana, *Bast.*, Saucats, Léognan; C. tincta, *Lam.*, perfect resemblance to existing species, Saucats, C. leonina, *Bast.*, Léognan, Sau-

cats; C. undata, *Bast.*, Saucats, abundant at Mérignac; C. niti-
dula, *Lam.* (Venus transversa, *Sow.*), Grignon, Barton, Saucats.
Cyprina Islandicoides, *Lam.* (Venus æqualis, *Sow.*), Suffolk,
Placentine, Antwerp, Dax, Bordeaux.
Venerupis Faujasii, *Bast.*, Bordeaux.
Petricola peregrina, *Bast.*, in large madrepores, Mérignac.
Saxicava anatina, *Bast.*, in the holes which it has bored in the
fresh-water limestone, when the latter was covered by the waters
of an ancient sea, Saucats.
Clotho? unguiformis, *Bast.*, in holes which it has pierced in
the marine and fresh-water limestones, Saucats.
Corbula revoluta, *Sow.*, Barton, Italy, Dax, Léognan, Méri-
gnac, Saucats; C. striata, *Lam.*, Grignon, var. Angers, var.
Bordeaux.
Mactra striatella, *Lam.*, analogous to the existing species? Sau-
cats; M. deltoides, *Lam.*, Grignon, Saucats; M. triangula, *Broc.*,
analogous to the existing species, Placentine, Saucats.
Lutraria Sanna, *Bast.*, Saucats. Mya ornata, *Bast.*, Dax.
Panopæa Faujasii, *Mesnard de la Groye*, Parma, Sienna, Pisa,
San Miniato (Reggio), Placentine, Piedmont, Léognan.
Psammobia Labordei, *Bast.*, approaches Ps. vespertina, *Leach*,
a living species, Saucats.
Solen strigillatus, *Linn.*, analogous to the existing species,
Placentine, Piedmont, Vienna, Grignon, Léognan, Dax; S. Va-
gina, *Linn.*, analogous to the existing species, Placentine, Gri-
gnon, Saucats; S. Legumen, *Linn.*, Saucats.
Pholas Branderi, *Bast.*, in the rolled stones and corals, Tou-
raine and Mérignac.
Clavagella coronata, *Desh.*, Meaux, Pauliac, nine leagues
from Bordeaux.

D. *Cretaceous Rocks at Stevensklint, Seeland.*

The description of these rocks by Dr. Forchhammer (Brewster's
Edinburgh Journ. of Science, vol. ix. 1828,) is more particularly
interesting when connected with the zoological passage of the cretaceous into the supercrefaceous rocks, noticed at p. 253. It appears
that the base of the cliff at Stevensklint is formed of chalk with
beds of nodular flints. Upon the chalk, which is represented to
have an undulated surface, rests a thin bed (about six inches thick)
of a bituminous clay, containing a *Zoophyte, Sharks' teeth,* a *Pecten,* impressions of a bivalve, and traces of *vegetable remains.* Incumbent on this is a hard yellowish white limestone, containing
the remains of the genera—*Patella,* 1 species; *Cypræa,* 2; *Fusus,*
1; *Cerithium,* 2; *Ampullaria,* 1; *Trochus,* 1; *Dentalium,* 1; *Arca,*
1; *Mytilus,* 1; *Serpula,* 1; *Spatangus,* 1; *Favosites,* 1; and *Turbinolia,* 1; with *Fishes' teeth* and undeterminable univalves, bivalves,
and corals. This limestone contains green grains, seldom exceeds

three feet, and is sometimes only a few inches, in thickness, but it is nowhere entirely wanting. It is covered by another limestone, from thirty to forty feet thick, almost entirely composed of fragments of corals, and forming the upper part of the cliff. This is divided by corneous flint into many beds, the flint beds being bent and curved. It is remarkable that the organic remains of this deposit are such as are considered characteristic of the chalk, consisting of *Ananchytes ovata, Ostrea vesicularis, Belemnites mucronatus,* &c. Dr. Forchhammer observes that the remains of *Ananchytes ovata* are occasionally so abundant, that the limestone almost entirely consists of them.

It would appear from the above, that there is an apparent alternation of the fossils usually considered cretaceous, with those more commonly referred to the supercretaceous period;—a circumstance so far remarkable, that it shows a state of things somewhat different from the more gradual mixture and change supposed to have taken place in the Alps and Pyrenees, and at Maestricht; being one in which the conditions were alternately favourable to the presence of animals supposed characteristic of two different classes of rocks.

E. *On Geological Maps and Sections.*

It is of the very first importance that the geologist should, before he proceeds to the examination of a country, be provided with the best physical map that can be procured, so that his observations may be recorded on that which will not deceive him. It must be admitted that such maps are sufficiently rare; but this defect may now be considered as being gradually removed, for in many of those recently published in this country and on the continent, much attention has been generally paid to the real physical features of the country, more particularly to the exact delineation of the mountain masses. Formerly, geographers contented themselves with running a range of high land between all the water sheds, with little regard to relative heights; so that a real depression between two ranges of mountains was not unfrequently converted into a high connecting ridge, merely because the streams of water flowed different ways in consequence of a very trifling degree of elevation.

Respecting the maps of our own country, too much praise cannot be given to the late sheets of those published by the Ordnance, remarkable not only for their general fidelity, but also for the shading of the hills; in this last respect very superior to the earlier sheets. With these maps in his hands the geologist feels that his time is not thrown away, and by noticing various minute circumstances upon them, he is subsequently enabled to soar, as it were, above the country he has examined, and by combining his various observations, he may arrive at general conclusions, with which he might not otherwise feel satisfied, and to which he might never have been led without an exact document of this nature *.

The value of exact geological maps is daily becoming more apparent, and it is by no means difficult to foresee, that many geological problems will eventually be solved by their accumulation. Already has a great change been effected in this department, but much more remains to be accomplished, and it is exceedingly desirable that even general lines in sketches should not be hastily run. The best general geological maps, which have been, or will soon be, published, are Greenough's Map of England and Wales, second edition; Elie de Beaumont and Dufrénoy's France; Hoffman's North Western Germany; and Oeynhausen, La Roche, and Von Decken's Rhine. Smaller maps of greater or less interest are sufficiently common, and will be found in various scientific works, more particularly in the Transactions of the Geological Society of London.

As the direction of faults and mineral veins, and the dip of strata, are daily becoming of greater importance, the student should be careful always to represent them on his map when possible. The former are certainly often very difficult to trace, yet much may be accomplished by diligent research, and the singular lines frequently obtained will amply repay the patient investigation of the observer.

With respect to geological sections, too much stress cannot be laid on the importance of rendering them as conformable to nature as circumstances will admit; that is, the perpendicular elevations and base lines should be as much as possible in proportion to each other. Without this necessary precaution such sections are little better than caricatures of nature, and are frequently much more mischievous than useful, even leading those who make them, to false conclusions, from the distortion and false proportions of the various parts; gentle sloping valleys being converted into deep ravines, moderate mountains into enormous elevations, and the possibility of conjecturing the kind of surface upon which any particular deposit was thrown down, and the relative importance of the deposit itself, being entirely destroyed. It will at once be admitted that the proportional thickness of a deposit is sometimes so trifling when compared to its length, that it could not be conveniently represented on paper; but as the relative importance of such a deposit is precisely one of the circumstances that should be

---

* The student must carefully guard himself from considering all well engraved maps exact; for very frequently a hard unpromising engraving far exceeds in value one finely got up, which may falsely represent minute undulations of land, and all its varied characters, with the greatest apparent precision; while such physical features only exist in the imagination of the artist.

exhibited in a section, it will be obvious that though it may be necessary to quit exact proportion, the section should be kept as nearly to it as possible. The cases, however, in which exact, or nearly exact, proportion can be kept are sufficiently numerous; and, unless it be desirable to convey false impressions, it is clearly in the interest of science that sections should be what they pretend to be, miniature representations of nature *.

F. *Tables for calculating Heights by the Barometer.*

The following Tables are those of M. Oltmanns, which are generally admitted as among the most convenient hitherto published. Being calculated for the metrical barometer, they were useless to persons employing that graduated according to English inches and their decimal parts. To render them applicable to our barometers, a table (A) has been prefixed, in which the equivalent of every millimetre of the metrical barometer is given in English inches and the thousandth parts of inches.

To reduce the metres used in these tables into English feet, a table (F) is appended, where the number of English feet corresponding to any number of metres up to 10,000 will be immediately obtained.

Abstraction being made of table A prefixed, and table F appended, the march of operations is as follows:

Let h be the height of the barometer at the lower station expressed in millimetres; h' that of the higher station; T and T' the temperature of the barometer at the different stations according to the centigrade thermometer; t and t' that of the air.

We search in table B for the number which corresponds to h; let us call it a: we likewise search in the same table for that which corresponds to h'; let this be named b: let us call c, the generally very small number which, in table C, faces T—T'; the approximate height will be $a—b—c$. (If T—T' is negative, it should be written $a—b+c$.) In order to apply the correction necessary for the strata of air, it will suffice to multiply the thousandth part of the approximate height by the double sum $2(t+t')$ of the detached thermometers; the correction will be either positive or negative, according as $t+t'$ is itself either positive or negative.

The second and last correction, that for the latitude and the diminution of weight, is obtained by taking, in table D, the number which corresponds vertically to the latitude, and horizontally to the approximate height: this correction, which can never exceed 28 metres, is always added.

In those very rare cases where the lower station is itself consi-

---

* For the differences between correct and caricature sections, see Sections and Views illustrative of Geological Phænomena, pl. 2.

derably elevated above the sea, it will be necessary to apply a small correction to be found in table E.

In order to understand the calculation of a height by means of these tables, and those prefixed and appended, let us suppose that in latitude $= 44°$ we had, at the level of the sea, the barometer $= 30·040$ English inches, temperature of the instrument $= 22°·5$ centigrade, and of the air $= 22°$. At the top of a mountain, the barometer $= 26·575$ English inches, temperature of the instrument $= 17°·5$, and of the air $= 17°$.

In order to obtain the equivalents of the English inches in millimetres, search in table A ; where the number of millimetres corresponding to 30·040 inches observed at the sea will be 763, and that of 26·575 observed on the mountain will be 675. Having obtained these equivalents, the calculation proceeds :

	Mill.	Metres.	
Barometer at sea level	= 763	= 6182·0	} Table B.
Barometer on the mountain	= 675	= 5206·1	
		975·9	
Diff. of attached thermometers . . . . = 5° =		7·4	Table C.
Apparent height		968·5	
Double the sum of the detached thermometers } multiplied by the thousandth part of 968·5 .		75·5	
		1044·	
Correction for latitude		3·1	Table D.
Height of the mountain		1047·1	
Height in English feet		3435	Table F.

When the height of the barometer, graduated according to English inches and their parts, does not precisely correspond with a certain number of millimetres, and when great accuracy is required, it will be obvious, that instead of taking the next nearest number to it in the tables, as might otherwise be done, it will be necessary to calculate the difference.

TABLE A.

Inches.	Mil.	Inches.	Mil.	Inches.	Mil.	Inches.	Mil.	Inches.	Mil.	Inches.	Mil.
14·56	370	16·457	418	18·347	466	20·237	514	22·126	562	24·016	610
·606	371	·496	419	·386	467	·276	515	·166	563	·055	611
·646	372	·536	420	·425	468	·315	516	·205	564	·095	612
·685	373	·575	421	·465	469	·355	517	·244	565	·134	613
·725	374	·614	422	·504	470	·394	518	·284	566	·174	614
·764	375	·654	423	·543	471	·433	519	·323	567	·213	615
·803	376	·693	424	·583	472	·473	520	·363	568	·252	616
·843	377	·733	425	·622	473	·512	521	·402	569	·292	617
·882	378	·772	426	·662	474	·551	522	·441	570	·331	618
·921	379	·811	427	·701	475	·590	523	·481	571	·370	619
·961	380	·851	428	·740	476	·630	524	·520	572	·410	620
15·000	381	·890	429	·780	477	·670	525	·559	573	·449	621
·040	382	·929	430	·819	478	·709	526	·599	574	·489	622
·079	383	·969	431	·859	479	·748	527	·638	575	·528	623
·118	384	17·008	432	·898	480	·788	528	·678	576	·567	624
·156	385	·047	433	·937	481	·827	529	·717	577	·607	625
·197	386	·087	434	·977	482	·866	530	·756	578	·646	626
·236	387	·126	435	19·016	483	·901	531	·796	579	·685	627
·276	388	·166	436	·055	484	·945	532	·835	580	·725	628
·315	389	·205	437	·095	485	·984	533	·874	581	·764	629
·355	390	·244	438	·134	486	21·025	534	·914	582	·804	630
·394	391	·284	439	·174	487	·063	535	·953	583	·843	631
·433	392	·323	440	·213	488	·102	536	·992	584	·882	632
·473	393	·362	441	·252	489	·142	537	23·032	585	·922	633
·512	394	·402	442	·292	490	·181	538	·071	586	·961	634
·551	395	·441	443	·331	491	·220	539	·111	587	25·000	635
·591	396	·481	444	·370	492	·260	540	·150	588	·040	636
·630	397	·520	445	·410	493	·300	541	·189	589	·079	637
·670	398	·559	446	·449	494	·339	542	·229	590	·119	638
·709	399	·599	447	·488	495	·378	543	·268	591	·158	639
·748	400	·638	448	·528	496	·418	544	·307	592	·197	640
·788	401	·677	449	·567	497	·457	545	·347	593	·237	641
·827	402	·717	450	·607	498	·496	546	·386	594	·276	642
·866	403	·756	451	·646	499	·536	547	·426	595	·315	643
·906	404	·795	452	·685	500	·575	548	·465	596	·355	644
·945	405	·835	453	·725	501	·615	549	·504	597	·394	645
·985	406	·874	454	·764	502	·654	550	·544	598	·433	646
16·024	407	·914	455	·803	503	·693	551	·583	599	·473	647
·063	408	·953	456	·843	504	·733	552	·622	600	·512	648
·102	409	·992	457	·882	505	·772	553	·662	601	·551	649
·142	410	18·032	458	·922	506	·812	554	·701	602	·590	650
·181	411	·071	459	·961	507	·851	555	·741	603	·630	651
·221	412	·110	460	20·000	508	·890	556	·780	604	·670	652
·260	413	·150	461	·040	509	·929	557	·819	605	·709	653
·299	414	·189	462	·079	510	·969	558	·859	606	·749	654
·339	415	·229	463	·118	511	22·008	559	·898	607	·788	655
·378	416	·268	464	·158	512	·048	560	·937	608	·827	656
·417	417	·307	465	·197	513	·087	561	·977	609	·866	657

Table A. (*continued.*)

Inches.	Mil.	Inches.	Mil.	Inches.	Mil.	Inches.	Mil.	Inches.	Mil.	Inches.	Mil.
25·908	658	26·772	680	27·639	702	28·504	724	29·371	746	30·237	768
·945	659	·811	681	·678	703	·544	725	·410	747	·276	769
·985	660	·851	682	·718	704	·583	726	·449	748	·316	770
26·024	661	·891	683	·757	705	·622	727	·489	749	·355	771
·063	662	·931	684	·796	706	·662	728	·528	750	·394	772
·103	663	·970	685	·836	707	·701	729	·567	751	·433	773
·142	664	27·010	686	·875	708	·741	730	·607	752	·473	774
·182	665	·049	687	·915	709	·780	731	·646	753	·512	775
·221	666	·089	688	·954	710	·819	732	·686	754	·552	776
·260	667	·128	689	·993	711	·859	733	·725	755	·591	777
·300	668	·167	690	28·032	712	·898	734	·764	756	·630	778
·339	669	·206	691	·071	713	·937	735	·804	757	·670	779
·378	670	·246	692	·111	714	·977	736	·843	758	·709	780
·418	671	·285	693	·150	715	29·016	737	·882	759	·749	781
·457	672	·324	694	·189	716	·056	738	·922	760	·788	782
·496	673	·363	695	·229	717	·095	739	·961	761	·827	783
·536	674	·403	696	·268	718	·134	740	30·000	762	·867	784
·575	675	·442	697	·308	719	·174	741	·040	763	·906	785
·615	676	·482	698	·347	720	·213	742	·079	764	·945	786
·654	677	·521	699	·386	721	·252	743	·119	765	·985	787
·693	678	·560	700	·426	722	·291	744	·158	766	31·024	788
·733	679	·600	701	·465	723	·331	745	·197	767	·064	789

Table B.

Mil.	Metr.	Mil.	Metr.	Mil.	Metr.	Mil.	Metr.	Mil.	Metr.	Mil.	Metr.
370	418·5	392	878·5	414	1313·3	436	1725·6	458	2117·6	480	2491·3
371	440·0	393	898·8	415	1332·5	437	1743·8	459	2135·0	481	2507·9
372	461·5	394	919·0	416	1351·7	438	1762·1	460	2152·3	482	2524·3
373	482·9	395	939·2	417	1370·8	439	1780·3	461	2169·6	483	2540·8
374	504·2	396	959·3	418	1389·9	440	1798·4	462	2186·9	484	2557·3
375	525·4	397	979·4	419	1408·9	441	1816·5	463	2204·1	485	2573·7
376	546·6	398	999·5	420	1427·9	442	1834·5	464	2221·3	486	2590·2
377	567·8	399	1019·5	421	1446·8	443	1852·5	465	2238·4	487	2606·6
378	588·9	400	1039·4	422	1465·7	444	1870·4	466	2255·5	488	2622·9
379	609·9	401	1059·3	423	1484·6	445	1888·3	467	2272·6	489	2639·2
380	630·9	402	1079·1	424	1503·4	446	1906·2	468	2280·6	490	2655·4
381	651·8	403	1098·9	425	1522·2	447	1924·0	469	2306·6	491	2671·6
382	672·7	404	1118·6	426	1540·8	448	1941·8	470	2323·6	492	2687·9
383	693·5	405	1138·3	427	1559·5	449	1959·6	471	2340·5	493	2704·1
384	714·3	406	1157·9	428	1578·2	450	1977·3	472	2357·4	494	2720·2
385	735·0	407	1177·5	429	1596·8	451	1994·9	473	2374·2	495	2736·3
386	755·6	408	1197·1	430	1615·3	452	2012·6	474	2391·1	496	2752·3
387	776·2	409	1216·6	431	1633·8	453	2030·2	475	2407·9	497	2768·3
388	796·8	410	1236·0	432	1652·2	454	2047·8	476	2424·6	498	2784·4
389	817·3	411	1255·4	433	1670·6	455	2065·3	477	2441·3	499	2800·4
390	837·8	412	1274·8	434	1689·0	456	2082·8	478	2458·0	500	2816·3
391	858·2	413	1294·1	435	1707·3	457	2100·2	479	2474·6	501	2832·2

TABLE B. *(continued.)*

Mil.	Metr.	Mil.	Metr.	Mil.	Metr.	Mil.	Metr.	Mil.	Metr.	Mil.	Metr.
502	2848·1	551	3589·8	599	4254·9	647	4868·7	695	5438·7	743	5970·4
503	2864·0	552	3604·2	600	4268·2	648	4881·0	696	5450·1	744	5981·2
504	2879·8	553	3618·6	601	4281·4	649	4893·3	697	5461·5	745	5991·9
505	2895·6	554	3633·0	602	4294·7	650	4905·6	698	5472·9	746	6002·5
506	2911·3	555	3647·4	603	4307·9	651	4917·8	699	5484·3	747	6013·2
507	2927·0	556	3661·7	604	4321·1	652	4930·0	700	5495·7	748	6023·8
508	2942·7	557	3676·0	605	4334·3	653	4942·2	701	5507·1	749	6034·4
509	2958·4	558	3690·3	606	4347·4	654	4954·4	702	5518·4	750	6045·1
510	2974·0	559	3704·6	607	4360·5	655	4966·6	703	5529·8	751	6055·7
511	2989·6	560	3718·8	608	4373·7	656	4978·7	704	5541·1	752	6066·3
512	3005·2	561	3733·0	609	4386·7	657	4990·9	705	5552·4	753	6076·9
513	3020·7	562	3747·2	610	4399·8	658	5003·0	706	5563·7	754	6087·5
514	3036·2	563	3761·3	611	4412·8	659	5015·1	707	5575·0	755	6098·0
515	3051·7	564	3775·4	612	4425·9	660	5027·2	708	5586·2	756	6108·6
516	3067·2	565	3789·5	613	4438·9	661	5039·2	709	5597·5	757	6119·1
517	3082·6	566	3803·6	614	4451·9	662	5051·2	710	5608·7	758	6129·6
518	3097·9	567	3817·7	615	4464·8	663	5063·3	711	5619·9	759	6140·1
519	3113·3	568	3831·7	616	4477·7	664	5075·3	712	5631·1	760	6150·6
520	3128·6	569	3845·7	617	4490·7	665	5087·2	713	5642·2	761	6161·1
521	3143·9	570	3859·7	618	4503·6	666	5099·2	714	5653·4	762	6171·5
522	3159·2	571	3873·7	619	4516·4	667	5111·2	715	5664·6	763	6182·0
523	3174·4	572	3887·6	620	4529·3	668	5123·1	716	5675·7	764	6192·4
524	3189·7	573	3901·5	621	4542·1	669	5135·0	717	5686·8	765	6202·8
525	3204·9	574	3915·4	622	4554·9	670	5146·9	718	5697·9	766	6213·2
526	3220·0	575	3929·3	623	4567·7	671	5158·8	719	5709·0	767	6223·6
527	3235·1	576	3943·1	624	4580·5	672	5170·6	720	5720·1	768	6234·0
528	3250·2	577	3956·9	625	4593·2	673	5182·5	721	5731·1	769	6244·4
529	3265·3	578	3970·7	626	4606·0	674	5194·3	722	5742·1	770	6254·7
530	3280·3	579	3984·5	627	4618·7	675	5206·1	723	5753·1	771	6265·0
531	3295·3	580	3998·2	628	4631·4	676	5217·9	724	5764·2	772	6275·4
532	3310·3	581	4011·9	629	4644·0	677	5229·7	725	5775·1	773	6285·7
533	3325·3	582	4025·6	630	4656·7	678	5241·4	726	5786·1	774	6296·0
534	3340·2	583	4039·3	631	4669·3	679	5253·2	727	5797·1	775	6306·2
535	3355·1	584	4052·9	632	4682·0	680	5264·9	728	5808·0	776	6316·5
536	3370·0	585	4066·6	633	4694·5	681	5276·6	729	5819·0	777	6326·7
537	3384·8	586	4080·2	634	4707·1	682	5288·3	730	5829·9	778	6337·0
538	3399·6	587	4093·8	635	4719·7	683	5300·0	731	5840·8	779	6347·2
539	3414·4	588	4107·3	636	4732·2	684	5311·6	732	5851·7	780	6357·4
540	3429·2	589	4120·8	637	4744·7	685	5323·2	733	5862·5	781	6367·6
541	3443·9	590	4134·3	638	4757·2	686	5334·8	734	5873·4	782	6377·8
542	3458·6	591	4147·8	639	4769·7	687	5346·4	735	5884·2	783	6388·0
543	3473·3	592	4161·3	640	4782·1	688	5358·0	736	5895·1	784	6398·2
544	3487·9	593	4174·7	641	4794·6	689	5369·6	737	5905·9	785	6408·3
545	3502·5	594	4188·1	642	4807·0	690	5381·1	738	5916·7	786	6418·5
546	3517·2	595	4201·5	643	4819·4	691	5392·7	739	5927·5	787	6428·6
547	3531·8	596	4214·9	644	4831·7	692	5404·2	740	5938·2	788	6438·7
548	3546·3	597	4228·2	645	4844·1	693	5415·7	741	5949·0	789	6448·8
549	3560·8	598	4241·6	646	4856·4	694	5427·2	742	5959·7	790	6458·9
550	3575·3										

TABLE C.

Deg.	Metre.	Deg.	Metre.	Deg.	Metre.	Deg.	Metre.	Deg.	Metre.
0·2	0·3	4·2	6·2	8·2	12·1	12·2	17·9	16·2	23·8
0·4	0·6	4·4	6·5	8·4	12·4	12·4	18·2	16·4	24·1
0·6	0·9	4·6	6·8	8·6	12·6	12·6	18·5	16·6	24·4
0·8	1·2	4·8	7·1	8·8	12·9	12·8	18·8	16·8	24·7
1·0	1·5	5·0	7·4	9·0	13·2	13·0	19·1	17·0	25·0
1·2	1·8	5·2	7·6	9·2	13·5	13·2	19·4	17·2	25·3
1·4	2·1	5·4	7·9	9·4	13·8	13·4	19·7	17·4	25·6
1·6	2·3	5·6	8·2	9·6	14·1	13·6	20·0	17·6	25·9
1·8	2·6	5·8	8·5	9·8	14·4	13·8	20·3	17·8	26·2
2·0	2·9	6·0	8·8	10·0	14·7	14·0	20·6	18·0	26·5
2·2	3·2	6·2	9·1	10·2	15·0	14·2	20·9	18·2	26·8
2·4	3·5	6·4	9·4	10·4	15·3	14·4	21·2	18·4	27·1
2·6	3·8	6·6	9·7	10·6	15·6	14·6	21·5	18·6	27·4
2·8	4·1	6·8	10·0	10·8	15·9	14·8	21·8	18·8	27·7
3·0	4·4	7·0	10·3	11·0	16·2	15·0	22·1	19·0	28·0
3·2	4·7	7·2	10·6	11·2	16·5	15·2	22·4	19·2	28·2
3·4	5·0	7·4	10·9	11·4	16·8	15·4	22·7	19·4	28·5
3·6	5·3	7·6	11·2	11·6	17·1	15·6	22·9	19·6	28·8
3·8	5·6	7·8	11·5	11·8	17·4	15·8	23·2	19·8	29·1
4·0	5·9	8·0	11·8	12·0	17·6	16·0	23·5		

TABLE D.

Appr. Ht.	0°	5°	10°	15°	20°	25°	Appr. Ht.	0°	5°	10°	15°	20°	25°
	m.	m.	m.	m.	m.	m.		m.	m.	m.	m.	m.	m.
200	1·2	1·2	1·2	1·0	1·0	1·0	3200	19·1	18·9	18·7	18·0	17·0	15·7
400	2·4	2·4	2·4	2·2	2·0	2·0	3400	20·5	20·3	20·1	19·3	18·4	16·9
600	3·4	3·4	3·4	3·2	3·0	2·8	3600	21·8	21·7	21·4	20·4	19·6	18·0
800	4·5	4·5	4·5	4·3	4·1	3·8	3800	23·1	22·9	22·6	21·6	20·6	19·1
1000	5·7	5·7	5·7	5·3	5·1	4·8	4000	24·6	24·4	24·0	22·9	21·9	20·3
1200	7·0	7·0	6·8	6·4	6·0	5·8	4200	25·9	25·7	25·3	24·3	23·0	21·6
1400	8·2	8·2	8·0	7·6	7·1	6·7	4400	27·5	27·3	26·8	25·8	24·3	23·0
1600	9·2	9·2	9·0	8·8	8·2	7·6	4600	28·9	28·7	28·2	27·1	25·6	24·3
1800	10·4	10·4	10·2	9·8	9·4	8·6	4800	30·4	30·2	29·6	28·4	27·0	25·5
2000	11·6	11·5	11·3	11·0	10·4	9·6	5000	31·8	31·6	30·9	29·8	28·4	26·7
2200	12·8	12·6	12·6	12·1	11·4	10·6	5200	33·0	32·8	32·1	31·0	29·7	28·0
2400	14·0	14·0	13·8	13·3	12·5	11·6	5400	34·3	34·1	33·5	32·4	30·8	29·2
2600	15·2	15·2	15·0	14·4	13·6	12·6	5600	35·7	35·5	34·8	33·7	32·1	30·2
2800	16·6	16·5	16·4	15·6	14·8	13·6	5800	37·1	36·9	36·1	35·0	33·2	31·3
3000	17·9	17·7	17·6	16·8	15·8	14·6	6000	38·5	38·3	37·5	36·3	34·3	32·3

TABLE D. (*continued.*)

Appr. Ht.	30°	35°	40°	45°	50°	55°	Appr. Ht.	30°	35°	40°	45°	50°	55°
	m.	m.	m.	m.	m.	m.		m.	m.	m.	m.	m.	m.
200	0·8	0·8	0·6	0·6	0·6	0·4	3200	14·6	13·1	11·5	10·1	8·6	7·0
400	1·8	1·7	1·4	1·2	1·0	0·8	3400	15·7	14·1	12·4	10·9	9·2	7·7
600	2·6	2·4	2·0	1·8	1·6	1·2	3600	16·7	15·0	13·4	11·6	9·8	8·2
800	3·5	3·1	2·8	2·4	2·0	1·7	3800	17·7	15·9	14·3	12·4	10·5	8·7
1000	4·3	3·8	3·4	3·1	2·6	2·2	4000	18·7	17·0	15·1	13·1	11·2	9·4
1200	5·1	4·6	4·2	3·6	3·1	2·6	4200	19·9	18·0	15·9	14·0	12·0	10·1
1400	6·1	5·4	4·8	4·2	3·6	3·0	4400	21·1	19·1	16·9	15·0	12·9	10·8
1600	7·0	6·2	5·6	4·8	4·1	3·4	4600	22·3	20·3	18·0	15·9	13·6	11·5
1800	8·0	7·0	6·3	5·4	4·6	3·8	4800	23·4	21·3	19·0	16·7	14·3	12·1
2000	8·8	7·8	7·0	6·0	5·1	4·2	5000	24·6	22·3	19·9	17·4	15·0	12·7
2200	9·7	8·6	7·6	6·6	5·6	4·6	5200	25·7	23·3	20·8	18·2	15·7	13·3
2400	10·6	9·4	8·4	7·2	6·1	5·1	5400	26·7	24·3	21·7	19·1	16·4	13·9
2600	11·6	10·5	9·2	8·0	6·8	5·6	5600	27·8	25·3	22·6	19·9	17·2	14·5
2800	12·6	11·4	10·0	8·8	7·4	6·2	5800	28·9	26·3	23·6	20·7	17·8	15·1
3000	13·6	12·2	10·8	9·4	8·0	6·6	6000	30·0	27·3	24·6	21·5	18·5	15·7

TABLE E.

h	Metres.	h	Metres.
400	1·71	600	0.63
450	1·39	650	0.42
500	1·11	700	0.22
550	0·86	750	0.03

Let, for example, the height of the barometer at the lower station be = 600 millimetres; the difference of level = 1500 metres, we have 1000 : 0·63 = 1500 : 0·95, and the difference of the level corrected = 1500·9 metres. This correction is always added.

TABLE F.

Reduction of Metres into English Feet and Inches.

Metr.	Feet.	Inches.	Metr.	Feet.	Inches.	Metr.	Feet.	Inches.
1	3	3·370	50	164	0·514	900	2952	9·261
2	6	6·740	60	196	10·217	1000	3280	10·290
3	9	10·111	70	229	7·920	2000	6561	8·58
4	13	1·481	80	262	5·623	3000	9842	6·87
5	16	4·851	90	295	3·326	4000	13123	5·16
6	19	8·222	100	328	1·029	5000	16404	3·45
7	22	11·592	200	656	2·058	6000	19685	1·74
8	26	2·963	300	984	3·087	7000	22966	0·03
9	29	6·333	400	1312	4·116	8000	26246	10·32
10	32	9·702	500	1640	5·145	9000	29527	8·61
20	65	7·405	600	1968	6·174	10000	32808	6·90
30	98	5·108	700	2296	7·203			
40	131	2·811	800	2624	8·232			

Reduction of Decimetres, Centimetres, and Millimetres, to English Inches.

Dec.	Inches.	Cent.	Inches.	Milli.	Inches.
1	3·937	1	0·393	1	0·039
2	7·874	2	0·787	2	0·078
3	11·811	3	1·181	3	0·118
4	15·748	4	1·574	4	0·157
5	19·685	5	1·968	5	0·196
6	23·622	6	2·362	6	0·236
7	27·559	7	2·755	7	0·275
8	31·496	8	3·149	8	0·314
9	35·433	9	3·543	9	0·354
10	39·370	10	3·937	10	0·393

G. *Comparison of English and French Measures.*
(From Baily's Astronomical Tables.)

	Eng. Inches.		Fr. Metres.
French Metre ...... =	39·37079	French Toise......... =	1·949036
———— Toise ...... =	76·739400	———— Foot .... =	0·324839
———— Foot .... =	12·789900	———— Inch .... =	0·027070
———— Inch .... =	1·065825	English Foot......... =	0·304794
———— Line .... =	0·088819	———— Inch.......... =	0·025399

Constant logarithms (always additive) for converting

French Toises into Met. 0·2898200 || French Ft. into Eng. Ft. 0·0276860
———— Feet into Metr. 9·5116687 || ———— Met. into Eng. Ft. 0·5159929
———— T. into Eng. Ft. 0·8058372 || Millimet. into Eng. In. 8·5951741

INDEX.

ABEL, Dr. Clarke, on the bank thrown up at the Cape of Good Hope, 78.
Aix in Provence, supercretaceous rocks of, 221.
Alps, Austrian and Bavarian, supercretaceous rocks of, 207 ; cretaceous rocks of, 259 ; lias of, 313.
Arago, M., on the temperature of the globe, 5.
Arkose, 309.
Atchafalaya, raft of, 65 ; section of the alluvial banks of, *ib.*
Attraction, local, of the magnetic needle, renders the determination of currents doubtful, 100.
Ava, organic remains found in the kingdom of, 239.

Baculite limestone of Normandy, 263.
Badku, gaseous exhalations of, 133.
Bagshot sands, 235.
Barometer, Tables for calculating heights by, 519.
Basaltic dyke, converting chalk into granular marble, N. of Ireland, 250.
Basin, on the term as applied to supercretaceous deposits, 186 ; of Paris, 223.
Basins, rock, termed Kettle and Pans, St. Mary's, Scilly, 43.
Basterot, M. de, on the supercretaceous rocks of Bordeaux, 222.
Beaches, shingle, 72 ; travel in the direction of the prevalent winds, *ib.* ; bar up the mouths of valleys, forming lakes, 74; sandy, 76 ; raised at Plymouth, 149 ; raised in the Island of Jura, Hebrides, 151.
Beaufort, Capt., on under-currents in the Mediterranean, 105 ; on the gaseous exhalation of the Yanar, 132 ; on a consolidated beach, Karamania, 78.
Beaumont, M. Elie de, on the erratic blocks of the Alps, 166 ; on the gravels of the Lyonais, Dauphiné, and Provence, 189; on the supercretaceous rocks of the Pertuis de Mirabeau, 219 ; on the epoch of the cretaceous deposit, 261 ; on the oolite of Burgundy, 305 ; on the lias of the Alps, 313 ; on the Grès de Vosges, 381 ; on the elevation of mountains, 496.
Bertrand de Doue, M., on the ossiferous beds of the Velay, 243.
Beudant, M., on the volcanic rocks of Hungary, 249.
Bigsby, Dr., on the erratic blocks of North America, 164.
Black Head, (Babbacombe Bay, Devon,) on the trap and limestone of, 478.
Boase, Dr., on the submarine forest in Mount's Bay, 147.
Boblaye, M., his account of the course of the Meuse, 54 ; on the oolite of the N. of France, 305.
Bore of the Ganges, 89 ; of the Maranon, *ib.* ; of the Aruary, 90.

2 A

530   *Index.*

Forests, submarine, 143.
Freshets of rivers, 58.
Fresh-water formations of the Isle of Wight, 235.
Fuchsel, M., his geological researches, 181.
Fundy, Bay of, great tides in, 85.

Gaseous exhalations, 131.
Gault, 255.
Geological terms, explanation of, 507.
Geysers, eruptions of, 19.
Glaciers, 59; advance of, *ib.*
Globe, changes on the surface of the, 32.
Gneiss, 463.
Gorges and ravines, 27; gorges cut by the drainage waters of lakes,
    50; Clifton gorge, near Bristol, 54.
Gosau, valley of, 251.
Gosse, Dr., on the baths of Vignone and San Filippo, 136.
Granite, 468; veins of, 473; superposition of, in the Alps, 472; rest-
    ing on chalk at Weinbohla, 262.
Gravel, superficial, in Devonshire, 157; of Central England, 160;
    rounded in caves, 176, 177.
Grauwacke, 433; arrangement of the laminæ in the slates of the, 434;
    remarks on the limestones of the, 435; red, 437; remarks on the
    organic remains of, 451.
Green-sand, upper, 255; analysis of green grains in, *ib.;* lower, 256.
Greenstone and other trappean rocks, 469.
Grès de Vosges, 381.
Group, modern, 40; Erratic block, 155; Supercretaceous, 181; Creta-
    ceous, 254; Oolitic, 304; Red Sandstone, 376; Carboniferous, 398;
    Grauwacke, 433; Lowest fossiliferous, 456.
Gulf stream, 93.
Gypsum, ossiferous, of Paris, 226.

Hall, Sir James, on the transported gravel near Edinburgh, 162; on the
    fusion of limestone beneath pressure, 479.
Hall, Capt. Basil, his plan for ships' tracks on charts, 101.
Harris, Mr., on the blocks moved during gales at the Breakwater, Ply-
    mouth, 75.
Hastings sands, 293.
Hennah, Mr., organic remains in his cabinet from the Plymouth lime-
    stones, 451.
Hibbert, Dr., on the erratic blocks of Shetland, 162; on the rocks of
    the Velay, 245; on the passage of granite into basalt, 471.
Hitchcock, Mr., on the carboniferous deposits of Connecticut, 422.
Hoffmann, M., on the Valley of Pyrmont, 29, 137; on the coal mea-
    sures of Thuringia, 414.
Hornblende rock and slate, 462.
Horner, Mr., on the submarine forest in Somersetshire, 146.
Hugi, M., on granite resting on lias in the Alps, 472.
Humboldt, M. von, on the temperature of the atmosphere, 26; on

LONDON:

PRINTED BY RICHARD TAYLOR,
RED LION COURT, FLEET STREET.

Printed in the United States
By Bookmasters